MINERALOGY

Mineralogy

JOHN SINKANKAS
Captain, U. S. Navy, (Ret.)

LINE DRAWINGS AND DIAGRAMS BY THE AUTHOR

VAN NOSTRAND REINHOLD COMPANY
New York   Cincinnati   Toronto   London   Melbourne

Van Nostrand Reinhold Company Regional Offices:
New York   Cincinnati   Chicago   Millbrae   Dallas

Van Nostrand Reinhold Company International Offices:
London   Toronto   Melbourne

Library of Congress Catalog Card Number 75-20850
ISBN 0-442-27624-9

Published by Van Nostrand Reinhold Company
A Division of Litton Educational Publishing, Inc.
135 West 50th Street, New York, N.Y. 10020

16   15   14   13   12   11   10   9

To
the late Peter Zodac
for a lifetime devoted to the encouragement of amateur
interests in earth sciences

Acknowledgments

I am deeply grateful for the unstinting help given me by various individuals during preparation of this book. The first part of the manuscript, dealing with mineralogy, was carefully examined by Dr. Harry F. Miller of Oceanside, California, who offered numerous helpful suggestions toward its improvement; these were unhesitatingly incorporated. The entire manuscript, Parts I and II, were similarly examined by Dr. Gustaf O. Arrhenius, Professor of Geology, Scripps Institution of Oceanography, La Jolla, California, whose helpful suggestions were also incorporated. To both of these gentlemen, I wish to express my appreciation, as much for encouragement to proceed with the work as for ideas for improvement.

The major portion of the photographs were made in the studios of O. D. Smith, Photographer, La Jolla, California, under the personal direction of the author, using specimens primarily from the author's collection, but also from the collections of Alice J. Walters and Elbert H. McMacken of Ramona, William C. Woynar, San Diego, and William Schneider, Poway, California. Additional photographs were provided by the late Raymond E. Schortmann of Schortmann's Minerals, Easthampton, Massachusetts, by Paul E. Desautels and Dr. George Switzer of the Smithsonian Institution, Washington, D.C., and by David New and Scott J. Williams of the Southwest Scientific Company, Inc. of Scottsdale, Arizona.

To aid in the description of species, the collections of the following individuals were examined: Alice J. Walters, and Elbert H. McMacken, Ramona, Carl E. Stentz, Newport Beach, Peter Bancroft, El Cajon, William C. Woynar, San Diego, Norman and Violet Dawson, San Marcos, California, and many other outstanding private collections, in part exhibited at the shows of the California Federation of Mineralogical Societies, at shows of Arizona societies and State Fairs at Phoenix and Tucson, and at the outstanding shows of the annual San Diego County Fair at Del Mar, California. In addition, the collections of the Smithsonian Institution, Washington, D.C., American Museum of Natural History, New York, and the Mineralogical Museum of Harvard University, Cambridge, were also examined.

vii

Much useful information on sizes and qualities of specimens offered in the market during the past ten years, was gleaned from the sales catalogs and bulletins of Southwest Scientific Company, Inc., Scottsdale, Arizona, Minerals Unlimited, Berkeley, California, Ward's Natural Science Establishment, Rochester, New York, Schortmann's Minerals, Easthampton, Massachusetts, Burminco, Monrovia, California, Filer's, Redlands, California, Shale's, Los Angeles, California, E. M. Gunnell, Denver, Colorado, Specimen Minerals, Ltd., Woodville West, South Australia, Gregory, Bottley and Company, London, England, and others.

Throughout the writing, editing, and typing of the manuscript, my wife, Marjorie, has been of inestimable value, while assistance was also furnished by my daughters Sharon and Marjorie, as well as by my mother-in-law, Emma Jane McMichael.

Preface

When he lays aside the badge of beginner, the amateur in mineralogy soon finds himself wandering unhappily in a no man's land. He earnestly wishes to know more about minerals, but is prevented from doing so unless he is able to enroll in courses taught at college level. His plight is emphasized by the books available for private study, which fall into two widely separated classes, those for the beginner, and those for the college student. Beginner's books are factual, easy to read, and provide valuable introductories to the fascinating world of minerals. However, they must necessarily be largely descriptive with minimum emphasis on theory. On the other hand, college texts, as simple as they may seem to the professional mineralogist, prove to be stumbling blocks to many amateurs because they assume prerequisite courses in mathematics, physics, and chemistry, which many amateurs have not had. Between these extremes of literature there is little specifically aimed at the interests of amateurs advanced beyond the stage of beginner. It is the aim of this book to provide a bridge between these extremes.

Once it is granted that the amateur's pursuit of mineralogy is because of personal tastes and interests, that he has no intention of becoming a professional, it is at once easier and more difficult to write for him. It is easier because of his spontaneous enthusiasm, in itself a powerful motivation to learn, but on the other hand, the writing cannot be technical for the reasons stated before. These considerations have led to a simplified presentation of numerous ideas ordinarily reserved for introduction to students of mineralogy at college level, but the presentations are not so elementary that they repel the intelligent reader. Greatest emphasis is laid on descriptive drawings and photographs and the reader is constantly referred to them in the text. The subjects chosen are primarily of interest to the amateur, and not necessarily those which would be considered fundamental for the student intending to become a professional mineralogist.

The subject matter in the main portion of the book begins with the atom because this is the logical beginning of mineralogy. By demonstrating how atoms form distinctive combinations, many puzzling features of minerals fit more logically into larger patterns which, when explained, enliven the amateur's interest and excite even keener appreciation. Following chapters take up the effects produced by the regular array of atoms in mineral crystals, and relate external appearance and properties to this cause. All the means available to the amateur for identifying minerals are emphasized, including the optical techniques developed by gemologists; these seem strangely neglected in books meant for amateur consumption. A number of gemological tests, easily understood and applied, are not only ingenious but virtually infallible. When their results are considered along with other bits of evidence, these tests often enable the amateur positively to identify minerals which had baffled him before.

The last portion of the book deals with the descriptive mineralogy of a selected group of species representing those most likely to be encountered and collected by the amateur. Special pains are taken to point out characteristic or distinctive properties and habits of each species, again with the view that personal identification of minerals is one of the most rewarding aspects of mineralogy. The book concludes with appendices containing identification tables, and a bibliography covering amateur interests in further reading.

J. S.

San Diego, California

Contents

PART II — DESCRIPTIVE MINERALOGY

APPENDICES

PART 1

Mineralogy

1

Introduction

The finest mineral specimens are at once so colorful, so complex in their crystallizations, and so far beyond ordinary experience that few persons can resist being attracted to them. Practically all amateur mineralogists trace their entry into the hobby from the moment their sensibilities were first devastatingly yet pleasantly assaulted by the vision of a superlative specimen. This experience is almost always followed by a burning desire to have similar specimens for one's own, and so another mineral collector is born.

Although many beginners in the hobby remain collectors only, most are impelled by healthy curiosity to learn more about minerals. It is to the credit of mineral collecting that it persuades most beginners to climb successively through the stages of beginner, collector, to amateur, and even to professional. Some amateurs will lose interest in mere collecting, particularly of common species acquired chiefly for their attractiveness, and attempt to get those which are mineralogically interesting, though to the eye they may be as drab as pieces of old brick. These specimens are then examined carefully, subjected to a variety of tests, and often described in scientific papers submitted to amateur and professional journals.

THE COLLECTING HABIT

The collecting habit comes early and stays late. It is a deep, undying instinct that emerges as soon as we are able to wrap our fingers around an object that fits them, and changes only in refinement as we grow older.

When it manifests itself in its worst form, as in the much-publicized case of two miserly brothers of New York City who collected all sorts of trash until their home had no room left, it seems to serve no useful purpose and indeed is a hindrance to doing something worthwhile. On the other hand, when it leads to the preservation of great works of art or literature, of inventions, or of objects which tell future generations what has happened before and so gives them targets for bettering their own efforts, then the habit is worth the time, trouble, and expense involved.

Some collectors spend lifetimes in the unswerving pursuit of bigger and better objects to put in their collections, or add varieties to make collections as comprehensive as possible. Others decide to specialize, and collect only within narrower fields. But even restricted collections may become too broad, and we hear of collectors who collect paintings by only one artist, or books dealing with one specific subject, and in minerals, of those who collect only crystals, or one species only, or even specimens only from one mine. Whatever limits the collector imposes on himself, he, in common with thousands of others, sniffs out new "finds" with all the determination and ingenuity that he can muster. Needless to say, when such efforts succeed, the fruits taste sweet indeed.

But collecting merely for the sake of possession is apt to be sterile pleasure unless more is known about the specimens themselves. A famous New York art dealer of some years ago was very aware of this and used it to great advantage in disposing of many fine paintings, which now hang on the walls of most of our prominent museums. He knew that his clients behaved like customers anywhere: they were reluctant to buy unknowns, but were happy to buy famous works of art if the price was within reach. If a prospective customer knew nothing about the paintings of an obscure artist, or conversely, an obscure work by a great artist, the art dealer took the trouble to point out the characteristics which made the painting desirable, and by the time that he was through, had so whetted his customer's appetite that when an example appeared, as it had a habit of doing soon afterward, the sale was usually made. Some of our better mineral dealers are also well acquainted with the need to educate customers, and their catalogs and sales bulletins are colorful and exciting descriptions of new acquisitions. Not much is usually said about specimens from classical localities because these sell themselves to knowing collectors at the mere mention, but much space is devoted to new offerings, telling what is found, how the material compares to specimens of the same species from better known localities, or otherwise pointing out something unique or different which makes the specimen worth buying.

In time each collector realizes the need for knowledge, if for no other reason than to guard himself against fakes, imitations, and the multitude of specimens of indifferent quality which may be found in every corner of the mineral collecting field. Truly fine specimens are never easy to obtain even for the well-to-do, for there is no guarantee that only the best will be offered nor even that the best are available at the time a collector is willing to buy them. Every experienced collector ruefully recalls instances when he was "stuck" with trash, probably because he bought without really being sure of quality. Or he may have passed up something really choice because he was unable to recognize a good thing when he saw one. In this class fall mineral specimens, often neither colorful nor flashy, from old localities which produced very few in number, or which are now unproductive, or specimens with special features which make them highly prized by the educated collector and connoisseur. In the latter group may be, for example, alexandrite twins from the Urals, really rather dull objects all things considered, or Japanese twins of quartz which, to the beginner, may look little different from hundreds of other specimens of rock crystal.

The application of knowledge to mineral collecting spells the difference between the mere amasser of bulk and the amateur connoisseur who tastefully assembles a suite of specimens each of which, regardless of size, is representative of the highest quality or of some unique feature which sets it aside from thousands of others. Unlike paintings, books, tapestries, and a host of other objects which man collects, minerals occur in a vast range of sizes and qualities, and perhaps more than is possible in other hobbies, collecting them can be gratifying to persons from all walks of life and of all incomes.

DISTINGUISHED AMATEURS

The list of amateurs in mineralogy is studded with distinguished names, some for accomplishments in mineralogy and others for fame in far different fields. It is the best evidence that an amateur need not stop at collecting, nor feel that he has nothing to contribute to the science of mineralogy.

An abundant iron mineral, *goethite,* is named after Johann Wolfgang von Goethe (1748-1832), better known to modern high school and college students as a great German author, but recognized in his day as a skilled amateur scientist with a strong interest in minerals. A contemporary, the Englishman James Smithson (1765-1829), whose name now is borne by the zinc mineral *smithsonite,* was also an amateur of the sciences, particularly mineralogy. The general public knows him as the benefactor of the Smithsonian Institution in Washington, D.C., for the establishment of which he left to the United States a huge sum of money. Appropriately enough, the Smithsonian now houses not only splendid examples of smithsonite, but also the finest collection of minerals and gems in the world.

Another Englishman, John Ruskin (1819-1900), whose literary works are required reading for today's scholars, wrote mainly on subjects far removed from mineralogy. Though famed throughout his life as an essayist and art critic, Ruskin early gained an interest in minerals which he kept until his death. His love of minerals and crystals is clearly shown on every page of his "Ethics of the Dust," a small volume of ten lectures on matters mineralogical. Beginning in 1850, he gave numerous specimens to the British Museum

(Natural History), among them the Edwardes Ruby and the magnificent yellow Colenso diamond, a beautifully formed octahedral crystal of 133 carats weight!

The best known American amateur is Colonel Washington A. Roebling (1837-1926), whose collection of over 16,000 specimens was bequeathed to the Smithsonian along with a substantial sum of money to acquire and maintain specimens in his name. Colonel Roebling earned his rank in the Civil War, and afterwards, rejoined his father in the bridge-building enterprises which were the engineering specialty of this distinguished family. His father conceived the Brooklyn Bridge but never lived to see it completed. This honor, as well as the difficulties attendant to a project so advanced for its day, fell upon the shoulders of his son, and when problems seemed unsolvable and responsibilities too much to bear, Roebling obtained relief by spending time with his beloved minerals. He was unable to collect personally in the field, because of poor health in the years after the Civil War, coupled with an attack of bends as a result of staying too long in the underwater caissons of the Brooklyn Bridge. Yet this handicap did not stop him from amassing a tremendous fund of knowledge and what was probably the best personal mineral collection in the world at the time.

The same period in American mineralogical history, often called the "golden era" of American collecting, brought forth two other important collectors equipped with the knowledge and the means to demand the best: Clarence S. Bement (1843-1923), a wealthy manufacturer of Philadelphia, and Frederick A. Canfield (1849-1926), of Dover, New Jersey. Bement's collection was eventually bought by J. P. Morgan through the services of G. F. Kunz, and was presented to the American Museum of Natural History in New York City. Canfield's collection was already rich in "old timers" from the famous zinc mines of Franklin, New Jersey, which he inherited from his father and continually enlarged with more specimens from this highly mineralized deposit. At the time of his death no other collection was comparable in richness, size, and perfection of crystallizations from this famed locality. It was bequeathed to the Smithsonian, where selections from it may be seen in the Mineral and Gem Hall.

Other famous names from the same era are William Earl Hidden (1853-1918), for whom the emerald-green spodumene from North Carolina, *hiddenite,* is named; George Frederick Kunz (1856-1932) an ardent and skilled collector of gems and minerals as well as a prolific writer, and known best by his mineralogical namesake, the pink-to-lilac spodumene gemstone, *kunzite;* and George Vaux, Jr. (1863-1927), of Philadelphia, a lawyer with a strong interest in minerals who succeeded in collecting an imposing total of 10,000 specimens in his lifetime, and whose name appears in the minerals *vauxite, paravauxite,* and *metavauxite.*

In this same period lived George Letchworth English (1864-1944) whose name is familiar to every amateur mineralogist from his well-known book *Getting Acquainted with Minerals,* lately revised by David E. Jensen. English began as an insurance man but soon gave up this business for the far more exciting career of a mineral collector and dealer. Before his colorful life closed, he had traveled one end of the globe to the other in search of the finest specimens which personal diplomacy could uncover and money could buy. *Englishite,* a rare phosphate mineral, honors his name. Lazard Cahn (1865-1940), another well-known and respected mineral dealer, and whose name is given to *cahnite,* a very rare boro-arsenate of calcium from Franklin, New Jersey, also began as an amateur mineralogist but became so engrossed that he took up formal studies of mineralogy in American and German universities a considerable number of years later.

In more recent years the name of Samuel George Gordon (1897-1952) and Harry Berman (1902-1944) still are recalled by every professional mineralogist, but amateurs may take pride in the fact that both "backed into mineralogy" the hard way from amateur status. Gordon developed a keen interest in mineralogy even as a small boy but circumstances prevented him from completing formal schooling in this subject. Nevertheless, mineralogy was his chosen field, and he found ways to stay in it. Before he died, he contributed many scientific papers and described nine new minerals. Berman's education was also interrupted, and after one year of mathematics and engineering at the Carnegie Institute of Technology, he was forced to leave. Nevertheless he went on to mineralogical positions in the United States National Museum and Harvard University and improved his knowledge during work in those institutions. In addition to writing several dozen scientific papers, he is also known as the designer of an extremely sensitive specific gravity balance used regularly in many mineralogical laboratories today.

There are at present here and abroad many well-known amateur mineralogists but perhaps it would be best to close this discussion with one whose name is still frequently mentioned in California and other western states, Magnus Vonsen (1879-1954), formerly of Petaluma, California. Vonsen's interest in minerals developed during World War I and grew steadily thereafter with specialization in the borate minerals of the California desert regions. His study and collection of borates led to the discovery of *teepleite.* Although he never had formal training in mineralogy, and indeed devoted the major portion of his time to a thriving business, Vonsen privately studied mineralogy and became highly skilled in chemical work and the use of the mineralogical microscope. At the time of his death his collection was rated as the finest private collection in the United States. It was bequeathed to the California Academy of Sciences and is housed in their museum in Golden Gate Park, San Francisco.

THE FUTURE OF THE AMATEUR

Although the amateur mostly collects and studies minerals for his own amusement, increasing experience and knowledge place him in a position where he can make valuable contributions to mineralogy far beyond mere donation of specimens. The field of mineralogy has plenty of room in it for amateur and professional alike, in fact, professionals frequently acknowledge that the lively curiosity, boundless enthusiasm, and energetic field-tripping of the amateur result not only in an abundance of fine specimens, but also the discovery of new localities, and—not so rarely as one would imagine—the detection of a new species.

Without the long academic training and familiarity with the professional's specialized laboratory equipment, however, the amateur can seldom hope to equal him in technical skill. But in the field, and at home where many happy hours are spent examining finds, the amateur becomes an invaluable extension of the busy professional. The preliminary time-consuming screening which the conscientious amateur gives to his specimens not only tells him more about them but reveals oddities or unusual features which may be of great significance to the professional. Perhaps when screening is over, and dozens of specimens have been examined carefully, nothing remarkable will show up, but the amateur will have the satisfaction of tracking everything down until there are no "unknowns" left.

In places where mineral deposits are being opened or systematically worked, as in mines and quarries, the amateur is far better able to keep track of progress, and by his collecting efforts, discover new or unusual minerals almost as quickly as they are blasted out. Frequently it is of great importance to mineralogy and geology, especially in respect to the formation of minerals, that the exact circumstances under which the minerals were found be noted. The combination of minerals, or *associations,* tell much about how they came into existence. Such information may seem insignificant at the time, but no one can predict when it will suddenly become of grave importance. For example, only a short time ago the small bits of knowledge assembled about radioactive minerals over a period of many years seemed far less valuable as a whole than the knowledge about just one important ore mineral of a useful everyday metal. Yet had the assembly of this information been neglected, atomic energy programs might have waited for years before they could be sure of the necessary raw material. Every scientist dreams of discovering some principle so basic and far-reaching in its consequences that it will take the scientific world by storm. But extremely few ever do so, and they must be content to add a bit here and there to the slowly rising edifice which is their science. In this process the amateur can help.

In still another field, that of education, the amateur becomes extremely important, perhaps more so in some respects than the professional. Certainly he educates himself whenever he collects and studies a specimen, or looks up the data on it, but it is assumed he will do this in any case. It is his influence on others that counts. If he displays enthusiasm for his hobby, invites others to look over his collection, participates in club or society educational programs and public displays, and patiently answers the questions put to him by children who may be seeing minerals for the first time as something more than "rocks," then he may inspire that one youngster, out of hundreds of prospective college students, to go on to a career in mineralogy. Unfortunately, the doors of Mineralogy Departments in colleges are not being battered down by eager applicants, particularly when the word among the students is that mineralogy courses are far from being easy. Yet the impressive number of college graduates in other fields who are now enthusiastic, devoted, and really scholarly students of amateur mineralogy points to a lack of previous exposure and motivation, which, had it been given at an earlier appropriate moment, might have resulted in far different careers. There is now an urgent need for professional mineralogists, while indications are that the special proficiencies called for will be even more useful in the future as man enters space. Again, the amateur can indirectly help the future of mineralogy, if not directly help its present.

THE DEVELOPMENT OF MINERALOGY

The earliest treatises on mineralogy were written when little was known about minerals except what they looked like, and consequently the pages were filled with descriptions and it still is true that the easiest way to identify a mineral is merely to look at it. If you have seen a certain species before, the chances are you will recognize it again in another specimen. Sight recognition works well if representative specimens have been handled at close range, but it works far less well if one has to depend on written descriptions and match those against a strange specimen. Therefore the usefulness of early mineralogy books depended entirely on how skillfully the author transferred his impressions to paper. If he did a good job, readers made sense out of it and found that they could recognize the mineral when it was encountered in specimens. All too often, it was the other way around, and readers were left confused and disheartened.

To sort order out of confusion in *descriptive mineralogy,* as this part of the science is now called, a splendid book was written in 1774, by the noted German geologist Abraham Gottlob Werner (1750-1817), attempting and succeeding admirably, in standardizing the terms and methods used to describe minerals on the basis of outward appearance. So well did Werner do his job that if the 1962 translation by Albert V. Carozzi is consulted, it will be seen that many of the terms we use today were first used by Werner.

The science of mineralogy did not stop at mere descriptions, and hundreds of years ago, students were already speculating about the inner make-up of minerals, causes of color, why some minerals were hard, others soft, and why some formed certain crystals and not others. Close on the heels of purely descriptive mineralogy arose *physical mineralogy,* in which important physical properties were described as they were discovered. Differences in hardness among the various species were known at an extremely early date but a systematic scale to which every mineralogist could refer was not devised until Friedrich Mohs (1773-1839), a German mineralogist, proposed that ten minerals, ranging from talc, the softest, to diamond, the hardest, be used as points one to ten respectively in a hardness scale, with other standard species placed between. The scale, the *Mohs Scale* of *hardness,* is still in use. Another important physical property discovered early and even of greater importance today, is that of cleavage. Other physical properties were also discovered, adding more certainty to the identification of strange minerals, and encouraging further investigation.

The discovery that certain transparent minerals affected light rays in characteristic ways led to still another division of mineralogy—*optical mineralogy.* The many methods of optical mineralogy are extremely complicated, difficult to understand, and require long training periods as well as specialized and costly equipment to carry out. Nevertheless, in hands of experts, mineralogical optical equipment can be used to identify many species quickly and certainly. One of its strongest attractions is that bits of material too small to test in other ways, or even to see well with the naked eye, can still be identified.

The chemical properties of minerals, at least a small number of them, must have been known by various civilizations many thousands of years ago. By the time of the Renaissance, metallurgy and other chemical arts had advanced to an amazing degree. Methods for smelting, refining, making acids, and other useful chemical products, are described in detail in the *Pirotechnia* of Vannoccio Biringuccio (1480-1539), translated by C. S. Smith and M. T. Gnudi, and in *De Re Metallica* of Georgius Agricola (1494-1555), translated by Herbert C. Hoover and Lou H. Hoover. Both volumes show that efficient chemical processes, many still in use today in more refined form, had been hit upon either by accident or design throughout previous centuries.

Since the Renaissance, awareness grew slowly but surely that minute particles of matter, now called atoms, comprise all solids, liquids and gases. Every schoolchild knows that there are many kinds of atoms, or *elements,* which, in their pure form, are very seldom seen except in the case of certain metals such as gold, silver, iron, etc. Even these are really never pure because they would be too weak to be useful without carefully controlled dosages of other elements. John Dalton (1766-1844), an English chemist, was the first to announce distinctly an atomic theory based on the existence of such particles.

Confirmation of Dalton's theory was not long in coming from all quarters, but perhaps the most valuable long-range discoveries supporting his theory came from mineralogists who were interested in why crystals grew as they did and concluded that it must be because the various atoms formed regular building blocks and were not merely haphazard mixtures of various elements. The idea of structure within a crystal was not a new one by any means, but had been advanced most clearly and strongly by Rene-Just Haüy (1743-1822), a French mineralogist who came to believe that the regularity of mineral crystals could arise only from an orderly stacking of minute components, all of the same size and shape. Haüy's ideas revolutionized the study of crystals, and led to many interesting and wonderful discoveries about their geometry, from which were drawn shrewd conclusions about atomic arrangements. The rapid advancement of science and technology in the last hundred years is due in no small measure to this important beginning. The developing science of crystal geometry came to be known as *crystallography,* and now more than ever before, it is of greatest importance to the scientist in mineralogy and in other fields dealing with the nature of matter.

Confirmation of orderly inner structure in crystals did not take place until much later, although enough convincing evidence had been accumulated to make it very unlikely that there was any other reasonable explanation for the outward forms of crystals. In 1895, the German physicist, Wilhelm Konrad Röntgen (1845-1923), discovered x-rays and the fact that they penetrated solid matter. In 1912, Max von Laue, his student, conceived the idea that x-rays, passed through a crystal, might furnish regular reflections from planes of atoms, and accordingly began experimentation toward this end. The experiments eventually succeeded but proof of cause and effect, and development of the most advantageous methods for using this powerful new scientific tool, had to wait for the extensive work of a pair of English physicists, William Henry Bragg, and his son, William Lawrence Bragg. Later work by the Bragg team attempted to show exactly where atoms were located in any crystal framework, and how they were attached to each other, and what effect all this had upon the properties of the crystals themselves. These relationships were not long in appearing, and it is now recognized that *every* property of matter is due to its atoms, their arrangement within the matter, and the nature and strength of the attractive forces which keep them together. The modern study of mineralogy begins from the study of the inner structure of crystals.

AMATEUR STUDY OF MINERALOGY

Where should the amateur begin *his* study of mineralogy? Certainly a good beginning is always to collect specimens, to observe them carefully and to test them with whatever means are at hand. But sooner or later certain features about them arouse curiosity and pose questions which cannot be answered from inspection alone. For example, the question of why malachite is green is usually answered by saying that it has copper in it. But when the intelligent amateur recalls that vivid *blue* azurite also has copper, and is very close to malachite in chemical composition, he finds small solace in this kind of an answer. As interest in the hobby grows, he develops an increasing interest in the *whys* rather than the *wherefores*, and the simple though truthful answers dating from the traditional study of mineralogy are not good enough.

This is basically the reason for beginning mineralogy with atoms and the atomic architectures known as crystals: it is to answer more of the *whys* which amateurs are posing today. Beginning at this point has the great advantage of leading logically into clearer understanding of what minerals *really* are. Such simple-sounding properties as hardness, specific gravity, and even the colors of the copper minerals malachite and azurite mentioned above, will take on new meaning and lend even greater fascination to the hobby than is now provided. The chapters to follow will begin with the atom, lead into crystals, then into their properties and tests, and will finish with descriptions of the species the amateur is most likely to place in his collections.

SOME PRELIMINARY DEFINITIONS

Before going on, it is a good idea to define mineralogy, minerals and crystals, although these definitions will not mean nearly so much as they will later. Let us be content to say now that *mineralogy* deals with the study of the typical stony substances in the earth's crust which comprise the components of rocks, and the products of their decay such as gravel, sand and soil. These components, or *minerals*, numbering now nearly 2,000, can be distinguished from one another by individual characteristics which arise directly from the kinds of atoms they contain and the arrangements these atoms take inside them. Practically all minerals show an orderly inner arrangement of atoms, and are then called *crystalline* minerals, while any part of a mineral in which the orderly pattern is the same from one end to the other is called a *crystal*. The pattern of atoms in any crystal is regular and repeating, like the small decorative designs which are printed on wallpaper. However, there is an important difference, the pattern repeats itself in *all* dimensions and there is no limit to how far it can go. For this reason, a crystal may be so small that it can't be seen, or so large that it takes a crane to lift it. Large or small, it is a crystal if the atomic pattern is the same in every part of it. When a crystal grows in an open cavity or in a liquid which does not hinder its outward development, it frequently is covered with shining flat faces which sometimes look as if they were cut by the lapidary. This is the popular concept of a crystal, but the real concept is the internal pattern, which if it were not there, would never permit the exterior flat faces to develop. As we will see later, the flat faces are *because* of the pattern, and not the other way around.

2

Atoms and Minerals

When we talk about atoms we must adjust our thinking to very small terms because atoms are very small objects. The head of a common nickel-plated dressmaker's pin seems small to many of us, indeed we may have to put on spectacles to see it well or even use a magnifying glass, but it is tremendously large compared to atoms because upon its curved surface there are about *200 quadrillion* nickel atoms exposed to view! If this seems a staggering number, consider how many atoms there must be in an ordinary coin. If we tried to write down the number, we would put down a lonely and, really, a rather foolish-looking digit at the head, followed by a procession of zeros covering solidly several sheets of paper. In terms of minerals, even the smallest of visible crystals, much smaller than the head of a pin, is still a vast complex of atoms.

Exactly what is an atom? The chemist defines it as the smallest part of matter which combines with others like it to make ordinary compounds, as salt, sugar, aspirin, and even mineral compounds. To the physicist, the atom is but one form of matter, a small one to be sure, but by no means the ultimate smallest particle. He is interested in the atom because it is a storehouse of energies and a steppingstone to his understanding of the universe, of which our world is but a small speck in a vastness which seems to have no end. The electronics scientist regards the atom as a source of negatively-charged motes called *electrons*, which ordinarily cling tenaciously to atoms, but can be lured away and made to scurry tirelessly around closed tracks in the astonishing electronic devices which we have come to regard as commonplace fixtures in our homes. The mineralogist is also interested in the atom, not merely because all minerals are made from atoms, but because the vast majority of minerals are crystals, which, as mentioned previously, are regular stackings of atoms in three dimensions, following rules set by the atoms themselves. To understand the rules, and thus understand better the nature of crystals, it is necessary to look inside atoms and see what they are composed of.

14

ATOMIC PARTICLES

When it isn't squeezed out of shape by neighbors, an atom is a ball-like object about one one-hundred-millionth of an inch in diameter, more or less, composed of particles arranged concentrically like a miniature solar system. The central ball of matter, comparable to the sun, is surrounded at relatively great distances by electrons whirling in orbits corresponding to planets. The heart of the atom, the small dot of matter in the center known as the *nucleus*, consists of even smaller particles clinging tightly to each other with tremendously strong forces. As physicists bombard atomic nuclei, causing them to split, they discover more and more particles, the last count listing about 30. No one knows where this digging into the inner atom is going to end, but for our purposes, we need only consider the two major nuclear particles called *protons* and *neutrons*. As the names imply, *pro*tons are positively-charged particles, while *neu*trons carry no charge at all. Thus the central cluster in every atom is also positively charged, according to how many protons it has in it. The neutrons, which weigh about the same as protons, contribute mass to the nucleus but do not influence the electrical charge.

If we think that an atom is small, we are sure to be overwhelmed by the smallness of atomic nuclei! From bombardment experiments, designed to burst open nuclei and observe the actions of the separate particles, it has been discovered that nuclei, even the large complex ones of uranium, are very small targets. Indeed it is estimated that nuclei are only from one ten-thousandth to one twenty-thousandth the diameter of the whole atom. Another astonishing discovery is that practically all the weight of any atom is contributed by its nucleus, painting a picture of nuclear particles as being not only extremely small, but extremely dense. One physicist estimates that if electrons could be stripped away from atoms, and a cubic inch of nuclei collected, this small volume of atomic cores would weigh nearly fifteen tons! When we consider that a cubic inch of lead weighs only 0.41 of a pound, while atomic nuclei weigh 30,000 pounds for the same volume, we, like the nuclear physicists, are irresistibly drawn to the conclusion that atoms, even lead atoms, must consist mostly of empty space! Perhaps this isn't the impression we receive when we drop a lead weight on our toe, but apparently it is true.

Atomic nuclei are seldom found alone in nature because each speedily attracts electrons to neutralize the positive charges of its protons, one electron for each proton. These minute particles then act like continuously patrolling sentries around a fortress, fending off other atoms and preventing them from drawing too close. It is the electron swarm which determines the atom's *size,* and depending on what kind of atom it is, it also determines the relative weight of any substance composed of a large number of closely packed atoms. Thus the weight of a block of lead, or any other solid substance, is primarily the weight of the atomic nuclei—*as modified by the fending-off distances established by the electrons*. It is somewhat like filling a basket with ping-pong balls, each of which has a small lead shot in it corresponding to a nucleus. The basket would not weigh much as a whole, but would weigh enormously more if the balls were discarded and just lead shot used to fill it. What we see of atoms in the ordinary course of events is comparable to the ping-pong ball representation; we can neither see nor touch the perfectly protected nuclei.

Turning to electrons, previously mentioned as very small negatively-charged particles, it should be said that not very much is known about their exact nature despite the fact that because every atom is enveloped in them, it must be electrons we feel every time we pick up something, and electrons we see every time we open our eyes. Some physicists believe electrons to be vague force fields, which, when surrounding a nucleus, form an approximately spherical blob of negative electricity somewhat like the drawing in Figure 1. Others believe them to be discrete particles, or possibly intensely concentrated force points, and even estimate them to be about one two-thousandth the diameter of a proton. How they shield nuclei is a puzzle because if they consist of particles, they must orbit around the atomic cores so rapidly that they are, in effect, everywhere at once. Nevertheless, most representations of atoms are similar to those in Figure 2, which show electrons as if they were fixed in position instead of moving rapidly. Such representations have the great advantage of making it easier to explain how atoms join to form crystals.

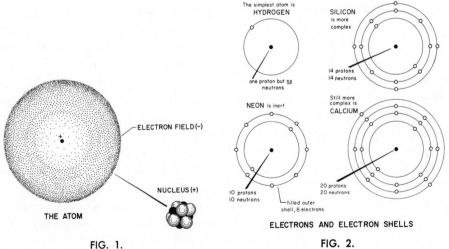

THE ATOM

FIG. 1.

ELECTRONS AND ELECTRON SHELLS

FIG. 2.

Because electrons are negative, and nucleonic protons are positive, every nucleus attracts electrons in number equal to the protons. Thus a ten-proton nucleus captures ten electrons, an eleven-proton nucleus attaches eleven electrons, etc. How the electrons are arranged around nuclei is an interesting matter because it is far from haphazard, and indeed, follows a rigid set of rules whose workings largely determine the chemical behavior of atoms. More will be said of this shortly, but for the moment we should consider the kinds of nuclei found in nature, for it is these extremely small dots of matter which attract electrons to create the numerous varieties of atoms known to exist.

ATOMIC COMPLEXITY

The latest ideas on matter and the nature of the universe, conceive of all nuclei as being manufactured by fusion in the intensely hot interiors of stars, using single protons and neutrons, plus other particles as raw material. The hydrogen nucleus is the simplest of all, consisting only of one proton. When it captures an electron, it becomes the completed hydrogen atom. Fusion results in agglomeration of more protons and neutrons, and the attraction of more electrons, until a large variety of atoms is created, the last natural one being the enormously complicated uranium atom, containing 238 particles in its nucleus, of which 92 are protons and the remainder neutrons. The count of protons alone is called the *atomic number,* and of course, because protons are balanced by electrons, it is also the count of the electrons.

As previously mentioned, the weight of an atom is the weight of its nuclear particles, the electrons being a very small and practically negligible part. Although one may expect that as nuclei grow in complexity, neutrons will appear in proportion to protons, this is not the case as examining the number of protons and neutrons in the uranium nucleus shows. Early in the atomic scale, additional neutrons are added to the nuclei such that the *atomic weight* rises in jumps until the uranium nucleus contains almost twice as many neutrons as it does protons. This will be clear in the table of common mineral-forming atoms which appears later in this chapter. It may be mentioned now that the *kinds* of atoms we speak of are solely due to the number and variety of particles in the nuclei, inasmuch as electrons can always be captured because of their abundance in almost any circumstance. The customary name given to an atom variety is *element.* Thus, as astonishing as it seems, especially when we consider the enormous differences in properties and appearance between elements as gold and oxygen, or carbon and lead, all are brothers under the skin, basically differing only in the number and variety of the *same* particles in their nuclei!

ELECTRON SHELLS

To complete the present theoretical picture of the atom in its numerous varieties, we now turn to the way electrons adhere to nuclei when captured. In general, although negatively-charged electrons are strongly attracted toward nuclear protons of opposite charge, they do not plunge into the nucleus to actually make contact. Instead, they are attracted only to certain levels outside the nuclei, at which points they are not permitted to draw closer. This curious attraction-repulsion poises the electrons in well-defined layers or *shells,* each shell, furthermore, being permitted to contain only so many electrons. At times, electrons can be knocked out of these shells, but extra ones cannot be jammed inward.

A few electron diagrams of typical elements are shown in Figure 2 to illustrate how the shells are formed. Beginning with the hydrogen atom with one proton, a single electron orbits the nucleus. The next more complex atom, not shown here, is helium whose nucleus contains two protons and two neutrons. It therefore has an atomic number of two, and an atomic weight of four. The innermost shell apparently never can contain more than two electrons, because a new shell, somewhat farther out, is formed as more protons in the nucleus acquire more balancing electrons. Thus, lithium, with three protons, contains two electrons in the innermost shell and its third electron in an outer shell. Next comes beryllium, which differs from lithium in having an atomic weight of 9 instead of 7, and one additional electron in the outer shell. This progression of electron shells follows until the complex and heavy atoms of thorium and uranium are reached, whose cross-section electronic diagrams look like slices of onion.

Note that in Figure 2, the second shell of the element neon is filled with exactly eight electrons. For some reason, *this is all that any outermost shell can contain,* although in very complex atoms, shells beneath the outermost shell can contain as many as thirty-two electrons. It doesn't make any difference whether shells beneath are partly or wholly filled, so long as the outermost shell contains no more than this magic number of eight! A few elements naturally contain exactly eight electrons in their outermost shells, and strangely, they seem completely satisfied with this arrangement because they refuse to have much to do with other elements and usually even refuse to enter into combination with atoms of their own kind. Consequently they are found as single atoms, floating free of any attachments, usually in the atmosphere and very sparsely in the earth's crust. These are the *inert* elements, and include neon, argon, xenon, krypton, and radon. Helium, which has a filled inner shell of only two electrons, behaves in precisely the same way and is also inert. All of these elements are gases, usually found in the atmosphere in very small quantities, but sometimes found in the mineral kingdom although even here they are not chemically combined in minerals. Helium, which results from the radioactive decay of uranium and thorium, is sometimes discovered in measurable quantities and thereby affords a means for estimating ages of rocks because more of it means a longer period of time has elapsed since the radioactive elements began decaying.

ELEMENTS FOUND IN COMMON MINERALS

The table on p. 20 lists elements found in common minerals; several are left out because they are so rare that they are not of much interest in mineralogy. The atomic number is the number of protons, but the atomic weight, as noted before, jumps considerably above the atomic number because more and more neutrons are attached to the nuclei as they grow larger. To find the number of neutrons, merely subtract the atomic number from atomic weight. Obviously it is impossible to weigh atoms separately, but means have been found to calculate the number of atoms in any compound, and from noting the proportion of one element as compared to another (usually oxygen), it is possible to state the weights *as compared to oxygen*. Thus the weight of hydrogen, given as 1, means that its atom weighs one-sixteenth as much as oxygen with atomic weight 16. The cadmium atom, of atomic weight 48, weighs three times as much as oxygen, etc. These are not *actual* weights, but weights *proportional* to oxygen.

Many element names are very old, particularly of early discovered metals such as gold and copper. Other names are newer because the elements themselves could not be isolated until much later. Instead of writing names in full, chemists and mineralogists use symbols to save time and space. Standard symbols are shown in the table on the next page and are usually straightforward borrowing of one or more letters from the full name. Each symbol is as brief as possible, yet clear and helpful to the memory. Some symbols, derived from Latin, Arabic, and German names, are completely unlike their present names and must be memorized if they are to be used. These are:

Antimony	Sb	—from *Stibium*		Potassium	K	—from *Kalium*
Gold	Au	—from *Aurum*		Silver	Ag	—from *Argentum*
Iron	Fe	—from *Ferrum*		Sodium	Na	—from *Natrium*
Lead	Pb	—from *Plumbum*		Tin	Sn	—from *Stannum*
Mercury	Hg	—from *Hydrargyrum*		Tungsten	W	—from *Wolfram*

LOSS OR GAIN OF ELECTRONS

As we have seen, the size of every atom is determined by the electrons which surround the nuclei, and more specifically, by the electrons in the outermost shell. From this we would assume that the hydrogen atom,

TABLE OF COMMON MINERAL ELEMENTS

Atomic Number	Name	Symbol	Atomic Weight
1	Hydrogen	H	1
3	Lithium	Li	7
4	Beryllium	Be	9
5	Boron	B	11
6	Carbon	C	12
7	Nitrogen	N	14
8	Oxygen	O	16
9	Fluorine	F	19
11	Sodium	Na	23
12	Magnesium	Mg	24
13	Aluminum	Al	27
14	Silicon	Si	28
15	Phosphorus	P	31
16	Sulfur	S	32
17	Chlorine	Cl	35
19	Potassium	K	39
20	Calcium	Ca	40
22	Titanium	Ti	48
23	Vanadium	V	51
24	Chromium	Cr	52
25	Manganese	Mn	55
26	Iron	Fe	56
27	Cobalt	Co	59
28	Nickel	Ni	59
29	Copper	Cu	64
30	Zinc	Zn	65
33	Arsenic	As	75
34	Selenium	Se	79
35	Bromine	Br	80
38	Strontium	Sr	88
40	Zirconium	Zr	91
41	Niobium	Nb	93
42	Molybdenum	Mo	96
47	Silver	Ag	108
48	Cadmium	Cd	112
50	Tin	Sn	119
51	Antimony	Sb	122
52	Tellurium	Te	128
53	Iodine	I	127
55	Cesium	Cs	133
56	Barium	Ba	137
58	Cerium	Ce	140
73	Tantalum	Ta	181
74	Tungsten	W	184
78	Platinum	Pt	195
79	Gold	Au	197
80	Mercury	Hg	201
82	Lead	Pb	207
83	Bismuth	Bi	209
90	Thorium	Th	232
92	Uranium	U	238

being the simplest of all, should be smallest, while the uranium atom, being the most complex, should be largest. Unfortunately, such a steady progression in atom diameters does not occur for several reasons, although there is certainly a general tendency for size to increase as atoms contain more nuclear particles and are enveloped in more electron shells. It has been found that as nuclei become more complex, and thus much heavier, internal attractive forces increase, tending to draw in all electron shells somewhat closer than is true for atoms near the beginning of the scale. In effect, it seems as if all atoms *try* to be about the same diameter. The greatest variations in size, however, occur when electrons are either lost or gained, because this materially affects the outermost shell and directly affects the atomic diameter. Thus an atom with a single electron in an outermost shell shrinks considerably if this particle is lost, while another atom, should it gain an electron, can be expected to swell in diameter. Thus atoms do not stay fixed in diameter, but can exist in several sizes depending upon what has happened to their outer electrons.

The loss or gain of electrons is very common, in fact, if it were not so, we would have very few compounds of any kind, and our world would become essentially a collection of native elements. The loss or gain of electrons depends on how strongly the nuclei attract electrons to begin with, and also upon how many electrons are present in outer shells. A number of filled shells tend to "screen" the attractive forces emanating from the nuclei, leaving little attraction for outermost electrons, especially when they are very few in number, as one or two. On the other hand, if an outer shell lacks just one or two electrons for completion of the eight-electron shell which every atom seems to want, then strong attractive forces develop which reach out to draw in electrons to fill the gap. Because most electrons are ordinarily attached to atoms to begin with, any atom seeking electrons usually takes them away from another which does not attract them strongly enough. This brisk trading of electrons follows the rules mentioned, that is, an atom with one or two outer electrons loses them easily, while an atom which needs one or two to fill an outer shell to eight, seizes them. When three outer electrons are involved, or a gap of three, the trading is less brisk. When just four electrons are involved, which is exactly half of the eight-electron configuration, the trade can be very complicated and may go either way, that is, an atom can gain four electrons or lose four.

Atoms which lose electrons, in effect shed them to drop down into a lower, filled-shell configuration, while those which gain them, attempt to fill a shell. However, giving or gaining electrons upsets the neat balance of electrical charges, an atom losing electrons becoming more positive by exactly the number of electrons lost, while an atom gaining electrons becomes more negative according to the number gained. When this happens, as it most commonly does, each atom becomes an *ion,* which simply means it is a charged atom. The process is called *ionization.* Each ion is not only charged but becomes, as mentioned before, a smaller or larger atom depending on whether it has lost or gained electrons.

IONS AND VALENCE

Because ions are no longer electrically neutral, oppositely charged ions draw toward each other with strong electrostatic attraction, combining into compounds as a result. The loss or gain of electrons during ionization results in an electrical *charge,* either plus or minus, the number of the charge being called the *valence.* Thus a single lost electron means a charge of plus one, a gain of a single electron means a charge of minus one, or using the term valance, a valence of plus one and valence of minus one respectively. Valence is indicated next to the element symbol by placing the sign and number in the upper right hand corner as in the following examples: oxygen, O^{-2}, potassium, K^+, etc. If the charge gained or lost is one, the number is not written, only the plus or minus sign. Practically all elements in minerals are ionized; their relative sizes and valences vary as shown in Figure 3.

Some elements can give up more electrons than appear in their outer shells and therefore have several valences. For example, iron and manganese commonly occur in mineral compounds in either of several valence states, and by doing so, drastically affect the appearance and properties of the several minerals formed by them even though the element is exactly the same. This explains why two iron oxides such as hematite and magnetite, both compounds of iron with oxygen, are vastly different in certain properties. In iron and manganese, and several other elements resembling them in this respect, the extra electrons are removed from an inner shell where they happen to be weakly bound by nuclear attractive forces.

ATOMIC BONDS

Aside from the few inert gases mentioned before, all other elements combine in various proportions to form an enormous range of compounds. The attractions between them which cause combination are called *atomic bonds.* One type of bond has been briefly touched upon before: the attraction of oppositely charged ions. A number of bond types have been discovered among minerals, the most important of which are the *metallic bond,* the *covalent bond,* and the *ionic bond.* Sometimes one bond prevails throughout a mineral crystal, but often atoms in a single crystal are held to-

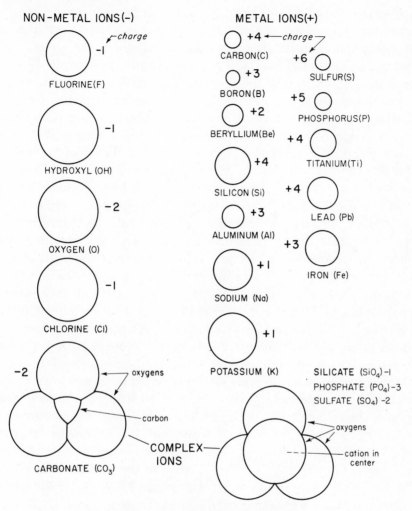

NON-METAL IONS (−)

METAL IONS (+)

FLUORINE (F)   −1   ← charge

CARBON (C) +4 ← charge

SULFUR (S) +6

HYDROXYL (OH)   −1

BORON (B) +3

PHOSPHORUS (P) +5

OXYGEN (O)   −2

BERYLLIUM (Be) +2

TITANIUM (Ti) +4

SILICON (Si) +4

CHLORINE (Cl)   −1

LEAD (Pb) +4

ALUMINUM (Al) +3

IRON (Fe) +3

CARBONATE (CO₃)   −2

SODIUM (Na) +1

POTASSIUM (K) +1

oxygens

carbon

COMPLEX IONS

SILICATE (SiO₄) −1
PHOSPHATE (PO₄) −3
SULFATE (SO₄) −2

oxygens

cation in center

SIZES AND CHARGES OF IONS

FIG. 3.

gether by several types in *combination bonds,* one type linking certain atoms, and another furnishing links elsewhere. The kinds of bonds present in crystals are nearly as important in their influence on the appearance and properties of minerals as the atoms themselves. Thus the element carbon produces two characteristic natural crystals, graphite and diamond, both comprised solely of carbon atoms, yet so different in all respects that it is scarcely believable that both are basically the same substance! Graphite is soft, greasy to the touch and opaque black, while diamond tends to be transparent, colorless and so hard that no other mineral approaches it in this property.

The metallic bond is characteristic of ordinary metals and imparts many of the properties for which we prize them. Bonds between metal atoms are peculiar in that outer electron shells overlap and merge until the entire mass can be thought of as many atomic nuclei spaced at regular intervals in a solid swarm of electrons. This is shown schematically in Figure 4. Electrons are free to drift about, and by their movements, enable metals to carry electricity and to rapidly transmit heat. The slackness of attractive forces between atoms allows entire blocks to slip easily in opposite directions when a strong force is applied, but the atoms tend to re-bond as soon as the force is removed and the metal mass remains in one piece. As a result, metals are noted for their ability to be beaten into thin sheets, or to be bent or twisted repeatedly without breaking. *Malleability* and *toughness* are therefore valuable identification properties of native metals. The mobility of electrons also allows maximum interference with the passage of light such that even extremely thin sections fail to transmit light. The typical "metallic" luster which needs no description is also due to this cause. Native metals are considerably heavier than most other minerals, and again this is due to the metallic bond which permits heavy metal atoms to approach each other as closely as possible. On the other hand, the looseness of the metallic bond also encourages metals to combine easily with other elements and accounts for the general rarity of native metals in the mineral kingdom. A few native metals, particularly gold and the platinum metals, are nearly chemically inert and seldom combine with other elements except in alloys.

As remarked in the previous section, atoms often attract electrons in order to fill gaps in their outer shells, the object being to surround themselves with an envelope of exactly eight electrons. Sometimes this is done by "sharing" outer electrons so that neighboring atoms have the services of eight electrons although some electrons must do double duty in the process. This will be made clearer by referring to the center drawing in Figure 4. At the left are shown carbon atoms, each having only four electrons in its outer shell and thus lacking four for completion. However, if forces are joined as shown, each atom will have a total of eight electrons although a pair on each of the four sides must be shared with neighbors. Figure 4 also shows how this scheme works in three dimensions in the case of natural carbon crystallized as diamond. Each carbon atom is now situated with four others around it. The pattern repeats itself throughout the crystal. Because atoms *cooperate* in sharing outer or *valency* electrons, the name *covalent* is given to this type of bond. Mineral crystals covalently bonded may be very strong as in diamond. In general, they do not melt or dissolve readily, nor do they conduct electricity. Many strongly affect light to the extent that they become opaque or exhibit a peculiar silvery luster.

The *ionic bond,* as its name implies, is the bond between oppositely charged ions. For example, in the mineral halite or common salt, sodium

METALLIC BOND

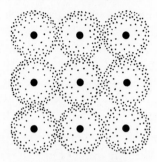 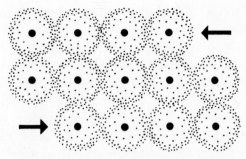

Atomic nuclei surrounded by clouds of mobile electrons.

Shear forces easily slip atoms because bonds are not very strong, but atoms quickly re-bond in the new position. This results in malleability.

COVALENT BOND IN CARBON (DIAMOND)

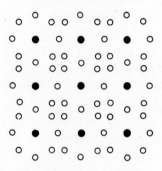 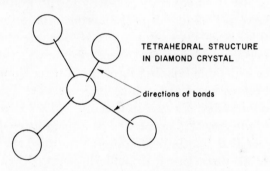

TETRAHEDRAL STRUCTURE
IN DIAMOND CRYSTAL

directions of bonds

Schematic of electron arrangement around bonded carbon atoms
showing how each atom is surrounded
by eight electrons.

Actual arrangement of atoms in three dimensions.
The outer atoms are bonded to others, repeating the
tetrahedral pattern to the limits of the crystal.

IONIC BOND IN CALCIUM FLUORIDE (FLUORITE)

CUBIC STRUCTURE IN
FLUORITE CRYSTAL

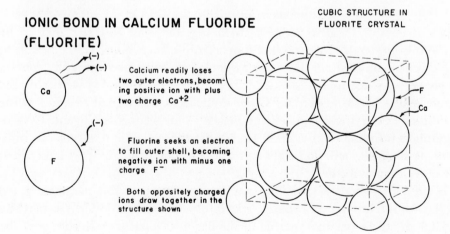

(−)
(−)

Ca

Calcium readily loses
two outer electrons, becoming positive ion with plus
two charge Ca^{+2}

(−)

F

Fluorine seeks an electron
to fill outer shell, becoming
negative ion with minus one
charge F^-

Both oppositely charged
ions draw together in the
structure shown

F

Ca

FIG. 4.

atoms lose single outer electrons to become positive ions and then are able to combine with chlorine atoms which, gaining single electrons, are negative in charge. This process is illustrated in the lower part of Figure 4. Forces between ions are of a far different nature from those in metallic or covalent bonds and the properties of the minerals also vary greatly. The ionic bond is very common in the mineral kingdom, more so than either of the others. Ionically bonded minerals are moderately hard, are poor conductors of heat and electricity, have medium-high melting points, and many dissolve easily in water or acid. Most tend to be colorless when pure.

Many minerals are formed through the *combination bonds* previously mentioned, usually involving the covalent bond with the ionic bond. In some cases, no clear-cut bonds exist between certain atom layers and only very weak residual or "left over" attractions hold layers together. This is true in graphite and mica where strong bonds exist between the atoms in the exceedingly thin layers but very little attraction is left over to hold the layers together. Consequently it is easy to rub graphite between the fingers and observe the smudge from numerous minute bits detached from the main portion, or in the case of mica, to separate the sheets by slipping a fingernail between them to peel them off like pages of a book. Other minerals exert considerable bonding force in all three dimensions and are far less likely to be torn apart easily, but where decided differences in degrees of attraction occur, they too can split more easily in certain directions than in others. Many peculiar properties of minerals are due in no small measure to the fact that bonds do not stretch outward with absolutely uniform attraction in all directions within the mineral. When crystal structures are discussed later, these effects will be more clearly understood.

THE DESIGN OF CRYSTALS

Now that we have atomic "bricks" to build with and several types of bonding cement to hold them together, just how are they to be made into minerals? It is too bad that we can't have atoms large enough to directly fit them together, but if we assume them to be like miniature spheres in shape, why not use substitutes which should do almost as well? Colored glass marbles, the kind used for Chinese checkers, make good "atoms," while a tube of fast-drying Duco cement makes a fine "bonding" agent. Crystallographers building small atomic models use more elaborate means but ordinary glass marbles will do nicely even though they may fall apart later when the glue becomes too dry.

If we take a single marble and see how many others like it can be glued to it without leaving any spaces, inevitably we will find that six marbles can surround the original marble in the flat hexagonal pattern shown at the top in Figure 5. But this is only the beginning because we have cemented them together in a flat or two-dimensional pattern, and we need three dimensions as in real atomic structures. Now, as shown at B in Figure 5, we can fit a marble in each of the corners marked with dotted circles. Only three will fit and still touch all the rest. We now have a hexagonal ring surmounted by

a trio of marbles. In a few minutes when the cement is dry, we can flip over the model and cement three more marbles to the other side. This results in the cluster shown in D. There are now six marbles in a ring around the original marble, and three on each side for a total of twelve marbles surrounding the original! No matter what arrangements we try, we discover that for marbles of the same size the maximum number of contacts with other marbles is exactly twelve.

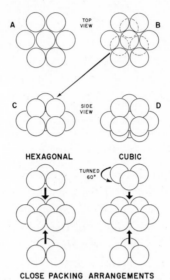

CLOSE PACKING ARRANGEMENTS

FIG. 5.

However, suppose that when the last three marbles were added, they were turned by sixty degrees or one third of the way around. What would happen? We would find that twelve marbles still surround the center marble but the

pattern has changed in a minor but very important respect. The difference is shown in the bottom illustration in Figure 5. In one pattern, called *hexagonal close packing*, the arrangement always leads to a honeycomb pattern, one layer on top of the other, with the pattern repeating every other layer. However, in the diagram showing *cubic close packing*, each layer skips two before it repeats and thus is different from the first. If a small model of each is made and turned about to provide views from all directions, it will be seen that the cubic model presents a square outline as shown at the bottom of Figure 5, but no matter how the hexagonal model is turned, it really appears balanced only when viewed from the top. Filled out models which show the developed form better are sketched in Figures 6A and 6B.

By now one fact should have impressed itself as we look upon the atomic models in Figures 5 and 6. It is the beautifully precise spacing of atoms which, as we can easily imagine, could go on forever if we continued to add more marbles to the arrangement that we started with. The even spacing not only takes place in one dimension, but in all three dimensions. In fact, if we took a model such as one shown in Figure 6, and turned it about in our hands, we would be astonished to find a distinctive pattern appearing in one direction, then another as the model was turned to another viewpoint, and still others!

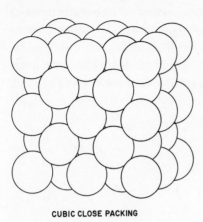

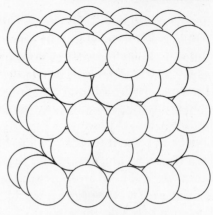

CUBIC CLOSE PACKING   HEXAGONAL CLOSE PACKING

FIG. 6.

Though the patterns seem to change, this is only because we look at them from different directions; however, the basic arrangement is fixed and must always repeat itself in space. By creation of such simple models we have also created the essence of the atomic arrangements found in minerals, which are better known as *crystals*.

Thus a crystal is essentially a simple arrangement of atoms in a pattern which repeats itself in three dimensions. As we will see later, there are variations in the pattern, but in every case, all atoms in the structure of a crystal attempt to draw as close to each other as possible. This in itself strictly limits the ways in which they can fit together, and therefore limits the number of patterns which are possible. Variations occur because all atoms are neither the same size nor of the same degree of attraction. Even when atoms are exactly the same, they do not always form crystals with the same internal atomic arrangements, as noted before in the case of diamond and graphite (carbon). In the native metals, held together by the metallic bond, cubic close packing occurs in gold, silver and copper, but hexagonal close packing prevails in the metals of the platinum group. A few *semi-metals,* so named because they have some properties of metals, but lack others, form curiously warped crystal frameworks which are essentially close packed but tipped to one side, as if the entire crystal had been squeezed upon diagonally opposite edges after it had grown. These elements are antimony, arsenic and bismuth. The tilting is due to bonding which exerts slightly greater force in certain directions, thus internally pulling the framework out of perfect alignment.

BASIC ATOMIC PATTERNS

Hexagonal and cubic close packed patterns work well when all atoms are of the same or nearly the same size, but do not work when there are great differences in size. In such instances, large and small atoms in the same crystal must work out other arrangements in order to permit them to draw as closely together as possible and to spread bonds as uniformly as possible. As the difference in size decreases, that is, as all atoms become nearly the same size, each atom is able to surround itself with more neighbors. The simplest pattern is shown in the sketch at the top of Figure 7. It is a dumb-

ATOM PATTERNS IN CRYSTALS

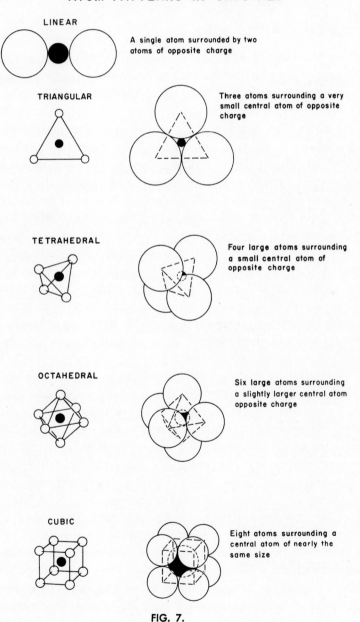

LINEAR

A single atom surrounded by two atoms of opposite charge

TRIANGULAR

Three atoms surrounding a very small central atom of opposite charge

TETRAHEDRAL

Four large atoms surrounding a small central atom of opposite charge

OCTAHEDRAL

Six large atoms surrounding a slightly larger central atom opposite charge

CUBIC

Eight atoms surrounding a central atom of nearly the same size

FIG. 7.

bell of two large atoms with a small one between. Below this sketch is shown another common arrangement in the form of a triangle with a small atom nestled between three large atoms. Next on the scale of complexity is a four-atom tetrahedral arrangement, with others following which permit a larger number of atoms to fit around a central atom until the cubic and hexagonal close packing schemes previously described are reached. All of these patterns appear in crystals.

RADICALS OR COMPLEX IONS

A number of elements of greatly different size sometimes form tight clusters of atoms called *radicals* or *complex ions,* the latter name being used to distinguish them from ions which contain only one atom. Attractive forces within radicals are so strong that for all practical purposes, each behaves like a single-atom ion. Calcite furnishes the prime example of a mineral containing a radical, in this case, the flat triangular *carbonate* radical, shown schematically at the bottom of Figure 8. Although calcite is composed of three elements, namely calcium, carbon and oxygen, the crystal structure really contains only two ions: calcium and carbonate. The carbonate radical consists of three oxygen ions surrounding the smaller carbon ion which fits snugly between them. Within the radical, four outer carbon electrons merge with outer electrons of the oxygens in a strong covalent envelope. Because the charge of four on carbon (C^{+4}) is not enough to balance the total of six negative charges on the oxygen (O^{-2}), there are two negative charges left over. These are neatly disposed of by ionic bonds to the calcium (Ca^{+2}) ions, which space themselves evenly throughout the structure as shown in Figure 8. Incidentally this is a good example of combination structures and combination bonds all within the same crystal.

The strength of carbonate radical bonding is shown by what happens to the radical when calcite dissolves in water to later form stalactites and stalagmites as in caverns. When the calcite dissolves, ionic bonds between calcium and carbonate are broken. Both ions drift in the water until a favorable site is reached for redeposition, as upon the walls of an opening. Here they rejoin to form calcite which eventually creates the beautiful formations for which limestone caverns are noted. During the watery trip, the carbonate radicals *do not break down* but remain together as units. Other radicals are also abundant, notably the *sulfate* and *phosphate* radicals. Each has a similar arrangement of large oxygen atoms surrounding smaller sulfur or phosphorus atoms. However, because the inner atoms are somewhat too large to allow the oxygens to form a flat radical, like the carbonate radical, the next best arrangement is taken, namely a *tetrahedral* arrangement as shown in Figure 7. Because all radicals are considerably larger than single-atom ions, their

CALCITE CRYSTAL STRUCTURE

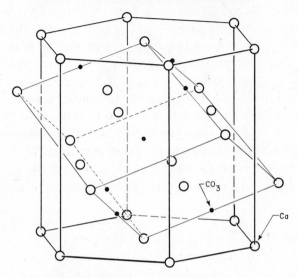

○   CALCIUM ATOMS
●   CARBON ATOMS marking the centers of each carbonate
    radical. One of the latter, greatly enlarged, is shown
    below. Each radical lies in horizontal position within the
    structure shown above.

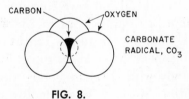

FIG. 8.

presence in any crystal influences strongly the geometry of the structure.

When written in chemical formulas, radicals are commonly enclosed in parentheses to show that they are to be treated as single ions: carbonate—(CO_3); sulfate—(SO_4); phosphate—(PO_4); etc.

WATER AND HYDROXYL

Water molecules sometimes fit into crystal structures which contain large enough spaces to accommodate them. However, they are weakly attracted as a rule, and only a modest degree of warming is enough to drive off the water. In other instances, water is more firmly bound in the crystals and actually forms an essential part of the crystal structure and when driven off, as by vigorous heating, results in destruction of the crystal. A good example of weakly-bound water is found in the calcium sulfate, gypsum, a mineral which provides the raw material for plaster of Paris. Two water molecules per calcium sulfate unit are present and occupy positions between layers of calcium and sulfate ions. Attractive forces are extremely weak and layers can be separated with ease into perfectly flat sheets. When heat is applied to gypsum, layers are disrupted and some of the water escapes. In the plaster industry, this is called "burning" gypsum. In powder form, burnt gypsum is sold as plaster of Paris and all that is needed to cause it to harden is to add water. When this is done, water molecules take their original positions between the calcium sulfate layers and gypsum is reconstituted. If the original gypsum is heated too strongly, all the water is lost and the resulting mineral is anhydrite which, as its name implies, is the waterless, or *anhydrous,* sulfate of calcium.

The hydroxyl ion, written (OH)−, contains hydrogen and oxygen as in water but contains one less hydrogen. Unlike water, which is usually weakly bound, hydroxyl is strongly bound within whatever crystal structure it happens to be because of its single negative charge. Its behavior is similar to that of other closely-knit radicals.

SILICATE TETRAHEDRA

Oxygen and silicon have a unique and enduring relationship because they combine in complex ions in which a central silicon atom is closely surrounded by four oxygen atoms. The fit is neat and precise, forming the *tetrahedral* arrangement shown at the top of Figure 9, in which atoms are cemented very strongly with covalent bonds. Because of the shape of the silicon-oxygen building block, it is commonly referred to as the *silicate tetrahedron.* While silicon has a valence of four, only one charge from each of its surrounding oxygens is neutralized, leaving another charge from each oxygen extending outward from the corners of the unit. In turn, the corner charges are used either to bond to other oxygens on corners of neighboring units, or to other ions. In either case, great structural variations are possible although the essence of all of them is the basic silicon-oxygen tetrahedral unit. The minerals arising from the combination of these units with other elements are called *silicates* which because of the great abundance of oxygen and silicon in the earth's crust, form nearly 95 percent of all mineral matter.

The increasing complexity of silicate unit combinations is shown in Figure 9. The simplest structure is a single tetrahedron which by attachment to other atoms becomes part of a larger crystal structure in many minerals. The name *nesosilicate* is given to this class, *neso* being derived from the Greek for "island" indicating that the silicate tetrahedrons are separate structural units. Pairs of tetrahedra may link as also shown in Figure 9, forming

SILICATE STRUCTURES

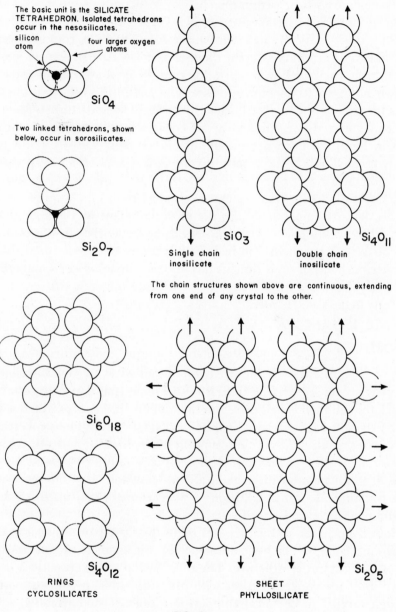

The basic unit is the SILICATE TETRAHEDRON. Isolated tetrahedrons occur in the nesosilicates.

silicon atom

four larger oxygen atoms

SiO_4

Two linked tetrahedrons, shown below, occur in sorosilicates.

Si_2O_7

SiO_3

Single chain inosilicate

Si_4O_{11}

Double chain inosilicate

The chain structures shown above are continuous, extending from one end of any crystal to the other.

Si_6O_{18}

Si_4O_{12}

RINGS
CYCLOSILICATES

Si_2O_5

SHEET
PHYLLOSILICATE

FIG. 9.

sorosilicate structures. *Soro* is derived from the Greek word for "group." Pairs of tetrahedra are linked by one oxygen ion which acts as a common corner for each of the tetrahedra. The third method of linking is by lengthening the double-unit group to form chains. The Greek word for "chain" or "thread" is *ino,* and the class is known as the *inosilicates.* Although bonds are strong within the chains, cross-bonds may not be as strong, leading to more or less easy splitting between chains. Double-chain inosilicates, shown schematically in Figure 9, are cross-linked by common oxygens.

Tetrahedral units curved upon themselves form ring structures, or *cyclosilicates,* from the Greek word *cyclo* for ring. Rings may contain three units as shown in Figure 9, four units, or six. The *phyllosilicates,* from the Greek *phyllo* for leaf or sheet, clearly describes the sheet structure formed by a series of connected rings linked at all corners in a hexagonal network as shown in Figure 9. Each sheet contains one or more networks and is very strongly bonded *within* the sheet but very weakly bonded *between* sheets, as in common mica.

The last class of silicates is an extremely important one because two of its minerals, feldspar and quartz, are the most abundant of all minerals. This class is called *tektosilicates,* with *tekto* meaning framework in Greek. Instead of silicon-oxygen tetrahedral units being linked in chains or rings, the linkage pattern extends in three dimensions like the filaments in a sponge, except in perfectly regular fashion. Properties of silicate minerals will be discussed more fully in the next chapter.

ISOMORPHISM

Among the hundreds of specimens in large collections, the amateur is sure to see striking similarities in the crystal shapes of certain specimens although their labels proclaim them to be different species. For example, he may note calcite, rhodochrosite, and siderite specimens covered with sharp-edged crystals which aside from differences in color and size, appear to be identical. Is there a relation? A closer examination of labels shows that each is a *carbonate,* and that chemically at least, the only difference is that calcite contains calcium in addition to the carbonate radical, rhodochrosite contains manganese, and siderite contains iron.

As it so happens, all of them crystallize in the same pattern; that of calcite is shown in Figure 8. The basic difference is in manganese or iron taking the place of calcium in the structure. The relationship seems clear: it is possible to have similar atomic structures with differing atoms in similar positions. Although varying in chemical composition, calcite, rhodochrosite, siderite, and several other carbonates are *isomorphs,* and the relationship is called *isomorphism.* The terms are from the Greek, *isos* for same and *morphe* for form or shape.

Interestingly enough, the isomorphous carbonates just mentioned also have similar properties, a not unexpected result in view of the nearly identical structures. In all three species, crystals split in the same directions, all are about the same in hardness, and even the angles between similar crystal faces measure nearly the same. A further point of interest is that the angles change slightly but measurably as the kind of ion changes. This is also not surprising because ions *are* of different size and certainly the structure can be expected to expand or shrink slightly according to which ion is present. The strong likenesses between isomorphs provide valuable clues to identity. Thus the similarity between crystals of dolomite, calcite, siderite, etc., are so consistent that after learning a few of their chief characteristics, one can recognize carbonate crystals of this class in a strange specimen, although the exact place in the class may not be known until tests are made.

ISOMORPHOUS SUBSTITUTION

Because some ions resemble each other in size and electrical properties, they not only form similar crystal structures, but are interchangeable in any given crystal. For example, many calcite specimens, when analyzed, contain appreciable amounts of other elements, usually magnesium, manganese, and iron. These prove not to be mere "impurities" but actual substitutes for calcium in the places in the crystal structure which calcium ions normally occupy. Because the structure isn't changed at all in respect to the way ions are spaced, such substitution is called *isomorphous substitution* or *simple proxy*. Perhaps the crystal is not as perfect as it could be with just one kind of ion, but it manages to adjust its framework to accommodate the slightly larger or smaller ion. Where there isn't too much difference in size or charge, as between calcium and manganese ions, or between magnesium and iron ions, the growing crystal cannot tell them apart and takes in considerable amounts of both. However, when there are great differences in size, as between calcium and magnesium ions, appreciable substitution is no longer permitted and each ion must form its own crystal. Thus if calcites ($CaCO_3$) and magnesites ($MgCO_3$) are tested, it is found that while some calcites contain considerable manganese they contain practically no magnesium, while magnesites contain scarcely any calcium but may contain some iron because the latter ions are of nearly the same size as magnesium ions and therefore fit into the magnesite crystal structure.

The rules for such substitution require that before any ion can take another's place, it must be of the same or nearly the same size and comparable in charge. If the size varies much, the ion may not fit although a few may be accidentally attached and covered over by a quickly growing crystal, in which case they must usually squeeze themselves between other atoms in the structure. Ions of this kind are actually impurities because they are not proxying for another ion. The name given to this process, *interstitial stuffing*, indicates how it happens. In crystals with more space for adjustment, in cases of poor fits by atoms of considerably different size, the structures may accept a number of such foreign ions but may be warped slightly and may even fall into a different crystal structure class. Sometimes ions of suitable size but not of the proper charge are taken into the crystal in which event, the electrical balance is restored by taking in ions which can make up the difference. This happens extensively in the feldspars and accounts for some of the remarkable variety of compositions and properties noted in them.

Substitution is sometimes limited, as in carbonates of the calcite group where manganese can substitute for only part of the calcium in the case of calcite and is therefore termed *partial* substitution. On the other hand, full proxying is permitted in other minerals, in which case it is termed *complete* substitution, as for example between fayalite, an iron silicate, and forsterite, a magnesium silicate. Both species occur with all proportions of iron/magnesium, providing a string of gradually-changing species called an *isomorphous series*. The pure species, fayalite and forsterite, are called the *end members*. Substitution causes changes in appearance and properties. In some instances, as in the fayalite-forsterite series, fairly accurate estimates of substitution can be made on the basis of change in color or in other properties although chemical analysis is needed to determine the exact proportions of iron and magnesium.

In addition to the ion-for-ion substitution described, some species show substitution of radical or ion clusters, such then being known as *cluster proxy*. This type occurs in the scapolite series where various proportions of chlorine, with carbonate and sulfate radicals have been discovered.

MISSING IONS

When analyzed, some minerals never provide ideal proportions of elements. For example, the iron sulfide, pyrrhotite, with ideal formula FeS, indicating one atom of iron per atom of sulfur, always contains somewhat more sulfur than it should have according to this formula. At one time this was thought to be caused by extra sulfur ions dissolved in the crystal structure, but later work proved that it was just the other way around: some of the iron ions were *missing*. If pyrrhotite crystal structures could be examined on a minute scale, gaps would be found where iron should be but is not. Such defective structures are being increasingly found among metal, semimetal and sulfide minerals, and account for puzzling departures from ideal chemical composition, even in what appear to be pure crystallized samples.

SOLID SOLUTION

At higher temperatures, mineral ions have more freedom of movement because heat imparts energy to them causing them to vibrate and "sweep out" more room. The practical effects of this vibration can often be observed in ordinary metal bars which show decided lengthening when measured before and after heat is applied. As a matter of fact, the blacksmith given the task of sharpening points on steel tools, as upon a pick, uses this principle to cause rearrangement of iron and carbon atoms in the steel crystals at higher temperatures, permitting the metal to be beaten into shape without cracking. In order to bring back the internal arrangement to that necessary for proper strength, he dips the hot point in water or oil, watches for certain signs which tell him how the recrystallization process is coming along, and finally quenches the point to cool it completely when the proper signs appear. Definitely, the kind of crystals in the steel when cold and when hot are different. In somewhat the same way, certain mineral crystals exhibit similar behavior, consisting, at low temperatures, of a mixture of several kinds of crystals, really two or more mineral species to be exact, and at higher temperatures, consisting of only one kind of crystal. Such substances are called *solid solutions,* because the minute crystals forced to form at low temperatures are often so small as to be invisible and therefore the mix somewhat resembles a liquid solution.

An excellent example of *unmixing* from solid solution is furnished by some feldspars, notably a kind called perthite which consists of microcline and albite feldspars at low temperature but is a uniform crystalline mineral at high temperatures. All ions of the several species adjust themselves at high temperatures into the crystal structure without difficulty, spacing themselves and spreading electrical charges to make a large crystal. This is possible only because heat allows more freedom for all ions to do so. At lower temperatures, the entire framework shrinks with atoms drawing closer together. At this time, loose fits are no longer possible and ions must move about to find better arrangements. At this point, the process of *exsolution* or *unmixing* takes place. The results are often visible in perthite crystals as alternating thin layers or criss-cross streaks consisting of the two species named.

POLYMORPHISM

During solidification of the earth's crust, minerals formed under widely varying conditions. In some instances, the crystal structure assumed by a certain species proved to be best under conditions then prevailing, which is to say, that this structure was *stable* under the circumstances. However, under different conditions, another structure may have been better and crystal ions accordingly shifted slightly from former positions. Specimens of both may show that the overall pattern is nearly the same but that details differ. Sometimes this is manifested in slight changes in outward appearance and properties, sometimes by drastic changes. When no change in chemical composition is involved, the several distinctive crystal structures are called *polymorphs,* meaning "several forms."

Undoubtedly, the most striking changes are produced in the carbon polymorphs: diamond and graphite, both of which are stable in different regions of pressure and temperature. The drastic difference in properties has already been noted; however, their distinctive structures, shown in Figure 10,

CARBON POLYMORPHS

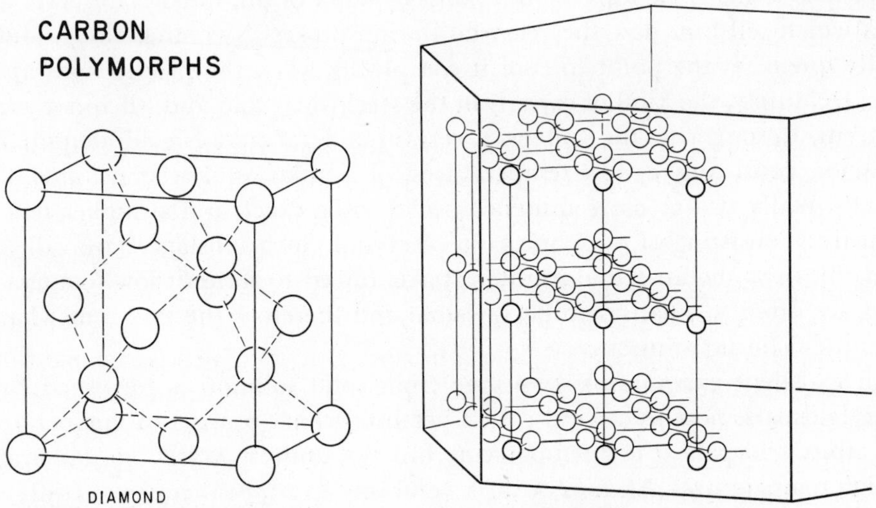

DIAMOND

Each carbon atom is at the center of a tetrahedron of four other carbon atoms. This arrangement continues throughout the crystal.

GRAPHITE

Carbon atoms are closely packed in sheets but sheets are separated by a considerable distance. Bonds are strong within sheets but very weak between them.

FIG. 10.

indicate why external appearances and properties are so at variance. The diamond structure is very uniform in respect to the way the carbon atoms are distributed in space. The distance between atoms is also uniform and the entire framework is extremely strong. On the other hand, the graphite structure shows atoms distributed in exceedingly thin layers with the distance between atoms least in the layers and greatest between them. Bonding forces are strong only within layers and practically no force holds the layers together. Graphite therefore flakes readily, is black in color, much lower in hardness and relative weight, etc.

Other well-known polymorphs are tridymite, cristobalite, and quartz, all of which are silicon dioxide, and rutile, brookite, and anatase which are titanium dioxide. It is known that some polymorphs grew under conditions of higher temperature and pressure than others, and by observing their presence in rocks, it is possible to make accurate estimates of pressure and temperature which prevailed at the time the minerals were formed.

NON-CRYSTALLINE MINERALS

Most minerals are crystalline but a few are not. Opal, for example, is silica (SiO_2) with a little water; as such it has practically the same composition as quartz, yet it always occurs in formless masses which, when broken, show the same kind of curving fractures noted in ordinary glass. Still, it is *not* a glass because it is not a product of melting; instead it formed from a solution of silica in water in which the silica divided into very small particles, just touching each other, causing the entire mass to become a jelly-like substance known as *colloidal solution*. When the water evaporated, the entire mass consolidated into a relatively hard and uniform material. Although each original particle may have been a minute group of crystals, or perhaps each still is, the properties of the whole are far from crystalline. It is known that opal is porous and relatively lighter than an equal mass of quartz. It must therefore contain numerous minute pores. Naturally it never forms distinct crystals.

While on the subject of glass-like materials, it may be wise to mention that a melted and solidified igneous rock, obsidian, is commonly mistaken for a mineral. Actually it is a mixture of several silicate minerals which developed into extremely small, partly-crystallized clusters, all of which together form the rock mass. Obsidian, opal, and window glass all lack a continuous crystal structure but may be thought of as mixtures of exceedingly minute crystals or crystal groups too small in size to be detectable even with the use of x-rays. Because of the lack of obvious crystallinity, such materials are called *amorphous,* from the Greek, meaning "no form."

Some originally crystalline minerals become amorphous through destruction of their crystal structure by radioactive bombardment. Sometimes the mineral happened to grow next to a species containing uranium or thorium, both of which are fissionable and shoot out high velocity particles capable of knocking atoms out of position, or even splitting them and creating new elements. In other instances, the mineral contained one or both of these radioactive elements as regular constituents, in which case, the internal destruction is nearly complete, although the crystal, when removed from the ground, appears quite undamaged on the outside. Zircon is commonly affected by thorium, as is monazite and other minerals containing rare earth elements. When such crystals are broken open, the fracture surfaces resemble dark brownish or black glass, while other tests quickly prove that they are dissimilar in many ways from normal crystals of the same species. Amorphous minerals of this kind, originally crystalline, are called *metamict*. For the reasons given, their testing and identification is often difficult and time-consuming.

SUMMARY

Without becoming too detailed, this chapter has attempted to show the relationships between atoms which force them to combine only in certain ways. With 92 kinds of atoms or elements, the possible combinations could go into the millions but fortunately for mineralogists, professional and amateur, mineral species are presently limited to a number approaching 2,000. However, new mineral species are being reported at about the rate of twenty-five per year but most are obscure compounds, usually very rare, and certainly unimportant as compared to the handful of species which are actually common enough to be found everywhere and to be actively studied by amateurs.

In summary, the atom itself dictates how it will join others. All atoms are created from exactly the same raw materials and it is a source of wonder and astonishment that merely changing the number of particles in the nucleus and in the surrounding swarm of electrons, spells the sole difference between a heavy metal such as gold and a light invisible gas such as helium. Because atoms vary in size and in degree of attraction for each other, only a limited number of geometrical patterns can be used when joining. In general, the patterns are dictated by the large non-metallic ions of relatively few but very abundant elements, of which oxygen is by far the most abundant. Some patterns are simple, others very complex, but whatever pattern is eventually adopted by a swarm of combining atoms, it is regular and repeating, and extends unbroken throughout the mass. Such a mass is called a *crystal,* to which there is no limit in respect to size so long as the pattern from one end to the other is uniform in spacing and direction.

Practically all minerals are *crystalline,* while the few which are not, the *amorphous* minerals, may be composed of crystals which are so small and so turned about in all directions that no external evidence can be gained as to their existence. Others which appear to have been crystals at one time, have been disturbed by radioactive bombardment to such an extent that the regular internal arrangement has been partly or wholly destroyed.

Some atoms commonly found together in nature have such a strong attraction for each other that they frequently form small clusters of characteristic geometry which retain their identity within larger crystal structures of which they may be part. Such *radicals* or *complex ions,* strongly influence the final structure of the crystal.

Because the variety of structural patterns is limited by considerations of ion size and charge, similar ions tend to form *isomorphous crystals* though they may not be related chemically. In some crystals, it often happens that a position normally occupied by ions of one element may be occupied instead by ions of another element providing they are of nearly the same size and charge. *Atomic substitutions* of this type significantly affect its composition and properties but do not destroy the structure although the latter changes slightly in geometric proportions. Substitution is so common that minerals nearly pure according to an ideal formula are seldom found.

Temperature and pressure around a growing crystal often determine the geometry of the crystal structure so that one structure in preference to another is chosen as being more stable under the environment at the time. Although composed of exactly the same elements in the same proportions, several structures may therefore be found for a given chemical compound and each is then considered to be a *polymorph* and a different mineral. Polymorphs usually display widely varying properties although they are composed of exactly the same elements in exactly the same proportions.

Elements in the earth's crust therefore *prefer to form crystals* of characteristic internal geometry according to the elements involved. The definition of a *mineral* may now be modified to say that *it is almost always a crystalline solid of characteristic internal atomic structure, composed of a limited but not exclusive number of elements combined in specific proportions.* This definition covers what is considered to be a mineral *species*. Within each species may be found numerous varieties, some of which seem to be strikingly different in appearance, but are essentially identical in terms of composition and internal atomic structure.

3

Classification of Minerals

As with other objects from the world of nature, minerals are classified with the idea of placing related species together. By doing so, it is easier to point out likenesses and dissimilarities within each class and to point out how one class differs from another. Classifying minerals is not easy because of the great number of species and the considerable variety within many of them. From the earliest days of mineralogy, this problem has occupied the attention of many mineralogists, each of whom attempted to develop some system which would make recognition of minerals a simpler and surer task.

Before anything was known about atomic structure and modern chemistry, the obvious method for classifying minerals was on the basis of outward appearances and such external tests as could be applied to minerals without involving chemical tests or analyses. Georgius Agricola, writing in the sixteenth century, used external characters in describing minerals in his *De natura fossilium* (the nature of "fossils," as minerals were then called). By the eighteenth century, schemes had been devised to classify minerals in much the way botanists classify plants, complete with Latin terminology; this was called the "natural history" method. The chief early proponent of this method was Carolus Linnaeus (1707-1778) of Sweden who published his suggested system in 1735, followed by Friedrich Mohs of Germany, who published his in 1820. In 1758, Axel Frederic Cronstedt (1702-1765), a Swedish chemist, suggested that minerals could be sensibly classified according to their chemical composition, but the strongest support and convincing reasons waited until 1814, when another Swedish chemist, Jöns Jakob Berzelius (1779-1848), published the first classification based on extensive mineral chemical analyses. A "mixed" system combining both natural history and chemical features was proposed in 1774 by A. G. Werner, as previously mentioned in Chapter 1. With the work of Karl Wilhelm Scheele (1742-1786) in Sweden, and in Germany by Martin Heinrich Klaproth (1743-1817), the merits of chemical classification became even more convincing. Particularly persuasive was the series of brilliant experiments clearly demonstrating isomorphism which

42

were carried out by Eilhard Mitscherlich (1794-1863) in Germany; these did much to explain hitherto puzzling variations in composition and properties of many minerals.

It is interesting to see the impact of all this upon the development of the Dana *System of Mineralogy* which is now in its seventh edition and is regarded as the most authoritative compilation of mineral data in existence. The first edition of the *System*, by a then very young author, James Dwight Dana (1813-1895), appeared in 1837 and followed the natural history scheme. However, as the advances previously described made themselves known, Dana threw out the natural history scheme and, in his third edition, adopted an approach that was the forerunner of the one now used in the seventh edition. Practically every book on mineralogy, and this one too, uses this system or one like it. As pointed out so long ago by Berzelius, the system has the merit of putting minerals together which are not only alike chemically but also alike from other standpoints. There are some places where the progression of chemical composition and crystal properties are not as smooth as one would like, but on the whole the scheme is the best that can be devised.

CLASSIFICATION BY IONS

The Berzelian system places minerals into large classes according to negatively-charged ions or *anions*. These are placed to the *right* in mineral formulas. Thus we find large classes called "oxides," "carbonates," "sulfates," etc., which reflect an apportionment according to each of these anions. If this is thought to be an arbitrary method, it is pointed out that because anions gain electrons in many instances, or form complex ions, they become the largest units in the crystal structure and therefore dictate the kind of structure to be adopted by the growing crystal. This problem in crystal architecture is similar to making a monument primarily out of boulders with small pebbles used as fillers, and properly most mineral crystals should be thought of as structures framed from large anions, as oxygen and oxygen-containing radicals, with smaller metal ions fitting between them. It is no accident that when large radicals as the carbonate or sulfate anions are present in crystals, basic patterns are inclined to be the same in the several species containing them. This is why a "carbonate group" of related species means far more than a convenient listing of names.

Some of the important building-blocks of mineral crystals are furnished in the following table of common mineral ions. Note the wide variation in charge or valence. No doubt if all atoms were limited in their ability to cast off or gain electrons, there would be far fewer species. Thus we discover that relatively few atoms, as shown at the top of the table, have fixed valences, while the greater number shown below have variable valences. The latter consequently contribute greatly to the wide variety of species. The positively-charged ions, or *cations,* appear at the left because this is the place reserved for them in mineral formulas. The far fewer anions, appearing at the right, may seem grossly outnumbered but, as pointed out before, carry much weight when it comes to forming crystals. Furthermore, large anions such as oxygen are incredibly abundant, although when one observes most rocks they seem far removed from the gaseous state of oxygen with which we are familiar. Oxygen is so abundant that it comprises over 62 percent of *all* atoms in rocks, followed by silicon with a little over 21 percent. Between them, over 83 percent is accounted for! The remaining percentage is divided almost completely among the few elements following: aluminum (6.5 percent), sodium (2.6 percent), calcium (1.9 percent), iron (1.9 percent), magnesium (1.8 percent), and potassium (1.4 percent). The remainder of elements comprise only about 1 percent! Thus the table below contains only one cation of great abundance, silicon (Si), while the anion side contains oxygen (O) as being also truly abundant. The role of oxygen is also predominant in the

COMMON MINERAL IONS

FIXED VALENCE

Charge	Cations	Charge	Anions
+1	Na, K, Li, Ag, Cs, H	−1	F, Cl, Br, I
+2	Ca, Mg, Ba, Zn, Be, Sr, Cd	−2	O
+3	Al, B		*Radicals and Groups*
+4	Si, Zr, Th	−1	(OH)
+5	Ta	−2	(CO_3), (SO_4), (WO_4), (MoO_4), (CrO_4)
+6	W	−3	(PO_4), (BO_3), (AsO_4), (VO_4)

VARIABLE VALENCE

Charge	Cations	Charge	Anions
+1,+2	Cu, Hg	−2	S, Se, Te
+1,+3	Au	−3	P, As, Bi, Sb
+2,+3	Fe, Ni, Co	−4	C
+2,3,6	Cr		
+2,3,4,6	Mn		*SILICATE GROUPS*
+2,+4	Pb, Sn, Pt	Charge	Group
+2,4,6	S, Te, Se	−2	SiO_3 (inosilicate)
+3,+4	Ti, Ce	−2	Si_2O_5 (phyllosilicate)
+3,4,6	Mo	−4	SiO_4 (nesosilicate)
+3,+5	Sb, As, Bi, Nb, V, P	−6	Si_2O_7 (sorosilicate)
+4	C	−6	Si_3O_9 (cyclosilicate)
+4,+6	U	−6	Si_4O_{11} (inosilicate)
		−8	Si_4O_{12} (cyclosilicate)
		−12	Si_6O_{18} (cyclosilicate)

radicals listed at the right and finally, with silicon, is responsible for the vast realm of silicate minerals which owe their existence to the linked oxygen-silicon tetrahedrons forming the structures shown at the bottom of the table. More detailed remarks on the formation of minerals will be furnished in Chapter 9.

MINERAL FORMULAS AND CHEMICAL COMPOSITION

Complex mineral formulas appear to be meaningless jumbles of letters and numbers when one sees them for the first time. In reality, as mentioned above, a definite scheme is followed in their preparation which not only results in stating the chemical composition but also furnishes information on crystal structure. In effect each chemical formula states that if a sample of the mineral were taken apart, atom by atom, and all atoms counted, the sums would be in definite proportions. The small numbers placed at the lower right of element symbols show these proportions. A simple example is that for quartz, SiO_2, meaning that each atom of silicon is attended by two atoms of oxygen. In calcite, $CaCO_3$, there are three atoms of oxygen for each carbon in the carbonate radical, and one carbonate radical for each calcium atom. This formula could be written 1Ca,1C,3O, but certainly this would not show the important relationship between the carbon and oxygen atoms in the carbonate radical. It is therefore not as good a way of putting it. Other formulas, unfortunately, are not as meaningful; the quartz formula, for example, really tells us nothing about structure. Nevertheless, the formula system is again the best that can be devised without becoming overly complex.

Figure 11 shows the hematite formula which is not as simple. As pointed

CHEMICAL FORMULA-HEMATITE

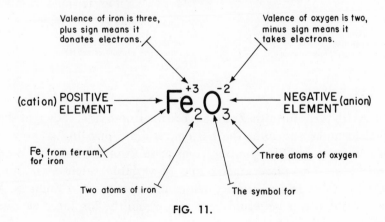

FIG. 11.

out before, the formula, Fe_2O_3, places the *cation* (Fe) to the left and the *anion* (O) to the right; both will be found in the same relative positions in the ion table furnished earlier. The numbers at the lower right of each ion indicate the combining proportions, that is, two irons for three oxygens. Now let us check the ion table for valences to see if this compound does provide the necessary balance of electrical charges. Because oxygen is a *fixed valence* ion, we first refer to the upper right portion of the table and note that its valence is minus two, that is, it will take two electrons. Because there are three oxygens, each taking two electrons, a total of six must be furnished to the oxygens by the irons. The left side of the table shows that iron is listed as a *variable valence* ion of either $+2$ or $+3$ charge, that is, it is capable of giving up *either* two or three electrons. Obviously because the formula shows only two iron ions, each must have given up the higher number of electrons, namely three. Thus the balance is struck: two 3-charge irons equal three 2-charge oxygens, or six equals six.

In case substitution takes place, as really happens more often than not, formulas become more complex because substituting ions must be shown in proper position. In Figure 12, such an example is given, namely, the mineral

CHEMICAL FORMULA-FRANKLINITE

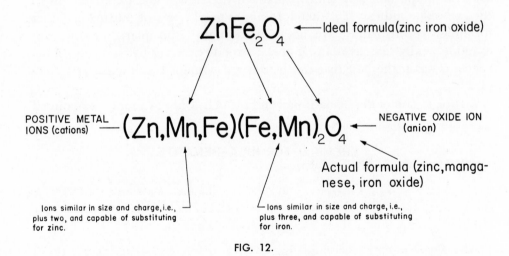

FIG. 12.

franklinite, with *ideal* formula $ZnFe_2O_4$, or zinc iron oxide, shown at the top. The *actual* formula, shown at the bottom, is more puzzling in appearance although not really so. The complications arise from the fact that a double substitution has taken place among ions of variable valence, in this case, manganese and iron. In franklinite, manganese displays two valences, Mn^{+2} and Mn^{+3}, while iron is Fe^{+2} and Fe^{+3} also. From the ion table we discover that zinc is a fixed valence ion of $+2$ charge, and therefore when substitution occurs for zinc, it must be by divalent manganese and iron. In the simplified formula, iron must have a valence of $+3$ in order to balance charges. Therefore both iron and manganese ions proxying for each other in the second part of the more complex formula must be triple-charge or trivalent. It is to be noted that when such substitutions occur, they are always indicated by placing the exchangeable ions together in the same set of parentheses.

In interpreting any mineral formula, the ion table furnished above is useful, particularly the portion showing ions and radicals of fixed valence. For example, to determine the valence state of copper in two common oxides, cuprite, Cu_2O, and tenorite, CuO, one need only remember that oxygen is fixed in valence at minus two. Thus the valence of copper in cuprite must be plus one, while in tenorite it must be plus two. Complexities and curious inconsistencies arise in some sulfides where simple ionic bonding is *not* involved, while metallic and covalent bonding may be. In them, some formulas do not balance at all neatly, as for example in realgar, AsS, as compared to orpiment, As_2S_3. The latter can be easily balanced by taking sulfur as divalent, and arsenic as trivalent, the total of six sulfur charges being balanced by a like number of arsenic charges. On the other hand, in realgar, taking either $+3$ or $+5$, the common valences of arsenic, does not result in an even balance against sulfur which is usually -2 in valence in metal compounds. Because analyses of realgar show consistent proportions of arsenic against sulfur, the formula must be correct, but some feature of the crystal structure is causing a left-over arsenic charge to be used up in some way, possibly by covalent bonding. In any event, the compound does not seem to be a lasting one, because realgar decomposes or *alters* when exposed to light, into orpiment, also absorbing oxygen from the atmosphere to produce an additional compound, an arsenic oxide.

MINERAL COMPOUND NAMES

The way to *say* and *spell* the chemical names of mineral compounds is puzzling unless the system used to coin such names is understood. By general agreement, the following rules have been adopted:

- Two-element compound names end in *ide. Examples:* silver sulf*ide*, molybdenum sulf*ide*, sodium chlor*ide*, etc.
- If the positive element (left-hand side of formula) exists in two valence states, a suffix is often used to show which is which. *Examples:* Iron has common valences of 2 and 3, therefore, two-valent or *divalent* iron has *ous* added to its name as in FeO, ferr*ous* oxide; *trivalent* iron has *ic* added as in Fe_2O_3, ferr*ic* oxide. Expect such suffixes in elements as tin (*stannous, stannic*), chromium (*chromous, chromic*), etc.

■ Prefixes are sometimes used with negative elements (right-hand side of formula) to indicate the number of atoms. *Examples:* realgar, AsS, arsenic *mono*sulfide; rutile, TiO_2, titanium *di*oxide; orpiment, As_2S_3, arsenic *tri*sulfide.

■ Compounds end in *ate,* when it is supposed that they have been formed from an acid which ends with the suffix *ic.* Often the acids are hypothetical, as the so-called *silicic* acids which at one time were thought to react with elements to form silic*ates. Examples:* Carbonic acid = carbon*ates,* sulfuric acid = sulf*ates,* phosphoric acid = phosph*ates,* vanadic acid = vanad*ates,* etc.

While the above rules are frequently followed in the naming of mineral compounds, some are disregarded without causing any real confusion. For example, one could call molybdenite, MoS_2, molybdenum *di*sulfide because it has two atoms of sulfur per atom of molybdenum, but inasmuch as there is only one molybdenum sulfide in the mineral list, it seems hardly worth the bother and certainly omitting the *di* cannot cause confusion in this case. In other instances, however, as with *polyvalent* elements as manganese, which can combine with other elements in several ratios, it may be helpful to be more specific. For example, manganosite, MnO, reflects manganese combining with a valence of two, and one could call this manganese *mon*oxide to distinguish it from pyrolusite, MnO_2, in which manganese is tetravalent and where one properly calls it manganese *di*oxide. To avoid confusion where such might exist, or to be exact in indicating which chemical compound is meant, one should use either the chemical formula or the species name. In some instances, as in *series,* it is necessary to use both because the chemical composition can change from one end member to the other.

MAJOR MINERAL CLASSES

Traditionally, the Berzelian system places the elements first for simplicity, followed by more complex compounds of two or more elements, then those containing radicals, and finally brings up the rear with the large and important class of silicates. Within each of these major classes are divisions according to increasing complexity in crystal structures. Thus all carbonates, for example, are found together, but even here are subdivided according to crystal structure into the *calcite group, aragonite group,* etc. Within groups may be found the series discussed in the previous chapter, if such exist, and finally species. The latter are placed under their own headings and if important varieties are known, these are given under the appropriate species. The following classes will be used in the portion of this book devoted to descriptive mineralogy:

Major Mineral Classes

Class	Remarks
Elements	Metals, alloys, semi-metals, and non-metals
Sulfides	Sulfur is the principal anion; includes tellurides and arsenides
Sulfosalts	Sulfur and a semi-metal (As,Bi,Sb) are anions
Oxides	Oxygen is the principal anion; includes hydroxides in which the hydroxyl radical (OH) acts as an anion
Halides	Halogen elements are anions, namely F, Cl, Br
Carbonates	Carbonate radical (CO_3) is the principal anion
Borates	Borate radical (BO_3), and borate groups are anions
Sulfates	Sulfate radical (SO_4) is the principal anion
Phosphates, chromates, arsenates, vanadates	Anions are principally PO_4, CrO_4, AsO_4, and VO_4
Tungstates, molybdates	Anions are principally WO_4, MoO_4
Silicates	Silicate groups as anions combined with other elements; includes silica (SiO_2)

CHEMICAL CLASSES SUMMARIZED

As remarked before, placing chemically related minerals together generally results in putting similar species in similar classes. Naturally, formulas are also similar, the pattern being nearly the same in each class and serving, when such formulas are seen by themselves, to refresh one's memory as to the features of the class. Because likenesses within each class of minerals are often consistent and striking, knowing the chief characteristics is of great help in identification. Thus the metallic luster typical of sulfides as a whole, suggests that a strange mineral belongs in this class if it also has such luster, while the near-perfect angular cleavage of calcite, if seen in an unidentified specimen, similarly suggests that a rhombohedral carbonate is at hand, though it may not necessarily be calcite. Conversely, seeing either the symbol S or radical symbol CO_3 in a mineral formula, evokes a mental image of properties associated with all sulfides and carbonates.

In respect to class formulas, the clue to class is provided by the right-hand part of the formula, that is, the part belonging to the negative ion or anion. Thus the oxygen symbol to the right, alone or modified by a number at its lower right, indicates an oxide. If it is included as part of a radical, this is sometimes difficult to discover unless the radical as a whole is enclosed in parentheses. Sulfides will show the sulfur symbol to the right, silicates the silicate radical, etc. These points will become clearer when the last part of the book dealing with descriptive mineralogy is consulted.

Elements. The letter symbols, as Au for gold, Ag for silver, etc., are usually written alone because impurities are ordinarily present only in small amounts. However, natural alloys are shown with two or more symbols because alloying proportions are fixed and the results are true chemical compounds. Among the elements are included the *metals,* the *semi-metals,* and the *non-metals.* As may be guessed, the metals are chiefly recognizable because they look like metals, and behave as we believe metals should. They are heavy, soft, malleable, and when freshly cut, metallic in luster. Gold and platinum often occur as nuggets because their malleable nature permits them to survive stream pounding without disintegration. Semi-metals display some properties of true metals, that is, they are metallic in appearance, are heavy and soft, but are also brittle and therefore lack the malleability so characteristic of the true metals. The non-metals include diamond, graphite, and sulfur; each of these species is distinctive in appearance and properties. On the whole, native elements are rare because they readily combine with other elements to form compounds. About 30 species are known in this class, but only gold, copper, silver, sulfur, graphite, and diamond are familiar to amateur collectors.

Sulfides. Sulfides are mainly compounds of metals and semi-metals with sulfur. The formulas always show the sulfur symbol S last, but one or more metals may appear before this symbol. In instances where two metals can substitute for each other in the crystal structure, as when iron (Fe) takes the place of zinc (Zn) in sphalerite, the formula is written to show substitution as follows: (Zn,Fe)S, thus making the relationship clear. Placing the zinc first indicates that it usually is found in greater quantity than iron. It is to be noted that iron and zinc are exchangeable for the same positions in the sphalerite crystal structure, however, a more complex class of sulfides, as represented by cobaltite, contains *two* metals each of which occupies its own positions in the cobaltite crystal structure, and therefore are *essential* constituents. The formula for cobaltite is accordingly written as CoAsS and without parentheses. Whereas iron could be completely absent in sphalerite, and the resulting mineral would still be sphalerite, the removal of either cobalt or arsenic from cobaltite would result in an entirely different compound and a different mineral.

Compounds containing tellurium, arsenic, antimony and selenium, occupying the negative element position of sulfur, are similar in properties and are usually included with the sulfides. These are known as *tellurides, arsenides, antimonides,* and *selenides.* All of them are uncommon to rare.

Sulfides, and the other species mentioned, are compounds of small metal ions and consequently provide weighty hand specimens which fact, in itself, is a valuable recognition feature. Almost all are opaque, either very dark in color or metallic in luster. Pyrite, for example, looks like freshly polished brass, chalcopyrite looks like brass that has discolored a bit, while galena resembles tarnished lead. On the other hand, bornite displays a beautiful iridescence on fracture surfaces resembling some of the deepest blue or purple hues of peacock's feathers, while covellite discolors to rich purple, and chalcocite is so black that it looks as if it had been coated with soot from a candle flame. Sulfides commonly occur mixed together in large masses and then are distinguished with difficulty. Transparent sulfides, such as sphalerite and cinnabar are still somewhat metallic in luster as if they had been partly silvered and not a very good job done of it. All sulfides are very weak, most are brittle but a few can be cut with a knife to produce shavings. Sulfides are common and make excellent, showy, mineral specimens.

Sulfosalts. The name *sulfosalt* is derived from hypothetical acids of sulfur which, in equally hypothetical reactions with metals, produce *salts* of sulfur. Although it is doubtful that such acids did create the sulfosalt minerals, the name is a convenient one to distinguish these compounds from sulfides. As a class, sulfosalts differ very little in properties and appearances from sulfides, with which they are easily confused. The chief difference, chemically, lies in the fact that semi-metals, as arsenic, antimony and bismuth, join with sulfur to form negative ions. Formulas therefore show As, Sb, or Bi joined with S on the right side. As an example, enargite, copper arsenic sulfide, is written as Cu_3AsS_4. Copper is the positive ion while arsenic and sulfur are negative. There are many complex sulfosalts but on the whole, few are important mineralogically except to collectors who prize fine specimens of enargite, pyrargyrite, tetrahedrite, bournonite, etc.

Oxides and Hydroxides. In formulas of this class, the oxygen symbol O, or the hydroxyl symbol OH, appear at the right in the negative ion position. In some instances, O and OH appear together, e.g., in manganite, $MnO(OH)$. A number of oxides contain water, H_2O, and a few rare species contain carbonate, CO_3. *Simple* oxides are compounds of a metal and oxygen, sometimes containing water. *Multiple* oxides are compounds of *two essential metals* with oxygen, as in spinel, $MgAl_2O_4$. In older mineralogical literature this particular oxide was called magnesium *aluminate,* indicating the belief that the aluminum and oxygen atoms were linked in a radical or complex ion similar to carbonate, sulfate, phosphate radicals, etc., but it has now been shown that the bonds between oxygen and the two metals concerned are alike and therefore no radical exists. This has also been shown to be true for other species supposed to contain radicals, as the manganates, columbates, tantalates, and titanates.

Oxides as a whole are strong minerals of medium to heavy weight. Some such as cuprite (Cu_2O) are relatively soft, but a number in the spinel and hematite groups are extremely hard and consequently are used as gemstones when transparent and flawless. The hydroxides are generally much softer than the oxides, and because a number form from the alteration of other minerals, they are inclined to grow as dense masses of minute crystals rather than large single crystals. Goethite often appears in dark brown cubes which look exactly like pyrite in form and markings because all the pyrite has altered into goethite without disruption of the pyrite form.

Because oxygen is so active chemically, and so universally abundant, oxides and hydroxides represent the end-products of the attack of oxygen upon other elements or minerals and are therefore stable compounds surviving under conditions which would cause other minerals to alter. It is for this reason that rutile, spinel, chrysoberyl, corundum, ilmenite, and others, are often found in gravels and sands, not much worse off except for the wear. Oxides are abundant and yield excellent, showy specimens.

Halides. The name *halide* is derived from the group name for the *halogen* elements: fluorine (F), chlorine (Cl), bromine (Br), and iodine (I). In halide formulas, one of these symbols will be placed at the right in the negative ion position. By far the greater number of halide compounds are chlorides, reflecting the abundance of this element, followed by scarcer fluorides and the rare bromides and iodides. The *normal* halides consist of a metal and a halogen, sometimes with water. *Oxyhalides* are principally oxychlorides, with characteristic formulas showing oxygen between positive metal ions and negative chlorine ions. *Hydroxyhalides* are similar but include the hydroxyl radical (OH). Some halide minerals, notably the *alumino-fluorides* cryolite, chiolite, ralstonite, etc., contain structural groups similar to silicate groups, but consisting of aluminum ions surrounded by six fluorine ions. Many species in this and other halide groups are rare.

By far the most common halide is ordinary salt, or *halite* (NaCl). Despite its abundance, or perhaps because of it, it is not collected very enthusiastically. Its tendency to absorb water from the atmosphere requires special storage precautions, a point which is not in its favor. On the other hand, the common calcium fluoride, fluorite (CaF_2), occurs in stable, colorful, and often large crystals which are much prized and actively collected. All halides are weak, brittle, and soft minerals, generally pale in color and sometimes, as noted in the case of halite, readily dissolved in water.

Carbonates. The carbonate anion (CO_3) is the distinctive unit, largely dictating the crystal structures of nearly 70 species. However, few species are abundant, the remainder being rare to very rare. The *normal carbonates* are most familiar to collectors, comprising three groups, the calcite, dolomite and aragonite groups. Of far lesser importance are carbonates containing water or hydroxyl, although within the latter sub-class fall the well-known and popular azurite and malachite, and several other colorful species familiar to collectors specializing in copper minerals. The calcite group formulas contain a single metal cation of small size, which together with the large carbonate radical, results in characteristic closely related structures exhibiting very similar properties. The dolomite species are similar but structures vary because two metal ions in the cation positions are involved, resulting in layered crystals, with every other layer containing ions of one metal only. Crystals of the aragonite group contain larger single metal cations, causing considerably different internal structures and requiring placement into a distinct crystal class.

All carbonates are relatively weak and fragile minerals, many fracturing easily and perfectly along characteristic crystal directions. The majority dissolve more or less readily in acids, giving off "fizzing" bubbles of carbon dioxide (CO_2) as carbonate radicals are detached and destroyed. The vigorous reaction to acids is a valuable identification test. A number of carbonates, particularly those containing water, dissolve in water. Calcite and dolomite are so abundant that in places upon the earth's surface, they form large expanses of rock consisting essentially of one or the other species. Cabinet specimens are readily available from among the common species and are prized by collectors for the great variety and beauty of crystallizations. Although some carbonates are vividly colored, the majority incline toward white, or colorless, and transparent.

Borates. Of the approximately 45 species in this class, the best known to collectors are the *hydrated borates* from sedimentary deposits at the bottom of playas in the arid regions of Southern California. Were it not for the dryness of this and similar regions, few species would be preserved because of their very easy solubility in water or decomposition when the water in their environment is changed in some respect. On the other hand, some borates, notably the *anhydrous borates* and the *hydroxyl-halogen borates,* are extremely stable, some also being very hard. The less stable borates contain the borate ion (BO_3) linked to others in flat networks interspersed with easily lost water groups, accounting for their general softness and solubility in water or acids. The more durable borate species contain the (BO_3) group in separate units in crystal structures. Formulas show the (BO_3) radical or sometimes more complicated groupings as (BO_5) and, when sheets of (BO_3) radicals are involved, or other complex linkages, the formulas merely show the count of boron and oxygen atoms without furnishing clues as to structure.

The common hydrated borates are weak, soluble, unstable minerals, mostly white to colorless, sometimes occurring in large, handsome crystals and furnishing good cabinet specimens. Many must be carefully preserved to prevent alteration due to loss or gain of atmospheric water.

Sulfates. The sulfate radical (SO_4) is the unfailing hallmark of formulas of this class, and consistently appears at the right in the negative ion position. There are several hundred sulfates divided according to chemical and structural similarities. Thus there are classes of *anhydrous* sulfates, or those without water, *hydrated* sulfates, or those with water, and further classes of each containing the hydroxyl (OH) ion or an ion of one of the halogen elements such as chlorine. Formulas in each of the classes reflect the composition, the hydrated sulfates showing (H_2O) attached to the body of the formula with the usual period. A small class of *compound sulfates* contains both carbonate and sulfate radicals.

Out of the hundreds of sulfates, only a few are familiar to amateur collectors. Because many are susceptible to gaining or losing water, they must be carefully protected in collections or suffer destruction. Most amateurs feel that this is too much trouble and don't bother to collect them. On the other hand, the members of the barite group are stable compounds yielding splendid crystals, sometimes of strong coloration and therefore much prized for collections. Gypsum, a hydrated sulfate, is relatively stable despite its water content and is also extremely popular. Except for a few copper sulfate compounds, and some others, the majority of sulfates are inclined to be white or colorless or only faintly tinged with color. All are very soft and fragile.

Phosphates, Chromates, Arsenates, Vanadates. Nearly 250 species are included in the species classes named in the above heading, but collectors seldom see more than a few of them. The reason is that many are rare or found in such small inconspicuous masses or crystals that they easily escape notice during collecting trips. A large number of phosphates occur in very small crystals and ferreting out their identity is a chore much esteemed by those who like to study minute crystals under magnification, the "micromounters." Formulas in these classes show the appropriate radicals in the anion position as before, namely: (PO_4), (CrO_4), (AsO_4), and (VO_4). Anhydrous and hydrated species are further modified by the addition of (H_2O), while other classes contain the hydroxyl radical (OH) and/or a halogen ion.

The phosphates are marked by many species yielding excellent colorful specimens, sometimes comprised of large crystals. Some of the better known are the apatite group, the pyromorphite series, turquois, wavellite, etc. The chromates are rare, and only one species, crocoite, is consistently sought for by collectors; specimens of crocoite are highly prized and costly. Of the arsenates, only mimetite, adamite, and erythrite are commonly collected, the other species being relatively rare. The vanadates are also rare, with species very low in number as compared to phosphates. However, colorful specimens, often splendidly crystallized, are furnished by vanadinite. In general, species included in the classes named above are inclined to be soft, brittle, and medium-to-strongly colored. Many are readily attacked by acids.

Tungstates, Molybdates. Very few species are represented in this division but several provide colorful specimens much prized by collectors. The tungstate radical (WO_4) is again at the right as in the wolframite group where a series exists between huebnerite, $MnWO_4$, manganese tungstate, and ferberite, $FeWO_4$, iron tungstate. Wolframite, which occupies a position between the two end-members, is written $(Fe,Mn)WO_4$, to show that iron and manganese can substitute for each other. Another tungstate, noted for its rarity in well-formed crystals and its bright blue ultraviolet fluorescence is scheelite, $CaWO_4$. The sole molybdate (MoO_4) seen to any extent in collections is wulfenite, $PbMO_4$, or lead molybdate, a mineral which frequently forms beautiful vivid orange crystals of glistening luster.

Tungstates and molybdates are soft and brittle, dark or vividly colored, and common in good crystals from a number of deposits here and abroad. Of the two classes, however, tungstates are less abundant as specimens because of their tendency to form masses rather than crystals.

Silicates. Formulas in the silicates range from simple to very complex but all show the familiar silicon-oxygen combinations, as simply SiO_2 in the silica group, or more complex linked groups as shown in the table earlier in this chapter. Water is present in a number of species, especially in the zeolite group where it is sometimes so loosely bound that it escapes spontaneously when specimens are taken from the ground. In other species it is held strongly and cannot be removed without destructive heating. Similarly, hydroxyl (OH) is also present in many species, but is essential to crystal structures and not easily removed.

Silicates are sub-classified according to the number and linking of silicate tetrahedrons because such smaller groups influence structures and properties of crystals more than other ions which may be present. The first sub-class is the *nesosilicates* in which tetrahedrons are isolated groups. Structures and properties vary widely because the isolated groups cannot exercise a consistent influence on the host structures. Thus a relatively soft species, such as olivine, is classed with hard species as those in the garnet group and the aluminum silicate group (andalusite, topaz, etc.). It is to be noted that wherever aluminum joins with silicon-oxygen tetrahedrons, as in the groups mentioned, stronger crystals result, furnishing hard, stable minerals, generally pale in color. They are important accessory minerals in rocks, particularly metamorphic rocks.

Double-tetrahedron structures, or *sorosilicates,* generally impart physical properties not greatly different from those of the previous class. Hard and strong species are found among the epidote group where aluminum coupled with silicon-oxygen tetrahedrons results in powerful bonds.

Ring structures, or *cyclosilicates,* form extremely strong minerals, as beryl and tourmaline, whose crystals reflect the neat stacking of tetrahedron rings. The internal structure of axinite, included in this sub-class, contains rings but these are not as symmetrically arranged and the crystals are therefore sharp-edged, wafer-like individuals, differing markedly in properties and appearance from other cyclosilicates.

The *inosilicates,* or linked tetrahedral chain structures, either single or double chain, impart the property of easy splitting in one crystal direction because bonds within chains are strong but are weaker between them. Crystals often develop as long needles or felted masses of minute interlocked crystals. Rocks formed from such masses are often exceedingly tough, as for example, jadeite and nephrite. In some instances, as in tremolite asbestos, the fibers are parallel and separable and it is easy to strip off long slender crystal fibers with the fingers. Aside from fibrous types, most species are relatively hard, of medium weight, and often strongly colored. They are important rock-forming minerals.

The *phyllosilicates* form sheet structures, the most familiar being found among mica group species. The continuous linking of hexagonal groups of silica tetrahedrons results in very easy splitting into exceedingly thin sheets. This property may not be apparent in some species, as chlorite or clay minerals, where sheets are often so small that they cannot be seen clearly even under magnification. The principal recognition clue to species in this class is the silvery glimmering of flat flakes easily dislodged from the crystals. All species are abundant, some being important rock-formers.

The most common minerals, the feldspars and quartz, are included in the sub-class of *tektosilicates* in which tetrahedrons are linked in three-dimensional frameworks tending to impart uniform properties throughout crystals, as in quartz, which breaks almost as readily in one direction as another. In the feldspar group, uniformity within crystal structures is not as great and its crystals split more readily in certain directions than in others. The zeolite group species are included in this sub-class for structural reasons, although compared to the others they are very weak, easily disintegrated, and chemically unstable. The framework within them is so open that water and ions of various elements can easily enter and can be just as easily removed. The majority of specimens in the tektosilicates tend to be pale in color, but striking exceptions may be noted such as vivid blue in sodalite, green in microcline, and purple in quartz. Because quartz is strong and resistant to chemical attack, it lasts when other minerals are destroyed by weathering of rocks, and is therefore the principal constituent of sands and sandstone. Feldspar species, however, easily alter into clay minerals.

Owing to the preponderance of oxygen and silicon, silicates are the most abundant minerals in the earth's crust, being found everywhere in all types of environments and generally characterized by hardness, chemical resistance to change, pale coloration, and translucency.

4

Crystal Growth

Crystals take places of honor in every collection. The better they are, the more glistening and elegantly shaped, the more highly they are prized. Without doubt, seeing a beautiful cluster of crystals does more to inspire a beginner than hours of lecturing. Even the seasoned mineralogist is moved by the sight of a splendid specimen, perhaps more so than the beginner because he appreciates the miracle which made all of it possible. The development of crystals is of intense interest to modern scientists as well as to collectors of minerals. Many of our electronic devices performing feats of near-magic, owe their operation to synthetic crystals grown in laboratories to precise specifications. Transistorized radios, TV sets, and far more complicated pieces of apparatus, are made smaller and more efficient by semi-conducting crystals, while LASER crystals, used to send intense beams of light over tremendous distances, are synthetic ruby crystals. Much has been learned about how crystals grow, but no doubt more still awaits discovery.

Mineral crystals form from molten solids, from solutions in which the necessary ions are dissolved, and from vapors carrying ions intermingled with the ions of gases. All of these growth processes are extremely important in nature, but generally speaking, our finest crystals are those grown from solutions, the next best from molten solids, and the least in quantity though sometimes very fine in quality, those grown from vapors.

CRYSTALS FROM MELTS

The commonest examples of crystals grown in molten material may be seen in igneous rocks, such as ordinary building granite and in metals. It is easy to see separate crystals in granite because they range from about $\frac{1}{8}''$ to $\frac{1}{4}''$ in diameter, sometimes larger. The gray crystals are quartz, the white or pink crystals are feldspar, and the black specks, if present, are biotite mica crystals. Metal crystals are not as easy to see because most metal objects are carefully smoothed and polished, effectively preventing detection of separate crystal grains. However, large flat crystals can be seen on slightly corroded brass objects or as star-shaped flat crystals on sheets of iron coated with zinc and popularly known as galvanized iron. Smaller crystals appear as sharp projections on surfaces of freshly broken steel and cast iron. In all of them, metals and rocks, the predominant feature is that few if any of the crystals show uniform geometrical outlines but, instead, are extremely irregular.

As heat dissipates from a mass of molten rock, various ions join in microscopic colonies to form the first regular frameworks characteristic of all crystalline minerals. Each colony draws nourishment from the molten material around it until none is left. At this time, the rock, though still very hot, freezes solidly. Figure 13 shows schematically the texture assumed by most igneous rocks and metals, sometimes on a scale large enough to see with the naked eye as noted before but often requiring the assistance of a microscope. At the bottom are shown several crystals still surrounded by molten material. Some show regular outlines but as boundaries grow outward to touch those of neighbors, all regularity will be obliterated and a mosaic pattern identical to that in the already frozen material results. Irregularity of all grains indicates that the several minerals froze at about the same time, but it is possible in some igneous rocks, that one or more minerals crystallized sooner, in which event their outlines will show the flat faces characteristic of crystals. This is why many diamond and sapphire crystals show faces even though they were formed in solid rock: they grew earlier and therefore developed perfectly before other minerals could interfere. Although found frozen in solid rock, diamond, corundum, garnet, kyanite, and staurolite commonly occur as well-formed crystals.

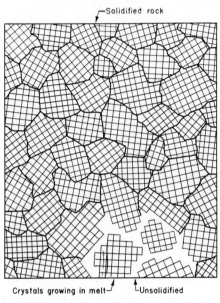

Solidified rock

Crystals growing in melt — └Unsolidified

CRYSTALLIZATION OF A MELT

FIG. 13.

CRYSTALS GROWN IN SOLUTIONS

Water is the universal solvent in the earth's crust, and most minerals lining cavities in rocks are grown from water solutions. A mineral such as halite, the basis of common household salt, dissolves easily in water as everyone knows but it stretches credulity to be told that something as seemingly imperishable as quartz also dissolves in ordinary water in substantial quantities. Yet this is proved every day by the production of synthetic water-grown quartz crystals. Many years ago it was found that quartz dissolved readily in water under high pressure and temperature, but this discovery was not put to much use until it was necessary to find alternate sources of quartz for electronic crystals rather than to depend on natural deposits. The process is simple but demands certain precautions: steel vessels must be made strong enough to prevent explosions and must be carefully sealed to prevent disastrous leaks. Quartz "seeds" in the form of thin wafers, cut in the proper manner for desired electronic properties, are fastened in a wire frame and lowered into the vessels which are about half-filled with water. In the bottom of each vessel is placed crushed quartz or quartz-glass to provide the nourishment. The vessels are sealed and lowered into explosion-proof pits where they are heated on the bottom. After a number of weeks, they are cooled, uncapped, and the contents removed. The thin seed crystals are now developed into flawless crystals, sometimes as much as a pound in weight.

Geologically speaking, a few weeks are as nothing compared to the many thousands of years of rock history. Nature's growing conditions may be poor but they are still fast enough to account for the many crystals which we find all over the world, sometimes in fantastically large sizes. Undoubtedly, the conditions for growing quartz crystals in the laboratory are still being duplicated deep in the earth's crust where pressures and temperatures are suitable. Perhaps someday these too, will be near the surface to delight the hearts of collectors.

The solubility of mineral ions in water depends on what kind they are, and as we have seen from the example of salt dissolved in water as compared to the solution of quartz in water, much also depends on pressure and temperature in the environment. This explains why ordinary quartz found on the surface of the earth may be millions of years old, yet shows little sign of solution, while the first rain shower quickly destroys substantial parts of a salt deposit. Mineral crystals grown from solutions are therefore found in geological situations which provided proper environments for their development, and later, for their survival. Many ore veins exploited in depth show a regular sequence of minerals from top to bottom, each association of minerals pointing to pressure-temperature conditions suitable for growth of the particular species involved. Among the species grown from solutions are quartz, calcite, zeolite minerals, clay minerals, and numerous others.

CRYSTALS FROM VAPORS

The growth of crystals from vapors is exemplified by sulfur, whose atoms easily escape to form a gas when massive sulfur is carefully heated. The process of vaporizing and condensing directly to the solid state is called *sublimation,* and may be seen actively in progress in the openings of fumaroles in volcanic regions. Vaporized elements and compounds are caught in streams of various gases, including water vapor, and remain dissolved until they encounter cooler surroundings, such as walls near vents. At this time the minute particles deposit directly upon the walls to form crystals. Aside from sulfur, there are few minerals that the collector is likely to obtain which provide impressive crystals produced in this manner.

HOW A CRYSTAL GROWS

Every crystal begins as a small cluster of atoms, of which there must be enough of the right kinds to establish the pattern and composition of the crystal. The seed cluster grows into a recognizable crystal as more and more atoms are attracted by the electrical forces of those already in the crystal structure. The regular pattern in space which every crystal must have is established at the very beginning, perhaps in a process similar to that shown in Figure 14. Not much is known about the exact mechanics involved in the building of crystals, but a number of ideas have been advanced of which this is one.

In this figure is shown a possible growth scheme for a crystal developing from a cube-shaped seed or block representing the smallest part of the crystal which has all its essential properties. Because the beginning block (A) is a cube, we can assume that all its faces are the same, and that atomic attractions on each of them is also the same. An unattached block drifts close and is attracted toward the seed, but having no preference for one face or the other, it takes the first at hand or the top face (B). Another is drawn to the broader side of the growing seed because now two faces exert more attraction than either of the single faces on the ends. At (C) the new block makes a "corner" which effectively focuses a stronger degree of attraction and causes the next block to drop into the corner in preference to any other place on the seed. At (D) essentially the same thing happens as in (B) where the broad face becomes more attractive than the edges. Again corners are formed as in (E) and are filled to make the larger cube (F). The process now repeats itself upon every face, all of which again have equal attraction.

SPIRAL GROWTH

From its smallest unit, the crystal grows larger so long as it receives more nourishment from its surroundings. Because the attraction of any corner or upraised part is so much stronger than elsewhere upon the face, incoming atoms are drawn to steps exclusively until a solid layer is

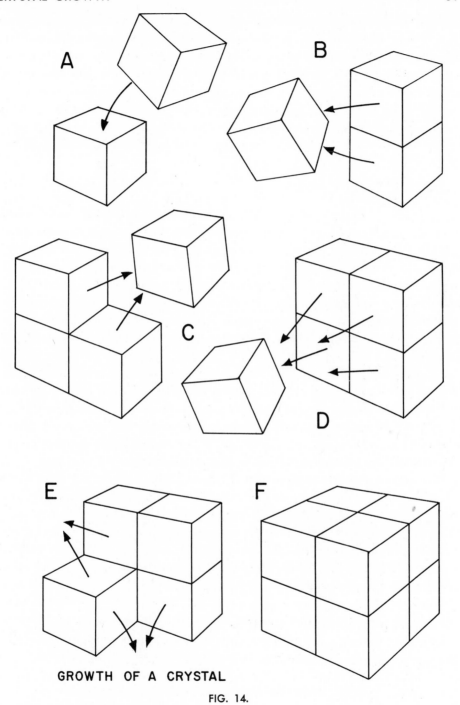

GROWTH OF A CRYSTAL

FIG. 14.

formed. Only then can others fall upon the new face to start a new layer. For many crystals this process is too slow and they take short cuts by developing imperfections which create artificial steps. Such an offset, or *spiral dislocation,* as it is called, is illustrated in Figure 15, and shows that one side of

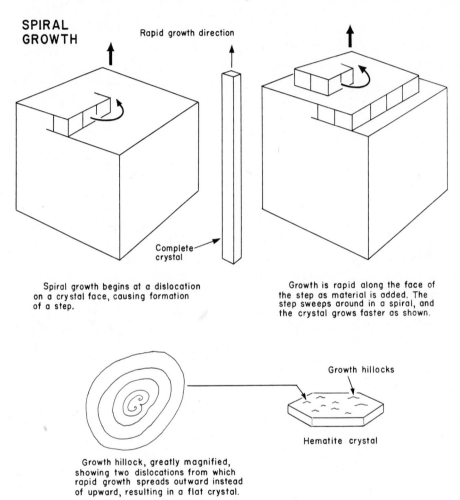

Spiral growth begins at a dislocation on a crystal face, causing formation of a step.

Growth is rapid along the face of the step as material is added. The step sweeps around in a spiral, and the crystal grows faster as shown.

Growth hillock, greatly magnified, showing two dislocations from which rapid growth spreads outward instead of upward, resulting in a flat crystal.

FIG. 15.

the crystal split and slipped so that the formerly smooth layer is offset by the thickness of one or more atomic layers. No one knows why offsets develop, but more and more of them are being found on crystals of various species and it appears that this is an important method of crystal growth even though it is an imperfection, technically speaking.

Figure 15 shows one spiral dislocation on top of a crystal, the arrow indicating how the small ledge attracts atoms or ions faster and consequently grows around in a spiral ramp which rises upward indefinitely. Although this crystal began as a cube, it is obvious that it is more likely to finish as a long slender crystal, perhaps like that shown in the center drawing. Spiral growth may account for the normal cubic crystals in a mineral such as cuprite developing into extremely long filamental crystals in the variety known as *chalcotrichite*. Double spiral dislocations cause equally interesting results as shown in the sketch of a hematite crystal at the bottom of Figure 15. A spiral starts at each dislocation and sweeps outward until it meets the other. Both fuse into a circular plateau which grows rapidly along its periphery, sweeping outward toward the edges of the crystal face. However, because the dislocations are still present, another cycle creates another plateau, and another, until a low mound or hillock is created. Such shallow bulges can often be seen upon the faces of tourmaline, apatite, and many other crystals.

GROWTH RATES

No crystal grows at exactly the same rate in every direction, although this seems a perfectly logical thing to happen. In Figure 16, for example, are two sketches, one of spherical growth, and one of cubic growth. We know of spherical growth outside of the mineral kingdom such as oranges, plums, etc., which receive their nourishment from *inside* and grow outward, but we know of no crystals which grow this way and certainly none which are spherical. Some real crystals are covered with so many faces that they *look* spherical, but of course this is only the general impression of their shape, and the fact remains that they are covered with flat faces. Thus, in every crystal, growth is the same only in certain directions, and is either slower or faster in others. Why should this be so? Why doesn't the crystal add layers of atoms on every part, keeping this up until it is fully grown?

These questions may best be answered by referring to the sketch at the bottom of Figure 16, showing the cross-section of a sphere, in which is fitted as closely as possible a number of cube units of some hypothetical crystal. Although the *general* cross-section is that of a sphere, there are still definite flat spots. As this almost-spherical crystal grows, all recesses or corners exercise the powerful attraction we have spoken of before, and attract atoms or ions to fill them. In a few more growth cycles, corners fill out as shown by dotted lines, and the crystal restores itself to cubic shape. Therefore attractions of ions already in the crystal tend to keep the exterior covered by flat planes although sometimes growth is so rapid that neat orderly allocations of space are not possible and "pile ups" occur on faces to make them rough. Many natural crystals show faces which have every appearance of being curved, but this is due to partial dissolution rather than growth. Severely etched crystals are most likely to display curved faces, but some are also noted

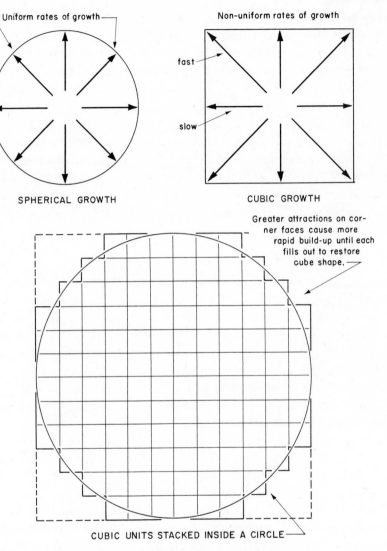

RATES OF GROWTH IN CRYSTALS

FIG. 16.

on crystals which develop so many faces, sometimes only a fraction of a degree apart, that the general effect is of curvature. Rounded and etched crystals are commonly observed in topaz and beryl.

A crystal growing in cubic arrangement can be expected to develop uniformly because equal attractive forces spread from all faces. It is not likely to develop into a stretched out box-like object, or some other form equally strange although this does happen occasionally. Even the spiral dislocations mentioned before tend to appear on *all* faces of a cubic crystal, rather than just on one as in the case of the *chalcotrichite* crystals. However, we may ask, what would be the result if the building block did not happen to be a cube, but one shaped like a brick as shown in Figure 17? What would the final crystal look like if it were allowed to develop perfectly?

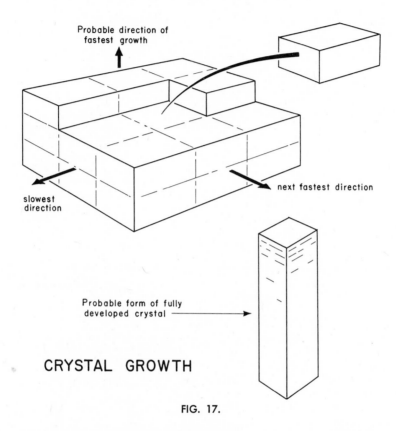

Probable direction of fastest growth

next fastest direction

slowest direction

Probable form of fully developed crystal

CRYSTAL GROWTH

FIG. 17.

Let us say that the greatest attractive force is on the largest face of the brick-like unit, less on the long and narrow face, and least on the ends. When each block draws close to the crystal, the strongest attractions force units to take the positions shown, and the finished crystal will most likely be a column stretching vertically. On the other hand, if the attraction were greatest on either of the other sides, the final shape could be vastly different, possibly producing a flattened or tabular crystal.

Growth, once begun, is not always steady nor even steady in any given crystal direction. Crystals are extremely sensitive to slight changes in their environment, as proven by countless experiments in the laboratory, and sometimes the slightest trace of an impurity in a solution may cause unequal growth, resulting in over-development of some faces and loss of others. This is made clear in Figure 18 where several examples are given. At the top is

CHANGES IN GROWTH RATE

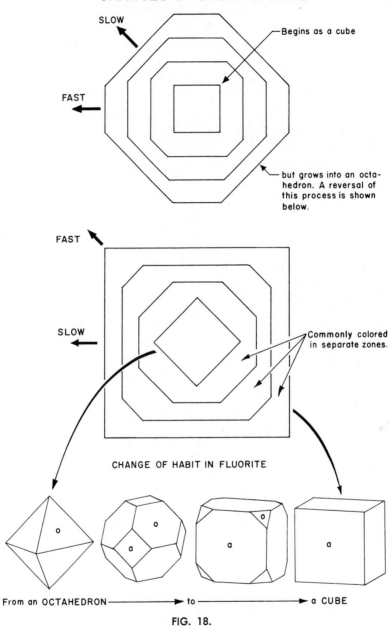

SLOW

Begins as a cube

FAST

but grows into an octahedron. A reversal of this process is shown below.

FAST

SLOW

Commonly colored in separate zones.

CHANGE OF HABIT IN FLUORITE

From an OCTAHEDRON ⟶ to ⟶ a CUBE

FIG. 18.

shown a crystal shaped like a cube which is rapidly changing its form because growth upon corners is slower. In time, its profile will be entirely different. At the bottom is a similar example in *fluorite,* whose crystals can be split open easily to display numerous zones of color furnishing evidence of the shape of the crystal in earlier stages. In this instance, the original crystal was an octahedron, but because of slower growth upon its corners, it developed truncating faces which gradually converted the crystal into a cube!

DEVELOPMENT OF FACES

One lesson that is obvious now is that the basic geometry of the smallest crystal unit tends to reflect itself in the finished crystal which, no matter how large it grows, is nothing more than a neat stacking of the smaller units. Sometimes the original shape of the unit reflects itself faithfully in the final crystal, the larger one being a greatly magnified version of the minute beginning unit, but at other times, as shown in the examples of Figure 18 where some faces develop at the expense of others, the final product bears little resemblance to the original seed shape. However, no matter how drastic the change, the final crystal form can always be related geometrically to the form of the smallest unit. This is to say, faces do not develop wherever they please but can only develop in certain directions. This can be seen clearly in Figure 19. Shown on the left are two networks of points corresponding to cross-sections of a square lattice as in a cubic crystal structure and a hexagonal lattice as in a hexagonal structure. Lines have been drawn to connect series of points along certain directions. As can be seen, each line represents a plane of atoms which can be transferred directly over to the crystal to which it belongs. Investigations of crystal structures show that all faces can be directly related to planes of atoms in every crystal, as shown diagrammatically in Figure 19.

In view of what we have learned about the way atoms fit themselves into the growing crystal structure, all of this makes sense. We have seen that atoms cannot attach themselves wherever they choose because forces from within the crystal cause them to drop into corners where attraction is greatest. When a crystal face develops, it therefore tends to remain flat, but this flatness must be related to an atomic plane; it cannot be otherwise. Early mineralogists soon discovered this interesting fact about crystals and one of them, the celebrated Dane, Nicolaus Steno, pointed out in 1669 that the angles between faces in quartz were always the same no matter where the crystal came from nor how misshapen it happened to be. Although Steno knew nothing about atomic structures, it is easy to see that he had to be correct because all faces on any crystal are related to the same atomic structure and therefore must always be related to each other in accordance with fixed rules.

Steno's discovery, called the *law of constancy of interfacial angles,* is illustrated by several examples in Figure 20. At the top are shown a variety of quartz cross-sections cut from crystal prisms. The perfect cross-section shown at the left is a regular hexagon with angles of 120° between each pair of faces. The others are distortions, but all produce the same angles between

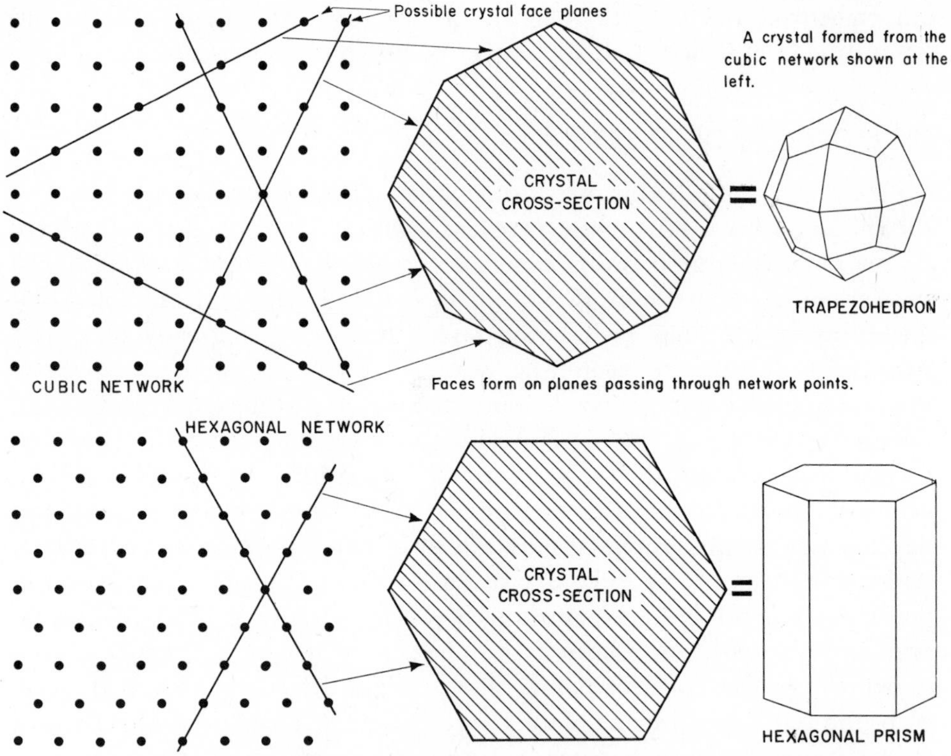

Possible crystal face planes

A crystal formed from the cubic network shown at the left.

CRYSTAL CROSS-SECTION

=

TRAPEZOHEDRON

CUBIC NETWORK

Faces form on planes passing through network points.

HEXAGONAL NETWORK

CRYSTAL CROSS-SECTION

=

HEXAGONAL PRISM

NETWORKS OF POINTS IN CRYSTALS RELATED TO CRYSTAL FACES

FIG. 19.

faces. The odd triangular cross-section at the right is from a crystal which developed alternate faces at the expense of the others and thus grew into triangular shape. However, note that its angles of 60° are exactly half of 120°. At the bottom is shown an ideal spinel crystal of perfect octahedral form, another distorted laterally, and still another so badly misshapen that it resembles a wafer; nevertheless, all angles between faces measure exactly the same.

PERFECTION OF FACES

Crystal faces vary tremendously in respect to flatness and luster from crystal to crystal. The so called "Herkimer diamonds" of New York State, are only quartz crystals but receive their misleading name from the nearly-perfect smoothness of faces. However, larger quartz crystals, usually grown under conditions less ideal than those which created the Herki-

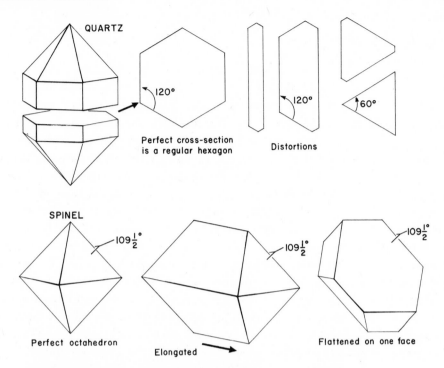

QUARTZ

120°

Perfect cross-section
is a regular hexagon

120°

60°

Distortions

SPINEL

$109\frac{1}{2}°$

$109\frac{1}{2}°$

$109\frac{1}{2}°$

Perfect octahedron

Elongated

Flattened on one face

CONSTANCY OF INTERFACIAL ANGLES IN PERFECT
AND IMPERFECT CRYSTALS

FIG. 20.

mer crystals, suffer far more accidents as they grow and are rougher and less lustrous. Imperfections on faces are traceable to growth conditions, the ideal circumstances being when only enough atoms are brought to a crystal as it can use, and this supply kept steady while temperature and pressure are kept steady too. All of this is a logistics problem which is so sensitive to disruptions that it is astonishing that we find as many good crystals as we do.

The delicate balance of conditions under which a crystal grows is well illustrated by the following incident which happened in a laboratory making electronic crystals for underwater sound equipment. The crystals were grown in a water bath, and all conditions around the tank were carefully adjusted to be sure that nothing went wrong. Nevertheless, the laboratory personnel were puzzled by the appearance of regularly spaced cloudy planes in their crystals as if some change occurred every few days to cause the crystals to grow poorly. The mystery was solved when it was found that over week-ends, with everyone at home, the janitor lowered the temperature in the laboratory to save on the heating bill. He turned it up again early on Monday morning and no one noticed this mistake except the crystals!

CRYSTAL IMPERFECTIONS

In view of the sensitivity of growing crystals to changes in environment it is not surprising that reasonably "perfect" crystals are extremely difficult to find in nature. Actually, the closest approach to perfection occurs in man-made crystals grown in the laboratory where environmental conditions can be closely controlled. On close examination, all natural crystals show mottlings, ridges, grooves, plateaus, pits, offsets, and a host of other defects which silently bear witness to the spirited battle every crystal went through as it strove desperately to keep some semblance of order in the face of all kinds of disruptions in environmental growth conditions. As small individuals, crystals are apt to be far closer to perfect, and it is for this reason that amateurs specializing in the study of minute crystals, the "micromounters," boast that their specimens are the finest obtainable. As crystals grow larger, opportunities for disruptions multiply, and large crystals are therefore seldom very good. For these reasons, large, excellent quality crystals are sometimes prized out of all proportion to size simply because the odds were so much against reasonably perfect development. On the other hand, many technically "imperfect" crystals are made far more desirable by their imperfections, as will be appreciated in the discussions which follow about inclusions in crystals.

Crystal imperfections may be conveniently divided into two broad classes: *external* and *internal*. Actually, it is impossible to draw the line between them because all are interrelated to some extent. Each type of imperfection will be remarked upon separately.

EXTERNAL IMPERFECTIONS OF CRYSTALS

Unequal Development of Faces. It is well known that many crystals prefer to grow faster upon certain faces and slower upon others. If a small, growing crystal is covered with several types of faces, one may develop faster than the other and literally grow itself out of existence. For example, in Figure 18, the triangular faces of the fluorite octahedron grow rapidly while the cube faces do not. In time, cube faces spread in area until they alone appear upon the crystal. Strictly speaking, this cannot be considered an imperfection except when some freakish circumstance causes a crystal to grow much more rapidly in one direction, resulting in grossly exaggerated individuals which little resemble others of their class. Many crystals which depart markedly from normal, are said to be "distorted," which lends the erroneous impression that the crystal was somehow twisted by external force until it yielded as if made of plastic. It is true that some crystals can be deformed in this manner, but by far the larger number appear distorted because of unequal face development, or in the case of the curious quartz crystals commonly noted in granitic pegmatite cavities, because the "seeds" from which they grew were badly misshapen to begin with. In pegmatite cavities, walls containing quartz, and sometimes feldspar also, commonly shatter during some stage in cavity mineralization, showering numbers of quartz and feldspar fragments into the cavity. Because these may be of almost any shape, subsequent light overgrowths of fresh material may do no more than smooth out irregularities without changing the basic irregular shape. If, for example, a thin sheet-like chip is lightly coated, it obviously very little resembles a "normal" quartz crystal and is therefore said to be "distorted." In other instances, when the supply of quartz from solution continues for a sufficient period of time, the shards may grow into complete crystals, sometimes with both ends terminated by faces, and of excellent "textbook" form. These are, in part, the *doubly-terminated* crystals highly prized by collectors. As mentioned earlier, other doubly-terminated crystals may form within solid rock because of having crystallized earlier than other constituents. In this connection, the terms *euhedron* or *euhedral* are often used to designate crystals faced on all sides, the prefix *eu* meaning "good" or "ideal." Crystals partly faced are then called *subhedrons,* and those with no faces, *anhedrons.*

Growth Hillocks. Spiral growths cause low mounds or hillocks on faces

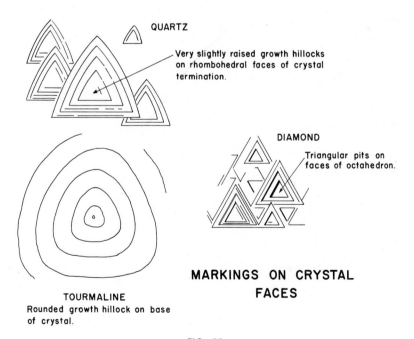

QUARTZ

Very slightly raised growth hillocks on rhombohedral faces of crystal termination.

DIAMOND

Triangular pits on faces of octahedron.

TOURMALINE
Rounded growth hillock on base of crystal.

MARKINGS ON CRYSTAL FACES

FIG. 21.

which are otherwise nearly flat. These do not stand very high and it may be necessary to cast a strong light from the side in order to shadow them. Similar low plateaus, sometimes of geometrical outline, also form on crystal faces because of excessively rapid growth on certain areas. Triangles with slightly curved sides are commonly seen on flat end faces of tourmaline crystals, while depressed triangles, of perfect equilateral outline, are exceptionally common on the octahedral faces of diamond crystals. A number of growths are sketched in Figure 21. In many instances, the *shape* of the growths is helpful in determining the internal geometry of the crystal, thus providing useful clues to identity.

Parallel Overgrowths. When growth hillocks or plateaus become overdeveloped, more atoms may be attracted to them and their growth made even faster, resulting in interesting overgrowths as shown in the *amethyst quartz* example in Figure 22. Because the overgrowth crystals began on the parent crystal, the atomic structure is continuous underneath, leading to parallelism in structure and of all faces. Growths of this kind are called *parallel growths*. Because all faces are parallel, turning the crystal to reflect

FIG. 22. Parallel growth of many smaller amethyst crystals upon larger amethyst crystals. Group of smoky purple crystals, 5″ by 4″, Iron Station, Lincoln County, North Carolina.

light from one face results in light being simultaneously reflected from all in the same position as can be seen in the photograph.

Another type of parallel growth appears in galena when for some reason, cube crystals stop growing as such and begin growing as octahedral crystals instead. Frequently the entire surface of a former cube may be covered by sprouting, sharp-cornered, octahedral crystals, roughly resembling the sole of a hobnail boot. Calcite also develops overgrowths, sometimes in strange ways! As shown in Figure 23, the earliest crystals were sharp-edged individ-

FIG. 23. Parallel growth on twinned calcite crystals. An earlier generation of calcite became coated by dark red-brown iron oxides, and was later enveloped by a second generation. Size of group 3″ by 1½″. Chihuahua, Mexico.

uals which got themselves coated with a dusting of brown iron oxide. A later generation of calcite was discouraged from taking up where the first generation left off by the presence of the iron mineral and began growing in parallel position at the base of the early crystals, finally overtaking and nearly enveloping them. Even more remarkable is the fact that the later calcite developed different crystal faces.

Scepter Overgrowths. Exceptionally interesting overgrowths are provided by symmetrical cappings of quartz atop slender prisms of an earlier quartz generation, as shown in Figure 24. Because of shape, such growths are called *scepters,* and obviously resemble those regal badges of office. Equally spectacular scepters are found on calcite and other species but quartz generally provides the best. The cause of this curious growth is not easily explained because additional silica ions should attach themselves to the crystal uni-

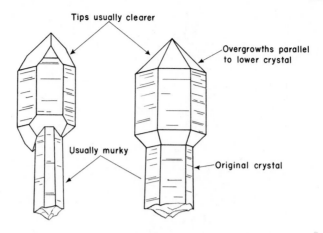

Tips usually clearer

Overgrowths parallel
to lower crystal

Usually murky

Original crystal

SCEPTER OVERGROWTHS ON QUARTZ
FIG. 24.

formly over all faces instead of selecting just the tips. Possibly the tips are
chosen because they protrude further into cavities and are better able to
obtain nourishment than other crystal parts. There is reason to believe that
scepter overgrowths usually occur when solutions contain little mineral
matter. Often, their quality indicates slow, steady growth as marked by transparency and freedom from internal defects, whereas in saturated solutions,
rapid and imperfect growth is likely to take place on all crystal faces. As
noted on previous overgrowth examples, scepters are structurally related to
supporting crystals but some are slightly tilted, as if the new growth did not
seat itself exactly in parallel although accepting the support provided.

Foreign Overgrowths. It seems perfectly reasonable that calcite overgrown
upon calcite, and quartz overgrown upon quartz, should grow in parallel
position, but it is noted that some species, either distantly related or not at all
related to host crystals, also form parallel overgrowths. For example, slender
crystals of rutile are commonly observed grown upon flat crystals of hematite
in such a way that they form star-like patterns, with legs 60° apart and
parallel to certain faces of the host. This cannot be an accident because
it is observed in too many instances. Another example is provided by
numerous minute chalcopyrite crystals growing upon sphalerite. When such
a crystal is turned, the many small chalcopyrite faces mirror the light
simultaneously, causing an unmistakable flash. Furthermore, when the position of the flashes from various faces is compared to the position of host
faces, again a relation is evident. Although neither rutile/hematite nor
chalcopyrite/sphalerite are capable of forming solid solutions, the influence
of atoms within surface layers of the host crystals must reach out to orient the

c

first seeds of the overcoating crystals. From that point on, additional material makes permanent this original guidance.

Selective Incrustation. Related to the previous type of overgrowth is another in which the host crystal exerts attractive force only from certain faces, resulting in attachment of small foreign crystals upon such faces, and not upon the others. In Figure 25, for example, alternate faces of quartz sometimes attract minute crystals of hematite, resulting in striking individuals with every other face colored vivid red. The attraction of quartz in this odd

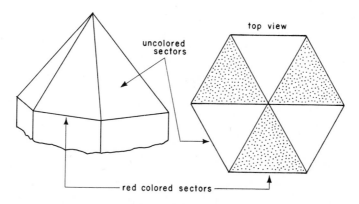

uncolored sectors

top view

red colored sectors

HEMATITE COATING QUARTZ

FIG. 25.

manner is related to its internal structure as will be described in a later section of this chapter. In other instances, as in the brilliant red, hematite-coated calcites from Cumberland, England, only one side of all crystals upon a matrix specimen is stained, the other side being white or colorless. Such coating is not due to any attraction from the calcite itself but is the result of hematite being carried along in a solution traveling steadily in one direction. Specimens of this sort are startling to the uninitiated because they seem to change color merely by turning them around.

Phantom Crystals. Sprinkling of foreign mineral crystals upon a host crystal during growth results in thin coatings which soon become inclusions. If such deposits are separated by time intervals, the crystal displays a number of ghostly inner crystals, called *phantoms,* each marking faithfully the size and shape of the crystal when it was younger. If the host crystals are transparent and the inclusions not too dense to see through, very beautiful specimens are the result. Quartz crystals again provide the best, but many must have their faces cut and polished to best show the effect. Phantom examples are sketched in Figure 26. It is to be noted that changes in color, commonly

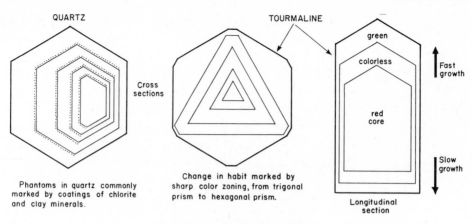

PHANTOMS IN CRYSTALS
FIG. 26.

accompanied by changes in habit, also cause phantoms, particularly in fluorite.

Offsets. An important cause of lack of good form and smooth faces on large crystals is the tendency of many to develop *offsets* during growth. One kind of offset, the spiral dislocation, has been mentioned, but others occur which cannot be assigned to the same cause. In general, offsets appear to develop early in the crystal growth cycle such that small portions of the crystal, possibly due to some ions attaching themselves a little out of alignment, grow slightly askew. In effect, the crystal divides into a number of closely intergrown individuals which, taken together, appear to be all one crystal. Whatever the cause, offset crystals are easily recognized by divergence of parts leading to considerable surface irregularity. Figure 27 illustrates several crystals marred by offsets, the galena showing faces of great roughness although the general form is unmistakably cubic, while the tourmaline, originally a single crystal, developed faster growth on some end face areas causing final development as a nearly parallel bundle of separate crystals.

Curved or "Twisted" Crystals. Offsets are responsible for the remarkable "twisted" crystals of smoky and colorless quartz from the Swiss Alps, consisting of series of individual crystals, each slightly offset from each other in corkscrew fashion. In what the Swiss call the "closed" form, the crystal resembles a single quartz individual, pressed flat on one prism face until it is much wider than thick, and then twisted along a lateral direction. In the "half open" form, the individual offset crystals are more distinct, so that their terminal points form a series of teeth along top and bottom. The "open" form consists of a twisted row of obviously distinct crystals although the faces upon front and back sides are still in curved alignment. An example is shown in Figure 238 (page 439). Quartz crystals recorded from Colorado

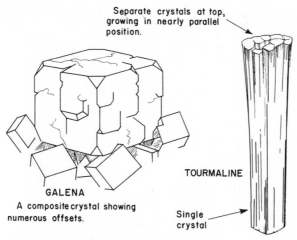

Separate crystals at top, growing in nearly parallel position.

TOURMALINE

GALENA

A composite crystal showing numerous offsets.

Single crystal

GROWTH IMPERFECTIONS

FIG. 27.

show twisting along a vertical rather than horizontal direction and, especially when slender, vaguely resemble a spiral wood-boring drill with very gently-curving flutes. Curved crystals are abundant in the dolomite group, some being so curved that the name "saddle crystals" is given to them. Somewhat similar curvature, again owing to development of many small crystals slightly offset from one another, is commonly seen in crystals of siderite, chabazite, apophyllite, heulandite, pyrite and arsenopyrite.

Cavernous and "Hopper" Crystals. Possibly because of defects developing along edges, crystals of certain species begin exceptionally rapid growth on sharp facial intersections so that decided hollows develop, while some even appear as if they had been bored neatly through. The *hopper crystal* of halite shown in Figure 28 indicates how fast growth on cube edges perpetuates itself until the crystal is deeply cavernous on all cube faces. Once rapid edge growth begins, more opportunity is afforded such areas to obtain ions from solutions and to grow even more exaggerated. Deep central cavities lined by faces parallel to prism faces are a common feature of pyromorphite crystals, as also illustrated in Figure 28. In some instances, outer edges develop slightly faster than those along inner cavities, resulting in a gradual curving over and development of barrel-shaped crystals.

Etching and Dissolution. Because each mineral grows best from solutions only in a narrow range of environmental conditions, changes may occur such that already grown crystals may dissolve instead of continuing to grow. In general, a great many crystals show light to moderate local dissolution, commonly called *etching,* usually evidenced by numerous small pits imparting a frostiness, like that observed on sand-blasted glass, or by distinct pits scattered here and there on otherwise smooth faces. Fascinating experiments designed

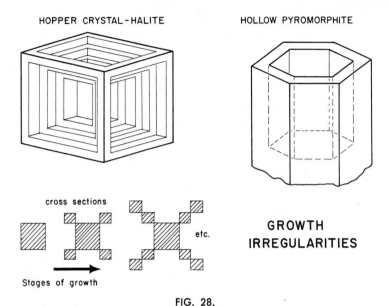

HOPPER CRYSTAL - HALITE   HOLLOW PYROMORPHITE

cross sections

etc.

GROWTH IRREGULARITIES

Stages of growth

FIG. 28.

to show the effects of dissolution are easily performed on fragments of calcite which can be immersed in dilute hydrochloric acid for short periods of from two to three minutes to ten minutes or more. Only a brief immersion is needed to show that the etch figures developed upon the exterior are related geometrically to the crystal itself. Similarly, close inspection of naturally dissolved crystals also reveals relationships to underlying atomic structure although severely corroded specimens, notably in spodumene, beryl, and to a lesser extent, in quartz, may prove puzzling to decipher especially if no signs of original crystal faces are present. Light etching results in characteristic markings related to the crystal structure as shown in Figure 29. The hexagonal pits on beryl exactly mimic the hexagonal outline of the crystal itself, and of course, their shape is regulated by the beryl structure. On the other hand, the boat-shaped indentations on the prism faces are not nearly so diagnostic. Severely corroded beryl crystals frequently reach the point where they resemble cigar-shaped objects with pointed ends and wrinkled sides, with all surfaces brilliantly reflective.

Because every crystal is a neatly balanced structure of like atoms in like positions, it follows that etch pits must look the same when present on identical crystal faces. In a salt crystal, for example, the same atomic pattern lies beneath all cube faces. If one face is partly dissolved and covered with small square pits, all other faces will develop exactly the same kind of pits if attacked by the same solution. This is of great importance in identifying like faces when examining etched crystals. By turning over a severely corroded specimen, it is sometimes possible to orient the crystal correctly from noting that pits of similar shape appear only in certain directions. In Figure 29 for example, if the beryl crystal were very badly corroded, it would still be possible to identify the end faces because the small hexagonal pits form *only on these faces.*

Oscillatory Striations. Crystals of pyrite, quartz and tourmaline, are commonly covered by numerous grooves or striations, sometimes resembling similar markings on phonograph records. In detail, each striation is caused by two narrow crystal faces forming the sides of a vee. Because several faces are trying to develop simultaneously upon the same crystal area, it is called *oscillatory growth* or *oscillatory combination.* In the examples shown in Figure 30, the general shapes are sometimes far different from what they would be if the competing faces grew normally. The competing faces are shown in the smaller

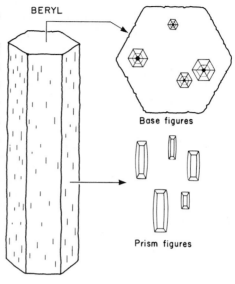

ETCHING FIGURES

FIG. 29.

sketches. All three species, quartz, pyrite, and tourmaline, can be identified with a high degree of confidence when these characteristic markings are present on crystals.

INTERNAL IMPERFECTIONS OF CRYSTALS

Liquid and Gas Inclusions. Rapidly growing crystals tend to attract so much growth material that uniform development becomes impossible, often to the detriment of quality. Milky or murky crystals, or those with numerous veils and wisps, prove to be filled with minute cavities in which are trapped liquids and gases. Enlarged examples of typical inclusions appear in Figure 31. Liquid droplets and gas bubbles adhere to growing crystal faces and are subsequently covered. In some instances, inclusions are so numerous that the crystals become nearly opaque. Veils and wisps sprinkled in planes may be phantoms of previous crystal faces, but if twisted or bent, they may mark cracks healed by additional growth material.

Examination of inclusions under low magnifications of 10 to 40 power, show many inclusion cavities bounded by faces corresponding to external crystal faces. This is not unexpected because growth at any place in the crystal conforms to the same rules. On the other hand, a large number of

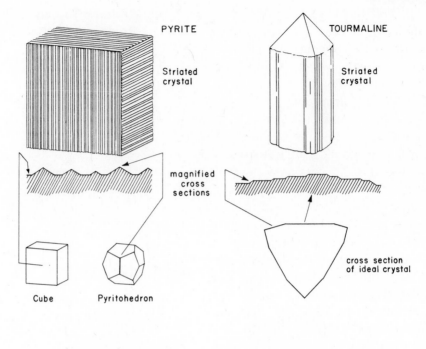

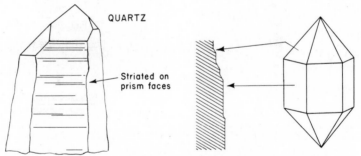

STRIATIONS CAUSED BY COMBINATION OF ADJACENT FACES

FIG. 30.

cavities are formless. Inclusions bounded by faces are sometimes called *negative crystals*.

Inclusions may be gas, liquid, liquid with a gas bubble, or even contain solid matter, usually as small crystals of a foreign mineral (Figure 31). Depending on which forms of matter are present, they are called *single, two, or three phase* inclusions. Many transparent gemstones display characteristic inclusions useful in their identification.

Solid Inclusions. Because many species develop nearly at the same time, it isn't at all surprising to find them intergrown in strange and wonderful

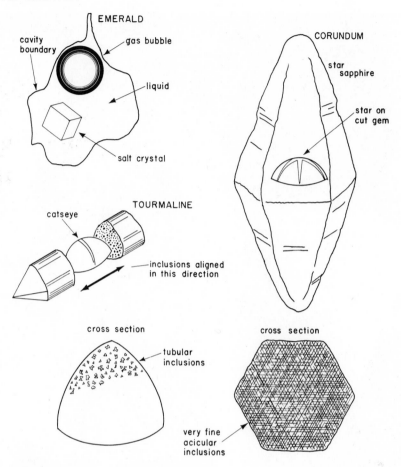

INCLUSIONS IN CRYSTALS

FIG. 31.

formations, such that one mineral is preserved inside the transparent crystal of another! Solid inclusions of this kind often make specimens of great beauty and value. Solid inclusions may be divided into two types: those formed earlier than the host crystals, and those formed at the same time. Inclusions in each class show important differences.

Early-formed inclusions ordinarily begin as growths upon cavity walls and are covered over by later minerals crystallizing in the same space. Slender sparkling needles of rutile in clear quartz provide a well-known example of this type of growth as shown in Figure 32. Rutile needles show no particular orientation in respect to the quartz crystal host, indicating that they grew first and were enveloped later. Some matrix specimens show rutile needles only partly enclosed by quartz with the remainder attached to cavity wall material. A tremendous variety of inclusions is found in quartz because this

FIG. 32. Slender coppery-hued rutile needle crystals in clear rock crystal quartz. Such crystals grew in a cavity before the crystallization of the quartz and were enveloped later. Size of specimen 6″ by 5″. Front face polished. Minas Gerais, Brazil.

mineral often forms last and therefore encloses earlier species. In addition to rutile, inclusions of zeolite minerals, epidote, tourmaline, clay minerals, calcite, pyrite, and many others, occur in quartz. A good example is shown in Figure 33.

Solid inclusions are sometimes so abundant that the crystals enclosing them are little more than a crystalline cement. Spectacular examples of calcite, forming distinct crystals in porous beds of sand, yield the interesting "sand calcites." When exhumed from the sand beds and loose material cleaned off, the crystal shapes of this species are easily recognized although the crystals are sand color, lack smoothness of faces, and otherwise scarcely resemble

FIG. 33. Inclusions in clear rock crystal quartz, possibly of anhydrite. Faces polished. Size of specimen 4¾″ by 2¼″. Itibiara, Minas Gerais, Brazil.

ordinary calcites. In some instances, sand calcites contain 60 percent sand by volume! Despite the difficulties, enough space is available between sand grains to permit crystal growth although the crystal must work its way around millions of small obstructions to develop. Similar crystals are furnished by the "sand euclases" of Brazil which enclose clay minerals; also by benitoite crystals enclosing crossite fibers.

Inclusions formed simultaneously generally appear as small crystals sprinkled in phantom fashion throughout the host crystal. Quartz again provides excellent examples. Former crystal faces are often marked by white or green chlorite forming mossy colonies, or numbers of hematite plates which, if thin enough, impart blood red color. Phantom layers of rutile and tourmaline needles are found in quartz also but were carried there by solution current from growing sites upon cavity walls. In such instances, phantoms sometimes appear on one side of the crystal only. Fluorite and calcite frequently exhibit sharp phantoms comprised of hundreds of very small sparkling sulfide crystals.

Inclusions provide a fascinating field of study, and indeed one could devote an entire lifetime to collecting examples of inclusions occurring only in quartz! Specimens containing beautiful inclusions in clear crystals are frequently cut and polished to make the inclusions easily visible, affording beautiful and valuable collector's pieces.

Aligned Inclusions. Curious and striking effects occur in some minerals because inclusions are forced by the host crystal to grow along specific directions. Aligned inclusions result from swarms of small crystals forming on growing crystal faces. Each minute individual is oriented by attractions between itself and the host crystal structure and is forced to grow in more or less parallel position. When eventually covered, all reflect a spangled light from parallel faces as the host crystal is turned. This effect, known as *aventurescence,* occurs in quartz containing small hematite plates, but is most beautifully pronounced in the oligoclase feldspar known as *sunstone.* When minute tubes or slender inclusions develop in certain directions in host crystals, another optical phenomenon is noted, namely *chatoyancy.* It is common in tourmaline crystals which, when properly cut and polished, provide single streaks of light reflecting from the tube walls, furnishing the gems known as *catseyes.* An example is shown in Figure 31. Chatoyancy occurs in any transparent mineral with suitable straight inclusions whether they are solid inclusions or slender cavities.

Two or more sets of chatoyant inclusions furnish *star stones* in ruby and sapphire, garnet and quartz. As before, inclusions are related to the crystal structure of the host but here result from the separative process called *exsolution.* Impurities which eventually form inclusions remain dissolved in the host crystal at high temperatures but as temperature drops, foreign ions can no longer remain in the host crystal as its own ions draw closer into the final low temperature structure. Because the host structure has some room for slender inclusions only in certain directions, foreign ions *exsolve* to form long needle-like crystals in these directions, and in the process, become directly related to the geometry of the enclosing crystal. In corundum, for example, better known to the public for its gem varieties, ruby and sapphire, rutile needles aligned in parallel sets throughout the crystal cause *asterism*, as the display is called, each set of needles being parallel to sides of the basal crystal face as shown in Figure 31. Interestingly enough, similar synthetic corundum gems use titanium dioxide (rutile) as a deliberate impurity. After the future star gem has been grown, it is reheated to temperatures just below the melting point in order to allow rutile impurities to crystallize into the necessary slender needles. Under controlled laboratory conditions, the results are usually far better than nature's products and synthetic stones can almost certainly be detected on the basis of unusual perfection.

Replaced Crystals. Chemical activity is not necessarily followed by permanent results even in such seemingly imperishable objects as minerals. Minerals once formed are subject to chemical attack and all that is needed to make the attack successful are suitable environmental conditions and suitable chemical agents. Some minerals resist chemical change for very long periods and it is these we find and preserve in our collections. But even here many of them must receive protection because our collecting activities remove them from environments in which they are stable and place them in those in which they are not. California borates often need protection from too-dry or too-wet atmospheres in homes, while deep red realgars from Nevada and elsewhere, must be shielded from daylight lest they convert into orpiment. Changes took place in certain minerals long before we ever found them and in many instances produced interesting and wonderful specimens. In general, two types of changes frequently occur in minerals (a) chemical alteration, and (b) dissolution; sometimes both.

A mineral alters when some influence in its environment causes it to change its chemical composition, either by losing or adding constituents. Losses or additions may be of water, of single elements, or of a number of elements. It may also alter non-chemically by changing crystal structure although all previous ingredients remain present in exactly the same proportions. Differing crystal structures of the same composition are called *paramorphs*. If a crystal changes chemically or structurally, yet keeps the shape of the original, it is called a *pseudomorph* or "false form"; it looks like a crystal of one species but is composed of another.

Pseudomorphs are common in the copper carbonates, malachite and azurite, which are closely related chemically and often occur together in the upper zones of copper deposits. Malachite is somewhat more stable than azurite. At times, a chemical reaction is triggered off in the latter mineral causing it to convert to malachite. Results are astonishing; some azurite crystals show patches of vivid green malachite dovetailed perfectly into otherwise deep blue azurite crystals! In other instances, alteration is complete and crystals are all malachite yet faithfully preserving the faces and angles of azurite. Paramorphic minerals may also exchange identities with equal fidelity, as when aragonite converts to calcite, or rutile to brookite. In each instance, the original faces are kept intact! A variety of interesting pseudomorphs is shown in Figure 34.

Pseudomorphs also occur when one species is replaced by another which is in no way related chemically. Beautiful "crystals" of fluorite have been obtained from European deposits which prove to be quartz in the form of chalcedony. From the copper deposits in Bolivia come splendid specimens of native copper perfectly replacing crystals of aragonite. There are many other instances of pseudomorphism, perhaps the best known to the amateur lapidaries being the replacement of wood, coral, and even dinosaur bones by chalcedonic quartz! Each cell, each pore of the original, is sometimes so faithfully preserved in the atom-by-atom replacement process that the original materials are easily and positively identified. It is customary to use the term "after" in connection with pseudomorphism, as "quartz after fluorite," or "native copper after aragonite," etc. Some important examples of pseudomorphism are given in the table below.

TABLE OF PSEUDOMORPHS

Replacing Mineral	⟶ *after* ⟶	*Replaced Mineral*
Quartz		Calcite, aragonite, fluorite, azurite, datolite, serpentine, asbestos; fossils (Figure 34)
Opal		Glauberite; fossils (Figure 34)
Goethite		Pyrite (Figure 164), siderite, magnetite, chalcopyrite, sphalerite
Pyrite		Fossils
Calcite		Aragonite, celestite; fossils (Figure 34)
Clay minerals		Feldspars, tourmaline, beryl, spodumene
Hematite		Magnetite (Figure 148)
Gypsum		Anhydrite, aragonite (Figure 34)
Copper		Cuprite, azurite, aragonite
Malachite		Cuprite, azurite
Chlorite		Garnet
Andradite		Biotite
Muscovite		Tourmaline, spodumene
Rutile		Brookite
Cassiterite		Quartz, feldspar (Figure 34)
Anglesite		Galena
Pyromorphite		Galena
Pyrolusite		Manganite
Serpentine		Olivine, calcite, pyroxenes
Talc		Quartz; pyroxenes
Dolomite		Aragonite
Magnesite		Calcite

FIG. 34. Pseudomorphs. *Top left:* gypsum replacing glauberite crystal; Camp Verde, Yavapai County, Arizona; size 2″ by 2″. *Top center:* calcite replacing interpenetrant sixling twins of aragonite, 3″ by 2¼″; Las Animas, Bent County, Colorado. *Top right:* Kaolinite after orthoclase feldspar Carlsbad twins; St. Austell, Cornwall, England. *Center:* precious opal replacing clam shell; Coober Pedy, South Australia. *Center right:* chalcedony replacing marine shell; New Port Richey, Pinellas County, Florida. *Bottom left:* talc replacing enstatite crystals; Odegarden, Bamle, Norway. *Bottom center:* chalcedony after sharp sixlings of aragonite; Chihuahua, Mexico, and pyrite replacing tabular crystals of pyrrhotite; Amelia, Amelia County, Virginia. *Bottom right:* cassiterite replacing Carlsbad twin of orthoclase; Wheal Coates, St. Agnes, Cornwall, England.

87

Another type of pseudomorphous activity also results in filling of empty spaces by minerals making *casts* of some original crystal or object. Tree trunks which have been enveloped by hot lava and then completely destroyed by burning leave cavities which may be later filled or lined by quartz. Because only filling is involved, the internal structure of the wood is not preserved as it is in replacement. In other instances, crystals coated with a more stable mineral, dissolve from beneath and leave cavities which faithfully mirror their form in the bottom of the more enduring species. If these should happen to be filled with another mineral, accurate casts result. Casts of this sort are easy to identify because fillings are granular or earthy rather than consisting of single crystals.

CRYSTAL HABITS

Because a crystal's structure determines size and shape of faces, the *habit* or *general shape,* is of great importance in identifying many species which prefer to grow in typical habits. For example, the cubic habit of galena is so consistent that the sight of lead-gray cubic crystals at once makes us suspect that they are this species and no other. The following terms are used to describe single crystal habits, with appropriate descriptive sketches furnished in Figure 35. Not much difference exists between some terms, it being a matter of choice which to use when describing crystals.

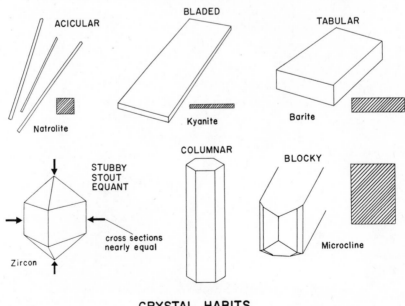

CRYSTAL HABITS

FIG. 35.

TERMS DESCRIBING CRYSTAL HABITS

(see Figure 35)

Term	Meaning	Crystal Examples
Acicular	Needle-like	Natrolite, scolecite
Capillary, filiform	Extremely slender, hair-like	Millerite, chalcotrichite
Columnar	Like a column; thicker than acicular	Beryl, tourmaline
Bladed	Flat and long, like a knife-blade	Kyanite, hemimorphite
Tabular	Like a tablet of paper	Barite, wulfenite
Stubby, stout, equant	Nearly equidimensional in cross-sections	Quartz, apatite
Blocky	*Same,* but crystals block-shaped	Feldspars

Familiarity with habits enables experienced mineralogists to make shrewd guesses as to identity. However, it must be remembered that habit implies the presence of characteristic faces, and when the expert examines a crystal, he is also mentally noting what kind of faces are present, how flat and smooth they are, whether they are striated or etched, and how they are placed in relation to each other. If he were to describe the particular crystal to someone else, he might very well add such terms as *prismatic,* because the crystal happened to be stretched in the direction of a prism as in the case of the columnar beryl shown in Figure 29, or he may say that it is an *octahedral* crystal, as is the spinel in Figure 20, or use other terms to point out some feature which is important. In any event, the selection of terms has as its sole purpose a description which is so clear that another person reading or hearing it, receives the correct mental image.

The shape of inner frameworks of atoms is often, but not always, reflected in the habit. Thus a cubic cell generally produces crystals of cubic shape, or if modified by additional faces, sometimes to a shape which approximates a sphere. Finding a crystal that answers this rough description should lead the amateur to suspect that it is a cubic structure species. Because species are also classified according to structure, this clue is helpful when taken with other

SILICATE CRYSTAL HABITS

Inner Structure	Habit	Minerals
Single tetrahedrons (nesosilicates)	Blocky to nearly spherical	Zircon, garnets
Rings (cyclosilicates)	Long hexagonal or trigonal prisms	Beryl, tourmaline
Chains (inosilicates)	Long prisms or fibers	Amphiboles, pyroxenes
Sheets (phyllosilicates)	Platey	Micas, chlorites
Frameworks (tektosilicates)	Blocky	Feldspars, quartz

data in finally determining the identity of the mineral. Similarly, it is likely that a mineral of hexagonal outline is structurally hexagonal. More will be said later about this aspect of examining crystals; for the moment we turn to some of the internal silicate structures which often reflect their geometry in crystal habits. Some of these structures and their reflected crystal habits are given above; others are listed in Table 2, Appendix A.

CRYSTAL AGGREGATES

The majority of crystals lack faces because they grew in groups competing for the same nourishment in the same cavity space. The absence of crystal faces on melt minerals, as in the granite example shown in Figure 13, is almost a foregone conclusion except when one species crystallizes early and retains its earlier euhedral shape. The term *massive* is used to describe close-knit growths of one or more species, of course implying that crystal faces are absent. Crystal aggregates vary widely from simple massive to characteristic groups of faced crystals grown in openings. In some instances, special hallmarks are so distinctive that they serve instantly to identify the species. A good example is pectolite which forms radiating growths of white, fibrous crystals tightly locked into tough masses (Figure 282, page 498). Very few other minerals resemble it, and well-developed specimens may be identified on the basis of its characteristic appearance in massive form. The following discussion begins with explanations of crystal aggregate formations.

CRYSTAL-LINED CAVITIES

Cavities occur in otherwise solid rock for many reasons. In some lavas, cavities are left by gas bubbles ranging in size from as small as a pea to some large enough to crawl into. In granitic rocks, cavities or "pockets," occur along centerlines of very coarse-grained bodies as *pegmatites*, and it is from them that many splendid gemstone crystals come. In soluble rocks such as limestone, water enters through cracks and dissolves the limestone, leaving openings suitable for later mineralization. Earth movements also produce fractures and fissures which may later fill with minerals introduced by rising solutions. Many are filled with blocks of detached stone with angular cavities between them which provide space for crystallization.

Gas bubble cavities in lava are shaped characteristically like biscuits or round loaves of bread. If filled completely by later mineralization, they provide *nodules*, commonly of agate. If empty spaces remain in their centers, as is frequently the case, *geodes* lined with sparkling quartz crystals, as shown in Figure 36, are the result. Pegmatite and vein cavities are ordinarily lined with crystals pointing inward. In pegmatites, the crystals are often broken or

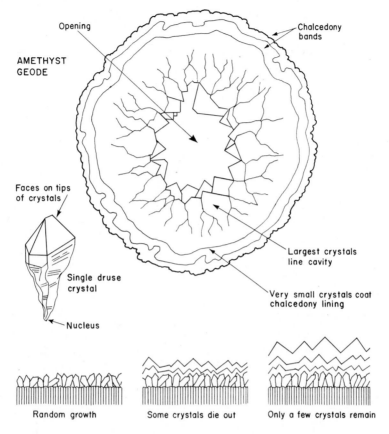

Opening

Chalcedony bands

AMETHYST GEODE

Faces on tips of crystals

Single druse crystal

Nucleus

Largest crystals line cavity

Very small crystals coat chalcedony lining

Random growth   Some crystals die out   Only a few crystals remain

GROWTH OF CRYSTALS IN DRUSES

FIG. 36.

loosely attached, while those in veins generally form more or less continuous blankets of numerous uniform crystals of small size. The adjective *drusy* is applied to formations of the latter type, and sections of cavity walls are called *druses*.

The development of a typical druse is interesting, the steps being illustrated at the bottom of Figure 36. As in many drusy formations, circulating solutions deposit numerous crystal nuclei upon walls. Many crystals begin but few survive; the results are layers of quartz crystals, individually much larger in size but also far fewer in number. Crystals that originally pointed toward the cavity opening were able to survive because quartz grows fastest in the direction of its crystal points and grows slowest to the sides. Any crystal pointed away from the wall, grows so fast that it keeps its head above water, as it were, but those lying on their sides collide with neighbors and are covered over. In any druse, the granular material nearest the wall consists of numerous small crystals; a little farther removed they are somewhat larger but fewer in number, while still farther out, they are least in number but greatest in size.

There is no real atomic structure continuity from crystal to crystal in drusy growths, and it is possible to loosen crystals easily from their neighbors. If a single druse crystal is removed and examined, it is seen to resemble a carrot in shape, that is, the upper euhedral part is broadest, but the lower part tapers toward the point where the original seed began its growth. This is shown to the left in Figure 36. The sides of "carrots" are covered with flat striated planes, often mistaken for real crystal faces, which represent compromises in growth between the crystal and its neighbors.

Sometimes cavities receive so much mineralization that they fill completely, but successive waves of mineralization usually leave "high water" marks as shown in Figure 37. This interesting formation shows drusy layers,

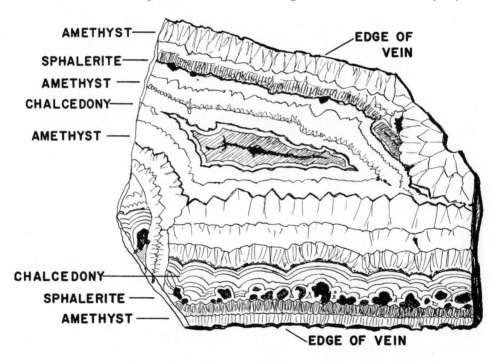

COMB STRUCTURE IN ORE VEIN

FIG. 37.

one deposited on top of the other from wall to wall, until all space is filled. Because of the distinctive toothed pattern, miners call this a "comb" or "comb structure."

Non-drusy crystal growth is also common in cavities, sometimes otherwise bare cavities showing single crystals or small groups sprinkled here and there upon the walls. At other times, later forming minerals deposit crystals on top of druses, providing splendid mineral specimens. Growths of the latter type are abundant in ore veins where sulfide crystals may be found perched on drusy quartz or calcite in single or clustered crystals. When any particular mineral deposit contains a number of species, the combinations of crystals may sometimes be as endless as they are interesting.

AGGREGATE TYPES

Inner crystal structure largely dictates habit; in turn, habit dictates how an aggregate is to grow. For example, malachite develops acicular crystals which, though sometimes developed as cottony tufts of brilliant green color, prefer to form as compact masses coating cavity walls in banded layers, from a fraction to several inches or more in thickness. A typical example is sketched in Figure 38. Close examination shows that the seemingly solid layers are composed of multitudes of tiny crystals, all growing away from the walls. All crystals are locked closely together in nearly parallel positions. Malachite begins growth in the same manner as drusy quartz, and again only those crystals which happen to be pointed upright continue to grow while others are suppressed. The result is near-parallelism, with crystals oriented at right angles to cavity walls, and grown to about the same length. One may say a malachite cavity lining is a druse carried to its ultimate conclusion.

The fibrous texture and mode of growth typical of malachite is duplicated in many other species, but some, similar in appearance, are believed to form from *colloidal gels*. Examples of gel growth are goethite and hematite which form smooth to glassy surfaced bulging masses. These show similar internal features although the mode of growth is not the same. Important features of the various aggregate types will now be summarized. Reference should be made to the appropriate sketches in Figure 38.

Massive or Compact. These terms imply many interlocked crystals, without faces, forming masses of considerable size. *Compact* further implies lack of openings or voids.

Granular. The common texture observed in granite and similar igneous rocks; also in marble. Grains may range from very small to very large in size. Lack of crystal faces upon grains is implied.

Fibrous. Crystals or crystal bundles are thin and long, like silk, and generally torn apart easily. Asbestos is *fibrous* but does not form in the same manner as malachite; it is a metamorphic mineral crystallized in thin seams under conditions of intense heat and pressure. The fibrous nature is a direct reflection of internal silicate chain structure.

Botryoidal—Mammillary—Reniform—Globular. These terms refer to the

GRANULAR MASSIVE   FIBROUS MASSIVE   BOTRYOIDAL, MAMMILLARY

RADIATE, GLOBULAR   MICACEOUS, FOLIATE, LAMELLAR   DIVERGENT

BLADED   DENDRITIC   ARBORESCENT

RETICULATE   PLUMOSE   STALACTITIC, COLLOFORM

AGGREGATE HABITS

FIG. 38.

outward appearance of fibrous aggregates of malachite, goethite, hematite, etc. *Botryoidal* means "grape-like," *mammillary* means "breast-like," *reniform* means "kidney-like," and *globular,* as in the example of wavellite shown in Figure 38, means that individual acicular crystals began growth from a common point, developed additional crystals, with all growing outward in sunburst fashion. When perched upon projecting points of rock, such growths furnish nearly perfect spheres. *Stalactitic* growths are also common, particularly of chalcedony, goethite, and sometimes malachite. The term is usually reserved for the downward-hanging calcite formations in limestone caverns, but similar growths form in various attitudes from colloidal gels. Because so many terms of essentially the same meaning are used to describe growth forms derived from gels, the present trend is to use *colloform* without further description. Colloform growths, and related drusy types, always display *concentric* banding. Shadings of color frequently mark concentric bands in malachite, smithsonite, and others, but are rarely observed in goethite and hematite. In the latter species, concentric bands or shells often split along the junctions between growth generations thus proving their concentric nature. Similar shells occur in wavellite and in other species.

Foliated—Lamellar—Micaceous. All terms refer to similar textures within massive material, namely thin, easily split sheets. *Foliated* means "leaf-like" as in a book, *lamellar* means "thin platey," and *micaceous* merely uses a mineral, mica, to describe the texture meant. Figure 38 also uses a single "book" crystal of mica to show its characteristic structure. In large masses, as in mica schist, numerous booklets of mica permit the rock to be parted readily along planes rich in this mineral. Micaceous texture is characteristic of mica minerals and chlorite minerals, but is also present in graphite, molybdenite, and a few other species.

Bladed. Aggregates of numerous thin lath-like crystals are termed *bladed*. Kyanite often appears in bladed masses of considerable size; stilbite frequently develops aggregates of bladed crystals which diverge slightly to form "wheat sheaf" bundles.

Plumose. Mica-like minerals sometimes form aggregates of numerous minute scales developed into spreading plume-like or *plumose* growths as shown in Figure 38. Occasionally the crystals are acicular rather than scaly, but the same descriptive term is properly used if the general form is that of a plume.

Divergent—Radiate. Both terms are used to describe long crystals which grow from common points but spread outward as single, euhedral crystals. *Stellate* is sometimes used to describe divergent crystals in star-like patterns.

Dendritic. Moss-like growths, often of great complexity and beauty, are termed *dendritic*. Chalcedony commonly contains dendrites of manganese oxides and extremely fine curved and branched filaments of silicate minerals which lend it the varietal name moss agate. Dendrites are common in very thin cracks in rocks along which manganese oxides crystallize in patterns similar to that shown in Figure 38. Native copper also forms interesting dendrites, some as thin as a leaf and complexly branched due to development of numerous lateral crystals.

Arborescent. Growths of this kind are "tree-like," that is, branched in systems which resemble the way a tree develops. It is common in native silver, copper, and gold.

Reticulated. Criss-crossed slender crystals form *reticulated* or "network" growths. Common examples are furnished by rutile and cerussite. Reticulated crystals usually bear a structural relationship to each other and consequently the angles between them are likely to be the same.

Cockscomb Crystals. Numerous crystals, offset slightly in relation to each other, sometimes form semi-circular fans or "cockscombs." Marcasite forms in this fashion and is easily distinguished from other sulfides on this account. Groups begin as several single crystals occupying nearly the same growth point on a cavity wall. With growth, they diverge in one direction only, creating characteristic aggregates. Similar growths are common in barite from England, hemimorphite from Mexico, and in autunite, torbernite, and columbite.

Barrel Crystals. Crystals sometimes grow freakishly into curved "barrels" which exhibit many separate crystal terminations at the ends of the "barrel" and more or less smoothly curved surfaces along the sides corresponding to the area in a real barrel covered by staves. Crystals of this kind are common in mimetite and pyromorphite.

Rosette Crystals. The celebrated "iron roses" of hematite are excellent examples of how numerous small platey crystals can grow in rosette fashion, sometimes perfectly circular in outline. Growth of a rosette begins with single crystals in the center from which others diverge outward, overlapping each other much like petals in a real flower. At the same time, each crystal slightly thickens toward the outer edges such that it becomes much thicker at the rim of the rosette. Perfect iron roses therefore show a decided depression in the center, with uniformly upraised rims formed by edges of many individual crystals. An iron rose from Brazil is shown in Figure 148 (page 327). Rosette growth of hematite is so characteristic that it can be used to identify this mineral at once. Similar but less perfectly developed rosettes are seen in pyrrhotite, but can be identified by the characteristic dull grayish-yellow metallic luster of this species. Aggregate habits are listed in Table 2, Appendix A.

OTHER DESCRIPTIVE TERMS

Less common terms include *pisolitic,* referring to concentric globular deposits, each about the size of a pea; *oolitic,* similar to pisolites but smaller; *concretions,* or rudely spherical masses formed by deposit of a mineral around a nucleus within enclosing rock and not in a cavity, and *amygdaloidal,* referring to nodules, usually in volcanic rocks, which are "almond-like" in shape.

TWINNING

Ordinary crystalline aggregates are essentially accidental in nature, representing numerous crystals which happened to choose the same time and place to grow. There is no real relationship between separate crystals except neighborhood, and it is usually easy to part them along junctions indicating that little holds them together except tight joints. However, crystals of many minerals *do* grow in fixed relationships known as *twins,* which as may be expected, are related to atomic structures.

Twinning begins in most instances, when crystals are very small. An ion arriving at some point on the growing crystal may take a position which is not geometrically perfect in relation to the existing crystal structure, but is nearly as good. If rapid growth takes place, it is possible for this ion to serve as a seed attractive to arriving ions which attach themselves and establish a new direction of growth. As more appear, the offset crystal enlarges and becomes half of the twin. Because twinning occurs early, it is usual to find twin parts nearly, if not exactly, the same size and exhibiting the same faces, markings, etch pits, etc. Fast growth promotes twinning for the reason mentioned; conversely, slow growth discourages twinning because ions have more time to shift to correct positions. Offset positions conducive to twin growths are *strictly limited in number* and *invariably consistent* from crystal to crystal. While crystals in druses may take any and all positions, showing no consistency in this respect, twins *always* take the *same* positions relative to each other. This is because some crystal structures possess certain points which will tolerate slightly offset ions while the ions beneath will exert enough attraction to keep offset ions in position. Thus twins may grow upon such points but only in certain directions. If other positions are chosen by incoming ions, the result will be formation of an aggregate. Attractive forces are usually very strong along twin junctions. For this reason, true twins do not break apart easily.

The consistent directions taken by twins are also evident by measuring angles between them. A good example is furnished by staurolite, a mineral found in some Appalachian mountain areas, in cross-like or *cruciform* twins, much prized as good-luck pieces. Each staurolite twin consists of two columnar crystals crossing at angles of close to 60° or 90°, the latter type illustrated in Figure 39. When faces are smooth and sharp, it is possible to measure the angle between twins and to see that no matter how many crystals are checked, the angle is always the same. *Constancy in twinning angles* is identical in nature to constancy of angles between faces of normal crystals; in both instances it is fixed by the crystal structure.

Features of Twins. Perhaps the easiest way to detect twins is to look for places where vee-shaped depressions mark the junctions between crystal pairs. Depressions of this sort are commonly called *reentrants,* and the angles, *reentrant angles.* A reentrant appears in the sketch of a twinned crystal of spinel in the lower part of Figure 39. A single crystal is on the left with a dotted plane passing through it representing the *composition plane,* or the

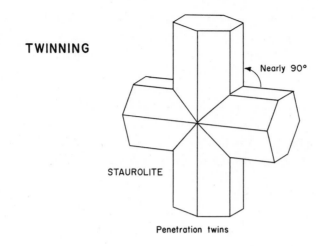

TWINNING

STAUROLITE

Nearly 90°

Penetration twins

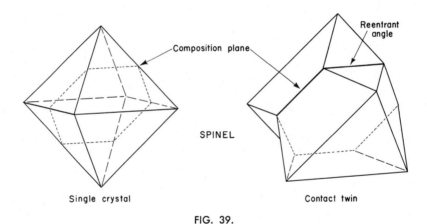

Composition plane

Reentrant angle

SPINEL

Single crystal

Contact twin

FIG. 39.

plane on which both halves of the twin meet. The twinned crystal can be visualized as a single crystal sliced in half, with the front half rotated to twin position. A twin of this kind is called a *contact* twin. Other examples are given in Figure 40.

Although reentrants are excellent clues to twinning, they are not always present and one may be forced to look for other signs. In sphalerite crystals, twinning commonly shows itself by a change in luster among narrow parallel bands which completely encircle the crystal. Band edges are very straight and sharp, but it takes a bit of turning to detect the slight differences in reflections which tell that each crystal actually consists of several crystals or *repeated twins*. A repeatedly twinned crystal of sphalerite can be compared to a loaf of bread in which every other slice is removed, turned around and replaced, retaining the general loaf shape but revealing differences along

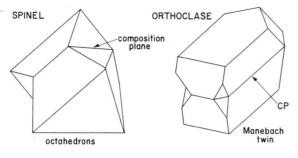

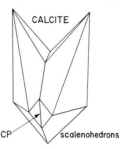

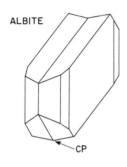

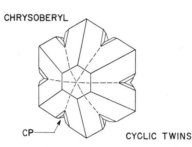

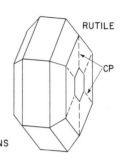

SPINEL

composition plane

octahedrons

ORTHOCLASE

CP

Manebach twin

CALCITE

CP

scalenohedrons

CONTACT TWINS

ALBITE

CP

CHRYSOBERYL

CP

CYCLIC TWINS

RUTILE

CP

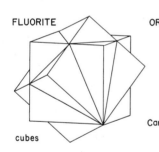

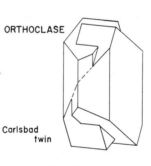

REPEATED TWINS

FLUORITE

cubes

ORTHOCLASE

Carlsbad twin

TETRAHEDRITE

tetrahedrons

PENETRATION TWINS

FIG. 40.

slice edges and in surface texture and color. Repeated twinning is called *lamellar* or *polysynthetic* twinning. It is so common in plagioclase feldspar, for example, that one can confidently distinguish between grains of this mineral and a similar potash feldspar because the characteristically fine straight twinning striations of plagioclase is not a feature of the potash feldspar. A simple twin of albite is shown in Figure 40 and a polysynthetic twin in Figure 249 (page 456).

Another form of the repeated twin is the *cyclic* twin. Several examples are shown in Figure 40. Cyclic twins develop when crystals twin on several sides at once, curving around to form disk-like or doughnut-shaped groups or groups of even greater complexity. Complete cyclic twins of this kind are much prized as specimens, particularly of the species illustrated.

Penetration twins occur in many species, several examples being given in Figure 40. All give the impression that two crystals were mysteriously pushed into each other as if each had no real substance. However, the twin relationship began when both crystals were very small and merely continued as they grew larger. Junctions between penetration twins are often irregular; if crystals are sliced on a diamond saw, decided differences in luster between individuals will be noted.

Cyclic and penetration twins produce misleading shapes, sometimes making one believe that they belong to one class of crystals when they really belong to another. Thus, in Figure 40, chrysoberyl cyclic twins are hexagonal in profile and one could easily believe that it is a single crystal belonging to a mineral with hexagonal structure. Another example is shown in Figure 41

ARAGONITE PENETRATION TWIN

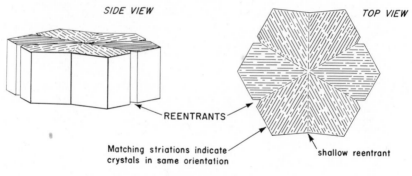

SIDE VIEW

TOP VIEW

REENTRANTS

Matching striations indicate
crystals in same orientation

shallow reentrant

FIG. 41.

where an aragonite twin is depicted. Again the impression is that a hexagonal crystal is being viewed. When the twin group is carefully examined, however, small reentrants are seen as well as large shallow reentrants on each pair of prism faces, making these faces slightly dished instead of flat. The upper face is also marked by striations as shown in the drawing, each set of markings belonging to a distinct crystal.

Quartz Twins. The popularity and abundance of quartz makes it deserving of special remarks concerning its twinning. Quartz twinning is so common that untwinned crystals are truly rarities! The most conspicuous twinning mode is called *Dauphiné*, after the numerous crystals obtained from localities in this area of France. By far the majority of crystals several inches or more in size from any locality show Dauphiné twinning by zig-zag offsets along prism and terminal faces. Also common is *Brazil* or *Brazilian* twinning, so named because it was noted in crystals from that country. Ironically, most quartz from Brazil is relatively free from this mode of twinning. There is no easy way to detect Brazilian twinning because its effects are internal and surface signs are usually absent. It is ordinarily detected by cutting crystals across and examining them optically. The third common quartz twin is known as the *Japanese* twin, because many fine examples came from the celebrated locality in Kai, Japan.

Twinning in quartz is complicated by its curious crystal structure composed of silicon-oxygen tetrahedra linked in spiral fashion. The structure can be likened to treads of a spiral staircase, each crystal composed of multitudes of staircases, oriented parallel to the direction from one termination to the other. As in staircases, spirals may turn to left or right. Quartz crystals commonly show outward evidence of spiraling direction and are then called *right handed* or *left handed*. Figure 42 shows examples of both. If crystals are

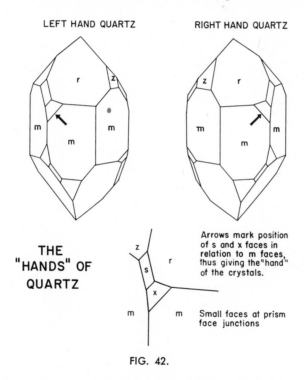

FIG. 42.

held in the positions shown, the handedness can be told by the location of the small auxiliary faces called *x* and *s* faces. If present at the upper left hand corner of the prism face marked *m*, the crystal is a left hand one and vice versa. Crystals of both are equally abundant in nature.

A further peculiarity due to structure appears in terminal faces, as distinguished by different letters in Figure 42. One set is marked with small *r*, the other set by *z*. Each set alternates on any given crystal, every other face is a *z* or an *r* face. One set may grow larger than the other, in which case the crystal terminations consist only of three large faces with three very small faces between them. In other crystals they are evenly developed. If it were possible to enlarge enormously the atomic structure beneath the *z* and *r* faces, we would see that the pattern is not the same, explaining why the faces are not the same. Notice how these faces also alternate on opposite ends of the crystal. Because of differences in atomic structure beneath these faces, etched crystals show similar etch marks only on similar faces, that is, if the termination is etched, the etch marks will look the same only on every other face. The alternate face coloration on hematite previously described in this chapter is due to differences in atomic attraction beneath these distinctive quartz faces.

In Dauphiné twins, two right hand or two left hand crystals are merged but one individual is rotated 180° as depicted in Figure 43. Prism faces,

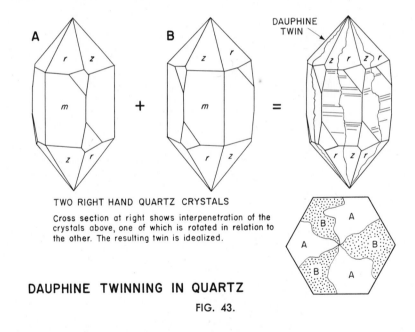

TWO RIGHT HAND QUARTZ CRYSTALS

Cross section at right shows interpenetration of the crystals above, one of which is rotated in relation to the other. The resulting twin is idealized.

DAUPHINE TWINNING IN QUARTZ

FIG. 43.

marked with small letter *m*, are usually striated horizontally as is true of most quartz crystals, but the striations are broken up by twin boundaries in a most irregular fashion as shown in Figure 44. In addition to offsets, zig-zag boundaries, etc., differences in faces are also revealed by etching, growth hillocks, or luster.

Brazilian twinning is extremely difficult to detect by external signs, but

FIG. 44. Dauphiné twinning in quartz. A 7″ rock crystal from the San Pedro Mine, Hiriart Hill, Pala, San Diego County, California, showing the many steps and offsets produced along the sides and top as the result of numerous dauphiné twins.

polished crystal cross-sections illuminated by polarized light show striking patterns of strong colors as indicated in the lower part of Figure 45. Similar cross-sections of amethyst display alternating purple and colorless segments due also to this mode of twinning. Unlike Dauphiné twins, internal portions of Brazilian twins form straight-sided triangular patches but this is not evident even when alternate patches are next to faces. Natural etching some-times shows patches of slightly different luster on some faces: when patch boundaries are irregular, the crystal is Dauphiné; if perfectly straight, Bra-zilian twinning is present. Most quartz crystals include both modes of twin-

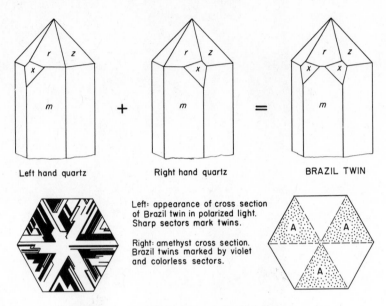

Left: appearance of cross section of Brazil twin in polarized light. Sharp sectors mark twins.

Right: amethyst cross section. Brazil twins marked by violet and colorless sectors.

BRAZIL TWINNING IN QUARTZ

FIG. 45.

ning. Thus the *s* and *x* faces may be present on *both* corners of the Brazilian twin prism faces as shown in the upper right sketch of Figure 45, but this is very rare.

Japanese quartz twins, illustrated in Figure 46, are simple contact pairs joined at angles of approximately 84°. Crystals are usually flattened with broad prism faces *m* forming continuous planes bisected by the zig-zag twin boundary.

Summary of Twin Features. *Position.* Twins show unvarying mutual posi-tions; angles between them measure the same from twin to twin.

Size. Because many twins begin growth at nearly the same instant, indi-viduals tend to be of the same size.

Reentrants. Shallow to steep angles between twins are usually present at one or more points or as deep depressions in the center of some cyclic twins.

Offsets. Marked offsets are usually present along twin boundaries but con-spicuous only in a few species.

Surface Indications. Close examination is necessary to detect differences in

FIG. 46. Japanese twin of rock crystal quartz. Size 3½" by 3½". From the famous locality of Otomezaka, Yamanashi Prefecture, Japan. The thickness of this crystal is only about ⅝", as is commonly the case in the Japanese twins of similar size from this locality. *From the collection of Alice J. Walters, Ramona, California.*

etch marks, pits, and luster between twin individuals. Striations meeting in vees usually indicate twin junctions.

Occurrence of Twins. Some species twin according to one mode only; others twin in several distinct modes. Very few species do not twin, and twins are therefore abundant although frequently overlooked.

5

Geometry of Crystals

Identification is usually made much easier if good crystals are at hand. Some species form such characteristic crystals that they can be identified on sight. Others are not so obliging, and for them, additional bits of information must be used before positive identification can be made. Despite wide variations in external crystal faces, as indicated in the previous chapter, clues provided by size and placement of crystal faces are of utmost value in making identifications. This can be appreciated from the experience of micromounters who cannot remove minute crystals and test them in ordinary ways, yet manage to identify accurately the majority of their specimens from the visual clues obtained through the microscope.

Although crystallography is extremely complex in its advanced phases, the amateur, fortunately, need not concern himself with more than the essentials. He will never be called upon to measure all the angles of a crystal, plot them on the special paper provided for this purpose, calculate missing elements of geometry, or attempt to establish atomic structure. These are tasks for the professional. The amateur should know how to recognize the distinctive crystal faces which reveal basic geometry, and from this, aid himself to recognize strange species without turning to others for assistance. Even these light requirements are not easily mastered because few crystals grow *exactly* like textbook illustrations! Actually, "textbook" crystals are so rare that many collectors look for them specially and show them with pride to fellow collectors. Despite the problems, it is possible to learn how to orient crystals so that their position in the fingers correspond to positions shown in crystal drawings. Once this knack is acquired, it is far easier to compare and identify faces, and ultimately identify the mineral itself.

DEVELOPMENT OF CRYSTALLOGRAPHY

The true meaning of crystals and crystal faces did not develop until relatively recent times although crystals were known to mankind for thousands of years. Rock crystal was known to the Greeks and Romans, indeed the word "crystal" is derived from the Greek *krystallos*, a term originally applied to ice, and indicative of the ancient belief that rock crystal was a kind of ice, frozen so hard that it could no longer melt. History records an abundance of Swiss crystals, probably as magnificently crystallized as those still obtained today, which were used by Roman artisans. Despite the evidence plainly exhibited on every crystal, the view was generally held that crystal faces were freaks of nature! It was not until 1669 that Steno discovered that quartz faces always met each other at constant angles. This important discovery passed almost unnoticed and for the next one hundred years, few advances were made in crystallography.

In 1783, Romé de l'Isle published a work on crystallography showing keen insight into the nature of crystals. This was followed in 1784 by the famous essay on crystals by a contemporary, the celebrated René Just Haüy, who published still another work on mineralogy in 1801. Haüy's studies revealed much about the consistency of crystals, perhaps most importantly pointing out that crystals of the same species were related through an unvarying internal structure despite numerous deceptive external features. The modern study of crystallography began with Haüy.

Haüy noted that calcite crystals always split along definite planes, and that no matter how the bits were further reduced in size, they invariably took the same shape. He believed that a calcite crystal could be split to an ultimate essential particle or, as he put it, the "integral molecule." He explained faces as being planes touching the tips of integral molecules in areas where crystals failed to develop fully, and made sketches to show this as in Figure 47, the roughness of such faces not being detectable because of the very small size of the particles. Although we know that crystals are not built from small "molecules" in quite the manner Haüy believed, his basic idea was so logical that it set off a train of investigations and discoveries which has not stopped to this day.

THE CRYSTAL CELL

Haüy's most important conclusion—that crystals were composed of many small identical "molecules"—is the ancestor of the modern idea that any larger crystal structure is composed of a regular stacking of much smaller building blocks. Each block, or as it is called, *unit cell,* is the smallest volume which keeps all the properties of the crystal: *chemical, physical,* and *geometrical.* This is to say that it must contain enough of the various atoms to establish the *chemical formula, the pattern of atomic arrangement,* and the *directions* and *proportions* of the crystal structure. Above all, each cell must be of such size and shape in relation to the atoms it contains, that

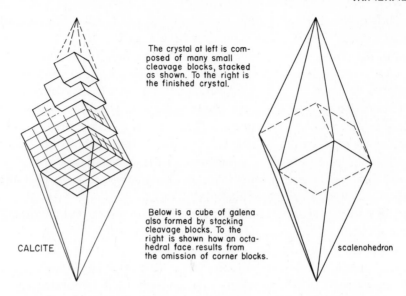

The crystal at left is composed of many small cleavage blocks, stacked as shown. To the right is the finished crystal.

Below is a cube of galena also formed by stacking cleavage blocks. To the right is shown how an octahedral face results from the omission of corner blocks.

CALCITE

scalenohedron

HAÜY'S IDEA OF THE INNER STRUCTURE IN CRYSTALS

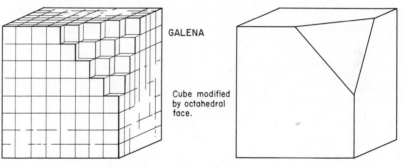

GALENA

Cube modified by octahedral face.

FIG. 47.

when many cells are stacked in three dimensions, every cell will be snugly surrounded by exactly similar cells with no gaps whatsoever between them.

The idea of cells in crystals is best understood by comparing a crystal to a large office building. While a building is being constructed, its steel framework is like the crystal lattice which crystallographers use to depict atomic patterns in space. Girders run vertically and horizontally, dividing the space into hundreds of rooms or "cells," much like similar divisions in a real crystal lattice. Even Haüy's idea of cells being left out systematically to create faces is duplicated in office buildings which are often stepped back to let in more light to the streets below. Unlike office buildings, crystal frameworks do not always grow at right angles: some are tilted in one or more directions, leading to crystals which depart from squareness. Nevertheless, all crystals, no matter how unsymmetrical they seem externally, still include the absolute essentials of all crystals: *the regular and repeated arrangement of atomic contents in space.*

CELL SHAPES AND SYSTEMS

One glance at a mineral collection rich in well crystallized specimens is enough to show a seemingly wild confusion of crystal shapes. It is true that there are many variations, but all of them can be related and placed into six *crystal systems* on the basis of like features of geometry. Because outer features of crystals depend on the regular build-up of unit cells, we are not surprised to learn that within each system there are basic unit cells with characteristic angles and proportions. The crystals we see and handle, are nothing more than greatly enlarged versions of these unit cells, complicated by development of numerous faces.

Nature always tries to solve her problems in the simplest fashion possible, and as we shall see later, the simplest crystal cell, the cube, is her attempt to pack the largest number of atoms in the smallest space. We may say that the cube cell is therefore the "ideal" cell, and if it were not for the large variety of individualistic atoms perhaps all crystals would belong to just *one* system, that with the cube as its basis! Fortunately, or unfortunately, depending on point of view, atoms vary so much in size and behavior that all cannot be accommodated within the simplicity of the cubic cell. There is a long gamut of changes in cell geometry, each departing a little more from the ideal. Within this long string, fences are erected to mark places where changes are decisive in character, the sections between being the *crystal systems.*

Let us then begin with the cube cell, and by pushing and pulling, distort its form to create the others. We must remember, however, that the "pushing and pulling" in real crystals is done from within by the atoms themselves which try to settle as best they can the conflicts between various atomic sizes and degrees of mutual attraction or repulsion which exist even when only one element is involved. The cube cell itself is the basis of one system, the *isometric. Iso* means "same," *metric* means "measure." This is to say, every *like direction* in a perfect cube measures the *same,* such as the lengths of edges, depths, and even the angles. This is shown in A of Figure 48.

A slight but important departure from cube perfection is shown at B in Figure 48. All we have done is pushed or pulled the cube along one direction without changing any angles, and created another unit cell—the *tetragonal.* The name is derived from *tetra* for "four," and *gonal* for "angular." It really means "four square," a good way to remember this cell. As slight as this change is, it has destroyed the squares upon the sides of the original cube cell, which are now rectangles. However, the end faces are still square and all angles are still right angles. The cell can be tall or squat.

THE CRYSTAL SYSTEMS

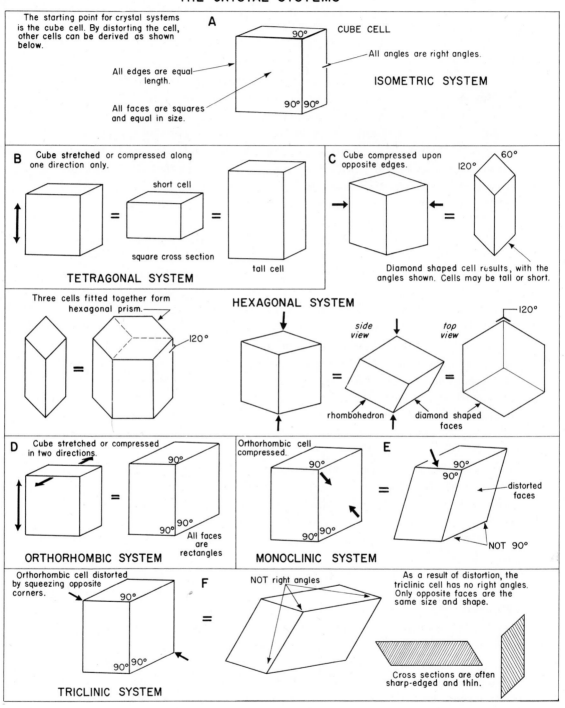

The starting point for crystal systems is the cube cell. By distorting the cell, other cells can be derived as shown below.

A

CUBE CELL

90°

All angles are right angles.

All edges are equal length.

All faces are squares and equal in size.

ISOMETRIC SYSTEM

90° 90°

B Cube **stretched** or compressed along one direction only.

= short cell = tall cell

square cross section

TETRAGONAL SYSTEM

C Cube compressed upon opposite edges.

60°

120°

= Diamond shaped cell results, with the angles shown. Cells may be tall or short.

HEXAGONAL SYSTEM

Three cells fitted together form hexagonal prism.

= 120°

side view

top view

120°

= rhombohedron = diamond shaped faces

D Cube stretched or compressed in two directions.

90°

= 90° 90°

All faces are rectangles

ORTHORHOMBIC SYSTEM

Orthorhombic cell compressed.

90°

90° 90°

E

90°
90°

distorted faces

NOT 90°

MONOCLINIC SYSTEM

Orthorhombic cell distorted by squeezing opposite corners.

90°

90° 90°

F

NOT right angles

As a result of distortion, the triclinic cell has no right angles. Only opposite faces are the same size and shape.

Cross sections are often sharp-edged and thin.

TRICLINIC SYSTEM

FIG. 48.

The next distortion calls for pressing the sides of the cube cell as shown by the arrows in C of Figure 48. Only enough pressure is applied to create angles of 60° and 120° within the end faces. As before, the cell can be tall or short. By placing three cells neatly together as shown at the left, we see that a six-sided cell is obtained. This is the basic form in the *hexagonal* system, the name being derived from *hexa* for "six." It is also possible to take a cube cell, as shown to the right in C, and by squeezing on opposite corners, distort it to the point where it also displays six-sided symmetry as viewed from the top. The angles are again 120°. The squashed cell is called a *rhombohedron,* and because it differs from the previous hexagonal cell, some prefer to place all crystals derived from the rhombohedron into a separate *rhombohedral* system. However, most mineralogists do not, but include them in the rhombohedral *division* of the hexagonal system.

At D in Figure 48, our distortion of the cube cell takes place in two directions at right angles to each other; one is to the side, the other vertically. The resulting cell looks like an ordinary packing carton with none of the sides being square but all angles remaining 90°. The proportions can be many, so long as the angles are right angles and no sides of the cell are square. Technically, the shapes are called *rhombs,* but because all angles are 90°, the system is called *orthorhombic,* from *ortho* meaning "at right angles."

The next cell is derived from the boxy orthorhombic cell by squeezing opposite edges as in E of Figure 48. Faces of one pair are changed into parallelograms, but the others remain rectangles. Only *opposite* pairs of faces match. Within two pairs, corner angles remain right angles, but angles within the parallelogram pair may be any value. Because the tilt is in one direction, the system is *monoclinic,* from *mono* for "one," and *clinic,* to "incline."

The last system is also derived from the orthorhombic cell, but the distortion is applied by pressing opposite corners as shown at F, resulting in the least regular of all shapes, with no right angles, and no square or rectangular faces. One may truthfully say that the result is the perfect opposite of the isometric system cell! The sketches in F also show cross-sections to indicate the sloped sides and parallelogram faces of the cell. Because all three sets of edges are now inclined to each other, crystals of this system are called *triclinic.*

Incidentally, it is helpful to the memory to put together initial letters as follows and then pronounce them: ITHOMT. It isn't easy to say ITHOMT but the mere effort is likely to engrave them on the memory. It is also helpful to remember that ISO means *everything the same,* TETRA means *four-square,* HEXA means *six,* ORTHO means *all right angles,* MONO means *one-tilt,* and TRI means *three-tilt.*

CRYSTAL AXES

It is much easier to describe features of crystals by referring them to imaginary lines called *axes*. Every geometrical solid has axes, even the office building in Figure 49. Although it is easier to visualize its axes as edges of the entire building, actually axes pass between every room. When it

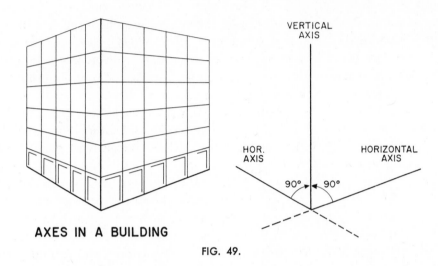

AXES IN A BUILDING

FIG. 49.

is considered that an office building is a composite of its rooms, this is the same as saying that the axes reflect the room edges. In crystals, axes similarly reflect cell edge direction. Because there are millions of cells in every crystal, there are also millions of axes, if we wished to look at it this way. Axes are therefore *directions only* and do not enter or emerge from crystals at given points. Drawings show them doing so only because it is easier to visualize how faces are placed in balanced arrangements around axes if just *one* set is shown. Without the use of axes, it would be nearly impossible to describe crystals and their properties in a sensible manner.

Figure 50 shows how axes are related to unit cell edges. The cells are placed to the left while the axes derived from them are placed to the right. Note that the operation is merely one of taking cell edges and extending them to make lines crossing at common points. Such figures are called *axial crosses*. In order to lend realism, axes and the faces placed upon them in completed drawings, are tilted forward slightly so that the reader's eye is a little above the horizontal. The axis running from left to right is also slightly tilted to heighten the effect. To avoid complications, all parallel lines in a real crystal are made parallel in drawings instead of being foreshortened as would be done in true perspective work.

The lettering of axes is uniform, following the scheme in Figure 50. In the isometric system, the edges of the cube cell are all of the same length, with all planes inclined to each other at right angles. The axes are lettered a, the forward pointing axis being a_1, the side axis a_2, and the vertical axis a_3. Note that certain ends are given plus signs, while the other ends, understood

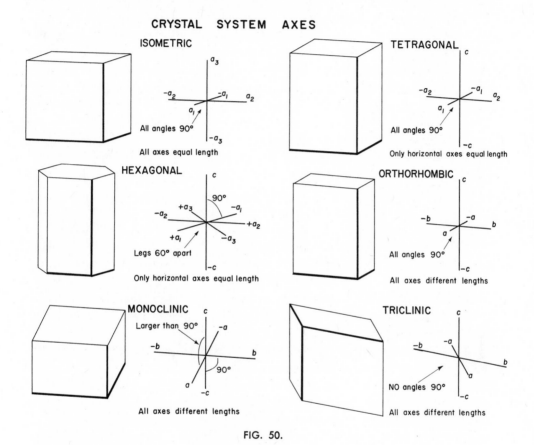

CRYSTAL SYSTEM AXES

FIG. 50.

to be minus, are unmarked. The reason for such marking will be clear later when the method of locating faces on crystals is explained.

In the tetragonal system, horizontal axes are the same in length, reflecting identical atomic structure directions. They are labeled a_1 and a_2. The vertical axis which reflects greater or shorter length of the unit cell along this direction is labeled the c axis. In the hexagonal system, all horizontal axes are again identical and uniformly labeled with a, plus a number. Note that plus and minus ends switch places around the crystal. Lateral axes in this system are 120° apart, or 60° apart between legs; their plane is at right angles to the c axis.

In the orthorhombic system, all axes are labeled separately because different cell dimensions and atomic structure patterns are encountered in all three axial directions. Thus the forward pointing axis is a as before, the lateral axis is b, and the vertical axis is c. All axes are perpendicular to each other but drawn to different lengths to reflect the varying cell dimensions. It is customary to place the shortest axis pointing toward the reader, the next longer to left and right, and the longest vertically. Monoclinic and triclinic systems also require axes of different lengths, inclined to reflect the shape of the cells. Monoclinic axes are drawn so that the axis which *does not* make a right angle points downward toward the reader and becomes the a axis. The other two, at right angles to each other, are drawn in the plane of the paper. In the triclinic system all axes are inclined to each other at angles other than right angles. The forward tilting axis is a; the other two b and c. Axes are also of different lengths.

The characteristics of the system axes are summarized as follows:

CRYSTAL SYSTEM AXES

Name	Axes	Remarks
Isometric	a_1, a_2, a_3	All axes perpendicular to each other and equal in length
Tetragonal	a_1, a_2	Horizontal, equal length; c axis vertical and *not* equal to horizontal axis length; all axes perpendicular to each other
Hexagonal	a_1, a_2, a_3	Horizontal, equal length; c axis vertical and *not* equal to horizontal axis length; horizontal axes 120° apart and their plane is at right angles to c axis
Orthorhombic	a, b, c	All axes unequal length and all perpendicular to each other
Monoclinic	a, b, c	All axes unequal length, two at right angles to each other (b, c), a axis inclined to their plane
Triclinic	a, b, c	All axes unequal length, none at right angles

AXIS LENGTH

One of the most confusing aspects of crystal study for beginners is to be told that axes are of certain *lengths*. Yet if they represent *directions*, as we have said they do, how can any direction have length? Actually, lengths refer to the *lengths of the unit cell* measured in the directions of the axes. Thus when it is said, *"The three axes of the isometric system are all at right angles to each other and of the same length,"* the more correct statement would be, *"The three mutually perpendicular edges of the isometric unit cell are of equal length."*

HOW FACES ARE RELATED TO AXES

Thus far, we have shown that the basic building block of any crystal, the unit cell, establishes the geometry of the crystal. It does so in much the same manner that the size and shape of the many rooms in an office building establish the size and shape of the finished building. Although it is easy to step into any building to see the proportions of the rooms, it is impossible to do so in the case of a crystal, and before the advent of modern x-ray methods for sending beams into crystals to determine from the reflections just how the atoms must be arranged, mineralogists had to depend on carefully measuring the faces of crystals to discover what the internal arrangements might be. As we shall soon see, much of this early work turned out to be very accurate, as verified by later mineralogists who could use x-rays to measure distances between atomic planes in crystals and discover the patterns of internal atomic arrangement.

To show how faces are related to axes, or to be more truthful, to the edges of the cells within the crystals, refer to Figure 51 where a kind of game involving the stacking of blocks has been depicted to show how many types of crystal faces can be built merely by changing the ways in which the blocks are stacked. At the top of the illustration are cube cells, at the bottom, orthorhombic cells with rectangles instead of squares for faces. We may imagine that we are looking down upon cross-sectional slices through crystals, each of which has the same kind of cells, but each of which has different faces. We shall see how the faces are related to the axes and to each other, and thus explain why any given mineral species can have more than one type of faces covering its crystals.

At the top, in the case of the cube cells, four are stacked along one crystal axis, and four along the other. Because there are four cells along each of the axes, the ratio is 4:4, or to put it simply, 1:1. If, as Haüy did, we draw a dotted line just touching the tips of the protruding cells, we can see that the line could represent an actual crystal face, and it could be wonderfully smooth because the cells are so small that our eyes would not be able to detect any of the surface irregularities which seem so pronounced in the drawing. In the next sketch, the number of cells along one axis is doubled. this immediately makes the corner angle a lower value, $26\frac{1}{2}°$ instead of $45°$, and at the same time, changes the proportions of cells along the axes to 4:8 or, simply, 1:2. The other examples along the top of Figure 51 show how other simple proportions can be obtained. Thus we can see why Abbé Haüy discovered how cells must be arranged inside crystals! For if we measure the angles on the outside of a crystal, and plot them on paper, much

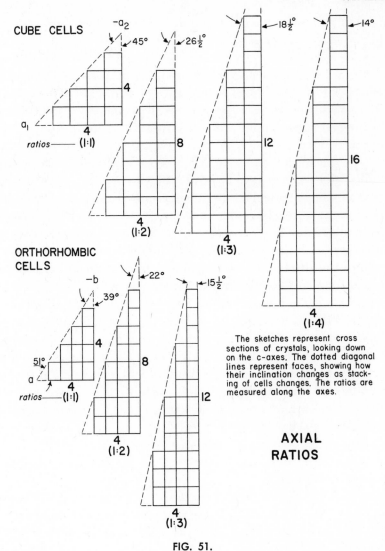

The sketches represent cross
sections of crystals, looking down
on the c-axes. The dotted diagonal
lines represent faces, showing how
their inclination changes as stacking of cells changes. The ratios are
measured along the axes.

**AXIAL
RATIOS**

FIG. 51.

as we have done in Figure 51, it is obvious that only cells of square cross-
section can furnish the angles shown at the upper corners, namely 45°, 26½°,
18½°, and 14°. But are such faces real? Do they really exist in crystals? The
answer is yes. The first face (ratio 1:1), belongs to the form known as an
octahedron in isometric crystals, the second face (ratio 1:2), belongs to the
dodecahedron, and the last two to forms known as *tetrahexahedrons.* All
of these forms will be depicted later in this chapter.

What becomes the effect when the proportions of cells are changed, as in
the lower part of Figure 51? Here we see that the rectangular cross-section of
the cells causes the slope of the faces to change in proportion to how much
the cell is stretched in one direction. The same number of cells are stacked

as before, and therefore give the same ratios of cells along the axes, but now the angles at which the faces meet are far different. In the first sketch where the ratio is 1:1, the angle at the top is no longer 45° but only 39°, in the second, the ratio 1:2 gives only 22° instead of 26½°, and so on. Again, if such a crystal were measured carefully and the angles and planes plotted on paper, the conclusion must be that the cells inside the crystal cannot be square in cross-section, but must be rectangular!

AXIAL RATIOS

Thus we can see that by measuring the angles made between faces on crystals, plotting them on paper, then measuring the distances along the lines representing the axes, we can arrive at what are known as *axial ratios*. Let us take a practical example of a real crystal to show how this is done, using a sulfur crystal which we know does not have cells shaped like cubes, but orthorhombic cells whose faces are rectangles. Figure 52 shows our

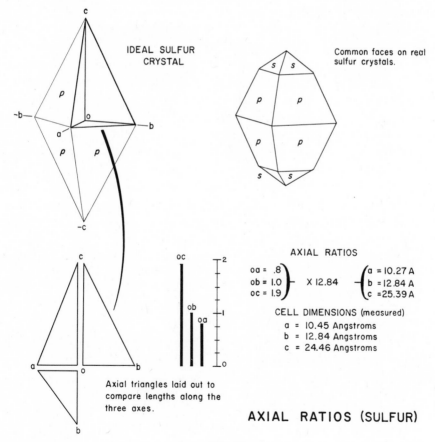

IDEAL SULFUR CRYSTAL

Common faces on real sulfur crystals.

Axial triangles laid out to compare lengths along the three axes.

AXIAL RATIOS

oa = .8
ob = 1.0   } X 12.84   { a = 10.27 A
oc = 1.9   { b = 12.84 A
                           { c = 25.39 A

CELL DIMENSIONS (measured)
a = 10.45 Angstroms
b = 12.84 Angstroms
c = 24.46 Angstroms

AXIAL RATIOS (SULFUR)

FIG. 52.

sulfur crystal, the sketch at the upper right being an idealized drawing of a crystal such as may be found in the Sicilian deposits, and the sketch at the upper left drawn to show just one set of faces, in this case, those belonging to the form known as the dipyramid and each face marked with the italic letter *p*.

By measuring the angles of the crystal, it is possible to discover what the cross-sections must be along each of the axes, almost as if we sliced the crystal first along the horizontal axes, then along the vertical axes. This process of measuring and plotting gives us the *interior* triangles as shown below the first sketch. Now it is easy to compare the lengths along each of the axes, using heavy ink to show them as bars stacked vertically against a scale. Using any convenient scale, and reducing the numbers to the simplest terms, the proportions, or cuts along the axes, turn out to be 1.9 for the longest cut from the center of the ideal crystal "O" to the tip at "C," then 1.0 for the distance OB, and finally .8 for the distance OA. Notice that the distance along the *b* axis is taken as one. This has been decided upon by all crystallographers to make matters simpler. Actually, if one measured a real sulfur crystal, perhaps one several inches long, the distance along the *b* axis would probably not be exactly one inch, but whatever it is, it is *assumed* to be *one* and the other dimensions changed accordingly.

Thus by measurements taken on real crystals as perfect as mineralogists could find, and by the use of some geometrical calculations, axial ratios were determined for many minerals much before anyone had any way of checking to be sure that such figures were correct. Just how correct were these early measurements? In Figure 52, at the lower right, the numbers are compared; the set of ratios at the left, namely .8:1:1.9, were determined before use of x-rays, while the set at the right, given in angstrom units by multiplying by a factor of 12.84, are very close to the latest measurements—shown just below. Each angstrom unit, incidentally, is one one-hundred millionth of a centimeter, and such units are commonly used to express very small dimensions as those of crystal cells.

From the above example, it is plain that when early mineralogists determined axial ratios by measuring the angles made between faces on real crystals, they were actually determining the proportions of the smallest building blocks in the crystals, the unit cells! They could not say what the actual dimensions were—that had to wait until the use of x-rays was discovered—but they could tell what the shape of the cell had to be, thus confirming the great idea first advanced by Haüy.

Sample axial ratios are given in the table below to show how they appear in mineralogical textbooks. Because the edges of the isometric cell must be equal in dimension since the cell is cube-shaped, there is no point in giving an axial ratio for it must always be 1:1:1. Similarly, in the tetragonal system, the cells are square in cross-section along the *a* axes, and therefore only a single "1" is used in the ratio number; the other number, which can be greater or less than one, is the proportion along the *c* axis. This is also the case in hexagonal crystals where the lateral *a* axes, representing like dimensions of the unit cells, are also given as "1," with the *c* axis proportion greater or less than one. In the other three systems, three numbers appear in the ratio, the middle one, representing the *b* axis proportion, is always given as "1."

Axial Ratios

Mineral	System	Ratios
Galena	Isometric	(none needed)
Idocrase	Tetragonal	a:c = 1:0.757 (*a* axes = 1)
Beryl	Hexagonal	a:c = 1:0.9975 (*a* axes = 1)
Topaz	Orthorhombic	a:b:c = 0.528:1:0.955 (*b* axis = 1)
Epidote	Monoclinic	a:b:c: = 1.592:1:1.812 (*b* axis = 1)
Rhodonite	Triclinic	a:b:c = 0.872:1:1.593 (*b* axis = 1)

LOCATING FACES ON CRYSTALS (Miller Indices)

To describe the shape and position of faces on a crystal, even a simple one as shown in Figure 52, would take many words and evoke more confusion than it eliminated. Just as a mariner at sea must use latitude and longitude to tell where his ship is, a crystallographer must use some equally accurate and logical method to describe how faces are placed about a crystal. This problem was solved many years ago when the method now universally employed was devised and published in 1839 by W. H. Miller of Cambridge, England. The numbers he devised have come to be called *Miller indices*. By using them, the position of any crystal face can be fixed because the numbers are derived from the axial proportions we have just discussed.

Each face on a crystal cuts one or more axes and some faces cut as many as three. In the examples shown in Figure 51, the sloped faces cut two axes, and as explained in the discussion accompanying this figure, we can use two simple digits to express just how many unit cells are stacked along each axis, and therefore just how the face must be tilted in relation to the axes. To carry this process to its logical conclusion and to show how all faces can be expressed in simple numbers, we turn to Figure 53 which shows three telephone poles rigged with wire braces, and from this common example, derive some Miller indices.

We imagine that the telephone linemen braced each of the poles differently. In the first example (A), they braced the pole 16 feet above ground,

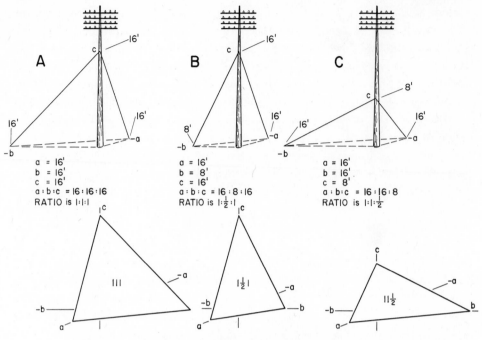

a = 16'
b = 16'
c = 16'
a:b:c = 16:16:16
RATIO is 1:1:1

a = 16'
b = 8'
c = 16'
a:b:c = 16:8:16
RATIO is 1:½:1

a = 16'
b = 16'
c = 8'
a:b:c = 16:16:8
RATIO is 1:1:½

The ratios given above are simplified by dropping the upper part of each fraction, thus converting all numbers into whole numbers.

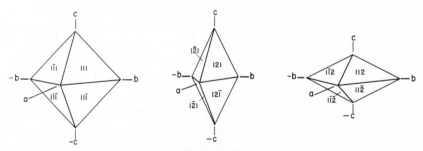

A minus sign is placed over each number representing the negative end of any axis. This helps to place any face in relation to others.

MILLER INDICES

FIG. 53.

and 16 feet along the street curbings. In B, the bracing wires were placed at the same height but the one to the left, along the *b* direction, was made shorter. In C, the vertical distance was changed. Note how the triangle behind the pole, made by the wires and a line drawn to their bases, changes shape accordingly. Thus it is obvious that this triangle depends for its shape on its inclination to the pole and to the ground. If the pole represents the *c* axis, and the street curbing the *a* axis and *b* axis, this is the same as saying that any crystal face also depends for its shape on the way it tilts in relation to the crystal axes. Instead of using feet, as the telephone examples do, we can express the same thing by using unit cells stacked along each axis just as was used in Figure 51. For the moment, let us stay with the telephone pole example and see how Miller indices can be derived.

In A, we note that the distances are all 16 feet, that is, from the place where the telephone pole enters the ground to the ends of the brace wires, the distances are the same. This is the same as saying that the ratio is 16 to 16 to 16, or, 16:16:16, or, to put it in its simplest terms, in the ratio 1:1:1. If we were to supervise the telephone crew, we could instruct the workmen to place the wires at "111" and they would understand that each wire must go up the pole and outward to the same distance. If it were a very tall pole, it might be 32 feet, or a very short pole, only 8 feet. In any case, the same shaped triangle would be created by the wire braces even though the triangle made by the longer distance would naturally be much larger than the one made by the shorter distance.

In example B, the problem is changed because we have made wire "b" only half the length of the other distances, namely 8 instead of 16 feet. This changes the proportions of the triangle, making the ratio $1:\frac{1}{2}:1$. Similarly, in C, the vertical cut is changed to 8 feet, making the ratio $1:1:\frac{1}{2}$. Just below each of the three examples given, we have turned the triangles around to make them face toward us, and have marked them with the ratios. Now it is clear that the triangles are of vastly different shape simply because the plane which creates each of them cuts the axes at different distances. Many other examples could be given to show how the shape of these faces changes with the way the planes cut the axes, but these will do for the moment.

Because fractions are troublesome things, and whole numbers much easier to use, we get rid of the fractions by discarding all the ones on top of the line and have left only whole numbers. Thus the ratio for B, of $1:\frac{1}{2}:1$, or $1\frac{1}{2}1$, becomes 121, and that for C, of $11\frac{1}{2}$ becomes 112. This is apt to be confusing because one may imagine that the larger digits mean that the cut along the axis is further out, instead of being closer in as is actually the case. It is very important to remember that in any set of Miller indices, the *larger* numbers always *mean* lesser distances. Note also that Miller indices are always given in the same order, that is, the first digit is for the cut along the *a* axis, the second for the cut along the *b* axis, and the third for the cut along the *c* axis. In isometric axes, these correspond to cuts on the first or a_1 axis, next on the a_2 axis, and the third cut on the a_3 axis. In tetragonal crystals it will be similar except that the vertical axis is the *c* axis. In hexagonal crystals which have three lateral axes, and a fourth vertical axis, there will be four digits in Miller indices, the first three in order, for axes a_1, a_2, and a_3, and the last for the *c* axis.

Referring back to Figure 53, we may now show how the planes creating the triangles between brace wires on the telephone poles can be converted directly into planes on crystals. At the bottom of the figure is shown how additional triangles are placed about the axes to "clothe" them and to completely enclose space. The same types of planes are used, except that a total of four must be used to make the crystal complete. Because each plane cuts each axis at the same distances, the Miller indices are the same but we show the position of each by putting small minus signs over the digits which represent the minus or negative end of the axes concerned. Thus in the first example at the bottom, the positive ends of the axes are all toward the upper right sector of space in the crystal and the face in the upper right which cuts nothing but positive ends of axes, has no minus signs over the digits of its Miller indices. However, the face just below cuts the negative end of the *c* axis, and this is shown by putting a minus sign over the last or *c* digit. The face to the upper left cuts the positive ends of the *a* axis and *c* axis, but cuts the negative end of the *b* axis. Therefore its center or *b* digit in the Miller indices is marked with a minus. This same system is used for all the crystals shown, and will be used for crystal drawings later in this chapter and elsewhere in the book.

In this connection, it should be pointed out that Miller indices in the hexagonal system are confusing because the positive and negative ends of the lateral axes alternate. This is shown in Figure 50. Thus the first digit may be positive, the second negative, and the other two positive, or some other arrangement may be shown, and one may have difficulty visualizing just where the face is located unless he remembers that the axes do alternate ends going around the crystal. The examples of faces given later with appropriate Miller indices should be studied carefully to fix these relationships firmly in mind.

Miller indices are spoken as follows: for (111), say "one,one,one"; for (12$\bar{1}$), say "one,two,*minus* one," etc. But again remember that the "two" really means "one half"!

Zeros in Miller Indices. Whenever a zero appears in a set of Miller indices, it means that the face does not cut one or more axes, or, in other words, it is parallel to them. In Figure 54, several examples of planes are drawn to show how they cut one or more crystal axes and how the indices vary accordingly. In the first example, marked *a*(100), the vertical plane is parallel to the *b* axis and *c* axis and therefore cannot intersect these axes no matter how large the crystal becomes. This fact is simply shown by using zeros for the *b* and *c* parts of the Miller indices. Because this plane cuts only the *a* axis, it is customary to mark the entire symbol with a small italic *a* as well as with the Miller indices. The use of small letters, with or without Miller numbers, will be explained in a later section. In the next example below, the plane cuts only the *b* axis and is therefore written as *b*(010); similarly, as

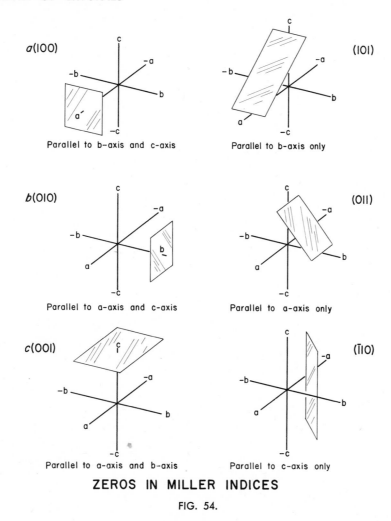

ZEROS IN MILLER INDICES

FIG. 54.

shown at the bottom, a plane which cuts only across the c axis is labeled $c(001)$.

In the right column of Figure 54 are additional examples of planes which cut *two* axes but not a third. At the top is a plane cutting the a axis and c axis, but not the b axis; the symbol is (101). The next symbol (011), shows that it is the a axis which is not cut, while ($\bar{1}$10) at the bottom, indicates that the vertical plane is parallel to the c axis, and strikes the a axis at its *minus* end and the b axis at its positive end. For this reason, a minus sign is placed over the a axis portion of the symbol.

Figure 55 shows additional complexity appearing in the hexagonal system because of the third horizontal axis. Hexagonal axes, as will be recalled, are labeled a_1, a_2, and a_3, with a plus in front to tell which end is which. Thus the Miller symbol for the hexagonal system always has four places in it, the

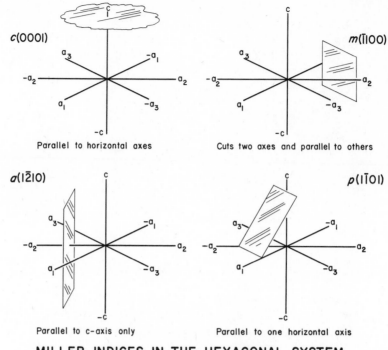

MILLER INDICES IN THE HEXAGONAL SYSTEM

FIG. 55.

first three for horizontal axes and the last for the vertical or c axis. The plane at the end of the c axis receives the indices (0001) because it is parallel to the horizontal axes and therefore cuts only the c axis. A vertical plane parallel to the c axis contains a zero in the c index place. Two vertical planes are also depicted in Figure 55, also a tilting plane, such as one would find on the end of a crystal, and which must, of course, intersect the c axis as well as at least two horizontal axes.

Axial Ratios and Miller Indices. A confusing feature of the Miller scheme arises from the very simplicity of its numerals which lead one to believe that faces with the same numbers should *look* the same. *This is true only if the faces belong to crystals in the isometric system.* To explain, in Figure 56 are shown two crystals which appear similar in a general way, but are considerably different in detail. Both show triangular faces marked with exactly the same indices (111), but one, galena, is faced with equilateral triangles, while the other, sulfur, is faced with narrow triangles which cannot be equilateral. Furthermore, cross-sections of each crystal show that galena is square while sulfur is diamond-shaped. Why should these faces, of different shape and of different inclination toward the crystal axes, bear the same Miller indices?

The answer lies in the shape of each unit cell. In galena, the cell is cubic, and because its dimensions are exactly the same along every edge, its outward

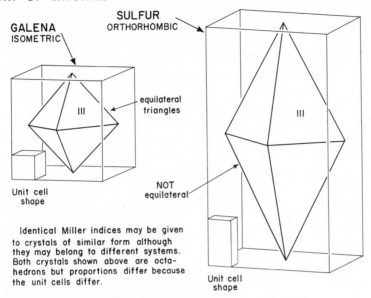

GALENA
ISOMETRIC

SULFUR
ORTHORHOMBIC

equilateral
triangles

III

III

Unit cell
shape

NOT
equilateral

Identical Miller indices may be given
to crystals of similar form although
they may belong to different systems.
Both crystals shown above are octahedrons but proportions differ because
the unit cells differ.

Unit cell
shape

MILLER INDICES

FIG. 56.

expression, the cube crystal, and any of its modifications such as the depicted octahedron, must also have equal angles along *all* edges. On the galena octahedron this means that triangular faces *must be* equilateral triangles, measuring 60° in the corners. However, the sulfur crystal, with nearly similar faces, arranged in similar fashion, *cannot* have equilateral triangles because its unit cells are rectangular instead of cubic. These proportions are shown in the sketches of Figure 56, where the cube enclosing the galena octahedron is assumed to have exactly the same number of cells as the rectangular solid enclosing the sulfur crystal. Because the cells are of different proportions, however, the crystals will also be of different proportions although the Miller indices for the faces are the same.

The Simplicity of Miller Indices. In an earlier chapter it was pointed out that crystals tend to grow in the simplest and most direct fashion possible, attracting additional ions and attaching them where the forces holding the crystal together can be most quickly satisfied. In effect, each crystal prefers to grow upon the least number of faces possible, and those usually the ones which allow the most uniform development. Consequently, when a large number of crystals of any given species are examined, it will be found that certain faces are more common than others, and that such faces will also have the simplest Miller indices. Thus in galena crystals, by far the commonest faces are those of the cube, $a(100)$, followed by faces of the octahedron $o(111)$. For galena, these faces represent the preferred *habit,* and as can be seen from the very simple numbers in the Miller symbols, presents a very direct solution to the problem of how to attach arriving lead and sulfur ions. On the other hand, the silicate species of the garnet group prefer to grow in dodecahedrons whose faces are given the symbol $d(110)$, again, a very simple one. In general, this same simplicity in Miller symbols is to be found in all species, the commonest faces always bearing the simplest numbers, and indicating the attempts of the growing crystals to solve growth problems in equally simple fashion.

Where faces are more complicated in respect to Miller numbers, as for example, a face labeled (561), they will also be far less common and the collector may go for years before a good example of a crystal bearing such faces is found. Collectors call such faces "rare," and some who specialize in the study of crystals bearing a variety of faces, prize such specimens highly.

Although most crystal faces will therefore be relatively simple so far as Miller indices are concerned, it is also to be noted that the numbers are always *whole numbers* and not decimals as 1.2 or .7. This follows when it is considered that a crystal must grow in orderly fashion from building blocks of specific size and shape. It is the same problem which faces a mason building a wall from ready-made bricks, all cast from the same mold. If he uses whole bricks, he must always use them in simple proportions, and no matter how he varies these proportions, whole numbers will be involved, that is, so many *whole* bricks laid this way and so many *whole* bricks laid that way. He cannot vary the whole numbers unless he breaks the bricks into pieces and only uses parts of them. Sometimes brick masons do this to change the shapes of walls but crystals do not, and for this reason, the digits within each Miller symbol are whole numbers.

Miller Summary. Miller indices are used to state how a face is positioned on any crystal in respect to any crystal axes. Numerals indicate the *relative length of cut* along any axis. All faces of crystals in the isometric system will be the same when marked with the same indices; angles between them will also be the same. In all other systems, this may not be true, and although a (111) face in the orthorhombic system is numerically the same as a (111) face in the isometric system, the inclination of faces against axes differs. This is caused by differences in relative axial lengths as stated by the mineral's axial ratio. The more closely the axial ratio of any crystal approaches 1:1:1, as found in the isometric system, the more closely will its faces resemble those found in that system. When there are large variations in axial ratios, as in sulfur, faces marked with the same Miller indices show marked differences despite the sameness of numerals.

There are three indices for all systems except for four in the hexagonal system. The last numeral is always reserved for the cut upon the vertical axis; the others are for cuts on the horizontal axes. Numerals and zeros may appear in any order. Zeros mean that the particular axis is not intersected by the plane of the face no matter how far it may be extended, hence the face is parallel to that axis. Two zeros mean a face is parallel to two axes. Three zeros occur only in the hexagonal system where (0001) means a face cutting only the c axis.

Numerals are proportional to cuts along the axes. The numeral "1" indicates a cut of standard distance, but *larger* numbers mean *lesser* distances. Thus, "2" means one half the distance, "3" means one third the distance, etc. There are no symbols such as "(333)" because the proportion is really one to one to one and is then written as (111).

It helps a great deal in understanding Miller indices to keep a mental image of the axial cross before you, and then apply each of the numerals to each of the axes. If you have paper handy, it is even better to sketch the axial cross, label each axis plus or minus, and then mark the points where the plane intersects each one.

AXES AND SYMMETRY

We are aware of *symmetry* in every object around us, although most of us do not use this term in ordinary conversation; we use the more familiar term *balance,* which essentially means the same. Symmetry is balance, perhaps the balance of like objects, as a pair of bushes, one on each side of a doorway, perhaps the balance of shape as when the same bushes are round, instead of one being round and one tall. Even our bodies display symmetry in disposition of limbs and features, and some artists and sculptors speak of the "symmetry of features" when describing the human face. Crystals also display symmetry, varying from the perfect symmetry of some isometric crystals to the chaotic appearance of crystals in the triclinic system which seem to have grown with little rhyme or reason. Symmetry isn't always obvious in crystals, even in those of the isometric system, as when some face is under- or overdeveloped, but it can usually be detected if the crystal is carefully turned about in the fingers and studied from all sides.

Symmetry is directly related to the atomic structure of crystals, the greater symmetry inside the crystal, the greater it is shown by balanced patterns of faces on the outside. Isometric crystals represent the highest symmetry, declining through tetragonal, hexagonal, orthorhombic, and monoclinic until the triclinic crystals are reached which, as mentioned before, frequently look completely unorganized. Typical arrangements of faces upon crystals from each of the systems are shown in Figure 57. By recognizing similar arrangements on strange crystals, it is often possible to determine the system, thus greatly aiding identification.

The first example is a complex garnet crystal from the isometric system

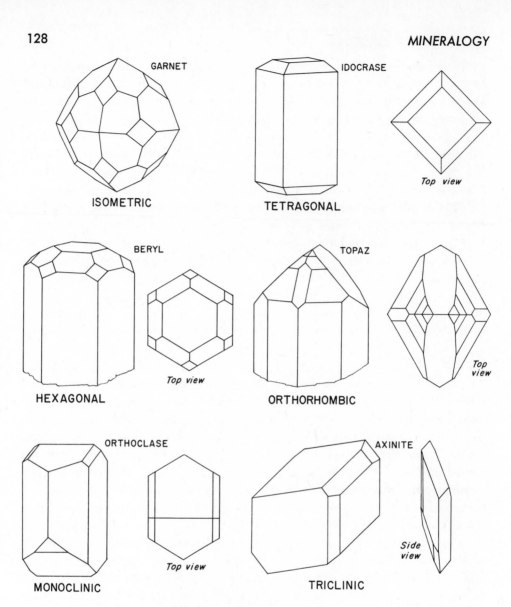

SYMMETRY IN CRYSTALS

FIG. 57.

which looks like a faceted ball. Note how the numerous faces arrange themselves in relation to the axes. This pattern can be followed completely around any fully-developed crystal. Furthermore, this crystal can be turned to various positions, other than the one shown, and more similarities can be observed. The highest possible symmetry occurs in many crystals of the isometric system, but some which do not seem symmetrical, as the peculiar crystals of pyrite shown in Figure 67 (page 142), owe their odd development to peculiarities in the atomic structure which tend to favor the development of alternate faces, and suppress those between. Nevertheless, they are all isometric because they can be related to the basic cube cell. Unidentified cube-like or ball-like crystals suggest the isometric system because of the tendency of its crystals to develop so symmetrically.

The idocrase crystal shown next in Figure 57, displays a square cross-section reflecting a similar arrangement of atoms in the horizontal plane. However, it is elongated vertically, and thus is far less symmetrical than crystals in the previous class. A top view shows how the terminal faces are neatly disposed around the vertical axis. Beryl crystals, one of which is shown in the next row, tend to be quite symmetrical, especially in their hexagonal cross-sections, and are easily recognized on this account. The hexagonal pattern of small terminal faces shows up well in the cross-section. In the rhombohedral division of the hexagonal system, terminations frequently show faces in sets of three, as upon calcite and tourmaline crystals. In the orthorhombic system, considerably less symmetry is shown in the topaz because faces are now disposed in rectangular fashion instead of in square, hexagonal, or triangular fashion as in the others before. Because faces are arranged in opposite pairs, one can easily imagine them as being divided into halves by planes passing through the crystal, two the long way, and one the short way.

Symmetry rapidly worsens in the monoclinic system, as shown by the orthoclase crystal of Figure 57. It is only when the crystal is held so that its narrow end points toward the eye, as shown by the top view, that some vestige of balance is seen. Rarely, orthoclase crystals are found equally developed on both ends, in which case additional symmetry is seen from the side view. The last crystal, an axinite, belonging to the triclinic system, is the "chaotic" crystal previously mentioned. In fact, it is this near-chaos which sometimes is the most useful identification feature of triclinic crystals because no matter how much they are turned about, only opposite pairs of faces seem to be parallel. Even this poor symmetry is often disguised by the fact that most crystals grow attached to walls, and faces which could show parallelism are obliterated. A good rule of thumb in examining crystals is this: if as you turn any crystal about in the fingers it does not show much symmetry, it is probably triclinic or monoclinic.

Symmetry features helpful in assigning crystals to a system are summarized in Figure 58. Some common sense must be used when applying them because it is easy to make mistakes if the crystals are examined carelessly. For example, in Figure 59 are shown two species which are often subjects for mistake, apophyllite, mistaken for an isometric mineral, and orthoclase, mistaken for tetragonal. Apophyllite commonly grows in pseudo-cubic crystals as shown, even to the extent of having small corner faces which the unwary assume to be octahedral faces found on true isometric crystals. However, the

SYMMETRY CLUES

GENERAL APPEARANCE	EXAMPLES	SYSTEM

Blocky, ball-like, similar appearance from many points of view.

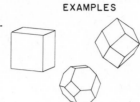

ISOMETRIC

Squarish cross sections, crystals often long, sometime very long and slender to acicular.

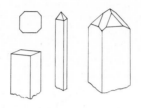

TETRAGONAL

Hexagonal or triangular cross sections, sometimes nearly round. Commonly short to long prismatic, columnar.

HEXAGONAL

Rectangular or diamond-shaped cross sections, stubby to short prismatic.

ORTHORHOMBIC

Blocky or stubby crystals with tipped faces which match only on opposite ends of crystals.

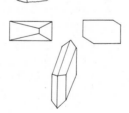

MONOCLINIC

Knife-edged, wafer-like crystals, only opposite faces match. Absence of right angles on faces or edges.

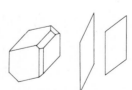

TRICLINIC

FIG. 58.

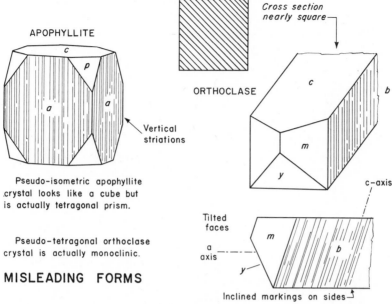

APOPHYLLITE

*Cross section
nearly square*

ORTHOCLASE

Vertical
striations

Pseudo-isometric apophyllite
crystal looks like a cube but
is actually tetragonal prism.

Pseudo-tetragonal orthoclase
crystal is actually monoclinic.

MISLEADING FORMS

Tilted
faces

a
axis

c-axis

Inclined markings on sides

FIG. 59.

masquerade is revealed by noting that apophyllite crystals show vertical markings along the four lateral faces, while such markings are absent altogether from the top face which instead shows a strong pearly luster. Remembering that tetragonal crystals are the same in crystal structure laterally, but different vertically, the strange crystal seems more likely to be tetragonal, especially when angle measurements along the side faces show them to be at right angles. The orthoclase example is superficially like a tetragonal crystal because, like so many of its kind, it is nearly square in cross-section. However, markings on the sides are slanted and the end faces inclined, all of which arouses the suspicion that this is not the neat lateral arrangement characteristic of tetragonal crystals. It is also unlikely that it is triclinic because of the square to rectangular outline; it must therefore be monoclinic.

REFLECTION OF CRYSTAL STRUCTURE IN FACES

It is important to point out that *true symmetry* is not only balanced placement of faces about crystal axes, but *likeness in physical properties.* For example, tetragonal apophyllite crystals furnish one of the most important clues that they belong to this system by developing striations on *every* lateral face as shown in Figure 59. In short, these striations show that the crystal structure, as exposed on each of these faces, is precisely the same around the lateral axes. We also confidently expect that the small "octahedral" faces on the corners will display an identical appearance when examined closely. Just as confidently, we may expect that any set of faces, exposing the same kind of atomic structure, will reval identical features when the faces being compared are equally developed. In isometric crystals, in which the highest symmetry exists, there are many faces which uncover the same structure and therefore bear the same markings, luster, general flatness, etc. In other systems, where symmetry steadily degenerates, fewer and fewer faces display similar appearances, until the point is reached in triclinic crystals where only opposite face pairs may closely resemble each other. The amateur should examine as many crystals as he can to detect similar faces and compare them to other sets, and finally, try to fit them into arrangements which suggest one crystal system.

ODDITIES IN SYMMETRY

After laying so much stress on the wonderful symmetry of isometric crystals, readers must be certain that all of them are perfect. Yet many are peculiar in appearance, as the sphalerite crystal shown in the upper part of Figure 60, which seems a far cry from a cube! How is one to explain the absence of cube faces or why it should not resemble a faceted ball, like a garnet crystal? There is an explanation, but it rests in the arrangement of atoms within the sphalerite unit cell.

It was previously pointed out that the general shape of any crystal can be thought of as the stacking of cells in three dimensions. However, if we think of the *contents* of each cell, we can easily see how it is possible to have different arrangements of atoms in the cell, although from the outside it looks cubic. For example, the upper right hand sketch in the same figure shows the sphalerite cell of cubic shape which we know that it must have in order to belong to the isometric system. But if we were to rearrange the "furniture" in this minute cubic room, we could find a better way than placing one pair of large sulfur atoms straddling the other pair. A much neater arrangement, and far more symmetrical, would be to place a sulfur atom in each corner of the cube cell. Unfortunately, the sulfur atoms would not like this arrangement and if they did, would have taken it in the first place. The arrangement *inside* the cell is tetrahedral, and not as symmetrical as if it were cubic. As the crystal grows, the arrangement of sulfur atoms causes attractive forces to be exerted unevenly against arriving ions so that the peculiar tetrahedral crystal shown at the left is formed in preference to a more symmetrical crystal.

The tetrahedral crystal also shown in this illustration encloses an octahedron to show the relationship between them. As is apparent, the tetrahedron is no more than space enclosed by alternate faces of an octahedron. Since

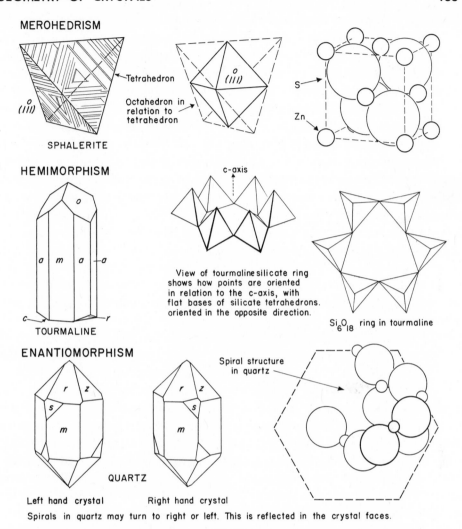

MEROHEDRISM

Tetrahedron

$\overset{o}{(111)}$

Octahedron in
relation to
tetrahedron

S

Zn

SPHALERITE

HEMIMORPHISM

o

a   m   a —a

c —————— r

TOURMALINE

c-axis

View of tourmaline silicate ring
shows how points are oriented
in relation to the c-axis, with
flat bases of silicate tetrahedrons.
oriented in the opposite direction.

Si_6O_{18} ring in tourmaline

ENANTIOMORPHISM

Spiral structure
in quartz

r   z
s
m

r   z
s
m

QUARTZ

Left hand crystal   Right hand crystal

Spirals in quartz may turn to right or left. This is reflected in the crystal faces.

FIG. 60.

the faces are the same in both figures, the same Miller indices are used (111).
Alternate face development of this sort is called *merohedrism, mero* meaning
"part" or "fraction," and *hedral* referring to faces. Later will be shown some
crystal drawings of pyrite, another isometric mineral which forms distinctive
crystals due to merohedral development. Its unit cell again displays less than
perfect symmetrical arrangement of its atoms.

   Doubly-terminated crystals of tourmaline show evidence of a peculiar internal structure which causes one pole of the *c* axis to be terminated by steep
faces, and the other pole by low faces as in the example shown in Figure 60.
When faces seem to be the same, differences in etch marks, growth hillocks,
luster, etc. can still be detected. The silicon-oxygen tetrahedrons of the

tourmaline framework are oriented along the *c* axis such that bases of tetrahedrons point toward one pole, but the upper ends point to the opposite pole. Even when both ends of the crystal are terminated by single flat faces of (0001), these are not really the same because different atomic patterns are exposed. Polar development is called *hemimorphism,* from *hemi* for "half," and *morph* for "form," in allusion to "half-development." Hemimorphism also occurs in hemimorphite, which receives its name from habitually forming crystals of this sort, and in other species.

Enantiomorphism is applied to crystals which show a right or left handedness, as in the quartz crystals of Figure 60. The term *enantio* means "opposite" in reference to the fact that some crystals develop atomic structures of opposite pattern. In quartz, the atomic pattern at the lower right of this figure shows how silicon-oxygen tetrahedrons link together on corners to make spirals, which can turn one way or the other, thus accounting for right or left hand crystals. Some physical and optical properties displayed by merohedral, hemimorphic, and enantiomorphic crystals will be discussed later; in many instances, these properties are of great value in scientific and technical fields.

FACES AND FORMS

As our appreciation of crystals grows, we realize that every crystal face strictly reflects the properties of the underlying crystal structure. Because atomic arrangements beneath geometrically similar faces sometimes differ, as in the crystals just discussed, only certain faces can be said to be alike in all respects. To sum it up, they must incline at exactly the same angles to the crystal axes, *and,* expose exactly the same kind of atomic pattern.

Faces which comply with both these requirements are said to belong to a *form*. For example, cube faces on galena crystals belong to the same form because they cut the axes at the same inclination and expose precisely the same pattern of lead and sulfur atoms. Similarly, the faces of the galena octahedron belong to a form, *but,* in the sphalerite crystal of Figure 60, though the faces are labeled with the same Miller indices, and though they may cut the axes at exactly the same inclination as the galena octahedron (111), they do not expose the same atomic pattern, *except on alternate faces.* Therefore what seem to be octahedral faces on sphalerite crystals, are actually tetrahedral faces, and properly belong to the *tetrahedral form.*

Some sphalerite crystals develop two sets of tetrahedral faces, both kinds showing marked differences in appearance. To distinguish them, one is called the *positive* form and the other the *negative* form. Similarly, quartz crystals at the bottom of Figure 60 display two distinct forms on each termination which are geometrically alike. One is the positive rhombohedron (r), and the other the negative rhombohedron (z).

Form Indices or Symbols. Whenever a *single* face on a crystal is being described, Miller indices are used as 111 or (111). However, if the entire *form* is being described, brackets are used, e.g. (111). To avoid complications, the form *symbol,* as it is called, is taken from the crystal face nearest the eye of the reader. This not only avoids the use of minus signs over numerals, but standardizes the procedure to be followed in making drawings and describing crystal faces.

Lettering of Faces. Small letters are used to indicate forms to which faces belong. The tendency is to discard letters because so many forms have been found that the resources of alphabets have been exhausted. Furthermore, many forms which are not really alike may bear the same letters and cause confusion. Miller symbols, with or without letters, sometimes further described by the geometrical form name, are increasingly used. The combination is often seen of letter and Miller indices such as $a(100)$, either a or the (100), signifying a cube face. One may find a crystal drawing with cube faces simply lettered a, or the Miller indices placed upon them. Because a and (100) mean the face cuts only the a axis, both symbols can also apply to crystals which are not cubes! This becomes clearer as we go along, but it is enough to say at this time that the same letters or Miller symbols apply to faces in all systems located in the same relation to axes. Thus the face $c(001)$ could be the face cutting the top of the c axis in the tetragonal, orthorhombic, monoclinic or triclinic systems. See also Figure 54 for other examples.

Faces in Zones. Faces parallel to the same crystal direction meet in parallel junctions to form *zones,* for example, the vertical sides of a building are *in* a zone. Referring to Figure 61, zones around the c axis are shown at A. At B, there are many zones because of the very high symmetry of crystals in the isometric system. The single outstanding feature of faces in zones is parallelism of edges. If a strange crystal is examined, it should be turned about until parallel edges are noted; the crystal is then oriented to a direction at right angles to observe the cross-section made by faces in the same zone. It is equivalent to looking down upon a skyscraper from an airplane to see the cross-section made by its sides. End views of typical crystal zones are depicted at C (see also Figure 58). Note how cross-sections can be extremely helpful in classifying the geometry of strange crystals and placing the crystal into its proper system.

Combined Forms. Very few crystals show faces belonging to *only* one form; by far the greater number show at least one other form *combined* in lesser or greater degree. When a slight change takes place, as in galena cubes whose sharp edges are cut off by another crystal form, the latter are said to *modify*

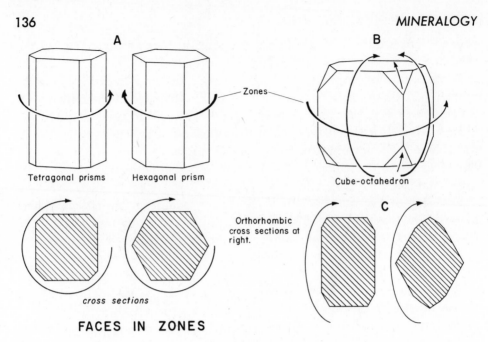

FACES IN ZONES

FIG. 61.

the basic cube form. Figure 62 shows *truncation* of a cube by an octahedral face, and the *bevelment* of an edge by a tetrahexadron. Modifications vary from scarcely perceptible truncations or bevelments, to developments equal in size to the basic form. Sometimes a special name is given to crystals which show nearly equal form development, as in the *cube-octahedron,* where faces

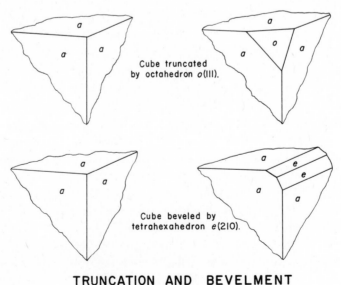

TRUNCATION AND BEVELMENT

FIG. 62.

of the cube and octahedron are nearly the same in area. In rare instances, as many as a dozen forms may be seen upon one crystal and very careful examination is needed to tell which is which. It is most helpful in solving such problems, to make sketches on paper showing faces and forms as each is recognized. Many face combinations will be shown in the last part of this chapter, and should receive the careful attention of the reader, particularly in respect to the patterns made by face intersections.

Closed and Open Forms. Whenever a form completely encloses space, as a cube or octahedron, such form is said to be *closed*. Because it is closed, it is entirely possible for its faces to be the only ones present upon crystals. On the other hand, there are forms which do not enclose space by themselves, and for this reason are called *open*. Both types appear in Figure 63. At the

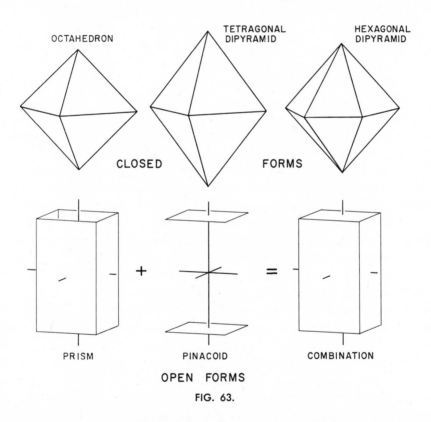

FIG. 63.

top are several forms differing in general shape but each complete in itself, and each capable of being found alone upon crystals. The lower illustration shows a *prism,* an open form requiring at least one more form in order to completely enclose space, in this case, a pair of parallel planes known as a *pinacoid.*

Every form in the isometric system is *closed,* while all forms in the triclinic and monoclinic systems are open. Open and closed forms are found in the tetragonal, hexagonal and orthorhombic systems, the last having as its only closed form, the *dipyramid.*

ISOMETRIC SYSTEM

Closed crystal forms are the hallmark of this system: every form is complete in itself. There are only seven basic forms; all others are derived from them by alternate omission of faces (merohedrism), or partial development. These seven forms are as follows:

ISOMETRIC FORMS

Name	Face Shape	No. of Faces	Forms	Angles
Cube	square	6	$a(100)$	fixed
Octahedron	60° triangle	8	$o(111)$	fixed
Dodecahedron	diamond	12	$d(110)$	fixed
Tetrahexahedron	triangle	24	$e(210)$	several
Trisoctahedron	triangle	24	$p(221)$	several
Trapezohedron	trapezoid	24	$n(221)$	several
Hexoctahedron	triangle	48	$s(321)$	several

The cube, octahedron, and dodecahedron are *fixed* forms, that is, they always show the same angles between faces. Thus a dodecahedron, regardless of where or how it occurs, or in what mineral species it occurs, always develops the same diamond-shaped faces containing the angles between them shown in Figure 64. The relationship of the cube, octahedron, and dodecahedron in the same figure, illustrate how any two, or even all three forms, may combine in crystals. The degree of combination varies from the merest trace of one form upon another, to equal face development.

The tetrahexahedron, trisoctahedron, trapezohedron, and hexoctahedron, are not fixed forms because several types are known, each displaying characteristic angles between faces. The *pattern* of faces remains the same for each and, if learned, is easily recognized when seen again. In Figure 65, one of the most common of complicated isometric forms, the trapezohedron, named after its peculiar trapezoidal faces, is shown in combination with the cube, octahedron, and dodecahedron. At the bottom is the tetrahexahedron and cube combination. Figure 66 shows a trisoctahedron and octahedron combination, and the very complex hexoctahedron combined with the dodecahedron.

Complicated forms are really variations of simpler forms. For example, the tetrahexahedron is basically a cube with four low triangles on each face; the trisoctahedron is an octahedron with three triangles on each face, and the

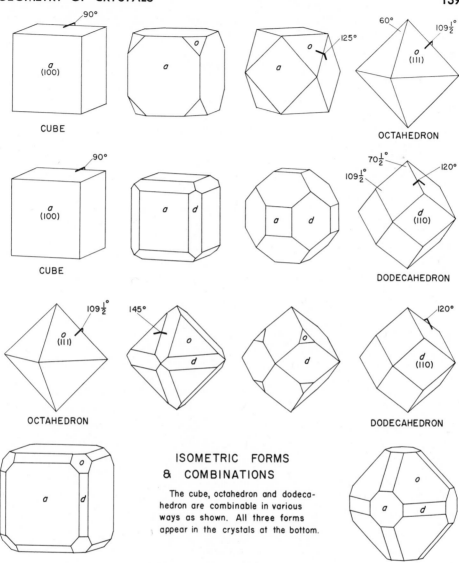

ISOMETRIC FORMS
& COMBINATIONS

The cube, octahedron and dodecahedron are combinable in various
ways as shown. All three forms
appear in the crystals at the bottom.

FIG. 64.

hexoctahedron is an octahedron with six triangles on each face. The trapezohedron is again an octahedron with three odd-shaped faces on each octahedral face. Form names provide useful aids to memory:

Hexa—means *six;* hence 6 faces (hexahedron = cube).

Octa—means *eight;* hence 8 faces.

Do-deca—means *two* plus *ten;* hence 12 faces.

Tetra-hexa—means *four* times *six;* hence 24 faces, four on each cube face.

Tris-octa—means *three* times *eight;* hence 24 faces, three on each octahedron face.

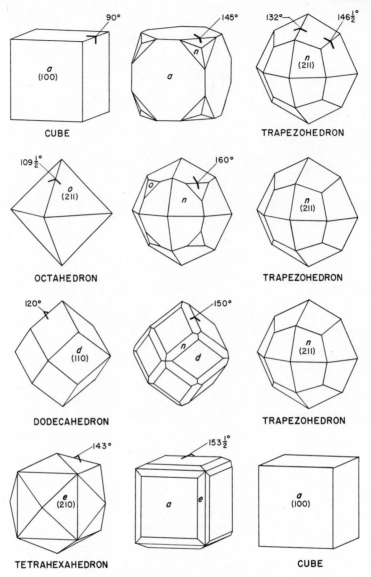

ISOMETRIC FORMS AND COMBINATIONS

FIG. 65.

Hex-octa—means *six* times *eight;* hence 48 faces, six on each octahedron face.

In the lower part of Figure 66 appear the merohedral positive and negative tetrahedrons previously mentioned. At the bottom, a tetrahedron and cube are combined. The pyritohedron shown in Figure 67 is extremely common in pyrite, from which it receives its name. The pyritohedron is derived

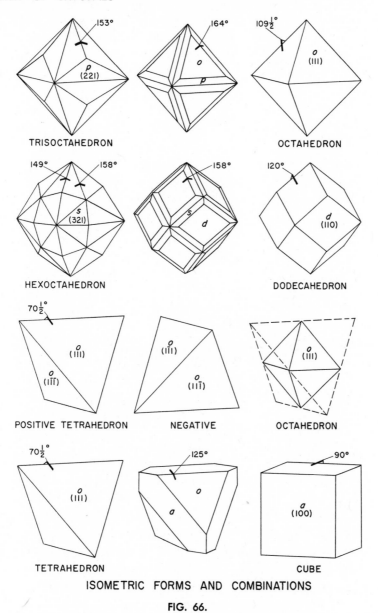

ISOMETRIC FORMS AND COMBINATIONS

FIG. 66.

from the tetrahexahedron shown in Figure 65. The two forward faces provide the necessary clue, it being seen that *alternate pairs* of the tetrahexahedron, grown fully at the expense of those between, form the pyritohedron. Figure 67 also shows combinations of pyritohedrons with other forms.

Isometric Crystal Clues. The best clues for recognizing isometric crystals are *cube-like to ball-like development,* plus *characteristic face shapes.* The

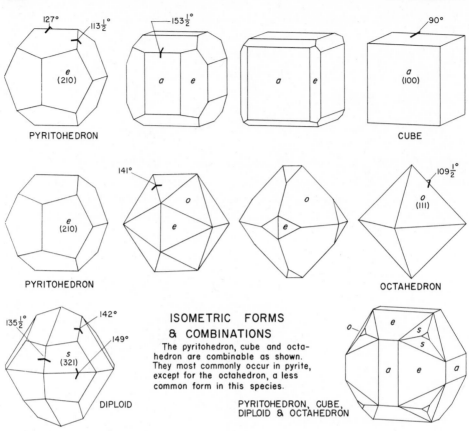

FIG. 67.

neat, even spacing of faces in drawings is seldom achieved in real crystals, but happens often enough to afford certain recognition in many instances.

Note how squares and triangles predominate, and how some corners show four-fold patterns, and others three-fold. Four-fold symmetry appears when looking down upon the axes of crystals, but three-fold symmetry appears when looking down on cube corner directions. It is very helpful to obtain and study garnet, pyrite, and galena crystals. Each should be rotated to place axes in the correct crystal drawing position, then turned to other positions to note how symmetry of faces changes according to direction.

Common Isometric Mineral Habits.

Cubes: galena, pyrite (often striated), fluorite, halite

Octahedrons: magnetite, spinel, franklinite, chromite, diamond; rarely, galena, pyrite, fluorite

Dodecahedrons: garnet, cuprite, magnetite; rarely, diamond

Trapezohedrons: garnet, leucite, analcime

Tetrahexahedrons: on edges of fluorite crystals

Pyritohedrons: pyrite, cobaltite
Tetrahedrons: tetrahedrite; sphalerite (usually distorted)
Hexoctahedrons: diamond (usually rounded)

TETRAGONAL SYSTEM

The forms of the tetragonal system are far less numerous than those of the isometric system but are rarely found alone and therefore often result in rather complex crystals. Discussion of this system introduces several *open* forms, namely, *prisms* and *pinacoids*. In Figure 68 examples are shown, including the tube-like prisms which consist of a fixed number of like faces all parallel to one axis, and the pinacoids, which consist merely of two parallel faces. Closed forms are double pyramids known as *dipyramids*. Another form, the *disphenoid,* is also a tetragonal form and is found on chalcopyrite.

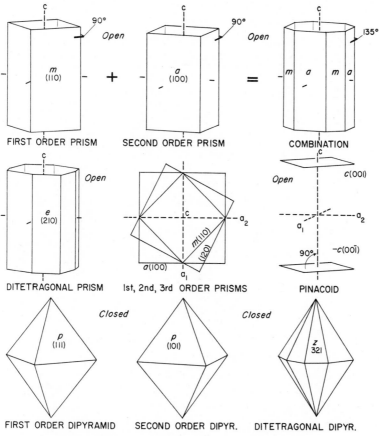

TETRAGONAL FORMS & COMBINATIONS

FIG. 68.

However, there is little point in showing it here because it closely resembles a tetrahedron (Figure 66).

Prisms of the *first, second,* and *third* orders differ from each other only in the angle at which they intercept the horizontal axes. The first order prism is that in which each face strikes two axes, and its symbol is therefore $m[110]$. The faces of the second order prism each intercept only one of the horizontal axes and the symbol is therefore $a[100]$. Note that the latter is exactly the same Miller symbol as for the cube in the isometric system. The three prisms are compared in the center of Figure 68. The *ditetragonal prism,* meaning a *doubled prism,* is shown in the center left, and consists of eight faces each intersecting both axes at the same angles. This form is frequently found as pairs of strip-like faces beveling otherwise sharp edges of the second order prism.

Prism plus pinacoid is a common form combination in tetragonal crystals. Because the only pinacoid in this system is across the *c* axis, as shown at right center of Figure 68, it is sometimes called the *basal pinacoid*. At the bottom of the figure are shown several dipyramids, and the ditetragonal dipyramid which corresponds to the ditetragonal prism in number and placement of faces. Several typical tetragonal crystals are shown in Figure 69.

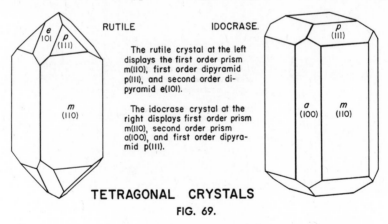

RUTILE   IDOCRASE.

The rutile crystal at the left displays the first order prism m(110), first order dipyramid p(111), and second order dipyramid e(101).

The idocrase crystal at the right displays first order prism m(110), second order prism a(100), and first order dipyramid p(111).

TETRAGONAL CRYSTALS

FIG. 69.

COMMON TETRAGONAL FORMS

Name	Face Shape (appr.)	No. of Faces	Forms	Angles
Prism, 1st order	rectangular	4	$m(110)$	fixed
Prism, 2nd order	rectangular	4	$a(100)$	fixed
Prism, 3rd order	rectangular	4	$g(120)$	fixed
Ditetragonal prism	rectangular	8	$e(210)$	several
Pinacoid	————	2	$c(001)$	fixed
Dipyramid, 1st order	triangular	8	$p(111)$	several
Dipyramid, 2nd order	triangular	8	$e(101)$	several

Tetragonal Crystal Clues. Valuable clues to recognition of crystals in this system are: first, to look for the typical square or octagonal cross-section, usually but not always, well formed; second, to observe dipyramidal cross-sections which must be square or octagonal when viewed from the top; and third, to note that crystals often incline to be *longer* along the c-axis than wide along the horizontal axes.

Common Tetragonal Mineral Habits.

Square prisms—idocrase (Figure 69), apophyllite (latter striated-vertically)

Square prisms with pyramids—zircon, rutile (Figure 69), scapolite

Dipyramids, usually without prisms—scheelite, cassiterite

Flattened square prisms—wulfenite, torbernite, some apophyllite

Pseudo-tetrahedral—chalcopyrite

HEXAGONAL SYSTEM

This system contains two divisions: *hexagonal* and *rhombohedral*. In the first, sixfold symmetry around the c axis is marked by appearance of hexagonal prisms, dipyramids, and other sixfold forms, while in the second division, the symmetry is threefold with appropriate forms. Hexagonal crystals vary remarkably in proportions, some calcite crystals exhibiting forms in which sometimes the crystals are nearly wafer-thin, while other forms create tapered crystals, or even slender prismatic crystals. Extremely elongated crystals may be found in beryl and tourmaline. In all crystals in this system, it is usually possible to observe sixfold or threefold repetition of faces by looking down upon the ends of crystals.

Common open forms are shown in Figure 70, and include first and second order prisms and the dihexagonal prism. These may be partly or wholly terminated by dipyramids, with or without the basal pinacoid (not shown). The *pedion* is a form consisting of one face only, c(0001), which can be plus or minus depending on the end of the crystal upon which it is placed in drawings. It occurs only in hemimorphic species, notably tourmaline (Figure 60) zincite (Figure 70), and greenockite.

The rhombohedron in Figure 71 is common in calcite, but this remarkable species which is known to have over 700 forms (!), has many other rhombohedrons, not to mention scalenohedrons such as the commoner types shown in the middle of the same figure, combined with other forms in crystals of great complexity. Also shown in Figure 71 are a prism and low rhombohedron combined in habits typical of prismatic calcites; also in the center, a calcite with prism and scalenohedron. Scalenohedral crystals frequently occur without modification, and create spectacular specimens of "dogtooth calcite" or "dogtooth spar." Doubly-terminated scalenohedrons show zig-zag waistlines.

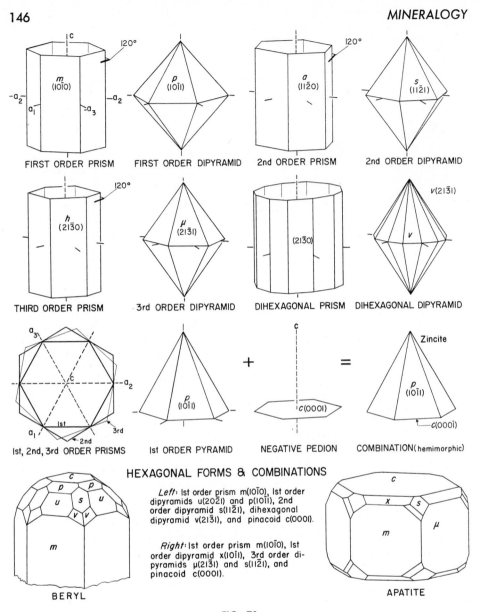

FIG. 70.

The hemimorphic nature of tourmaline is shown well by the very different character of faces on each end of the c axis in the lower drawing of Figure 71. One end is usually covered with steep pyramidal faces (*not* dipyramids), and the other by lower forms. Frequently the pedion, in this case the negative one, $c(000\bar{1})$, forms a small face at the apex of a pyramid, but in other instances, notably in tourmaline crystals from California, it may nearly cover one end of the crystal, leaving pyramidal faces present only in traces. Many

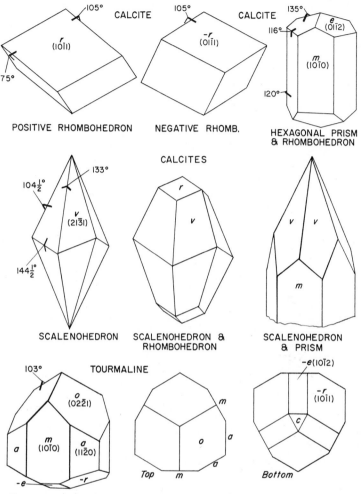

Hemimorphic tourmaline crystal showing pyramids, prisms and pedion -c(0001)

HEXAGONAL FORMS & COMBINATIONS
FIG. 71.

tourmaline crystals do not show good prism faces, the prism area being covered by oscillatory striations representing alternate-development of prisms *m* and *a*.

Hexagonal Crystal Clues. The most valuable clue to hexagonal crystals is three-fold or six-fold symmetry. Pits, etch marks, or growth hillocks, frequently afford other clues but must be located on basal planes to be indicative. As a rule, "dogtooth" type crystals and rhombohedrons usually indicate calcite or other carbonates. Crystals in the hexagonal division tend to form

COMMON HEXAGONAL FORMS

Name	Face Shape (appr.)	No. of Faces	Forms	Angles
Prism, 1st order	rectangular	6	$m(10\overline{1}0)$	fixed
Prism, 2nd order	rectangular	6	$a(11\overline{2}0)$	fixed
Dihexagonal prism	rectangular	12	$(21\overline{3}0)$	several
Dipyramid, 1st order	triangular	12	$p(10\overline{1}1)$	several
Dipyramid, 2nd order	triangular	12	$s(11\overline{2}2)$	several
Pinacoid	————	2	$c(0001)$	fixed
Pedion	————	1	$c(0001)$	fixed
Trapezohedron, trigonal	trapezoidal	6	$x(51\overline{6}1)$	several
Rhombohedron	diamond	6	$r(10\overline{1}1)$	several
Scalenohedron	narrow triangular	12	$v(21\overline{3}1)$	several
Pyramid, trigonal	triangular	3	$o(02\overline{2}1)$	several
Prism, trigonal	rectangular	3	$m(10\overline{1}0)$	two
Prism, ditrigonal	rectangular	6	$a(11\overline{2}0)$	fixed

straight-sided individuals, sometimes of considerable length as in tourmaline and beryl.

Hexagonal Crystal Habits.

Hexagonal prisms, long—beryl, tourmaline, willemite, calcite

Hexagonal prisms, short—calcite, apatite, vanadinite, pyromorphite, corundum, also beryl, tourmaline

Hexagonal prisms, tabular—morganite beryl, apatite, graphite, molybdenite, hematite, ilmenite, ruby corundum, pyrrhotite

Hexagonal prisms with pyramids and rhombohedrons—tourmaline, phenakite, willemite, dioptase, calcite, quartz, hanksite

Rhombohedrons—calcite, siderite, dolomite, rhodochrosite, chabazite

Scalenohedrons—calcite; rarely, other carbonates

ORTHORHOMBIC SYSTEM

Although commonly modified by several forms, crystals in this system tend to be "boxy" in general appearance, reflecting right angle axial intersections. Wedge-shaped terminations along any or all axes are common, however, by noting the mutual angles made by such faces, and comparing their position to adjacent faces, it will be seen that they "straddle" adjoining faces symmetrically.

There is only one family of closed forms in this system, the dipyramid shown at the upper left in Figure 72. There are many dipyramids with faces at different degrees of inclination to all axes. Two upon the same crystal are shown on sulfur in the lower part of the same figure. All other forms are open, as shown by the pinacoids and prisms of Figure 72. Several may be

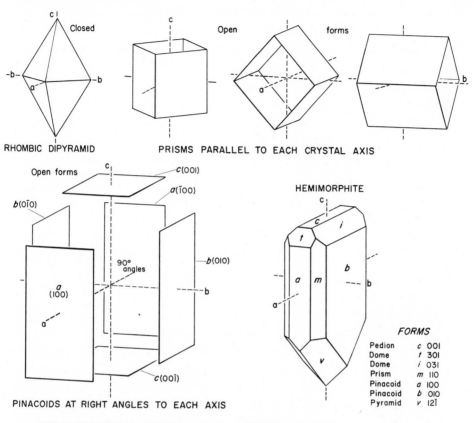

RHOMBIC DIPYRAMID   PRISMS PARALLEL TO EACH CRYSTAL AXIS

Open forms

PINACOIDS AT RIGHT ANGLES TO EACH AXIS

HEMIMORPHITE

FORMS

Pedion	c	001
Dome	t	301
Dome	i	031
Prism	m	110
Pinacoid	a	100
Pinacoid	b	010
Pyramid	v	12$\bar{1}$

ORTHORHOMBIC FORMS AND COMBINATIONS

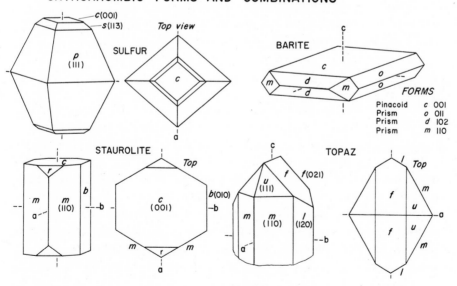

SULFUR   Top view

BARITE

FORMS

Pinacoid	c	001
Prism	o	011
Prism	d	102
Prism	m	110

STAUROLITE   Top

TOPAZ   Top

FIG. 72.

found on a single crystal, but their strict orientation, either squarely across an axis or parallel to same, results in crystal faces which are ordinarily neatly placed on all sides of crystals. Typical orthorhombic symmetry is shown in the bottom examples of Figure 72; hemimorphism occurs notably in hemimorphite, shown in the center right. The major portion of the hemimorphite crystal is enclosed by pinacoids and prisms, but the wedge-shaped end is enclosed by a pyramid v.

COMMON ORTHORHOMBIC FORMS

Name	Face Shape (appr.)	No. of Faces	Forms	Angles
Prism, 1st order	rectangular	4	$o(011)$	several
Prism, 2nd order	rectangular	4	$d(101)$	several
Prism, 3rd order	rectangular	4	$m(110)$	several
Dipyramid, rhombic	triangular	8	$o(111)$	several
Pinacoids	rectangular	2	$a(100)$	several

Orthorhombic Crystal Clues. Simple crystals tend to be boxy to rectangular-tabular. Faces commonly show decided variations in luster, etch marks, striations, etc., which help to identify forms. In topaz crystals, for example, it is often possible to find four faces with identical markings, one pair placed opposite another in neat balance. Profiles and cross-sections of prismatic crystals often approximate ovals or diamond-shapes, showing symmetry arising from axes at right angles to each other.

Orthorhombic Crystal Habits.

Prisms, very long—stibnite, natrolite

Prisms, short—staurolite, andalusite, topaz, danburite, columbite

Prisms, tabular—autunite, celestite, barite

Blocky or stubby—anglesite, chrysoberyl, olivine, topaz

Pseudo-hexagonal prisms—aragonite twins

MONOCLINIC SYSTEM

Principal forms are prisms and pinacoids as shown in Figure 73; there are no closed forms. A very common prism is $m[110]$ shown at the lower left. Two other prisms are also common and deserve a few words of explanation. One prism is parallel to the forward inclined axis (a axis). If it were present on the orthoclase crystal in Figure 73, it would truncate the edges between pinacoids c and b. The third prism is rather difficult to visualize except to say that it is like the prism $m[110]$ shown in Figure 74,

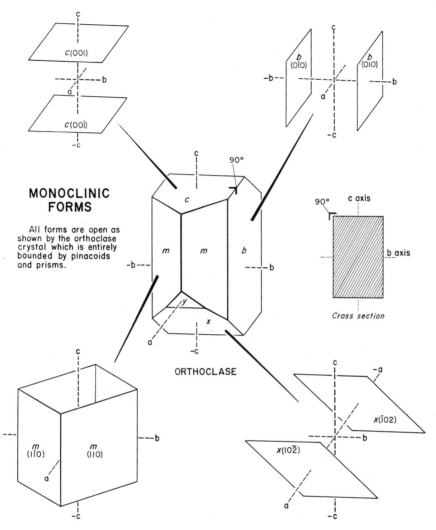

MONOCLINIC
FORMS

All forms are open as
shown by the orthoclase
crystal which is entirely
bounded by pinacoids
and prisms.

FIG. 73.

but tipped backward so that it not only strikes the *a* axis but also the vertical *c* axis. Its faces cut all three axes and there are therefore no zeros in its Miller symbol. A prism of this type may show a face such as *v*(111), indicating intercepts on all three axes. A number of pinacoids in this system strike two axes instead of one as previously noted in other systems, and their Miller symbols therefore show only one zero instead of two. For example, in Figure 73, the ordinary pinacoid face *c*(001) is shown atop the orthoclase crystal and is obviously parallel to both the *a* axis and *b* axis, and therefore cuts only the *c* axis. On the other hand, the pinacoid face *x*(10$\bar{2}$) cuts the *a* axis, is parallel to the *b* axis, and again cuts the *c* axis at its minus end.

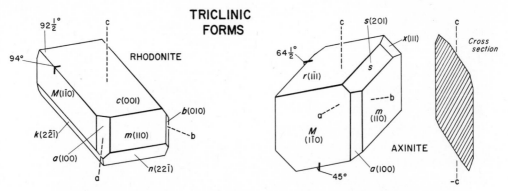

All forms are open and consist of pinacoids as shown here but pedions are found on crystals of other triclinic species as well as the usual pinacoids.

FIG. 74.

Common Monoclinic Forms

Name	Face Shape (appr.)	No. of Faces	Forms	Angles
Prism, parallel to c	rectangular	4	$m(110)$	several
Prism, parallel to a	rectangular	4	$n(021)$	several
Prism, cuts all axes	triangular	4	$u(111)$	several
Pinacoids, parallel to two axes	rectangular	2	$a(100)$	three
Pinacoids, parallel to one axis	rectangular	2	$y(101)$	several

Monoclinic Crystal Clues. Many crystals grow as stubby or blocky individuals, but exceptions are numerous. For example, mica group minerals are largely monoclinic, and certainly their chief characteristic is platey or foliated crystals of misleading hexagonal outline. It is worth remembering that crystals in this system do not differ greatly from many orthorhombic crystals except in respect to lacking squareness in one crystal direction due to the single tilting axis. As an example, some pyroxene crystals appear to be box-like in cross-section, suggesting an orthorhombic mineral; however, when closely examined, it is seen that the faces on the ends are tilted at angles which cannot be right angles. The orthoclase crystal in Figure 73 also suggests an orthorhombic symmetry if it is observed looking along the a axis, in which case, its cross-section shows parallel opposite faces as sketched to the right of the central figure. Yet if it is turned to the side the odd placement of faces is at once obvious.

Monoclinic Crystal Habits.

Prisms, long—gypsum, vivianite, spodumene, allanite, epidote, crocoite

Prisms, short—amphiboles, pyroxenes, orthoclase, phlogopite

Blocky numerous "odd" faces—datolite, borax, orthoclase, azurite, monazite

Platey—micas, chlorite, often of misleading hexagonal cross-section

TRICLINIC SYSTEM

The prominent feature of triclinic crystals is lack of symmetry due to inclination of all axes at angles other than 90°. A few species habitually grow in blocky crystals (microcline, rhodonite) but others develop tabular or platey crystals (axinite, kyanite). Blocky crystals are not always easy to distinguish from monoclinic crystals. In the case of microcline, whose name is derived from the fact that two of its axes incline *so little,* its crystals are extremely difficult to distinguish from monoclinic orthoclase, requiring other tests to do so with confidence.

From the viewpoint of symmetry, most triclinic crystals are bounded by pinacoids, and only pairs of faces opposite and parallel match each other. Pinacoids may intersect one or more axes; Miller indices may therefore contain one or two zeros. There are no closed forms. Pinacoid pairs are shown in the rhodonite and axinite crystals in Figure 74. Note that the blocky crystal of rhodonite is not much different in general shape from the orthoclase of Figure 73, nor really from some orthorhombic crystals which assume blocky shapes. On the other hand, the axinite crystal is definitely characteristic because of its wafer-like shape, sharp edges, and obvious lack of symmetry.

Triclinic Crystal Habits.

Platey, tabular—axinite, albite, kyanite

Blocky—microcline, rhodonite

THE CONTACT GONIOMETER

A goniometer is a crystal angle measuring device; two kinds are used. The first is the *reflecting goniometer,* a large, extremely accurate instrument used in mineralogical laboratories, but which is so expensive that it is completely out of reach of the amateur. It employs a narrow beam of light directed against a small crystal mounted in its exact center, from which face reflections are received in a telescope. By moving the telescope on an outer calibrated ring, degrees of arc between crystal faces may be accurately measured. The second instrument is a very simple device called a *contact goniometer.* Its use is shown in Figure 75. In essence, it is nothing more than a pivoted metal arm with a pointer which slides over a semicircular protractor, indicating angles as shown. It is very inexpensive.

Because few crystals have truly flat faces, especially those large enough to be held in the fingers and measured by a contact goniometer, considerable care must be used to obtain reasonably good readings. Carefully examine the crystal to determine the best pair of adjacent faces; place the card portion of the instrument on one face and bring around the movable arm to contact

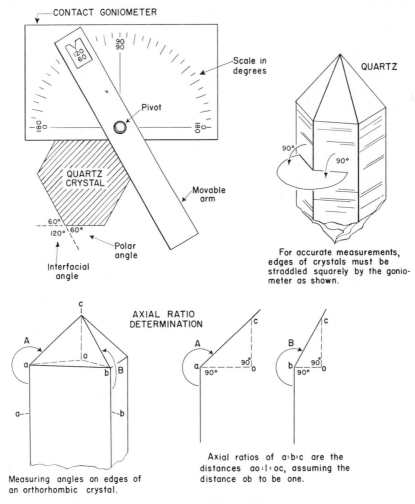

CONTACT GONIOMETER

Scale in degrees

QUARTZ

Pivot

QUARTZ CRYSTAL

Movable arm

60°
120° 60°

Polar angle

Interfacial angle

For accurate measurements, edges of crystals must be straddled squarely by the goniometer as shown.

AXIAL RATIO DETERMINATION

Measuring angles on edges of an orthorhombic crystal.

Axial ratios of a:b:c are the distances ao:l:oc, assuming the distance ob to be one.

USING THE GONIOMETER

FIG. 75.

the other face. Be sure to hold the goniometer so that it stands on the crystal faces at right angles to the edge between faces, as shown in the small sketch to the upper right. If it tilts, the reading will be in error; the more the tilt, the more the error. Next, raise both crystal and goniometer toward a light to see if any streak of light appears between the sharp edges of the instrument and the crystal faces. Adjust the instrument to either cut off all light or allow it to come through in uniform slits indicating that the angle taken is correct. Read off the angle and record.

The *interfacial angle* is the angle between faces; it is never greater than 180°. The *polar angle* is the complementary reading or, in the case of the reading shown in Figure 75, 180° —120° = 60°. Mineralogy textbooks usually furnish polar angles between faces. Their values may be confusing unless it is remembered that it is the complementary angle and not the usually larger interior angle which is being given. For practice, the amateur should obtain a set of good crystals, preferably no smaller than $3/4''$ in length, and take angular measurements upon them, recording results against a small sketch of each crystal. Results should then be checked against textbook values. Angles are frequently recorded as between (Λ) two Miller faces in the following manner:

$$(110) \wedge (1\bar{1}0) = 67° 54'$$

or as the angle between two form letters:

$$m \wedge m' = 67° 54'$$

The prism faces m and m' are located on the forward part of the crystal drawing adjacent to each other. The minus sign over the middle numeral in the second Miller symbol indicates that this face cuts the minus end of the b axis. The sign ($'$) next to the m similarly indicates that the face cuts the minus end. It is usually not possible to measure accurately to less than a degree with a contact goniometer unless the crystal is large and exceptionally flat-faced. In the case of the prism angle example given above as 67° 54' (67 degrees and 54 *minutes*), it is better to round it off to 68° inasmuch as 54 minutes is nearly another degree of arc.

The lower part of Figure 75 shows how several measurements can be used to obtain fair axial ratios, sometimes helpful in establishing the identity of an unknown crystal. The principle, the same as explained in a previous section, depends on placing the edges of the goniometer arms directly over the edges of prism and pyramid faces to measure interior angles. It is difficult to balance the goniometer on sharp edges, but it can be done with some juggling. After the angles are obtained, corresponding to the a axis and b axis of the crystal, the goniometer is laid flat on some paper and the angles drawn with pencil. Right triangles are marked off as shown by the dotted lines. Measure the dotted lines; assume distance bo as equal to one; from their lengths, obtain the axial ratio a:b:c.

This method works only when a face slopes to intersect all *three* axes in tetragonal, orthorhombic, and hexagonal crystals. It will not work with other systems. In the case of tetragonal and hexagonal crystals, where horizontal axes are of the same length, only one edge measurement is needed to obtain ratios. It may be used in isometric system crystals to *verify* that they do belong to this system but, as pointed out before, the axial ratios in this system must always be 1:1:1. Where one axis tilts, as in the monoclinic system, or all axes tilt, as in the triclinic system, the difficulties of plotting using the scheme just explained are too great to overcome easily, and trigonometric calculations are needed. As a word of caution, it should be mentioned that textbook values will be obtained only if the *correct* face happens to be present on the crystal as shown in Figure 52, where the *p* face of sulfur is the one used by all crystallographers by mutual agreement. Other faces also give axial ratios, but the numbers must be divided or multiplied by some simple number as two or three to convert them to textbook values.

6

Physical Properties

Considering the enormous quantities of ordinary rock which greet the collector in quarries, mines, and other collecting sites, it is obvious that euhedral crystals, such as every collector wants, are extremely rare. No one has ever estimated how much rock consists of well-formed crystals, but it must be a very small percent. The greatest number of specimens are broken masses showing little if any crystal face development. Naturally this makes identification more difficult in some ways, but considerably easier in others. Even the broken surfaces display signs reflecting inner crystal structure as surely as a set of crystal faces, and, when interpreted correctly, such subtle signs lead to identification with a certainty which sometimes is difficult to duplicate with crystals. For that matter, crystals are not always as useful as one may imagine because many adopt strange or misleading habits, others cloak themselves with foreign minerals, or become etched and disfigured to the point where they little resemble ideal forms. In these instances, the outward signs or *physical properties* must be taken advantage of in testing.

The term physical is derived directly from the realm of science known as *physics* which concerns itself with natural forces and their effects upon matter, including mineral crystals. Physical properties of greatest interest to the amateur, and particularly useful for testing, will be discussed in this and following chapters. The strength of crystals and related properties such as fracture, cleavage, and hardness are dealt with in this chapter along with minor properties such as heat, magnetism, and electricity. The relative weight of minerals, extremely useful in identification, is the subject of a separate chapter, as is the wide field of optical properties.

WHEN A CRYSTAL BREAKS

Why is the surface of a broken crystal important in identification? Only a glance over a collection of mineral specimens is needed to discover decided differences on exposed broken surfaces; some resemble broken glass, some are partly smooth, partly irregular, others differ in the way in which they reflect light. Decidedly, they are *not* the same! But just as decidedly, each surface is characteristic of the crystal structure beneath. It is these characteristics that we wish to investigate, and by doing so, relate them to separate species, thus aiding in their identification.

One of the most important features of any crystal structure is *direction;* it proved to be important in respect to creation and placement of faces on crystals, and it proves equally important in respect to the way any crystal breaks. When directional influences are at a minimum a crystal is said to *fracture.* If a more or less flat fracture is developed and is related consistently to certain planes in the crystal, it is said to be *cleavage.* A cleavage-like fracture due to separation between twins is *parting.* Beginning with fracture, it will be seen how the character of broken surfaces indicates the nature of the atomic structure beneath.

CONCHOIDAL FRACTURE

A solid composed of evenly-spaced and evenly-attracted atoms shows no tendency to fracture in any special direction, consequently its fracture is guided purely by mechanical considerations, as shown in the example of Figure 76 at the left. A homogeneous material of this sort fails in

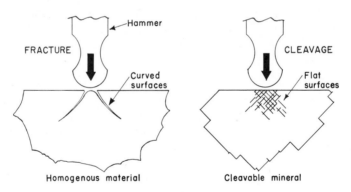

FRACTURES COMPARED

FIG. 76.

the manner shown when tapped by a ball-peen hammer. The impact momentarily pushes down a small cone of material whose sides slope in curved fashion and lead to the fracture being called *conchoidal,* or "shell-like." If the specimen is large enough to contain the crack and not let it pass through the sides, the conical surface immediately fills with air which keeps the sides from rejoining. Widely separated parts reflect light as silvery reflections, thinner parts interfere with light and develop vivid rainbow hues. In contrast, a cleavable material, shown at the right in the same figure, fails along approximately the same lines, but the smooth curve is replaced by numerous angular cracks directed by inner crystal structure. When turned, a specimen of this sort reflects light only from certain positions, indicating that the minute splits are oriented.

The smooth curvature of perfectly developed conchoidal fracture in any specimen immediately indicates that it is extremely homogeneous in internal structure. The finest examples of conchoidal fracture therefore occur in materials which are non-crystalline, as, for example, ordinary glass, the natural equivalents of glass as obsidian and tektites, also opal, amber, etc. In each of these substances, the constituents have been so thoroughly scrambled that atoms are distributed in very uniform fashion throughout. Indeed, when glass makers prepare glass batches for the furnace, they take great pains to select only those raw materials which will *not* crystallize when the melt is cooled. Conchoidal fractures of lesser perfection occur in very homogeneous, fine-grained rocks, as some rhyolites and basalts, or in minerals composed of many very small crystals, as in the chalcedony variety of quartz. However, these may be distinguished easily from non-crystalline materials by the lack of glassy luster on fracture surfaces. It is very useful to learn to recognize the characteristic conchoidal fracture of non-crystalline materials and minerals because it occurs in perfection only on very few species, notably opal and the volcanic melt-rock, obsidian.

Conchoidal Fractures in Crystals. Whenever crystal structures approach the internal uniformity of glass, good conchoidal fractures develop, as shown in Figure 77 where a large clear mass of rock crystal quartz is compared to a mass of obsidian. Even here the distinctions are plain, despite the fact that the quartz is an exceptionally perfect and flawless specimen and therefore expected to reveal the highest degree of conchoidal fracture possible in this species. By far most quartz specimens show distinct guidance of fractures, even in flawless specimens, as noted by surfaces which abruptly terminate in steps or pass into ripples due to internal twinning. Remarkable narrow ripples are a distinctive feature of amethyst fractures and are caused by repeated twinning characteristic of this variety of quartz. In other quartz specimens, fractures are distinctly guided by crystal structure so that attempting to break a crystal across usually results in a diagonal fracture approximately parallel to one of the terminal rhombohedral faces. As internal uniformity in crystals disappears, fracture surfaces are increasingly guided by the atomic structure, if not by planes of inclusions, bubbles, etc., until the highly perfect and completely directional fracture surfaces known as *cleavages* appear. More will be said about cleavages later. Because all fractures, regardless of nature, are indicative of internal structure, or its lack, as mentioned in the case of non-crystalline material, it is extremely important in

FIG. 77. Conchoidal fractures developed upon rock crystal quartz (*left*) and obsidian (*right*). The mass of quartz is cobbed from a much larger crystal and is virtually flawless. It measures about 7¾″ in diameter and originated from an unstated locality in Minas Gerais, Brazil. Conchoidal fractures are perfectly developed upon the obsidian specimen which comes from Coso Hot Springs area of Kern County, California. The latter specimen exposes a spherical inclusion of grayish-white cristobalite.

identification that such surfaces be examined most carefully. This will be appreciated more fully as the discussion develops.

Descriptive Terms. The ease with which any mineral is fractured is described by using terms as *tough, brittle, friable,* etc. *Toughness* is best exemplified by native metals and by granular or fibrous mineral aggregates, such as chalcedony or rhodonite, which are broken apart only with considerable difficulty. *Brittleness* is displayed by glass, quartz, opal, and most nonmetallic minerals. Considerable variation makes it necessary to use additional terms as when describing opal as *very* brittle. The term *friable* is also in common use, aptly describing the crumbly disintegration of earthy materials or crystals which are so penetrated by cracks that the fragments fall apart when rubbed in the fingers. Several terms describe less than perfect conchoidal fractures, as *sub-conchoidal* for surfaces which show decided conchoidal character but not so uniformly developed as upon glass or obsidian. More or less flat surfaces not, however, cleavage planes, call for use of such terms as *even, uneven,* or when more broken up, terms as *regular* or *irregular.* The jagged, torn surfaces of fractured metals are termed *hackly.* Some soft minerals, notably the native metals, are *sectile* because curved shavings can be turned up by the edge of a knife blade. Sectility is also observed in a few non-metal species, notably graphite, molybdenite, argentite, realgar, orpiment, etc.

Aggregate fractures receive their own descriptions when necessary to reflect the fact that numerous crystals are simultaneously fractured. When many slender crystals are broken parallel to their sides as in goethite, hematite, etc., the surfaces may be described as *splintery* or *fibrous.* Clayey or earth-like aggregates exhibit an *earthy* fracture, which also implies lack of luster. The terms *granular* or *sugary* are also appropriate when granular aggregates are involved.

Spontaneous Fracturing. Although the impression lent is that fractures are created by breaking specimens from cavity walls or by deliberate trimming, numerous instances are known where cracks appear spontaneously due to various causes. Opal and other amorphous minerals frequently contain so much water trapped in minute pores that its evaporation causes slight shrinking and "crazing" of outer portions much like those which appear upon mud when it dries. Internal stresses, sufficient to cause cracking, also commonly appear whenever sharp color differences are noted in crystals of tourmaline and beryl. Zones of color indicate changes in composition and slight structural strains from one part of the crystal to another. Such stresses may cause portions of the crystal to fail, sometimes leaving behind a more or less spherical nodule of great smoothness of surface and freedom from flaws. Such "nodules" are commonly observed in tourmaline from Maine, California and Brazil, while similar nodules are common in the morganite variety of beryl.

Spontaneous fractures also commonly occur around inclusions, even when inclusions are deeply buried. The host crystal ordinarily cracks along one plane, in which event each inclusion resembles a miniature saturnian planet, with the crack forming the rings. At other times, several rings surround inclusions if the host crystal is easily cleavable in several directions. Planes of inclusions, as those creating phantoms, also cause host crystals to crack along them. Still other cracks commonly radiate outward from large liquid or gas inclusions which, due to expansion, overstress the surrounding crystal and cause its rupture. Water inclusions are particularly destructive because of the tremendous expansive force exerted when frozen.

CLEAVAGE

A flat fracture guided by the atomic structure is a *cleavage*. It may be guided so well that it is unmistakable, and sometimes so easily developed, as in calcite, that it is nearly impossible to fracture its crystals *except* along cleavage planes. On the other hand, it may be present only in traces, and sometimes is absent in some specimens though present in others of the same species. A cleavage is always a more or less plane surface, often glassy smooth and sometimes absolutely continuous, as in mica, so that no imperfection mars its surface. Unlike ordinary fractures, which are unpredictable as to directions taken, cleavages can always be related to specific planes within the crystal and thence to crystal faces and axes. Cleavage planes are parallel to common faces in some species, serving equally well for identification purposes. Because any cleavage plane repeats itself as many times as the atoms which form its boundaries, there are multitudes even within the smallest crystal. In an easily cleaved mineral such as mica, it is theoretically possible to continue splitting to the point where nothing remains except the atomic structure characteristic of a single layer. This is impossible to achieve in practice because of the extreme thinness of such layers; even the thinnest sheet split by hand still contains thousands of structural layers.

A cleavage plane cannot pass through an atom, nor through a strongly-bound atomic cluster such as a radical group; it can only pass between or around them. Silicate structures as rings, sheets, and chains of SiO_4 tetrahedrons, also discourage cleavages passing across them although they may pass between. Thus the silicate rings of beryl and tourmaline prevent development of cleavages parallel to the c axis but do permit imperfect cleavages to develop between them, corresponding to the basal plane c (0001). Where sheet structures, as in mica, are so strongly bonded within sheets and so weakly between them, the difficulty of breaking across sheets is compensated for by the perfect and very easily developed cleavages passing between them. Sheet structures in graphite and molybdenite also produce perfect basal cleavages. In the case of crystals whose structures are more uniform, as in diamond (octahedral cleavage) and salt (cubic cleavage) directional weaknesses in bond strengths are not so greatly pronounced, and crystals do not split as easily along cleavage planes as mica, graphite, etc. Nevertheless they do split if the proper blow is skillfully administered. In the case of diamond, the hardest mineral known, a perfect cleavage occurs along octahedral planes because bonds are fewer between these planes. In salt, bonds between atoms are weakest across cubic layers, and this mineral cleaves accordingly.

Cleavages therefore vary from mineral to mineral, and occur only when atomic structure happens to display directional bonding weaknesses which, when greatly pronounced, result in perfect and easily developed cleavages, and conversely, result in poor cleavages when attractive forces across crystal planes are not greatly different. Cleavages do not imply that cleavable crystals are weaker than non-cleavable crystals. Diamond cleaves perfectly along octahedral planes, yet is the hardest mineral known, while bornite, displaying very poor cleavage, is very soft and weak. Numerous examples can be cited to show that cleavage only means that if a mineral must break, it may prefer to do so along planes of relative weakness. If atomic bonds across such planes are decidedly weaker than elsewhere, cleavage is more likely to occur than fracture.

Cleavage Surfaces. As shown in Figure 78, cleavage surfaces vary greatly from species to species, and often within any given species. *Perfect* cleavage is used to designate surfaces of exceptional flatness and smoothness, such as produced in diamond, topaz, and mica. *Good, fair,* and *poor* are also used to qualify character, while *interrupted cleavage* is used to describe surfaces covered by step-like or shelf-like cleavage projections. Interrupted cleavages

FIG. 78. Cleavages in Fluorite and Calcite. The inset shows nearly perfect octahedrons obtained by carefully cleaving larger cubic fluorite crystals. Below are cleavage rhombs of colorless calcite, several of which show doubling of images due to the strong double refraction of this mineral. *Inset photo courtesy Schortmann's Minerals, Easthampton, Massachusetts.*

are easily detected by holding the specimen under a strong overhead light and slowly moving it about until many points of light reflect at once, indicating parallel alignment. This test is effective on waterworn topaz, for example, which sometimes is so smoothly rounded that very little can be told about its pebbles from casual examination. However, its single prominent cleavage evidences itself on opposite ends of a pebble by numerous pinpoint reflections which suddenly light up when the pebble is turned and just as suddenly eclipse if it is turned too far.

Cleavage surfaces are made less perfect by twinning, as in albite, in which otherwise smooth surfaces are interrupted by many narrow parallel twinning striations characteristic of this and other plagioclase feldspars. Similar twinning bands are noted upon sphalerite and calcite cleavage planes. In fluorite (Figure 78) octahedral cleavage surfaces are covered by numerous irregular patches, somewhat offset from the general level, but together giving the impression of flatness. Each patch represents a portion of the crystal which grew slightly out of registry with other parts. A similar patched effect is also noted on broad cleavages of galena and sphalerite.

The ease of cleavage developments is an important property quite aside from the fact that a cleavage exists. For example, only the pressure of the fingernail is needed to cleave a mica crystal, while a rather sharp tap, directed along a specially cut groove, is needed to split a diamond crystal. For this reason, descriptions of cleavage often contain terms such as *easy* or *difficult* to indicate how readily a cleavage can be started.

Cleavages in the Crystal Systems. Because cleavage is guided by the atomic structure of crystals, it follows that if cleavage develops along any plane *every other plane like it in respect to the same crystal structure will also develop a cleavage.* In the most symmetrical of all crystals, those belonging to the isometric system, this maxim is shown beautifully by the cleavages noted in some of its species. For example, in Figure 79, three common isometric cleavages are shown. The cubic cleavage, corresponding to the faces of the cube form, is observed perfectly developed in galena and halite crystals. Although a specimen of galena may show only one shining cleavage, perhaps developed where the crystal was broken from its matrix, any crystal needs only be tapped lightly along its outer edges to develop cleavages parallel to every cube face. Similarly, all corners of a fluorite cube can be skillfully chipped off with a sharp-edged tool, as is the custom of miners in the fluorite mining districts of the Midwest, to develop more or less perfect octahedrons, shown in process in Figure 79, and photographed in finished form in Figure 78. The dodecahedral cleavage in Figure 79 occurs in several isometric species but is well developed only in sphalerite. With care, a large piece of sphalerite can be split into a dodecahedron, but it is far less easy to do than to make an octahedron out of fluorite.

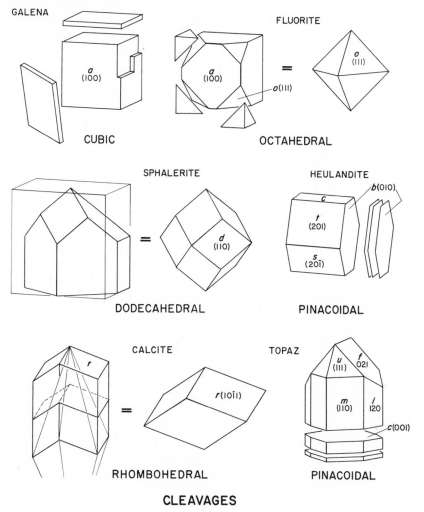

GALENA

FLUORITE

o (100)

o (100)   o(111)   =   o (111)

CUBIC   OCTAHEDRAL

SPHALERITE   HEULANDITE

b(010)

c

t
(201)

=   d
(110)

s
(201)

DODECAHEDRAL   PINACOIDAL

CALCITE   TOPAZ

r   u   f
                                          (111)   021

=   r(10ī1)   m   l
                                        (110)   120

                                                 c(001)

RHOMBOHEDRAL   PINACOIDAL

CLEAVAGES

FIG. 79.

Tetragonal species sometimes display excellent cleavages. Wernerite displays good prismatic cleavage, with 90° angles between planes in accordance with the symmetry requirements of this system. Apophyllite displays a nearly perfect cleavage parallel to the basal pinacoid; numerous partly developed cleavages usually impart a strong pearly luster which, with cleavage and square habit, makes recognition of this species a simple matter.

The hexagonal system is noted for providing perfect rhombohedral cleavages in the carbonate minerals as calcite, rhodochrosite, dolomite, etc. Calcite is favored for demonstrating this cleavage. Carbonate cleavage is guided by the large carbonate radical and is so easily developed that rarely is any other type of fracture surface found. The relationship of rhombohedral cleavage planes to a typical calcite crystal is shown in Figure 79, while cleavage rhombohedrons appear in Figure 78. It is to be noted that in aragonite, and other orthorhombic carbonates, cleavages lie not only in radically different crystal directions, but also indicate the lower symmetry of crystal structure in this system. As symmetry grows poorer, cleavages tend to become fewer in number and to show less resemblance to each other in respect to quality of surface, ease of production, etc. Before leaving the hexagonal system, it should be pointed out that a cleavage plane occurs in some species parallel to the basal pinacoid $c(0001)$, but is well-developed only in a few, notably graphite and molybdenite; traces are sometimes found in beryl crystals.

Basal pinacoid cleavage is characteristic of minerals in the orthorhombic and monoclinic systems, serving as valuable clues to identity. Micas are easily recognized by perfect basal cleavage. Topaz is probably recognized more times in poor crystals or irregular masses by its perfect basal cleavage than by any other property (Figure 79). Cleavages in two great mineral groups, the pyroxenes and the amphiboles, coincide with prisms $m[110]$, as shown in Figure 80. Although confusingly similar, they may be distinguished by the angles between cleavage planes: pyroxene cleavages are nearly 90° apart (actually about 87° and 93°); amphiboles about 56° and 124°.

Pinacoidal and prismatic cleavages commonly occur together as in barite, also depicted in Figure 80. A perfect basal cleavage crosses prismatic cleavage, contributing to the notorious fragility of barite crystals. Practically all barite crystals show them. Partly developed cleavages are observed in many species from all systems and, although marring the beauty of crystals, provide excellent recognition clues. They are commonly present in crystals of calcite, gypsum, sphalerite, heulandite, apophyllite, micas, anhydrite, celestite, spodumene, fluorite, etc.

Misleading Cleavages. A number of species develop cleavages which seem appropriate for a system other than the one to which they belong. The prismatic cleavage of pyroxene species, for example, resembles the square prismatic cleavages of tetragonal species. Similarly, the several pinacoidal cleavages of anhydrite, all at right angles to each other, yield fragments which look as if they must belong to an isometric species instead of an orthorhombic mineral. However, differences in luster and markings on the several cleavage surfaces show that they cannot belong to an isometric mineral. Cleavage surfaces must always be examined carefully to be certain that they are alike, or if not, just how they differ. Surfaces exactly alike *must* belong to the same crystal form, and this in itself may be a most important clue to identity. Cleavages are so important in this respect, that the amateur is well advised to assemble or purchase a set of specimens showing a variety of cleavages and to study them to be sure that the differences are understood.

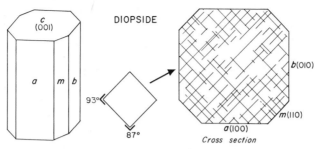

PRISMATIC CLEAVAGE OF PYROXENES

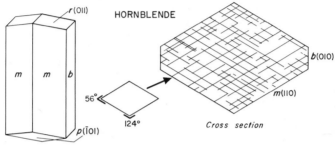

PRISMATIC CLEAVAGE OF AMPHIBOLES

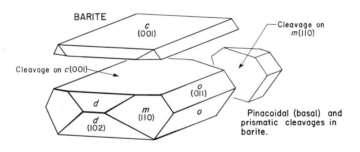

CLEAVAGES

FIG. 80.

PARTING

Confusingly like cleavage are flat, easily-developed fracture planes called *partings*. Sometimes, as in the corundum example of Figure 81, they are so smooth that they cannot be told from a true cleavage plane merely by casual inspection. No matter how closely they resemble cleavages, however, partings arise from different causes and cannot be the same. Partings develop along planes of weakness as between twins. If a crystal consists of many twins, there may be many partings; if only several twins are involved, then there can only be a like number of partings. Partings therefore occur in *limited number* in any crystal, and sometimes not at all in other crystals of the same species. Because this is not the case with cleavages, which can be developed in infinite number in any crystal, this limitation serves as a valuable means of identifying partings when seen in specimens. In the corundum crystal of Figure 81, for example, close examination shows numer-

FIG. 81. Parting along twinning planes in dark blue corundum from Mozambique. Specimen size 3″ by 1½″. The twinning joints can be seen at the top, crossing at angles of 60°. If this crystal were broken further, it would separate along such joints into squarish fragments. The top plane is c(0001); the side planes are r(10$\bar{1}$0).

ous partings, but gaps of from one-sixteenth to one-quarter inch exist between them. Small, nearly cubic blocks can be detached where several partings cross, but no matter how carefully these smaller blocks are tapped, they do not develop further flat planes, shattering only into glassy fragments covered with conchoidal fractures. This is proof that they are partings rather than cleavages.

The curious shape of the corundum in Figure 81 is due to a prominent parting developed on the basal plane c[0001], and additional partings developed on planes corresponding to the rhombohedron r[$\bar{1}$0$\bar{1}$1]. Large, impure corundum crystals, particularly those from localities in Georgia and North Carolina, frequently part readily on rhombohedral planes, yielding box-like forms which seem as if they should belong in the isometric system. Similar shapes are sometimes developed in hematite due to easy parting. Parting also arises from crystal deformation, as when a crystal solidly enclosed in matrix has been subjected to compression. Crystals of this type usually crumble into sand-like fragments along parting and cleavage planes when removed from matrix. Conspicuous fractures and cleavages are listed in Table 3, Appendix A.

HARDNESS

On a much smaller scale, *hardness,* or the resistance of any mineral to scratching, is indicative of the strength of crystals because even the creation of a small groove calls for separation of atoms. As may be expected from the wide variety of atoms joined in diverse combinations and structural patterns, hardness varies greatly from species to species. Measurement of hardness is of considerable value but results must be used with caution, as will be explained.

The earliest scale of hardness was devised by the German mineralogist Friedrich Mohs (1773-1839), who noted important variations in hardness and proposed his scale to provide a standard reference. Despite its antiquity the Mohs Scale is very much in use and indeed, when one considers the difficulties in trying to better it without expensive instrumentation, it is no wonder that everyone still refers to it.

MOHS SCALE OF HARDNESS (DESIGNATED BY H)

1. Talc (softest)	6. Feldspar
2. Gypsum	7. Quartz
3. Calcite	8. Topaz
4. Fluorite	9. Corundum
5. Apatite	10. Diamond (hardest)

According to this scale, all hardnesses are relative to each other and do not represent actual quantitative steps between mineral hardnesses. This is to say that a mineral of H 4 scratches another which is H 3. If a mineral lies somewhere between, its hardness may be designated by H 3½. It is not exact, but its lack of precision is more than made up in convenience. More scientific hardness tests have been devised, particularly for testing of metals, ceramics, and other hard substances used in industrial or scientific applications. For example, the Knoop indentation tester uses a small cut diamond point pressed by machine into the prepared surface of the specimen. Results are considered fairly accurate and consistent. To compare the accuracy of the Mohs Scale, corresponding Knoop values are given in the graph of Figure 82. Note how the curve of increasing hardness rises rather uniformly up to corundum (H 9), at which point it nearly vanishes out of sight before it stops at diamond (H 10). According to Knoop values, little difference in quantitative hardness appears among the softest minerals, but rather large steps are found between quartz and topaz, and between topaz and corundum. But the largest step of all, between corundum and diamond, is about three times the distance from corundum to the beginning of the scale! Thus diamond is vastly superior in hardness to its closest rival, corundum.

Hardness and Crystal Structure. In general, smaller atoms or ions promote greater hardness, while larger atoms or ions promote weakness. Thus many minerals with large ions such as the carbonate and sulfate radicals tend to be softer than other classes in which large atom groups are absent. The familiar silicate radical contributes considerable hardness to silicate minerals as a whole, all of which tend to be approximately the same in hardness. Aluminum and oxygen atoms combined in complex ions similar to the silicate ion also contribute hardness.

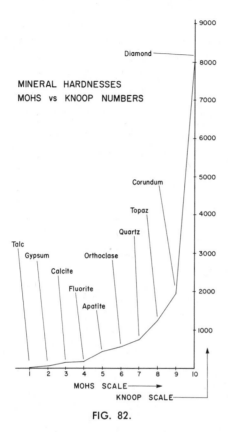

FIG. 82.

Marked differences are noted in polymorphs due entirely to structural reasons. The best known pair is diamond (H 10) and graphite (H 1-2), each on opposite extremes of the hardness scale! Whereas diamond is very uniform in respect to placement of atoms within the structure and the extension of bonds between them, graphite is marked by a sheet structure which is extremely weak in comparison. Another polymorphous pair also shows marked though less striking differences in hardness: calcite (H 2½-3) and aragonite (H 3½-4). Both are calcium carbonate, but differ in crystal structure.

Hardness and Crystal Directions. Because crystal properties vary with direction, hardness sometimes varies in spectacular fashion when measured in several directions on the same crystal. The best known example, kyanite, is a chain structure silicate in which a hardness of 7 is measurable *across* the chains, but a hardness of only 4½ *within* the chains. The flat cleavage, usually well developed upon most kyanite specimens, provides the necessary surface for demonstrating the profound effects of atomic structure upon this property. In other minerals, hardness variations can be easily noted, as upon calcite which is harder (H 3) upon the cleavage faces than upon the basal plane (H 2½). Variations exist in other species but are not always easily detected by the crude methods used to test them. Wide variations have been noted by gem cutters whose diamond-impregnated cutting laps provide a multitude of small uniform "hardness points." Changing the position of the gemstones from one orientation to another when cutting facets of the same size, provides indication of hardness variations. Hardnesses are listed in Table 4, Appendix A.

Testing for Hardness. For usual tests, such as the amateur is likely to employ, a set of *hardness points* is essential. They may be nothing more than small bits of suitable minerals with sharp points, or professional points, made by inserting pointed mineral fragments in small metal rods. Professional points are better because the metal rods provide a far better "feel" as slight vibrations are sent up the rods to the fingertips during scratching. Lightness of touch is important because most points are brittle and pressing too hard may cause tips to break off.

The test area on the specimen should be selected with care. It must be representative of the mineral in its freshest state and must not be coated with alteration products or other minerals, or be fissured, grained or ridged. If necessary, a small piece should be knocked off to expose fresh and smooth material. Testing begins using the hardest point, usually a sliver of corundum, and then working down until a point no longer "bites" into the surface but merely slides over it. Pressure must be light but firm. If the point slides rather than bites, it should be turned slightly to expose a sharper portion. To test directional effects, make scratches in several directions in star fashion. In a few materials, decided differences according to crystal directions provide useful recognition clues; in the case of kyanite, differences are sufficient to identify.

It is to be emphasized that hardness is subject to personal interpretation, and to surface conditions on the mineral being tested. It is not an accurate test but is most useful in placing unknown species into a *general* range of hardness. This is to say that it is easy enough to distinguish a soft mineral from a hard one, but much more difficult to distinguish between two which are nearly the same. Beginners frequently mistake quartz for topaz, or even for colorless corundum, because blunt hardness points are used on perfectly smooth surfaces, and light pressure merely results in sliding rather than scratching. For this reason, it is essential that only the sharpest of points be used, coupled with light application pressures to preserve sharpness.

A magnifying glass is indispensable for hardness testing. If the necessary light touch is used, the scratch made is so fine that it cannot be seen readily with the eye alone, nor is one certain—even when a slight "catching" is felt while using the point—that it is indeed the point which is doing the scratching and not the other way around. The experienced mineralogist therefore applies the point, feels the "catchiness" of a scratch being generated, wipes off the powder created, and finally observes the result using a magnifier. If a shallow "vee" appears in the test specimen, it is being truly scratched.

Useful hardness testing accessories are metal points which may be used in lieu of stone points, and have the merit of being less brittle and kept sharp easily. Copper points, preferably made from copper spikes or very heavy gauge wire, are generally H 3, and are useful accordingly. Mild carbon steel in penknife blades is about 5 to 5½. Window glass is about 5½. Steel files, much harder than ordinary steels, are about 6½; the business end only must be used, because the tang end, meant for insertion in a wooden handle, is annealed to softness. If possible, grind the tip of a three-cornered file to a point using a water spray on the grinding wheel to prevent burning and annealing; break off the file about half way to make it lighter and easier to use. Much better than any file, is a stout darning or sailmaker's needle inserted in a small holder of drilled wooden doweling. Needles are about the same hardness as files, but can be more delicately used and resharpened by a few twirling strokes on an oilstone. Under magnification, a well sharpened needle is very useful not only for hardness testing but for developing cleavages and thus adding to one's knowledge of the mineral being tested. In some instances, as when a specimen happens to be minutely cleaved, a rough hardness test gives false impressions of softness because numerous failed places beneath the surface rip up readily. The use of a magnifier and needle point shows if a clean surface is being scratched or whether an already weakened area is merely being plowed up.

A Summary of Strength Properties. In a previous chapter it was mentioned that the chemical-structural classifications of minerals generally proved to be the best method for putting like species together. When the strength properties of species within the classes are compared, the wisdom of using this method becomes apparent. There are exceptions of course, but by and large, properties displayed by species in each class show striking similarities as summarized below:

Native Metals. Gold, silver, copper, etc. Soft, malleable, sectile

Semi-Metals. Arsenic, bismuth, antimony. Soft, somewhat sectile, brittle

Native Elements. Sulfur: soft, brittle. Diamond: extremely hard, brittle. Graphite: extremely soft, somewhat sectile

Sulfides, Sulfosalts. Generally soft, brittle or sometimes somewhat sectile. Pyrite group *considerably* harder

Oxides. Hard to very hard, brittle. Exceptions: cuprite, zincite, which are soft

Hydroxides. Vary from soft to hard, brittle

Halides. Soft, brittle, outstanding cleavages

Borates. Soft, brittle, outstanding cleavages; a few species are hard

Sulfates. Soft, brittle, outstanding cleavages

Molybdates, etc. Soft, brittle; cleavages poor

Phosphates. Some soft, many moderately hard; brittle, cleavages common

Silicates. Great variations in hardness but average is hard to very hard; brittle; many species exhibit cleavages

From the above, and speaking generally, it can be seen that greatest hardness is encountered in oxides and silicates. Greatest softness and outstanding cleavages are met with in halides, carbonates, many borates, and sulfates. Malleability and sectility are outstanding properties of the native metals. Softness and weakness typify sulfides.

THE EFFECTS OF HEAT

The general effect of heat on solids is to set atoms in vibration, the intensity of vibration and the distance swept out by each atom during such movement being proportional to the intensity of heat (temperature). When temperature is sufficiently high, atomic bonds begin to yield and atoms move farther from each other until the formerly solid mass turns into a liquid and, if atomic agitation is severe, into a gas. Some solids, as lead, tin, gold, etc., suddenly slump into liquid because bonds break rapidly and uniformly however, other solids, as many silicate minerals, break bonds easily between the silica groups and the other ions, but reluctantly within the silica groups themselves. Consequently they slowly change from solids into very thick fluids, requiring prolonged high temperatures to do so. As may be expected from the large and varied assortment of atoms in minerals, and similar diversity of bonds, minerals behave very differently when subjected to heat. Sometimes this behavior is useful in identification as is explained in Chapter 9, where for convenience, fusion tests are discussed at the same time as other tests involving the use of a flame.

Melting Points. In general, melting occurs in fairly consistent temperature ranges for species within the same chemical class, although wide departures from this rule occur, as pointed out before and as will be noted in the table given below. Native elements range over the entire scale, from mercury, which is liquid at room temperature, to iron, platinum, and others, requiring very high temperatures to melt. The unique polymorphs of carbon and sulfur are in a class by themselves and cannot be compared to other elements. Sulfides and sulfosalts tend to be low upon the scale; notable exceptions are pyrite and sphalerite among common species. Halides melt at relatively low temperatures but oxides and hydroxides generally require very high temperatures; in fact, many oxides are used for refractories in high-temperature ovens and furnaces because of this property. Hydroxides lose water as heat is applied, but then becoming oxides, they resist further change. Carbonates are comparable to oxides in resistance to melting but are readily broken down chemically. Sulfates, borates, phosphates, etc., vary considerably in resistance to heat, but silicates as a rule withstand high to very high temperatures for the reasons previously given.

MINERAL MELTING POINTS

Mineral	M. P. Centigrade	M. P. Fahrenheit
Mercury	−39	−38
Sulfur	+113	+235
Bismuth	271	520
Orpiment	300	572
Stibnite	548	1018
Antimony	630	1166
Halite	801	1474
Silver	961	1762
Almandite	1050	1922
Gold	1062	1944
Copper	1083	1981
Albite	1100	2012
Pyrite	1171	2140
Cuprite	1235	2255
Fluorite	1360	2480
Iron	1535	2795
Barite	1580	2876
Quartz	1710	3110
Corundum	1774	3225
Zircon	2500	4532

Heat Transmission in Crystals. Heat transfer is rapid in some metals, as silver and copper, but slower in others, as iron. Extremely uniform structures, as glasses, obsidian, opal, etc., pass heat very slowly but uniformly. Crystals with decided differences in structure, as chain silicates, sheet silicates, etc., pass heat rapidly along certain crystal directions, but less quickly along others. To a lesser extent, this is true of all crystalline minerals although the effects in some species may not be readily apparent. Even isometric crystals which, as a class, are more internally uniform than crystals belonging to other systems, do not pass heat as uniformly as amorphous substances. For example, if a cube of halite is touched on a cube face by a hot wire to measure the rate at which heat is transmitted, it will be found that the outward spread of heat follows a cloverleaf pattern rather than a circular pattern as would be true if the same test were applied to an amorphous material as glass. The passage of heat in crystals is usually faster than in amorphous minerals because atoms are in contact and spread vibratory motions to neighbors more quickly than is possible in substances which are internally disorganized. The slowness of heat transmission in amorphous materials creates high local temperatures, which, causing expansion, result in the well-known easy cracking of heated glass, obsidian, etc. On the other hand, cryptocrystalline minerals, as chalcedony, can often be heated strongly without damage because the numerous minute crystals which comprise the whole allow for some adjustments between them to relieve stresses. However, large quartz crystals crack speedily because slight but important differences in rate of expansion according to crystal direction develop stresses which the crystal cannot resist. When a large quartz crystal of four or five inches in diameter is heated slowly and evenly as possible, it still cracks into blocks, each of which is then better able to resist further cracking.

Many minerals exhibit important chemical or structural changes when heated. The removal of water from gypsum, for example, converts it to anhydrite although the stress created by heating causes the entire sample to fall to powder. The "burning" of lime converts calcium carbonate (calcite) into calcium oxide by destroying the calcite structure and releasing carbon and part of the oxygen. Changes in structure without change in composition also commonly occur as may be noted in the case of quartz. The conversion of quartz into glass follows a series of steps, each changing the crystal structure in some respect. Ordinary quartz is not affected up to 573°C or 1063°F, but past this point, the linked silicate tetrahedrons alter position slightly in relation to each other but do not separate. In the low temperature region quartz is called *low quartz;* in the higher temperature region, with slightly different crystal structure, it becomes *high quartz.* At 867°C or 1593°F, the tetrahedrons separate and re-link in another pattern to form tridymite, but this new species also loses its identity at 1470°C or 2678°F when cristobalite is formed. At 1710°C or 3110°F, cristobalite becomes liquid. Curiously, as each of these steps is taken, the crystal structure becomes more symmetrical, until isometric cristobalite is reached.

MAGNETISM

The property of magnetism has been known since very ancient times and it is recorded that suitable pieces of highly magnetic iron ore, or "lodestone," were suspended on threads, free to turn, and thus used to indicate the magnetic pole to ships' navigators. Lodestones are exceptionally magnetic specimens of magnetite, which species, of course, takes its name from this property. Even weaker specimens are strongly attracted to magnets, and as such provide a very distinctive test for identifying this mineral. Magnetism has been intensively studied in modern times but only recently has the property been explained.

Magnetism is believed to arise from the joint influence of the small magnetic fields surrounding each electron. As electrons orbit around nuclei, they also spin on their own axes and by doing so within the larger atomic electrical fields, create miniature magnetic fields. Most electron spins are "paired," that is, each electron has an opposite spinning mate with a magnetic field canceling its own. However, in some atoms, spins occur in the same direction causing magnetic fields to arise from the atoms because electron fields support instead of oppose each other. A number of magnetic atoms in any crystal produce a general attraction for the magnet. If the separate fields happen to be aligned in one direction in the crystal, it becomes a magnet itself. Magnetite *lodestones* may have naturally formed in the way used to make artificial metal magnets, as follows. Iron, the chief magnetic element, is suitably alloyed and a bar placed within a very strong magnetic field created by electricity. Some sort of stress is applied to the bar, as by hammering sharply, upon which some electrons are momentarily jarred from fixed orientations and freed to tumble into new positions. They align themselves according to the outer magnetic field, and from this moment, make the bar a magnet.

Many minerals are not magnetic, in fact, they are repelled by magnetic fields and are therefore termed *diamagnetic*. In them, it is believed that all electron magnetic fields are canceled nearly completely, thus accounting for their repulsion. Those species showing some degree of attraction are called *paramagnetic,* while magnetite *lodestones* are called *ferromagnetic*. Ordinary magnets are useful for testing magnetism, but only magnetite and pyrrhotite display strong attraction, magnetite being considerably stronger than pyrrhotite. A number of species, particularly those containing iron, show slight but noticeable attraction. Specimens must first be crushed to small grains making it easier for the magnet to exercise its influence. A bar, rod, or horseshoe magnet, preferably Alnico, is passed near the grains while they are being observed with a magnifying glass. Sometimes grains jump to the magnet but at other times, only a slight stirring may be noted.

ELECTRICAL PROPERTIES

Electrical properties are seldom useful for practical identification but are interesting and deserve a few words of explanation. Such properties depend on the flow of electrons or their concentration in specific areas of the crystal. The steady movement of electrons, or electrical current, occurs only in crystals with the metallic bond, as in native metals and some sulfides. Most crystal structures lock electrons so firmly in place that they cannot move appreciably, hence cannot conduct current. In other crystals, some slight shifting is possible and temporary concentrations of electrons appear on points or areas which are then "charged" in relation to the rest of the crystal. Development of charged areas arises from distortions of the crystal structure when stress is applied, causing the electrical fields around atoms next to the surfaces to become lopsided, as it were. In turn, this forces electrons closer to the surface on some specific area of the crystal, but draws them deeper within the crystal framework at the opposite end. If the stress is induced by heat, the effect is called *pyroelectricity,* if caused by mechanical pressure, it is termed *piezoelectricity.*

Pyroelectricity is prominently displayed by tourmaline, which quickly attracts a disfiguring coat of charged dust particles as soon as its crystals are heated, sometimes greatly to the annoyance of jewelers and museum curators who must display cut gems of this mineral in showcases made warm by artificial light. Topaz also has this property but not to the degree noted in tourmaline. The lack of internal symmetry in tourmaline has been noted previously and is most obvious in doubly terminated crystals, where one end is usually capped by faces different in many respects from those on the other. Such hemimorphic development is proof of less than perfect symmetry. When it is observed in other species it indicates that these species can also display similar electrical properties, although for practical purposes few compare to tourmaline in strength of charge. The end of the crystal positively charged by heating is known as the *analogous* pole and is usually terminated by lower faces than observed on the opposite *antilogous* negatively-charged pole.

Of more practical value to science and industry is piezoelectricity, because suitably cut sections of piezoelectric crystals can be put into electronic circuits to regulate frequency with great accuracy. It was found many years ago that quartz crystals reacted to pressure or heating by developing opposite charges on alternate prism edges. Quartz is spiral in structure, and merohedral, which is to say that only alternate faces belong to the same crystal form. When a wafer is properly cut from a quartz crystal, it can be squeezed to distort the structure and made to develop opposite charges on opposite sides of the wafer. Where tourmaline shows this property only on ends of the *c* axis, quartz develops opposite charges at the ends of each of the *a* axes. Tourmaline has been used in high pressure gauges, as for measuring breech pressures of large guns, and is more satisfactory for this purpose than quartz. Tourmaline wafers are cut across the *c* axis, but those of quartz, though generally cut to intersect a line from one prism edge to the opposite, are sliced along a number of orientations to provide special electronic characteristics.

RADIOACTIVITY

The large nuclei of uranium and thorium, containing over 200 particles each, spontaneously disintegrate in the process known as *radioactivity.* Two varieties of uranium nuclei, known as *isotopes,* are important in minerals: U^{238} and U^{235}. In abundance the first outnumbers the second by 140 to one. The second, U^{235}, is of great interest outside the purely mineralogical realm because a critical quantity brought quickly together explodes with considerable violence, due to sudden fission. Only one variety of thorium is of interest mineralogically, Th^{232}.

The unwieldly nuclei of uranium and thorium continually disintegrate no matter how or where they happen to be included in minerals. As they do so, three types of radiations, *alpha, beta,* and *gamma,* are emitted. Alpha radiation consists of positively charged helium nuclei; beta radiations are electrons, while gamma radiations are very similar to x-rays. Alphas can be stopped by a few inches of air or several sheets of paper and their effect is strictly local. Betas are absorbed within twelve inches of air travel or by a thin sheet of metal. Gammas, in common with x-rays, are far more penetrating and travel several hundred feet through air or about one foot through rock or through $2\frac{1}{2}$ feet of water. Radiations are easily detected by scintillometers or Geiger counters, portable models of which are available for prospecting. By a long series of disintegrations, involving the formation of many radioactive elements of lesser atomic number in the process, lead is ultimately formed. If a rock containing these elements is analyzed, and lead, known to be formed through their decay, is detected, it is possible to make estimates of geological age by determining the ratio of lead to uranium or thorium.

Aside from their presence in minerals usually containing them, uranium and thorium are of great interest because radiations from them cause coloration of much smoky quartz and other minerals, as well as causing damage to crystal structures. Dark coloration of feldspar and quartz in the form of "haloes," particularly in pegmatite deposits, is commonly traceable to a small mass of a radioactive mineral at the center. Circular brown dots on quartz crystals are also caused by radiation damage. Zircon, monazite, allanite, and other species commonly containing small quantities of thorium are often so disrupted in structure that the previously crystalline material within them is reduced to near-glass, forming the *metamict* crystals previously described. Because radiations affect photographic emulsions, a flat slab containing radioactive minerals will "take its own picture" merely by resting on an unexposed film wrapped in light-proof paper.

7

Specific Gravity

The earth's gravity pulls every substance downward. Those with greater weight proportional to their volume, as rocks and metals, settle beneath soils and water, and the latter beneath atmospheric gases. The arrangement of matter upon our planet is in accordance with this scheme, and points to the fact that minerals are the heaviest of all substances. Although minerals seem nearly of equal weight in pieces of the same size, there are sufficient distinctions among them to make their relative weights most useful in identification. Relative weights are called *specific gravities* or *densities*. The first term, and the one which will be used here, refers to the weight of a mineral compared to an equal volume of water. As an example, if a quart of molten lead is poured into a mold, allowed to cool, and the casting placed on a scale, it will take a little over 11 quarts of water to balance it. Volume for volume, lead is about 11 times heavier, therefore its *specific gravity* is 11. Density means the same but requires that weights be stated as well as volumes, e.g., grams per cubic centimeter, or pounds per cubic foot. As will be seen later, it is not important for the amateur to use such units and hence specific gravity is used in preference. The symbol for specific gravity is G.

When it is remembered how widely the atoms of common elements vary in respect to size and number of subatomic particles, we gain an inkling of why minerals also range widely in gravities, from somewhat above water to as much as nineteen times that of water! Atomic weights rise steadily from hydrogen to uranium, but there are many differences in atomic *sizes* because it is the electron envelope which determines how large an atom is going to be in any particular mineral crystal. Even the same atom varies because of ionization. If it loses an electron it may effectively shrink in size; if it gains one, it grows in size. Many atoms, particularly those with higher atomic numbers indicating more particles in nuclei, do not grow in diameter in proportion to nuclear weight. These are the "heavy" elements and it is understandable why they contribute to rising specific gravity when present in any mineral species; it is not necessarily that many more atoms are packed

179

into the same space, but that those present contain heavier nuclei. Thus it is not surprising to find that small, dense metal atoms impart high gravities to native metals and sulfides. Conversely, silicates are not so heavy because most space in crystals is taken up by large and less dense oxygen atoms.

Because practically all minerals are in the solid crystalline state, atoms must be spaced at distances fixed by the atom sizes, the bonds, and the internal crystal structure. In view of the great variety of atoms and crystal structures known to exist in minerals, variations in gravities are also due to this cause. Of a polymorphous pair of minerals in which exactly the same atom is involved, one species will not weigh the same per unit volume as the other:

SPECIFIC GRAVITIES OF CARBON POLYMORPHS

Diamond —————————— $G = 3.5$
Graphite —————————— $G = 2.1 - 2.2$

The difference between diamond and graphite must be due to atom spacing in the crystal structure because the same atom is involved in both. On the other hand, if the same crystal structure is involved, and only one atom is exchanged, decided differences can occur, as shown in several rhombohedral carbonates:

SPECIFIC GRAVITY OF RHOMBOHEDRAL CARBONATES

Name	Atomic weight of cation	Cation radius(A)	G
Calcite	Calcium —— 40	.99	2.7
Magnesite	Magnesium — 24	.66	3.0
Rhodochrosite	Manganese — 55	.80	3.69
Siderite	Iron ————— 56	.74	3.96
Smithsonite	Zinc ————— 65	.74	4.43

The table above is very instructive because it shows a steady increase in relative weight as atomic weight and ion size decrease—*except for magnesium!* Although weighing much less than the other metals, the magnesium ion is considerably smaller; and, taking less space, the unit cell of magnesite becomes smaller than that of the other species. The result is that the specific gravity of magnesite falls in the order given above. The specific gravities shown are for pure or for nearly pure samples, and as may be expected, whenever isomorphous substitution takes place, as when iron, manganese or zinc substitute for calcium in calcite, specimens show increasing gravities. Despite variations, many species display reasonable constancy in specific gravity, making this property extremely valuable for identification purposes.

THE USE OF WATER IN DETERMINING SPECIFIC GRAVITY

Almost everyone is familiar with the tale of how Archimedes, the celebrated scientist of ancient Syracuse, determined whether his king's new crown was gold or was adulterated with silver, by the use of ordinary water. He hit upon the solution to the problem as he was preparing to take a bath in a brimming tub of water. He noted the water spilled over the moment he put in his foot, and it occurred to him that if he put equal weights of gold and silver into separate brimming containers of water, there *might* be a difference in the amount of overflow from each, and hence a way of telling if the crown were pure gold. His tests, carried on with appropriate drama before the court, clearly showed that an equal weight of gold displaced *less* water than an equal weight of silver. The overflow could be weighed and a proportion between gold weight and water weight obtained. He next immersed the crown and showed that the weight of water displaced by the crown was *not* in the same proportion, but was slightly *more,* indicating that the crown was only part gold and, as the rascally jeweler later confessed, adulterated with silver.

What Archimedes proved was that it was not necessary to cut out pieces of various substances into exactly the same volume shapes in order to obtain specific gravities. All that is needed is to place any solid in water and compare the weight of water it displaces to the weight of the substance. It can be of any size or shape, the only requirement is that it be fully submerged. Archimedes' principle is used today for determining the specific gravity of all types of solids, including minerals.

Reasonably pure water, or "fresh" water, is very constant in specific gravity. When not overly contaminated with soluble salts or other impurities, and when used at ordinary room temperature, it provides a cheap, reliable, and readily available yardstick. Everyone has agreed that the specific gravity of water is to be taken as *one,* and every other substance, including mineral matter, is to be compared to it. Distilled water, used at 4° Centigrade, is standard for accurate laboratory determinations, but for ordinary purposes, tap water at room temperature is good enough.

The arithmetic for determining specific gravity is simple, as the following example shows. Placing a cubic foot of lead in a brimming tub of water results in an overflow of exactly one cubic foot. If we weigh the lead and the displaced water, and divide the water weight into the lead weight, the answer is the specific gravity. This works as follows:

SPECIFIC GRAVITY OF LEAD

Weight of cubic foot of lead ———— 705 pounds
Weight of cubic foot of water ———— 62.4 pounds
Water divided into lead ———————— 11.3, specific gravity of lead.

We can also use the above method for finding the specific gravity of crystals, but for the amateur it becomes extremely awkward to fill a small glass beaker with water *exactly* to the brim, no more or no less, and then try to catch the overflow—whose volume must equal the volume of the specimen. Fortunately, there is a better and surer way of doing the job because Archimedes also found that anything immersed in water, including his king's crown, seemed to *lose* weight directly in proportion to the amount of water displaced. Take the case of the same lead block used before: If we suspend the block from the beam of a scale, it will weigh in air, the same as before, that is, 705 pounds. Now by putting a tub of water beneath the block so that the block is completely submerged, the scale will read a *loss* of weight equal to 62.4 pounds! As can be seen, this is exactly the weight of a cubic foot of water. In other words, the upward "push" of water always causes a specimen immersed in it to apparently lose weight exactly equal to the volume of water displaced. Now we may again calculate specific gravity by the new method as follows:

SPECIFIC GRAVITY OF LEAD
(LOSS OF WEIGHT METHOD)

Weight of cubic foot of lead ———————— 705   pounds
Weight of lead block in water ————————— 642.6 pounds

Difference in weight ———————————————— 62.4 pounds

Difference divided into weight of lead ——— 11.3, specific gravity of lead.

The latter method is the most useful to the amateur but requires a balance to measure weights. Sensitive laboratory balances are expensive but may sometimes be bought cheaply if they are second hand. Reasonably accurate balances, satisfactory for ordinary work, are available at modest cost from a number of mineral dealers and suppliers of mineralogical accessories. However, equally good results can be had on homemade balances for almost no cost; two types will be described. One is the *beam balance,* the other the *Jolly balance.* The first is no more than a light strip of balsa wood balanced on razor blades, while the second uses a spring which must be specially purchased. *Both are so easy to make and so easy to use that every amateur should have one.*

HOMEMADE BEAM BALANCE

Figure 83 shows details of a homemade beam balance constructed from scrap wood, razor blades, screws, and odds and ends likely to be found in a home workshop. Only the flat strip of balsa wood, preferably the "hard" type, must be purchased from a store selling model-making sup-

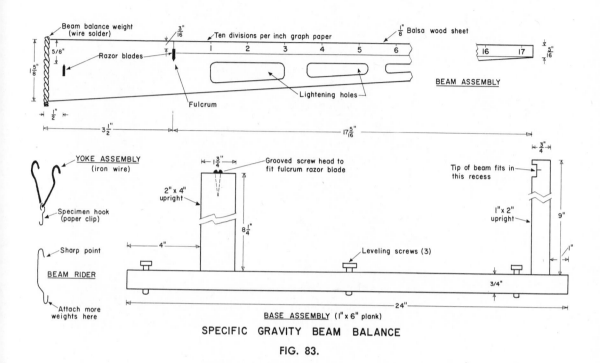

SPECIFIC GRAVITY BEAM BALANCE

FIG. 83.

plies. Because of the lightness of balsa wood, and the fact that no pans are used to hold the specimen, the scale is very sensitive. Consistent accuracies to the first decimal place are to be expected and sometimes reliable readings to the second decimal place. The greatest care is required only in moving the paper clip rider on the balsa wood beam and reading its indications on the scale as accurately as possible. The balance is shown in use in Figure 84.

For purposes of leveling the balance, use three bolts or machine screws, about 2 inches in length, turning them through slightly undersize holes. Spot them two on one side and the third in the back. By screwing up or down while a carpenter's level rests on the base, the balance can be brought to level for greater accuracy in operation. Once it is leveled on any work area, it can be left alone. The balsa wood beam is lightened by cutting holes, as shown in the illustration. Care must be used in inserting and cementing razor blades to be sure that they are exactly at right angles to the balsa wood sheet and are cemented firmly to prevent loosening during use. Slits for insertion of blades are cut with a sharp pointed knife, care being taken not to crush the wood. The upright post supporting the beam is slotted verti-

FIG. 84. Homemade specific gravity beam balance in use. A specimen suspended with thread is immersed in a beaker of water at the left. The author is adjusting the position of the paper-clip rider to achieve balance. The point of the paper clip is near the forward edge of the balsa wood beam, upon which is pasted a strip of ten-division graph paper, suitably calibrated, and from which the reading of weight-in-water is to be obtained.

cally; the slot must be wide enough to clear the beam but not so wide that the razor blade knife edge cannot straddle it. On each side of the slot insert two round-head wood screws. Using a three cornered file, enlarge the screw head grooves until they form continuous vees. Shape a piece of hard wood to the same profile and polish the grooves with jeweler's rouge. This prevents the razor blade from "hanging up" because of engaging its sharp edge in a file mark. When finished, the screw grooves are in line and the razor blade supporting the beam fits nicely in them, allowing the beam to tip back and forth in a very sensitive fashion.

Readings along the beam are obtained from a strip of ten-division-per-inch graph paper pasted to the balsa wood as shown. The strip begins at the balance point razor blade, or fulcrum, and extends toward the narrow end of the beam. Inch divisions are marked 10, 20, 30, 40, etc. Readings are obtained from shifting the position of the rider, a bent-out paper clip with one of its ends cut off to a sharp point with diagonal cutters. This sharp point is placed as close to the graph as possible when the balance is being used, to provide most accurate readings between graph paper divisions. It is *not slid along* but is *picked up and moved* from place to place until its weight balances that of the test specimen. *Because it rests upon a point instead of straddling the beam, much more accurate readings are possible.* Other paper clips, small pieces of copper or lead wire, etc., can be added to the rider hook to balance the weight of larger test specimens. Although larger specimens provide greater accuracy, it is best to limit size to no more than one inch in diameter for the particular balance described to prevent overloading. Conversely, specimens should not be less than about one half inch.

After the beam is made and the graph paper cemented on (see Figure 83) and allowed to dry thoroughly, the beam is placed in position and balanced by cementing soldering wire to its short end. The yoke assembly for suspension of specimens should also be in position at the same time. A small piece of paper-clip wire is used for making the final adjustment in addition to the solder-wire weight. This is necessary because the wood absorbs moisture or dries out, depending on atmospheric humidity, and some way is needed to make it balance correctly before testing is begun. When perfect balance is achieved, mark the stop post opposite the end of the beam. This serves as a reference when weighing specimens. The stop is very important because it keeps the beam from swinging wildly and damaging itself.

Operating the Beam Balance. Have ready a glass beaker or tumbler filled nearly full with water which has been allowed to stand for at least ten minutes to permit escape of trapped gases and to reach room temperature. Gases escaping during weighing form bubbles which cling to the specimen, buoying it up and giving it greater lightness than is really the case. If it isn't too much trouble, the water should be boiled to remove gases and then allowed to cool. A very small pinch of ordinary household detergent is always added to improve the "wettability" of the water and to prevent attachment of bubbles to the specimen as it is lowered into the beaker. Also have ready a pair of scissors, a pencil and some scratch paper. The first step is to fasten the specimen and hook it to the short end of the balance. Specimens are suspended by very thin nylon thread (Size 000) and the entire spool of thread is dipped beforehand in melted beeswax or paraffin to waterproof the thread and make it easier to tie. About 6 to 8 inches of thread are clipped from the spool, and a slip-knot noose made at each end; one passes around the specimen, the other around the small hook beneath the yoke. The noose around the specimen is carefully drawn tight and tag ends snipped off. The specimen is then gently lowered until its weight is borne by the beam. The paper-clip rider is placed at some point along the beam to achieve balance.

If it is run all the way to the end without striking a balance, additional paper clips are hung on its lower end and its position adjusted. It is desirable that balance always be achieved with the rider and its weight as far out on the beam as possible. When the beam stops oscillating and balance marks on the end match, read off the weight of the specimen in air opposite the sharp end of the paper-clip rider. Estimate the distance between ink marks on the graph paper as closely as possible. Let us say that the first reading lies between 14 and 15, and between the third and fourth ink marks on the graph paper. By estimating the distance between marks, it is possible to arrive at a reading such as 14.35. Ignore the decimal and make the reading 1435. This value is written down and marked "weight in air."

Cautiously place the water-filled beaker under the specimen until the latter is completely submerged; slip a block of wood under the beaker to keep it in proper position. Note that as soon as the specimen touches the water, balance is lost and the beam tips toward its narrow end. Run the slider *back* toward the fulcrum until balance is again achieved. Let us say that the reading is now 868. Write this under the previous reading and mark it "weight in water." Subtract the two readings and divide the difference into the weight in air number as follows:

BEAM BALANCE SPECIFIC GRAVITY DETERMINATION

weight in air ———————— 1435
weight in water ———————— 868

difference ——— 567 (weight of displaced water)

Divide 567 into 1435 as follows:

$$567\,\overline{)1435} = 2.51 \text{ Specific Gravity (G)}$$

Note that in obtaining readings, numbers on the graph paper mean absolutely nothing as far as *true* weight is concerned, and indeed they do not have to because all we are interested in is the weight of the specimen *relative to water*. This is what we have obtained. If one wishes, various gram weights can be used to calibrate the beam to provide an accurate weighing device for weighing small gemstones or bits of precious metal. The beam readings obtained are marked on paper opposite their gram or ounce weights. Thus if the weight of any specimen is desired, the appropriate riders are used and the beam reading is entered in the table to read off weight in ounces or grams.

HOMEMADE JOLLY BALANCE

The Jolly balance, named after its inventor, the German physicist P. von Jolly (1809-1884), uses the principle that a vertically-hung spring stretches proportionally to weights attached to its lower end. Thus in the homemade Jolly balance shown in Figure 85, a specimen attached by thread to the lower hook of the spring carries it downward a certain distance which can be measured. If the specimen is immersed in water, the spring will contract to a shorter distance which can also be measured. The difference is due to the buoyant effect of water and, as before, provides a means for quickly determining specific gravity.

The heart of any Jolly balance is the spring and the better the spring, the better the results. A spring can be made from piano wire but it is a great

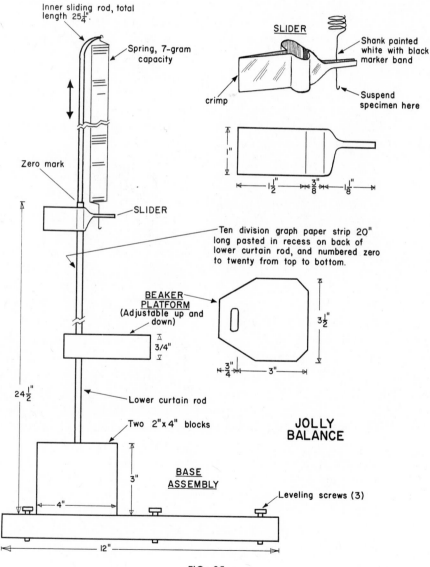

FIG. 85.

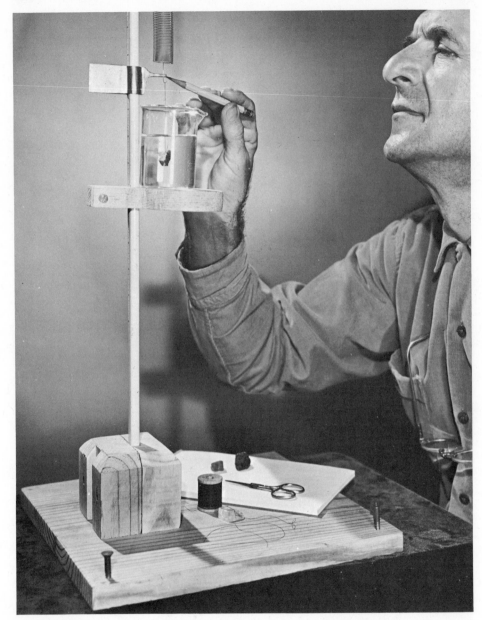

FIG. 86. Jolly balance in use. A specimen is suspended from thread at the lower end of the tension spring and is immersed in water for the weight-in-water reading. The author is checking the metal slider mark against the mark on the spring to see if marks match. The reading is taken from the strip of marked graph paper seen on the back side of the metal curtain rod.

deal of trouble and hardly worthwhile when a perfectly good one, such as shown in Figure 86, can be bought from suppliers of scientific apparatus for several dollars. Springs are available in various sizes but the one used by the author is a 7-gram capacity spring, excellent for weighing the majority of specimens likely to be encountered. Because the spring tends to unwind and spin the specimen as it stretches, the makers put a reverse loop in the center. Thus, as the spring stretches, half unwinds one way but is exactly opposed by the unwinding of the other half. The result is that the suspended specimen stays still.

The spring support is a telescoping metal curtain rod, one part fitting snugly into the other and expansible to various heights. The inner sliding part is drilled at the top with a small hole to support the spring; the lower part is fastened securely in a block of wood on the base. The vertical rod measurements used in Figure 85 have been calculated to prevent over-stretching of the spring, that is, if a specimen is *too heavy* for the 7-gram spring, the spring will stretch to the wooden beaker platform and no farther, preventing damage to the spring. The back of the lower curtain rod has a narrow recess and in this is cemented a long strip of marked ten-division graph paper, zero at the top, and numbered toward the bottom in the same manner as used in the beam balance. The slider shown at the top of the rod is made from a piece of sheet metal, cut out in the form shown, and crimped to make it grip the rod without slipping. Two arms come together in front to provide a slot in which the lower shank of the spring rides and to provide a means of reading the downward positions of the spring after it stretches. The upper edge of the slider at the back is used to read the graph paper markings. The beaker platform is of wood, cut to fit loosely around the lower curtain rod. When a beaker is placed on the platform, the weight jams it and prevents sliding. If preferred, a set screw can be put into the side to lock it in place. The platform is leveled by three bolts or machine screws as in the beam balance, except that the object here is to be sure that the spring rides parallel to the curtain rod and that the small lower shank of the spring does not rest against the slider but swings freely inside it. When not in use, the spring should be kept stored in the cardboard box in which it was mailed.

Operating the Jolly Balance. As before, set out some water ahead of time to have it ready for weighing. Install the spring at the top of the upper curtain rod and extend the rod until the small ink mark on the bottom shank of the spring is exactly at the edge of the slider arms. The slider must also be exactly on the zero mark at the top of the lower curtain rod. Because the spring flops about with every vibration or puff of air, try to work on a solid table and away from drafts. Patience is required for this initial but important step. Once the upper rod has been adjusted, it need not be readjusted unless some hours have elapsed or the room temperature has changed markedly. The spring is sensitive to temperature changes and when these occur, its initial calibration should be checked.

Next, suspend a specimen with thread as before, and allow to rest gently against the slider. Lower the slider slowly until the spring begins to support the specimen. Continue lowering until it is fully supported and the ink mark on the lower spring shank is again exactly opposite the edge of the slider arms. It doesn't make any difference whether the upper or lower edge of the slider arm is used for marking the position of the spring, but whichever it is, it should be used consistantly to avoid errors. When the specimen and spring have ceased motion, and the ink mark is exactly opposite the slider arm edge selected, turn to the back of the lower curtain rod and read off the distance at the slider edge. As before, record this as "weight in air" on a slip of paper.

Place the beaker of water on the wood platform below the specimen; immerse the specimen fully. Again the spring will contract because the specimen has apparently lost weight. Adjust the slider upward until the spring is taking the weight of the fully submerged specimen, then re-adjust until the arm edge is exactly opposite the mark on the spring shank. Again turn to the back of the rod and read off the distance on the graph paper scale. Record this as "weight in water" and, as before, find the specific gravity.

Insofar as accuracy is concerned, there is little to choose between the two balances described. The beam balance is considerably easier to operate but the Jolly balance is slightly more accurate. On professional Jolly balances, care is taken to allow smooth, easy screw height adjustments to initially calibrate spring position, while accurate means for reading spring positions are also provided. The homemade balance, without such features, requires care to be sure that consistent readings are taken. In any event, both devices are so easy to make and so inexpensive, that every amateur is strongly urged to provide himself with one and gain confidence in its use. Practice with clear quartz or calcite specimens because these species are almost constant in specific gravity and any error in determination must be laid to the instrument or technique used rather than to any unusual property of the mineral itself.

USE OF HEAVY LIQUIDS

It is sometimes convenient to use heavy liquids for approximately determining specific gravity, particularly of samples so small that they cannot be weighed with reasonable accuracy on balances. The following liquids are most useful:

Bromoform .... G = 2.86-2.90 (last figure is for
pure liquid)
Acetylene tetrabromide (tetrabromoethane) G = 2.96
Methylene iodide (very expensive) .... G = 3.3

Bromoform is a sweet-smelling, clear fluid which decomposes and darkens if exposed to daylight. It should be kept in a brown bottle away from light with a few pieces of copper placed in the fluid to keep it clear. It has been used as an anesthetic and should not be inhaled, nor should the other fluids mentioned. Acetylene tetrabromide is also a clear liquid with a peculiar odor reminiscent of moldy vegetation. It is preferable to bromoform in every respect. Methylene iodide is straw yellow in color with a decided iodine odor; it also decomposes in light and requires the same storage precautions as bromoform. Liquids are preferably kept in small wide-mouth brown glass bottles and poured when needed into small colorless glass beakers or tumblers where behavior of test fragments is easily observed. The above liquids can be obtained from druggists on special order or from dealers in geological or mineralogical supplies.

Heavy liquids are only approximate indicators of specific gravity unless portions are carefully diluted to provide a large range of liquids of varying densities. This is usually not practical for amateur purposes because considerable quantities would be required and each liquid must be tested separately to determine specific gravity. However, a useful liquid for indicating quartz is made by diluting bromoform or acetylene tetrabromide with acetone or xylene. Acetone is obtained at paint shops. A small piece of quartz, preferably colorless and flawless, is used as the indicator. It floats when placed in either fluid because its specific gravity is considerably less, that is, quartz G = 2.65, bromoform G = 2.86-2.90, and acetylene tetrabromide G = 2.96. If acetone or xylene is added drop by drop to a small portion of either fluid and the solution stirred to insure mixing, a point will be reached where the quartz barely floats. The jar containing this liquid is then marked and kept separately. Whenever very small mineral fragments thought to be quartz are placed in the jar, it is easy to tell if they are of nearly the same specific gravity by noting their behavior. Those which are heavier sink; those lighter, float. It is useful to note if a test fragment sinks slowly; if so, it cannot be much denser than quartz. Conversely, less dense minerals, of which there are many, will bob to the surface, quickly distinguishing themselves by their buoyancy.

With heavy liquids, it is possible to aid in identification by eliminating many species which from other properties, seem likely suspects. Whenever larger specimens are obtainable, the balance should always be used in preference to liquids.

SELECTION OF TEST SPECIMENS

Regardless of method used, it is very important that suitable specimens be selected for testing to avoid misleading results. The ideal specimen is that which is so large that it nearly taxes the capacity of the balance. Smaller specimens provide more room for error. The best are also as free as possible from cracks, flaws, inclusions, altered areas, and other defects. Cracks are specially to be avoided because they contain air which cannot be easily removed. When a cracked specimen must be used, it is weighed in air first, then placed in water and heated gently for some minutes to drive off the trapped air. The container is then allowed to cool to room temperature after which the specimen is quickly fastened with thread and weighed in water.

APPLICATIONS OF RESULTS

Assuming accurate determinations of specific gravity have been made, how shall they be used? As mentioned earlier, few minerals are constant in specific gravity because substitution of one or more elements for others so often takes place. If substituting atoms are heavier or lighter, the specific gravity also changes although the mineral remains classed in the same species. For example, sphalerite, ideally ZnS, seldom is without iron substituting for zinc. The range of G is therefore from 3.9 for zinc-rich varieties to 4.1 for iron-rich varieties.

Even larger variations are noted for minerals in a series, as in the olivine group where one end member, forsterite (Mg_2SiO_4), yields $G = 3.22$, while the other end member, fayalite (Fe_2SiO_4), yields $G = 4.39$. Ordinary greenish olivines yield $G = 3.3$-3.4. Similar wide variations are noted in many species and call for judgment in application of specific gravity values to the problem of identification. As with so many other mineral properties, the determination of a single value is not enough by itself to identify, but when coupled with others, it may be. Of all properties that the amateur can determine, however, specific gravity is one of the most useful and therefore should be taken whenever possible. A convenient listing of minerals according to specific gravity is provided in Table 6, Appendix A.

8

Optical Properties

Light does not pass freely through the atomic frameworks of crystals, but must conform to the rules imposed by the frameworks themselves. If we discover how light is affected by its passage, we can also learn much about the inner nature of crystals. In turn, this helps to classify any crystal and, in many instances, leads to quick identification of the species to which the crystal belongs. The influences on light are called the *optical properties* of the mineral, and it is the purpose of this chapter to explain some of them and point out how they can be used to advantage in identification. First it is necessary to examine the nature of light because how it behaves in any crystal depends as much upon its own nature as upon that of the crystal.

THE NATURE OF LIGHT

Light is intimately related to matter because it is created by the rapid vibrations of electrons. The exact nature of light is still not fully understood, some behavior suggesting that it is a wave motion, somewhat like the ripples on a pond which spread outward from where a stone has been dropped in, but other behavior suggesting that it consists of extremely small pulses of energy shot out like bullets from some tremendously fast machine gun. The theory of short bursts of energy is the *quantum theory,* the word "quantum" meaning packet of energy; in the case of light, this packet is the *photon.* The previous theory is called the *wave theory.* Both are now believed to be compatible and equally useful in explaining the behavior of light and related radiations caused by the vibration of atomic particles.

It is known that a tremendous variety of radiations and waves come from matter when enough energy is supplied to make atoms and electrons vibrate. Radio waves are caused by the back-and-forth rush of electrons through wires of a transmitting antenna, setting up electromagnetic waves which pass through the atmosphere and are received by the antennas of radio sets.

If the electrons are speeded up and made to oscillate even faster, the waves become shorter between crests, which is to say, shorter in wavelength, and their properties change. They no longer tend to flow around obstacles but travel in straight lines, severely limiting their range. Such waves are used for TV sets and explain why it is necessary to clutter the roofs of houses with special antennas, turned to special directions. Beyond these waves are extremely short radio waves used in radar, which are less easily deflected allowing them to be used in "beams." As waves grow shorter in length, a region just below visible light is approached where the waves no longer require elaborate electronic apparatus to create them; all one has to do to make them is to heat almost any substance, as a piece of metal for example. These are the "heat rays" or *infra-red* rays, familiar to us as those imparting the sensation of warmth even when we sense them from some distance away as from a fire, charcoal burner, or electric stove. By holding a hand before our face, we can instantly cut off the rays thus showing that they are very similar to light in their behavior. If more energy is supplied to the substance as by turning up a heating element, the first of *visible light* radiation appears—a dark red. More energy changes the color from dark red to light red to orange, yellow, and finally to a peculiar blue to violet light, such as seen in the arcs of electric welding torches.

Past this violet light, our eyes are no longer capable of sensing the radiations, but they exist nevertheless. In this region occur the shorter-than-light wavelengths of *ultra-violet,* followed by even shorter wavelength radiations which begin to behave more like streams of particles than waves. Here are found the highly penetrating x-rays, which pass so easily through flesh and even through minerals and metals, while further beyond occur the tremendously powerful cosmic ray particles which are known to pass easily through many feet of concrete.

This gamut of radiations, in the form of waves at one end, changing imperceptibly into particle or photon streams at the other, is called the *electromagnetic spectrum.* The waves at the extreme lower end measure many feet between crests; in the light region, they are very short and must be measured in correspondingly small units as angstroms (A), each of which is only one one-hundred-millionth of a centimeter. Beyond visible light, still other units must be used because the particle streams are quite unlike waves. Out of this entire spectrum, visible light occupies only a very small band.

THE LIGHT SPECTRUM

The easiest way to create visible light is to heat a suitable substance to the point where its atoms vibrate so vigorously that they impart energy to electrons, making them vibrate too. This is the principle of the electric light which merely feeds a stream of electrons into a thin tungsten wire, setting up a great internal agitation causing heat generation. However, in a solid or a liquid, atoms and electrons are constrained in their vibrations by neighbors and are not perfectly free to do as they please. Even when the tungsten wire is heated to incandescence, and the tungsten atoms are spread apart considerably, they remain in their original patterns. Thus the radiations which issue from the electrons are a complex jumble of all wavelengths which, when striking the eye, impart the sensation of "white" light. The same happens in the sun where all kinds of atoms are heated to tremendous temperatures, some even being torn apart and put back together in a process that has been going on for millions of years.

Thus complex mixtures of light wavelengths are sensed as white light, but narrow bands of only a limited number of wavelengths are sensed as color. The visible light spectrum, shown in Figure 87, is a smooth progression of hues ranging from the longest wavelengths at the red end to the shortest at the violet end. To prove this, individual waves, or *monochromatic light,* can be obtained from white light by repeating Sir Isaac Newton's (1642-1727) classical experiment using a triangular polished glass prism. If a small beam of white light, as from the sun or from a powerful slide projector lamp, is allowed to fall on the prism, the light waves separate or *disperse,* yielding an image with the same colors, in the same order as shown in Figure 87. More will be said about dispersion a little later, but for the moment we will be content with the fact that light rays of mixed wavelengths can be separated in this manner. Because the light is spread into a spectrum, the technique for doing so is called *spectroscopy,* and the prism is a simple *spectroscope.*

The value of a spectroscope is that it enables us to take what seems to be a hopelessly scrambled mixture of light radiations and sort them out for study. The importance of this study lies in the fact that certain atoms (heated to incandescence), when floating free and therefore not bothered by interference from neighbors, emit characteristic light wavelengths from their electrons. In fact, the wavelengths are unique to the elements concerned, and thus it becomes a simple matter to identify elements if a means can be found to heat some of the atoms in the free state. In practice, this is done by placing a compound containing the element in a little cup in one of the carbon rods of an arc lamp, turning on the current, and observing the appearance of the light after it passes through the spectroscope. A very pretty sight greets the eye because most of the field of view is black but interrupted here and there by brilliant narrow colored bands or *bright line spectra* of the elements present in the carbon arc cup. The spectroscopist usually takes a photograph of the display, then compares the lines to calibrated charts and unhesitatingly reports which elements are present. If iron

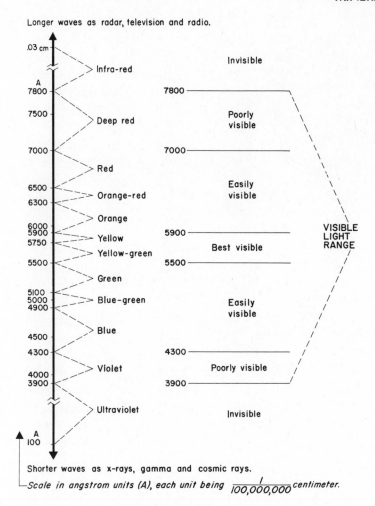

PORTION OF ELECTROMAGNETIC SPECTRUM

FIG. 87.

is present, for example, bright lines will be found only in certain positions on the spectrum scale, each line always at the same wavelength, as shown in Figure 88. It then becomes a question of matching these "spectral finger-prints" against those on file.

We now begin to see the origin of white light from our incandescent sun because with its internal temperatures of many millions of degrees, surely all elements are emitting light and other radiations at a furious rate. Because radiations overlap each other, the light as a whole impresses us as being white. However, refinements in the spectroscope some time after Newton's day, led to a curious discovery about the sun's light by a German physicist, Joseph von Fraunhofer (1787-1826). He found that its spectrum was not the continuous smooth band of colors previously supposed, but actually was interrupted by numerous narrow dark lines. Later work by other physicists soon proved that elements in the free state could not only emit characteristic light radiations, but could also absorb them, and that the dark lines noted in the sun's otherwise continuous spectrum were due to elements in the sun's outer brilliantly flaming envelope of gases, absorbing the light emitted by their fellows in the central portion of the sun. Thus the dark lines, called *Fraunhofer lines,* or *dark line spectra,* are the exact reverse of bright line spectra, but due always to precisely the same elements. Figure 88 shows not only Fraunhofer's lines, but also the bright line spectra of iron, showing how the latter exactly match dark line sodium spectra in the sun's light. Depending on how they are caused, the spectra just described are also called *emission spectra* and *absorption spectra.*

Spectroscopy is of value in mineralogy if the necessary apparatus is at hand. Gemologists are increasingly putting to use small hand spectroscopes through which the light from a transparent gem is observed. In some instances, excellent absorption spectra, as in the zircon example shown in Figure 88, can be observed, but many gemstones show nothing at all. Absorption occurs in much the same manner as noted in the gaseous envelope of the sun, that is, the electrons around certain atoms are partially free of restraint and can move to absorb certain wavelengths of light. However, their freedom is far more restricted in the majority of instances and thus most transparent crystals observed through the spectroscope show nothing but the smooth continuous spectrum of ordinary white light. General absorption of light over many wavelengths is far more common in crystals. Because it gives rise to color, it is of much greater interest, as the next section will explain.

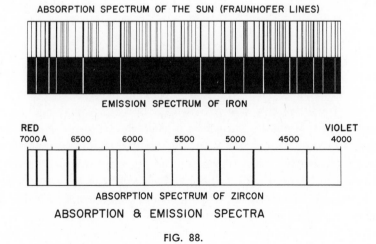

ABSORPTION SPECTRUM OF THE SUN (FRAUNHOFER LINES)

EMISSION SPECTRUM OF IRON

RED 7000 A 6500 6000 5500 5000 4500 VIOLET 4000

ABSORPTION SPECTRUM OF ZIRCON

ABSORPTION & EMISSION SPECTRA

FIG. 88.

ABSORPTION OF LIGHT IN CRYSTALS

Because light is an electromagnetic radiation and we know that atoms are fastened in crystals by intermeshed electrical and magnetic fields, the flood of light falling on any crystal results in an interaction between electromagnetic fields of light on the one hand and of the crystal on the other. Light does not pass *between* atoms, like raindrops working their way through tree leaves, but sets the crystal fields into vibration. If we imagined any crystal to be composed of numerous quivering gossamer soap bubbles, with each atom composed of one or more bubbles blown inside each other to correspond to electron shells, then the light photons can also be compared to much smaller soap bubbles hurled at tremendous speed at the interlocked structure, causing vibrations which spread according to the kind of bubbles and their distribution in the framework. If crystal fields are firmly intermeshed with strong bonds between them, it is quite possible that photons striking one side will send vibrations which emerge on the other not much changed in character. Our visual impression of such a crystal may very well be that it is "white" or "colorless." On the other hand, if certain fields in the crystal are free to vibrate and absorb light energy partly or wholly, the crystal may appear "colored" or even "black." As pointed out in the case of absorption spectra, it is the freedom of electron fields to interact with light fields which results in the maximum effects upon light. Thus light may pass into a crystal and be regenerated on the other side, unchanged in character, or it may be *selectively absorbed*. It is such partial absorption which gives rise to sensations of color because some wavelengths are removed but others allowed to pass, and when received by the eye, are interpreted as *color*.

SPECIFIC CAUSES OF COLOR

Many minerals are inherently colorless but commonly occur in several hues, for reasons which will be explained shortly. The fact that a number of colors are listed in descriptions shows that no *single* hue is inherent. Such species are *allochromatic,* from *allo,* meaning "other than usual." Species in which a certain hue is essential are termed *idiochromatic,* from *idio,* meaning "inherent." The true color of these species is sometimes masked or modified for various reasons, but is still as essential a part of the mineral as its own chemical composition. Despite the very wide variety of hues, as among those belonging to allochromatic minerals, definite color relationships can be pointed out which are fairly consistent and very helpful in making tentative identifications. As with any clue to identity, information provided by color must be checked against other bits of information before a positive identification is made. In a few instances, a specific hue strongly suggests the presence of a certain coloring agent which, when known to be present in only a limited number of species, whittles down the list of suspects considerably. Color and related properties of selected species are listed in Table 1, Appendix A.

Color characteristics of allochromatic species are often apparent at a glance. In quartz, for example, one may find amethyst crystals containing smoky or yellow parts among the normal violet colored areas, or even colorless portions as is commonly the case. On the other hand, idiochromatic species, of which malachite and azurite are beautiful examples, possess only one basic hue each. Sometimes the normal dark green of malachite is banded with light shades, but these are due to porous sections rather than to any real change in the basic hue. Similarly, azurite crystals are uniformly dark blue from tip to tip while masses may vary from pale to dark blue depending on porosity.

Color is also caused by inclusions, occasionally so fine that magnification must be used to see them. Minute scales of vivid red hematite commonly deposit in thin films on quartz and calcite crystals, and may be subsequently overcoated by more material disguising their true nature. Fortunately, most specimens of this sort can be readily identified either by direct observation or by looking for the very thin color bands which such inclusions tend to form.

Aside from coloring caused by mechanical inclusions, as the hematite mentioned above, by far the greatest portion of coloration is imparted by color-causing ions or *chromophores*. The term *color center* is also used to generally describe points within crystals from which color emanates regardless of how it is caused. In most instances, coloring ions are included in crystal structures as substitutes for other ions, the intensity and quality of color being then a good indication of the extent of substitution. Nearly pure sphalerite, ZnS, for example, is almost colorless or tinged very pale yellow or green. When iron substitutes for zinc, the color darkens to deep red, to reddish-brown, to nearly black, as iron content rises. Iron, plus a number of closely related elements, called the *transition elements,* are responsible for much color observed in both allochromatic and idiochromatic minerals.

TRANSITION ELEMENTS

The transition elements range from scandium, atomic number 21, to copper, atomic number 29. However, the great rarity of scandium does not merit placing it in this discussion and it is omitted from the following table. All transition elements are characterized by peculiar configuration of electron shells in which a fourth and outermost shell contains one to two electrons, but the third shell contains from nine to eighteen electrons. Because the ultimate capacity of the third shell is eighteen, it is reasonable to suppose that electrons added as new atoms are formed, should fill this shell to capacity. This is not the case; apparently nuclear attractive forces tending to draw electrons closer are not very powerful past the second shell, allowing electrons in the third shell to be readily lost. This accounts not only for the variable valences of these elements, but also for their coloring properties which arise from the ease with which the weakly attracted electrons vibrate under the impulses of light and absorb much of its energy. The exact wavelengths destroyed depend upon the environment in which the transition ion finds itself, and the final color lent to minerals varies considerably from species to species, and even within any species. In the event that two transitional elements are present in the same crystal, the effect of one influences the other. Further, it has been found that strongest coloration occurs when the same transition element is present in two valence states, as for example, when iron is present as Fe^{+2} and Fe^{+3}. Transition elements color strongly no matter where they find themselves, whether it be in glasses or in crystals, but their greatest influence is in crystals. An extremely small number of coloring ions is often enough to impart intense coloration. In some analyses, the quantity appears negligible yet is enough to color the mineral very noticeably and, at the same time, it scarcely affects density or other measurable properties.

COLORING PROPERTIES OF TRANSITIONAL ELEMENTS

Element, Atomic No.	Electron Configuration	Color Examples
Titanium, 22	2, 8, 10, 2	Blue—corundum, benitoite
Vanadium, 23	2, 8, 11, 2	Various
Chromium, 24	2, 8, 13, 1	Green, red—corundum (ruby), beryl (emerald)
Manganese, 25	2, 8, 13, 2	Pink, red—rhodonite, rhodochrosite
Iron, 26	2, 8, 14, 2	Green, red, brown, etc.—many species
Cobalt, 27	2, 8, 15, 2	Pink—erythrite; blue—glasses
Nickel, 28	2, 8, 16, 2	Green—garnierite
Copper, 29	2, 8, 18, 1	Green, blue—many species

COLORING AND CRYSTAL STRUCTURE

In general, as any crystal structure approaches simplicity and symmetry in respect to atomic spacing and bonding, light is less likely to be absorbed. In isometric minerals, as fluorite, halite, and others where very symmetrical arrangements are the rule, minerals tend to be colorless. However, pyrite is also isometric but is noted for its brassy metallic color;

the less than perfect arrangement of ions within its unit cell, as noted in a previous chapter, may be the cause here for strong coloration. In this instance, the large sulfur ion field is believed to interact easily with light fields, and there is also the strong coloring behavior of iron. Sulfur is a *latent color ion,* that is, it may influence light greatly when present in certain atomic structures but does not necessarily in all of them. Thus in sulfide minerals it has important coloring effects, but in sulfates it is too firmly bound with oxygens to have much influence, and many sulfates are pale in color or even colorless. Other latent color ions are arsenic, antimony, cadmium and iodine, and minerals containing them are often vividly colored. Elements resisting distortion of fields are sometimes called "colorless" ions, and include aluminum, barium, calcium, potassium, sodium and lithium. Their compounds tend to be colorless. Minerals in systems other than isometric can also be colorless, as for example, quartz, which in the vast majority of instances is colorless. It is largely a very pure species, but when it is colored, the explanation lies in the presence of foreign ions, substitutional ions, or defects in crystal structure.

Foreign ions can distort electrical fields in any crystal. They may be accidental inclusions during crystal growth, jammed between normal ions and distorting fields in their vicinity, or they may substitute for another element, often causing marked distortion because they are not exactly the same size. They may also be located along defects, as spiral dislocations or twin boundaries. "Extra" ions of any kind not only cause distortion in a physical sense but also upset the electrical balance of the structure, and this means imbalance in electrical fields and the likelihood of great influence upon light.

EXTERNAL INFLUENCES ON COLOR

Many crystals fade or change color when heated or exposed to light or other electromagnetic radiations. Sometimes the change reverses damage done while the mineral was within the ground, as when a crystal near a radioactive mineral is subjected to bombardment by fission particles. In some instances, original atoms are destroyed with formation of new elements, but mostly atoms and electrons are knocked out of normal positions and "hang up" until a disturbance sets them free to rejoin whatever group they were part of. Thus the smoky color of quartz caused by radiation is driven off when specimens are heated. In many instances, it can be restored by re-irradiation with x-rays. In pink-orange beryl or pale violet spodumene, exposure to daylight imparts enough energy to restore original internal conditions and beryl turns pink, its stable color, while spodumene turns colorless, also stable.

Heat is much used to change or drive off color in gemstones. Its effect may be simply to make all atoms vibrate faster and create more room in the framework for displaced particles to return to their original sites. Sometimes deliberate introduction of ions along with heat cures defective color by restoring normal balances between the number of cations and anions. This is done in the case of synthetic rutile boules which lose oxygens during flame-fusion, resulting in a surplus of titaniums. The absence of counterbalancing oxygens causes boules to be very dark in color. To restore oxygen, boules are heated almost to melting point in an oxygen atmosphere, at which time oxygens enter and migrate to sites where titaniums are in excess. After treatment, the color is very pale yellow.

Heat treatment is a very simple way to improve color in some gemstones but it cannot change color caused by atoms properly belonging in the crystal structure. In peridot, where increasing iron causes disagreeable brownish shades, no degree of heating short of melting will make iron atoms shift position or change their color-causing propensities. Heating results only in increased agitation of all atoms which, however, return to normal positions after heat is removed. It is interesting to note that during heating, color often changes with rise in temperature, usually for the better, but as the sample cools, the original undesirable hue returns.

STREAK

A useful color clue to identification is to scrape an unknown mineral upon a gritty white surface, as an unglazed porcelain bathroom tile, and note the color produced by the *streak*. Because the mineral is finely subdivided by such abrasion, many seemingly dark to black minerals prove to be lighter or actually different in hue. For example, hematite crystals appear quite black, but when rubbed across a porcelain streak plate, the characteristic deep red trace shows the true color. By this simple test, hematite is distinguished from minerals similar in appearance and color. The same result can be obtained by pounding a small bit of any mineral to powder and pouring it on a piece of white paper to check the color. Streak is most useful for dark to strongly colored minerals as sulfides, and less useful for many silicates, carbonates, and others, which produce white powders although they seem strongly colored in large pieces. Streak colors of selected species are listed in Table 5, Appendix A.

PHOTOLUMINESCENCE

The production of visible light by any means whatsoever is called *luminescence*. When produced by heat, as in incandescent electric light filaments, it is *thermoluminescence. Photoluminescence* is the process of creating visible light by bombardment with *photons,* hence the name. No real difference exists in light created by either means, except that thermoluminescence is usually accompanied by a great deal of heat and production of infra-red radiations. On the other hand, photoluminescence is generally free of heat and infra-red, and is therefore commonly called "cold light." This is the light given off by modern tube-lights and advertising or "neon" signs. It is also the light given off by certain minerals when flooded with ultra violet light. An older and still useful term, *fluorescence,* means the same as photoluminescence, and is derived from the mineral fluorite which spectacularly displays the effect.

Enough color and variety is displayed by fluorescent minerals to make them objects of special collections by amateur mineralogists, some confining their efforts to this field. Fluorescence is used by prospectors equipped with portable ultra-violet sources, or "black lights," to discover minerals fluorescing during darkness. Some minerals may be identified from their characteristic fluorescence, and for this reason, many amateurs obtain small ultra-violet lamps for use in home laboratories. The greater number of species do not fluoresce.

The production of visible light by incandescent filaments, as in ordinary electric lights, depends simply on putting enough electrical energy into the wires to create heat and make electrons emit light pulses. At best, it is a "shotgun" approach, wasteful of current. On the other hand, the photoluminescence process is more selective because the input photons strike electron fields more or less directly, and excite them to high enough energy levels to cause emission of light. For this reason, photoluminescent lamps are less wasteful of current than ordinary incandescent lamps.

According to Stokes' Law, named after its discoverer, all photoluminescent light must be of longer wave wavelengths than that of the light used to excite it. If *visible* light is used to excite electrons, photoluminescence will be in the infra-red and therefore *not visible.* On the other hand, if visible light is wanted, then shorter wavelengths must be used. This explains why the shorter wavelengths of ultra-violet light are the most popular means used to produce fluorescence. Being invisible itself, ultra-violet light does not distract the eye and permits excited minerals to glow in their true colors. Ultra-violet lamps produce their radiations in a combination process which will now be explained.

As pointed out before, atoms in gaseous state emit *line* spectra instead of a flood of radiations over the entire light spectrum. If an element can be found which emits powerful radiations in lines or narrow wavelength bands in the ultra-violet light region, this can be used for a source, after the visible light portions are screened off. Mercury proves to be most suitable because it is readily vaporized in sealed tubes, the vapors easily passing current which serves to intensely excite the electron fields around the mercury atoms. Several narrow bands in the ultra-violet are emitted; one centered at 3650 A is employed in the long wave lamp, while another at 2537 A is used in the short wave lamp. Both lamps are fitted with very dark purple glass filters to screen off practically all visible light but allowing ultra-violet to pass.

Mechanism of Photoluminescence. Although fluorescent minerals appear to glow immediately when irradiated with ultra-violet, the production of visible light is a roundabout process taking some time. Ultra-violet photons begin the process by imparting energy to movable electrons. Some are raised to orbits of slightly greater diameter, others may be knocked out of orbit and lodged in positions away from the atoms to which they belong. As soon as possible, each electron attempts to return to its normal position; those which do so immediately produce a very small pulse of light energy as they drop back into position. Those which delay in returning may fall back sometime later to cause an afterglow known as *phosphorescence*. In rare instances, electrons may be trapped away from normal positions for indefinite periods, but may be released by application of gentle heat to the crystal framework, in which event a long-delayed phosphorescence takes place. This has been called thermoluminescence but is more properly termed *thermostimulated phosphorescence*.

Impurities promoting fluorescence are called *activators*. Slight traces in the crystal framework are capable of introducing defects in the structure which serve as "traps" for electrons, making it easier for electrons to free themselves from parent atoms as well as introducing extra electrons which contribute to fluorescence. Minerals which do not fluoresce normally can be made to do so by introducing activators, as is done today in the production of commercial phosphors. Some examples of fluorescence are bright red in ruby corundum (activator: chromium), red in calcite (activator: manganese), and vivid green in willemite (activator: manganese). Too much activator prevents development of fluorescence; the proper amount must be properly distributed for the glow to appear. Some species which commonly fluoresce are listed in Table 8, Appendix A.

TRIBOLUMINESCENCE

Production of light by friction or abrasion, or *triboluminescence,* is noted in some minerals as sphalerite and quartz. The very faint glow is best observed in full darkness. It is due to development of charged areas on opposing faces of a fracture, the concentration of electrons on one face sometimes being sufficient to leap across the gap and create light in a manner similar to that used in the mercury vapor lamp. It is very commonly observed when quartz is sawed on a diamond saw, when a stream of "sparks" is seen to issue from the point of contact.

PATHS OF LIGHT

Now that some properties of light have been explained, it is time to turn to the paths taken by light when it enters substances denser than air. The simplest behavior occurs when a transparent substance with completely irregular internal structure is involved, as glass or liquids, or mineral crystals belonging to the isometric system, which, by virtue of having atoms in such regular arrangement, exert the least complicated influence on light. On the other hand, crystals belonging to the other systems strongly influence the paths taken by light passing within them; the resulting phenomena are bizarre and as interesting as they are perplexing. For the discussion immediately following it is best to assume that a simple substance such as glass is involved. Later sections will take up light behavior in various types of crystal structures.

Light bends sharply the moment it strikes a substance containing a greater concentration of atoms than that in which it was traveling. Thus it will bend only a slight amount when passing from the emptiness of space into the earth's atmosphere, but a great deal when it passes from air into water, glass, or any other substance of far greater density. Such bending, or *refraction,* is depicted in Figure 89, where various beams strike several glass shapes to show how paths differ in them. When any ray strikes a surface vertically, it is not refracted at all, but passes through undeviated. However, the slightest inclination from the vertical causes refraction, as shown by the arrow paths in the sketches. When the rays enter they are bent inward, as shown, but when they leave they bend away from the point of exit.

Light is also *reflected,* as shown in Figure 90, but here we discover that it divides so that part is reflected and part is refracted. Such division occurs at all times except in one special case when a light ray happens to be already inside a dense substance, as illustrated in the bottom drawing of Figure 90. A light ray, as B, may strike the inner wall at a steep angle, emerging as shown. Another ray, labeled A, strikes at a shallower angle and is reflected back into the medium instead of emerging. However, in doing so, it does not divide and *all* of it reflects back, or as it is said, suffers *total internal reflection.* The angle at which this curious event takes place is called the *critical angle.* It is of great interest to makers of optical

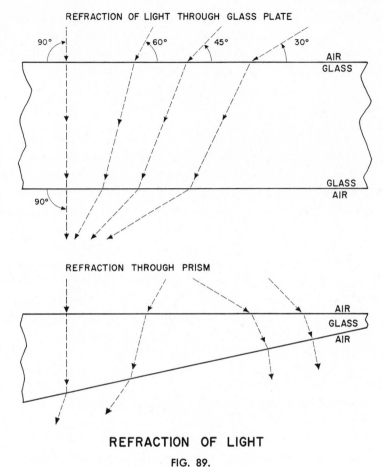

REFRACTION OF LIGHT THROUGH GLASS PLATE

REFRACTION THROUGH PRISM

REFRACTION OF LIGHT

FIG. 89.

instruments because by making use of this phenomenon, they can design prisms to turn light to any direction without having to bother silvering the backs.

REFRACTIVE INDEX

Figure 91 shows refraction as before, except that now the trouble has been taken to measure the angle at which the light ray falls, and the angle at which it bends after entering the test specimen. These angles are not accidental, as a learned scientist named Snell discovered many years ago. In fact, they are always related by a simple law which Snell formulated and which still bears his name. This law states that the amount of bending after a light ray passes into any given transparent substance depends on the angle at which it entered and that this relationship can be expressed by the sines of the angles concerned. For example, in

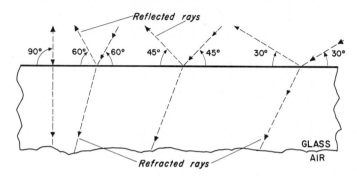

REFRACTION AND REFLECTION TAKE PLACE AT ONCE

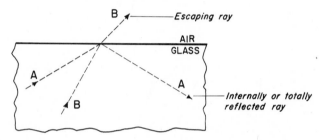

REFRACTION AND REFLECTION OF LIGHT
FIG. 90.

Figure 91, the angle of incidence (I) is measured at 38°, and the angle of refraction (R) is 27°. Looking up the sines of these angles in any trigonometry book, we obtain the values shown in the example. One value is simply divided into the other and the result is a number called the *refractive index.*

Perhaps this seems like a useless exercise in mathematics until it is learned that the refractive index *varies* from substance to substance, which is to say, each transparent substance has its own characteristic value. Immediately, refractive index becomes of far more interest because if it is measurable, it can be used to identify minerals. In fact, refractive indexes are used for this purpose regularly and have proved to be among the most dependable clues to identity that either the amateur or professional can use. To show how the bending of light, and also the refractive index, varies from substance to substance, the three sketches at the bottom of Figure 91 depict typical light rays passing into diamond, quartz, and water, showing clearly the variation in angles. However, these are simple rays of white light and really are not quite truthful representations of what happens because such rays do not stay "white," but disperse, as we have noted before, into spectral components. Therefore it is wise to discuss dispersion briefly, because it does have a bearing on refractive index.

DISPERSION

It was mentioned earlier that light passing through a prism divided into spectral colors, and indeed it was this effect which Newton studied so carefully, paving the way for Fraunhofer to discover the absorp-

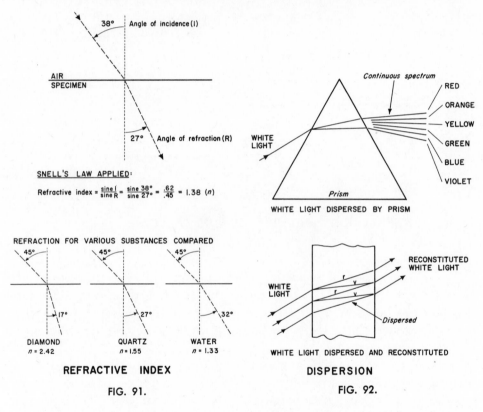

SNELL'S LAW APPLIED:

Refractive index = $\frac{\text{sine } I}{\text{sine } R}$ = $\frac{\text{sine } 38°}{\text{sine } 27°}$ = $\frac{.62}{.45}$ = 1.38 (n)

REFRACTION FOR VARIOUS SUBSTANCES COMPARED

DIAMOND $n = 2.42$ QUARTZ $n = 1.55$ WATER $n = 1.33$

REFRACTIVE INDEX

FIG. 91.

WHITE LIGHT DISPERSED BY PRISM

WHITE LIGHT DISPERSED AND RECONSTITUTED

DISPERSION

FIG. 92.

tion spectra in the sun's light. The division of light into colors by means of a prism, called *dispersion,* is shown in Figure 92. As will be noted, the longest wavelengths, namely those in the red end, are refracted least, while those in the short wavelength portion of the spectrum are bent most. It was not until considerably after Fraunhofer's time that an explanation was offered for this differential bending and then it was discovered that all light slows drastically when it enters a denser substance, but wavelengths of shorter length (violet) are slowed more than longer wavelengths (red), and therefore are bent more. All others between these are bent in proportion. To give some idea of what this slowing means, it may be recalled that the velocity of white light has been measured at about 186,000 miles per second when traveling in a vacuum. Its velocity is very slightly less in air, but much less in liquids and solids. Curiously, it was found that velocities in transparent substances are related to refractive index, such that in a substance of refractive index 2.0, light slows to half its speed in vacuum, or to about 93,000 miles per second. If the index is 3.0, it would only be one third, etc. It was also found that each wavelength of light therefore has its own refractive index, sometimes resulting in rather large spreads between the index of violet light and of red light, depending on the substances involved. For example, the indexes of diamond vary as shown below according to color wavelength:

REFRACTIVE INDEXES OF DIAMOND

Color	Wavelength	Refractive Index
Deep red	(6876 A)	2.4076
Yellow	(5893 A)	2.4175
Green	(5270 A)	2.4269
Blue violet	(4308 A)	2.4513

Even larger spreads in refractive indexes between extreme colors of the spectrum are noted in some minerals, and if these are capable of being cut into gems, they provide finished stones of considerable "fire." In the raw state, as in natural crystals, dispersion is seldom apparent because it is brought out only when all surfaces, as on gems and prisms, are carefully polished to perfect flatness. Dispersion is therefore due to the variation in velocity of the colored component of light but its development depends on the optical nature of the substance involved. It has little practical application in mineral identification except when refractive indexes are measured by means which will be described later in this chapter; it then becomes important because its effects, unless taken into consideration and corrected for, may result in considerable error.

Dispersion begins the instant a light ray strikes the entry surface but oddly it is not apparent unless the departure surface is tilted in relation to the first. When surfaces are parallel, as shown at the bottom of Figure 92, the light splits into colored components as before, but rejoins on leaving to reconstitute white light. The paths within the transparent medium show that each ray divides into colors, but the red of one ray rejoins the violet of its neighbor, and so on for all colors. In effect what one ray loses as it passes through is made up for by a gain from a neighboring ray. The net result is white light.

INTERFERENCE

The vivid hues observed on soap bubbles are due to a kind of dispersion called *interference*. Similar hues are sometimes seen on minerals when extremely thin films, usually caused by surface alteration, coat the surfaces. Figure 93 illustrates the creation of interference. Two white light rays are shown at left falling on a thin film, which here is much exaggerated in thickness. Both rays arrive at once, and both are partially refracted and reflected as explained before. Ray A reflects as shown, but part enters the film and follows a longer path before it again emerges.

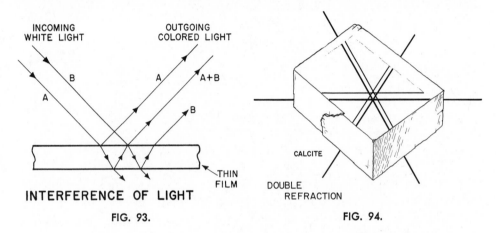

INTERFERENCE OF LIGHT

FIG. 93.

FIG. 94.

The film happens to be just thin enough so that the place where Ray A emerges coincides with the reflected portion of Ray B. Both travel upward merged as Ray A + B. However, because Ray B reflected instantly, while Ray A lost some time traveling through the film, its vibrations are not exactly in phase with Ray B. The vibrations mutually interfere, and the composite Ray A + B is no longer white light, but is colored, the film at the place of emergence seemingly colored too. The exact hue depends on the angle of incoming light and the thickness of the film. If the film is too thick, or too thin, no interference colors will be seen.

Cracks in quartz and other transparent minerals sometimes show the complete range of spectral colors because each crack begins as a very narrow wedge at one end and gradually opens toward the crystal surface, sometimes traversing a considerable distance. If the taper is very gradual, the colored bands spread widely, sometimes to as much as $1/4''$ to $1/2''$ in width. A similar effect can be gotten by resting a lens of very shallow curvature on a piece of clean plate glass. In the exact center will be a set of perfectly circular interference rings, but a magnifying glass may be

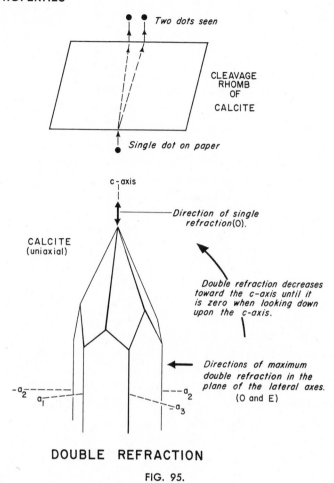

Two dots seen

CLEAVAGE
RHOMB
OF
CALCITE

Single dot on paper

c-axis

Direction of single
refraction(O).

CALCITE
(uniaxial)

Double refraction decreases
toward the c-axis until it
is zero when looking down
upon the c-axis.

Directions of maximum
double refraction in the
plane of the lateral axes.
(O and E)

$-a_2$ a_1 $-a_2$
                        $-a_3$

DOUBLE REFRACTION

FIG. 95.

needed to see them well. These are called *Newton's rings,* because it was he who first explained their origin.

DOUBLE REFRACTION

The use of glass to describe optical phenomena was purposeful; the complete lack of crystal order within this substance permits ordinary light phenomena to take place but bars others which are distinctive properties of all crystals other than isometric. Isometric crystals are so orderly in respect to atomic arrangement, that one may say they most closely approach the random disorder of glass and therefore behave like glass optically. But crystals of all other systems are less symmetrical in respect to internal atomic arrangement, and influence light in a tug-of-war game which causes some very peculiar results indeed! For example, Figure

94 shows how a transparent cleavage rhomb of calcite placed over ruled lines causes two images to appear instead of one in the effect known as *double refraction*. How this occurs is shown at the top in Figure 95.

If a cleavage rhomb is placed over a black dot on paper, the light ray traveling from the dot toward the eye splits in two as shown, furnishing two images. Knowing the position of calcite cleavage planes in respect to the crystal axes, it is discovered that double refraction occurs only along the directions shown in the lower drawing. The maximum doubling occurs along the plane of the horizontal axes, but diminishes steadily toward the *c* axis until none at all is seen when looking parallel to this axis. Because calcite cleaves so readily, it isn't possible to observe single refraction unless the trouble is taken to cut and polish a clear piece squarely across the *c* axis. Because this axis is the *only* direction along which single refraction takes place, calcite is said to have only one optical axis, and is therefore a *uniaxial* mineral.

Calcite belongs to the hexagonal system and it is to be noted that *all* hexagonal crystals display similar optical behavior. Here we may relate the crystal structure to the behavior of light, because in hexagonal system crystals, *and also tetragonal system crystals,* the patterns of atoms appear the same from the sides and is the reason for calling the lateral crystal axes "equal" in respect to direction and length. Thus all hexagonal and tetragonal crystals display the same type of double refraction and have only one optical axis of single refraction which coincides with the vertical crystal axis. Many species in these systems do not have the great strength of double refraction noted in calcite, but it is present nevertheless, and easily detected, as will be pointed out later.

The influence of crystal structure on optical properties is also shown beautifully by calcite and other rhombohedral carbonates, for they are all nearly alike in degree of double refraction. In transparent crystals it is so pronounced that when taken along with the inevitable easily developed cleavage, it is difficult to mistake these species for any others. Additional tests may be necessary to confirm exactly which of the several species is being examined.

Figure 96 is drawn to illustrate how crystals in the other systems, namely orthorhombic, monoclinic and triclinic systems differ optically from those just described. Instead of one optical axis, there are now two which cross each other at various angles, ranging from very slight to very great. Undoubtedly, the lack of symmetry in atomic arrangement causes such axial doubling, for as will be recalled in the crystals of these systems, the atomic patterns are not the same along the lateral axes, and it is for this reason that the axes are labeled with different letters and also said to be of different length.

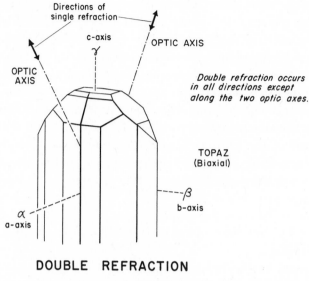

Directions of single refraction

c-axis
γ

OPTIC AXIS

OPTIC AXIS

Double refraction occurs in all directions except along the two optic axes.

TOPAZ
(Biaxial)

β
b-axis

α
a-axis

DOUBLE REFRACTION

FIG. 96.

The presence of two optical axes in *biaxial* crystals results in a complicated variation in double refraction, depending from which direction in the crystal it is viewed. In some directions it is very slight, in others great. In connection with double refraction as a whole, it should be mentioned that each doubly refracted ray travels at different velocity, as has been shown to be the case when rays are dispersed. The refractive indexes therefore also vary.

DOUBLE REFRACTION AND REFRACTIVE INDICES

As pointed out before, stronger refraction of light means slower travel and increasing refractive index. If it were possible to cut and polish spheres out of sample crystals, and to measure refractive indices at regular intervals over all surfaces, it would be found that tetragonal and hexagonal spheres produce only one reading at the ends of the *c* axis, but two readings gradually diverging as one axis pole is left and the equator of the sphere approached, as indicated in Figure 95. At the equator, the spread between indices is at a maximum. As the measurements are taken past the equator toward the other pole, they again draw closer until only one reading appears when squarely upon the opposite pole. This is true for all sides of the sphere. However, one of the values *never* changes and is therefore called the *ordinary ray,* designated by the capital letter O. The changeable one is called the *extraordinary ray,* designated by capital letter E.

Orthorhombic, monoclinic and triclinic crystals are more complicated because of the presence of two axes instead of one. Spheres cut from them produce four points of single refraction where optical axes emerge from opposite sides of the sphere. Between them will be found not two but three refractive index values according to direction as shown in Figure 96, labeled with the small Greek letters. Alpha α, represents the lowest value and gamma γ, the largest value, while beta β, is somewhere between them. Because light is singly refracted only along optical axes, measurements taken everywhere else produce two changeable values of refractive index which vary in value as different points on the sphere are selected. This is quite unlike the tetragonal and hexagonal case where only one ray out of a pair changes in value.

OPTICAL SIGNS

In uniaxial crystals, one index value always remains constant and the other changes according to the direction along which the indexes are measured. However, the changeable index can be higher or lower in value than the fixed index, and because some species follow one rule and others follow the opposite rule, it is helpful when measuring refractive indexes to determine which is which. By discovering this fact, an additional bit of information useful for identification can be added. Thus, in tetragonal and hexagonal crystals, if the changeable extraordinary ray E provides an index higher in value than the ordinary ray O, the mineral is said to be *positive in sign*. If the extraordinary ray is below the ordinary ray in value, the mineral is *negative in sign*.

A similar system is used for biaxial minerals, except that one notes whether the intermediate index β is closer to α or γ. If it is nearer to α, the mineral is *positive;* if nearer to γ it is *negative*.

As pointed out before, isometric crystals and amorphous transparent substances as glass, liquids, plastics, etc., do not doubly refract light and therefore furnish only one index of refraction no matter in which direction it is taken. Sometimes singly refractive and doubly refractive substances are distinguished by calling them *isotropic* and *anisotropic* respectively, the words being derived from "iso" meaning single, and "aniso" meaning *not* single, while "tropic" refers to "affinity."

POLARIZED LIGHT

Because ordinary light consists of electromagnetic pulses which interact with the electromagnetic fields of crystals, it not only doubly refracts when entering certain crystals, but each ray is forced to pulsate in opposite directions or poles, as shown in Figure 97. In effect, the original circular fields are split, then squeezed into ribbon-like fields, resulting in what is termed *plane polarization*. The entire crystal produces polarized rays in all directions except along optical axes where rays do not split and

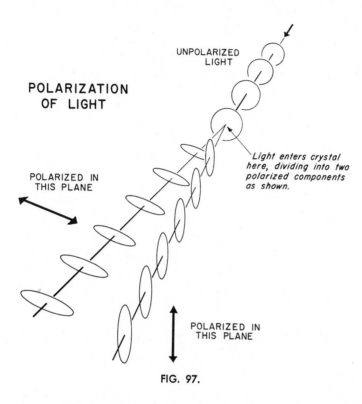

POLARIZATION
OF LIGHT

UNPOLARIZED
LIGHT

Light enters crystal here, dividing into two polarized components as shown.

POLARIZED IN
THIS PLANE

POLARIZED IN
THIS PLANE

FIG. 97.

therefore cannot become polarized. It is all very strange and puzzling that such rays can fill a transparent crystal but apparently they do. Actually it is fortunate that this occurs, because by analyzing the polarized light which issues from crystals we are able to discover extremely useful identification properties.

The easiest way to produce polarized light is to use sheets of the plastic material known as Polaroid. It can be obtained very inexpensively and, as will be seen, is very useful to the amateur in determining optical properties of minerals. Polaroid contains thousands of very small platey crystals of an iodine compound which produce polarized light by the process of absorption. Because all the crystals are aligned in the same direction, each sheet in effect becomes a gigantic single crystal of the iodine compound, except made extremely thin and quite transparent. The special ingredient crystals are actually deep purple in color if viewed in one crystal direction, but very pale gray when viewed through the extremely thin sides. In common with many doubly refracting natural crystals, the iodine compound crystals exercise a peculiar effect upon polarized light as soon as it is produced. One component is so strongly absorbed that for all practical purposes it is lost within the crystals, and only a very small amount, colored very deep purple, trickles out the other side of the Polaroid film. On the other hand, the remaining component, vibrating in an opposite field direction, is freely permitted to pass through the crystals, and it is this which we see imparting the smoky gray color typical of Polaroid. Because of the peculiar behavior of Polaroid, each sheet produces nearly perfect polarized light simply by absorbing one of the two polarized light components.

Even more strange than the nearly complete destruction of one component is the fact that when two sheets of Polaroid are crossed, as shown in Figure 98, the light passing through gradually dims and is almost completely absorbed when the sheets are overlapped at right angles to each other. The very minor trickle of purplish light mentioned before is seen only when the crossed sheets are held up directly before a strong light or toward the sun. For practical purposes, no light gets through when the Polaroids are in this position. Thus polarized light components are antagonistic toward each other. If each is allowed to vibrate in the same direction, as in Step B of Figure 98, each Polaroid sheet will permit the light of the other to pass through, but if crossed, as shown at the bottom in Step D, the light of one sheet is completely absorbed by its crossed companion. This peculiar behavior at once provides us with a simple and powerful means for determining the presence of similar behavior in natural crystals capable of producing polarized light.

PLEOCHROISM

The selective absorption of polarized light by colored crystals is called *pleochroism,* from the prefix "pleo" meaning several, and the root "chroism" meaning colored. When only two colors are noted, as in *tetragonal* and *hexagonal* crystals, which contain only two polarized light components, the term *dichroism* is used. Orthorhombic, monoclinic, and triclinic crystals have complex crystal structures which produce pairs of polarized light components that differ in character depending on direction. In these crystals *three* distinct hues may be observed and the term *trichroism* is used. It is emphasized that pleochroism can occur only in colored doubly-refractive crystals; it cannot occur in colorless crystals of any description, nor in isometric crystals or amorphous substances as opal, glass, etc. In fact, it is not always observable in colored doubly-refracting crystals. Some of

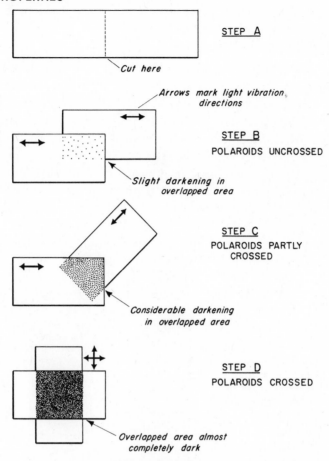

STEP A

Cut here

Arrows mark light vibration directions

STEP B
POLAROIDS UNCROSSED

Slight darkening in overlapped area

STEP C
POLAROIDS PARTLY CROSSED

Considerable darkening in overlapped area

STEP D
POLAROIDS CROSSED

Overlapped area almost completely dark

POLAROID SHEET EXPERIMENT
FIG. 98.

these though strongly colored, as green olivine, produce scarcely any pleo-chroic effect. In other words, in those colored doubly-refracting crystals capable of producing this effect, pleochroism still depends on the species concerned, and sometimes on the specimen concerned. In its favor, how-ever, it is pointed out that when the effect appears, it is proof positive that the mineral is doubly-refractive. In some instances the strength of the effect is so pronounced, as in colored tourmaline, that it becomes a valuable clue to identity. It is also pointed out that pleochroism does not occur along optical axes because light is not split along such axes and hence cannot produce two colored components. Furthermore, pleochroism is not to be confused with color zoning in crystals, or other forms of patchy color distribution.

DICHROSCOPES

A simple instrument called a *dichroscope* is used to determine pleochroism. Two colors may be seen in its eyepiece when a suitable specimen is examined, hence its name. Figure 99 shows methods for making two models which cost almost nothing for materials. The Polaroid dichroscope begins with a sheet cut into squares as shown; the arrows within the squares mark polarity direction. Take two of the pieces and turn one around so that the polarities are at right angles, as indicated by the arrows. Both sections are cemented between thin pieces of cardboard to hold them taut. If a razor blade is used to make the cut, a very neat joint without overlap is possible.

In use, a suitable transparent-to-translucent crystal is held toward a strong light and the dichroscope placed next to the crystal so that the light passes

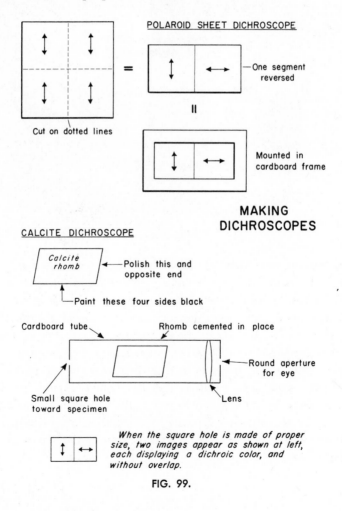

FIG. 99.

FIG. 100. An actual photograph of dichroism in tourmaline crystals. A small Eureka Polaroid sheet dichroscope has been placed before two slender pink crystals from the Himalaya Mine, San Diego County, California. Ordinary light is allowed to shine through the back. The segments of Polaroid are "crossed," the upper segment showing a very pale pink hue, but the lower showing a much darker pink, corresponding to the two dichroic colors observable in these crystals.

through the crystal, then through the dichroscope, then to the eye. This is shown in Figure 100, an actual photograph of dichroism in tourmaline crystals. One half of the Polaroid exhibits the color corresponding to one polarized light component while the other section displays the color of the oppositely polarized light; therefore both appear at once and can be directly compared. This test is very sensitive. If in doubt as to whether a difference in color exists, the frame should be rotated to various positions and also slowly waved back and forth to emphasize the change in hue.

Actually, it is not necessary to use two "crossed" Polaroids, one will do if it is turned to various positions. However, this requires that the eye "memorize" the various colors seen in the several positions and it may fail to do so if the change is slight. It is far safer to use the device described. Note in the illustration of the tourmaline crystals, that only two colors will been seen along the sides of the prism. Tourmaline is uniaxial, and therefore cannot display more than two hues. Furthermore, because the optic axis coincides with the crystal axis, viewing in the axis direction results only in seeing one color in both windows of the dichroscope because light is not doubly refracted in this direction. In the example shown, this color is deep pink. An important rule in the use of the dichroscope is therefore to test a crystal in several directions, trying to make them about 90° apart. This is even of greater importance in biaxial crystals, as will be pointed out.

For biaxial minerals, it is absolutely necessary that the specimen be examined in three directions at right angles to each other to insure that trichroism is not overlooked. As an example of how it can be easily overlooked, and the mineral therefore mistaken for a uniaxial species, consider the colors displayed by cordierite (iolite). In small transparent pieces, two colors are usually very obvious, even without the help of a dichroscope; they are dark purplish-blue and straw yellow. However, if the specimen is systematically examined, as recommended, a third color will be discovered, namely, pale blue.

The calcite dichroscope shown in the lower part of Figure 99 is superior to the Polaroid instrument because the clear colorless calcite used does not impart any color as does the Polaroid. All that is needed to make this instrument is a clear cleavage rhomb of calcite, large or small, with its ends polished in case they happen to be rough. If a good splitting job is done, even this is not necessary. The method of polishing is described later in this chapter in the section dealing with the use of the refractometer. The sides of the rhomb are painted black or covered completely with black masking tape to absorb stray reflections which may be distracting. The rhomb is inserted in a small cardboard tube and cemented in place. One end of the tube is fitted with an opaque cap of metal or plastic, through which is bored or punched a very small hole, filed to make it square. The opposite end is capped with a simple convex lens of from one to three power, which may be simply cemented in place, or cemented into another cardboard tube to slide over the main tube. Over the lens is placed another disk of opaque material with a peephole in the center. The trick in making this instrument is to shape the small square window on the end opposite the eyepiece in such a way that when it is seen through the eyepiece two small squares will be in view, each just touching but not overlapping the other. For this purpose, use a small jeweler's square file, enlarging the hole, from slightly less than $1/8''$ in diameter to begin with, until the desired result is obtained. This should be done before the disk is finally cemented in place.

To use the calcite dichroscope, place the specimen to be tested immediately in front of the small square window; hold both specimen and dichroscope toward a strong light until both images are clearly seen; observe color, rotating the dichroscope to obtain maximum color change. As remarked before, check at least three different directions in the specimen before being satisfied. When properly used with suitable specimens, dichroscopes give the following information:

When colors change, the mineral is doubly refracting; it therefore is *not* isometric, nor amorphous.

When colors do not change in one part of pleochroic crystal, the dichroscope is pointing along an optical axis.

When only two colors are observed, the mineral is probably uniaxial, therefore, tetragonal or hexagonal.

When three colors are observed, the mineral is probably biaxial, therefore, orthorhombic, monoclinic, or triclinic.

Cautions: Note the use of the word *probably;* the tests are strongly indicative but not absolute because colors sometimes change subtly from one part of a crystal to another, and in the case of biaxial minerals, sometimes two of the colors are not much different in quality and therefore can be mistaken easily for shades of the same color. It is strongly recommended that test specimens be obtained after a dichroscope is made, and practice gained on them. Excellent specimens are pink or green tourmaline (uniaxial), yellow Durango, Mexico, apatite (uniaxial), axinite (biaxial), cordierite (biaxial).

THE POLARISCOPE

The *polariscope* is merely a pair of crossed Polaroid sheets, *between* which is placed the specimen to be examined. The simplest kind is made from two disks of Polaroid sheet, about 1½″ to 2″ in diameter, mounted in frames and attached to a post, as shown in Figure 101. It is handy to be able to slide the top disk up and down, but it isn't necessary. If the disks are fixed in position, the space between them should be about 3″, because it will be necessary to hold specimens in this area while being tested, and some room is needed for the fingers. A microscope substage lamp is the best illumination source, the light passing upward from beneath the bottom Polaroid disk. However, any other lighting arrangement, such as a mirror casting light upward from an outside source, is equally satisfactory, providing that the light does not leak from around the disks and strike the eyes directly.

To use the polariscope, turn on the light source, and place a transparent or translucent mineral specimen between the crossed disks, in the manner shown in Figure 101, holding it with the finger tips so that it can be rotated to any position. Note that before the specimen is introduced, the field of view is quite dark except for the faint purplish glow produced by the Polaroid crystals. If the transparent specimen is amorphous, as opal, or ordinary glass, it remains generally quite dark except for vague streaks of bluish light which sometimes form cross-like figures. No matter which way

FIG. 101. Testing a thin cleavage flake of mica for a biaxial figure between the crossed polaroids of the Gemological Institute of America's polariscope. Light is provided from underneath by a small microscope substage lamp. The glass stirring rod ball-tip acts as a powerful but easily maneuvered magnifying glass to enlarge axial figures to the point where they can be clearly seen.

the specimen is turned, the same appearance will be noted. This at once indicates that the polarized light has the same effect on the specimen in all directions, and therefore the specimen cannot be doubly refracting. Try isometric specimens such as small pieces of transparent fluorite; note the same generally vague streaks of light against a dark background. However, if a piece of quartz is inserted, the results are startlingly different! As the specimen is turned, it "winks" dark and light, doing so abruptly in each 45° of rotation. Obviously, this specimen is decisively affecting the polarized light, and therefore must be a doubly refracting mineral. Thus the polariscope easily distinguishes between singly refractive and doubly refractive minerals. As an identification clue, even with small pieces no more than ¼″ across, this test may be decisive in distinguishing between two minerals that appear similar. Words are incapable of accurately describing these appearances, however, and the best way to gain experience in the use of the polariscope is to obtain samples of singly and doubly refractive minerals, and compare them, carefully noting how each differs from the other. It will be found that singly refractive substances, including all glasses, plastics, and isometric minerals, are seldom completely dark under the polariscope, but nevertheless furnish appearances so distinctive that they cannot be mistaken.

A distinctive appearance is also provided by translucent specimens consisting of multitudes of minute crystals, such as chalcedony, some jades, and other cryptocrystalline mineral varieties. It will be noted that light *always* passes through such specimens as a fairly strong glow no matter which way they are turned. This is explained by the fact that within such specimens, many crystals are always turned to the proper direction to pass light regardless of how the specimen happens to be oriented. When it is turned to another position, the original light-carrying crystals go dark, but others take their place to pass light and the glow continues.

The polariscope is also used to locate the optical axis in uniaxial minerals, and thereby obtain additional information about an unknown specimen which may be of value in identification. To learn the technique, begin with a specimen in which the direction of the axis is already known. For example, a slender crystal of quartz can be broken across the prism by light tapping, so that several of the prism faces are left to show in which direction the axis lies. By turning a suitable piece underneath the Polaroids, it will be observed that the light winks on and off when the prism faces are horizontal. However, when the specimen is turned upright so that the prism faces and the optical axis are also upright, the field of view remains generally dark, because it is in this direction that light in the quartz crystal is singly refracted. Again the distinctive appearance of the specimen should be noted and remembered against the day when a strange specimen, with no clue as to orientation of the axis, is examined.

Biaxial minerals are far less distinctive in this test because the spread between the two optical axes results in development of double refraction everywhere between them, in effect, everywhere throughout the crystal. Sometimes a definite darkening or lessening of the strength of "winking" is noted along the vertical crystal axis which lies between the optical axes (see Figure 96), but in most cases one cannot be sure.

AXIAL FIGURES

Under polarized light, a unique and colorful display of interference colors is seen surrounding the places on crystals where optic axes emerge. To see such displays, use small transparent crystal fragments, usually no more than $\frac{1}{32}''$ to $\frac{1}{8}''$ thick, and turn them between crossed Polaroids until some area is seen which suddenly flashes into strong colors. When the glass stirring rod of Figure 101 is placed close to the specimen over the colored area, the ball tip acts like a powerful magnifier, and produces an image on top of the ball similar to one of the figures drawn in the lower part of Figure 102. If the ball tip is moved about, and the mineral fragment tilted as necessary to put strongest colors face up, eventually a complete figure will be obtained. Such figures are called *axial figures,* and if consisting of a single set of closed rings, they belong to uniaxial crystals. If two sets of rings, usually connected by loops of color, are observed, each "eye" marks the emergence of one of the axes in a biaxial crystal.

It is far easier for the amateur to produce these figures than it is to describe them, and it is therefore recommended that for trial purposes, thin sections of clear mica, topaz, quartz, beryl, or tourmaline be obtained and tested. The most effective specimen is mica. Select a sheet and place it with cleavage plane parallel to the Polaroids, as shown in Figure 102. If colors do not appear, the sheet is probably too thick and needs to be split more. When the sheet is held horizontal, the axes are spread apart as shown to the right, and requires tilting of the sheet to one or the other side to bring the optic axis directly under the ball of the stirring rod. If the mica is perfect, un-

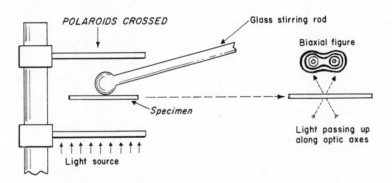

UNIAXIAL FIGURES

Normal uniaxial figure showing colored rings and black cross.

Quartz uniaxial figure showing absence of cross.

BIAXIAL FIGURES

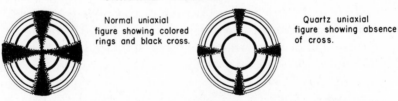

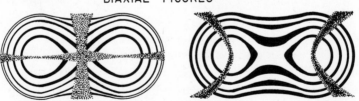

Depending on how the specimen is turned in relation to the Polaroids, either of the biaxial figures shown may appear, or only part of one as is usually the case.

USING THE POLARISCOPE

FIG. 102.

twinned, unriffled or otherwise marred with imperfections, it is easily possible to tilt it one way, to bring out an unmistakable axis figure of the kind shown in Figure 102, and then tilt it the other way to detect the other. It is a very pretty demonstration and easy to produce. A similar figure can be developed on a suitable cleavage plate of topaz.

While easily cleavable mica and topaz provide excellent specimens for display of biaxial figures, quartz, tourmaline, and beryl, and other species of both uniaxial and biaxial minerals may not cleave at all, or may cleave in the wrong directions. In such cases, the grains may be so rough on their surfaces that it becomes very difficult to develop an axial figure. In this case, it is often helpful to crush a sample to the size of sand, and place the grains on a slip of glass between the Polaroids. Usually one or more of the fragments will be in proper orientation and will announce itself by bright sparks of color. Sometimes it helps to place a drop of cedar wood oil on the grains, such oil serving to smooth out the surface irregularities and permit the colored rings to be seen more easily and with less distortion. In a later section dealing with the use of the refractometer, a list of fluids suitable for this purpose is furnished. If these fluids are not available, a small drop of salad oil is better than nothing.

The axial figures shown in Figure 102 present the ordinary appearance of uniaxial and biaxial crystals when properly oriented under crossed Polaroids and suitably magnified. The normal uniaxial figure shows a distinct maltese cross of black passing through and interrupting the brilliantly colored rings; it is quite unmistakable. On the right is shown the characteristic uniaxial figure of quartz and it will be noted that the black cross does not pass through the center. When seen distinctly this figure is certain evidence that the specimen *is* quartz. The absence of the cross in the center arises from the spiral structure of quartz, causing spiral or rotary polarization of light. Beneath is shown the usual biaxial figure, one part of which is all that is ordinarily seen, but this part is sufficient in itself to identify a biaxial mineral with certainty. The complete pattern shown at the lower right is sometimes seen in crystals whose axes cross at shallow angles. Note that only *one* dark line crosses the center of the colored rings. This is the most important means of distinguishing uniaxial and biaxial figures!

The polariscope, as simple as it is to make and to use, is an extremely useful instrument for determining characteristic optical properties of many species. When properly used, it will:

Distinguish singly refractive and doubly refractive substances
Distinguish cryptocrystalline minerals
Locate optical axes.
Develop axial figures, thus distinguishing between uniaxial and biaxial crystals

MEASURING REFRACTIVE INDEXES

One of the most powerful methods for identification is determination of refractive indexes. Unfortunately, the professional instruments used for this purpose are too expensive for the average purse: even the cheapest currently available refractometer costs at least ninety dollars. Because it is sometimes possible to obtain secondhand instruments, the methods of using them will be described later in this chapter. However, the first part of this section will deal with the means the amateur can use for which no special instruments are necessary to estimate refractive index.

All refractive index measurements compare the index of transparent substances to the index of a vacuum which is taken as 1.0. The index of air, however, differs so slightly from that of a vacuum, namely 1.00029, that measurements in air, as is the usual practice, are not corrected unless the most meticulous physical experiments are being performed. When an index is given without regard for crystal direction, as in isometric crystals, or for amorphous substances which can have but one index, the small italic n is used before the index value; e.g., the index of fluorite is $n = 1.434$. On the other hand, when polarized light components in uniaxial and biaxial crystals are measured, then the English and Greek letters previously mentioned are used, namely: for uniaxial crystals, O and E, corresponding to the ordinary and extraordinary rays respectively; and alpha α, beta β, and gamma γ, for the three rays in biaxial crystals.

ESTIMATING REFRACTIVE INDEX BY IMMERSION

The principle involved is very simply explained by referring to the sketches of Figure 103. Several transparent substances are placed in water to show how each varies in visibility depending on how closely refractive indices match that of the water. Ice, $n = 1.31$, very closely matches the index of water, $n = 1.33$, and explains why clear ice cubes put in water seem to disappear. The light rays simply are not deviated very much when they pass from water into ice and from ice into water. Unless bubbles are present in the ice, a lump cannot be clearly distinguished. To the right is shown a piece of clear quartz put in the same container, but now seen more easily because its index, $n = 1.55$, is enough above water to sharply deviate light rays and to make its outline more visible. A third specimen, that of the curious mineral known as cryolite, matches the index of water so closely that it is even more difficult to detect its presence than ice!

The lesson here is plain: if a liquid is found to match the index of a mineral, the latter disappears from view. Conversely, if the match is poor, the specimen fragment is easily visible. Naturally the most striking results occur when specimens are clear, colorless and flawless, but even when strongly colored, or filled with inclusions or cracks, the thin edges of fragments seem to merge imperceptibly into the fluid, the eye having considerable difficulty distinguishing the exact boundary between solid and fluid. The principle of immersion is most useful to the amateur, but calls for assembling a set of suitable fluids for carrying out tests.

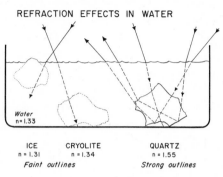

REFRACTION EFFECTS IN WATER

Water
n = 1.33

ICE
n = 1.31
Faint outlines

CRYOLITE
n = 1.34

QUARTZ
n = 1.55
Strong outlines

FIG. 103

Water is one of a number of fluids used for immersion testing, while other common fluids are obtainable from paint stores, druggists, and mineralogical suppliers. Some fluids are more difficult to procure and more expensive, but, in their favor, they last a long time. These must be ordered specially from drugstores or from large mineralogical supply houses. All liquids are stored in wide-mouthed colorless glass jars, which need not be over 1″ wide for the average run of test specimens of ¼″ to ½″ in diameter. Because some liquids are corrosive or attack ordinary liners inside lids, each jar should be fitted with a small square of thin polyethylene sheet, over which the lid is screwed down. Each jar is labeled as to contents and refractive index, and stored with others in a box. Many liquids can be diluted to lower their indices over a wide range, but then it is necessary that the amateur have access to a refractometer to determine indices of the resulting mixtures. Specimens are handled with a slender pair of tweezers; both specimen and tweezers must be cleaned and dried before immersion in another fluid. This prevents contamination of fluid. If possible, a number of pieces broken from the specimen should be used separately to avoid the inconvenience of cleaning.

The testing procedure requires chipping small pieces of the unknown mineral from the specimen, and placing suitable fragments in one fluid after another until a close match is obtained. It is usually possible to tell after several trials which way the index is going, and to test further accordingly. A magnifier of 5-10 power is useful for observing the edges of the specimen in the fluid. If the edges become blurred or difficult to see, a close match is indicated. Test specimens need not be colorless, nor completely transparent for that matter; all that is important is to observe the "disappearance" of the edges. When the closest possible match is made, the refractive index of the fluid is noted and checked against identification tables to select a list of minerals which could have that particular index. Usually on the basis of other conspicuous properties, many species in the list can be eliminated until only several serious contenders are left. Refractive indices and related properties of minerals are listed in Table 7, Appendix A.

Good practice is obtained by checking small pieces of known minerals against fluids to see how they look when immersed. For example, a clear piece of quartz is immersed in a variety of fluids and its appearance carefully noted. After some practice, one develops a keen sense of estimation, making it possible simply to glance at a specimen and often place it in the correct jar the first time. In this connection, a further test of great value is to use known species *with* unknown species when making tests. For example, if a small piece of quartz is placed with the unknown fragment in the quartz fluid, it will be instantly apparent if both are nearly the same, or considerably different. Suggested test specimens are included in the following list of fluids.

TABLE OF IMMERSION FLUIDS

Name	Refractive Index	Test Mineral (n)	Remarks
Methyl(wood) alcohol	1.33	cryolite(1.34)	Absorbs water; inflammable, keep stoppered
Water	1.33		Dissolves many low index minerals
Acetone	1.36	fluorite(1.43)	Inflammable; volatile
Kerosene	1.45	opal(1.44-1.46)	Check exact R.I.
Carbon tetrachloride	1.46		Volatile; do not inhale fumes
Turpentine	1.47		Inflammable
Olive oil	1.47		Wash off with lighter fluid
Glycerine	1.47		Absorbs water
Mineral oil	1.48		Wash off with lighter fluid
Castor oil	1.48		
Toluene	1.49		Inflammable
Cedar oil	1.51	gypsum(1.52-1.53) halite(1.54)	Wash off with lighter fluid
Clove oil	1.54	quartz(1.54-1.55)	Wash off with lighter fluid; strong aroma
Anise oil	1.55		Very strong aroma; wash off with lighter fluid
Orthotoluidine	1.57		Do not inhale fumes
Bromoform	1.59	colemanite(1.59-1.61)	Do not inhale fumes
Cassia oil	1.60	topaz(1.61-1.62)	
Monochlornaphthalene	1.63	apatite(1.63-1.64)	Do not inhale fumes
Acetylene tetrabromide	1.64		Unpleasant odor; do not inhale fumes
Monobromonaphthalene	1.66	phenakite(1.65-1.67)	Unpleasant odor; do not inhale fumes
Monoiodonaphthalene	1.70	spodumene(1.66-1.68)	Unpleasant odor; do not inhale fumes
Methylene iodide	1.74	spinel(1.72-1.73)	Protect from light
Methylene iodide and sulfur	1.78	spessartite garnet(1.79-1.81)	

Notes: A variety of mineral oils are available from automobile service stations for immersion purposes but require checking of refractive indices. Of the fluids listed, all those higher in index than orthotoluidine are not easily obtained except by special order through druggists. Many are expensive and prices should be checked before ordering. Most organic fluids listed can be mixed with acetone to lower indices but again require checking on a refractometer to determine new index values. Methylene iodide, monoiodonaphthalene, and bromoform should be kept in brown glass bottles to prevent adverse effects of light; small pieces of copper should be kept in the bottles to prevent decomposition. A number of the liquids listed are also useful for specific gravity determinations.

THE GEM REFRACTOMETER

The most useful refractometer for amateur identification purposes is the *gem* refractometer, specifically designed for use with the polished flat surfaces of gems. Similar use for mineral identification also requires the flatting and polishing of a small area on test specimens, a procedure which is not difficult and will be described shortly. The use of a Rayner refractometer is shown in Figure 104; it shows the refractometer, the eyepiece for viewing the calibrated index screen within the instrument, and the projection lamp throwing light into the instrument and from there to the eyepiece. At the top of the refractometer appears the flat polished surface of a small glass prism upon which is rested the polished surface of the test specimen.

Refractometers of the type described here are called *total reflectometers* because they measure the angle at which total reflection takes place *against* the test specimen. The principle of all these instruments is shown in Figure 105, and the vital part is a prism of very high refractive index glass shaped to half-cylinder cross section. The curved parts and top are polished. The polished test specimen, or gem, as the case may be, is placed on top but is separated from the glass prism by a very small droplet of high refractive index fluid. The purpose of this fluid is to provide optical contact because no matter how carefully the glass prism and test specimen are polished, they will never fit so precisely that all air is excluded from between them. In the example shown in Figure 103, the liquid is $n = 1.81$, but it could be lower providing it is always higher in index than the test specimen. Should the liquid be lower in index than that of the specimen, all that will appear on the scale of the instrument is its reading but not that of the test specimen.

Once the specimen is seated, light rays from the lamp are directed through the rear window as shown at the left in Figure 105. Rays which point too steeply upward pass completely through the glass, then through the liquid,

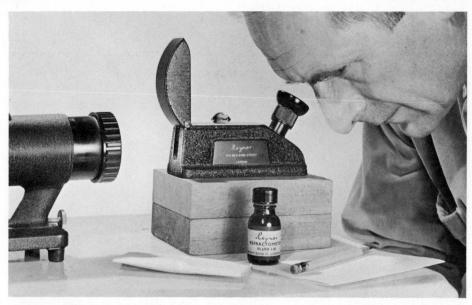

FIG. 104. The Rayner refractometer in use. The slide projector at the left shines a powerful beam of light into the refractometer window just beneath the raised hood. The light passes through a frosted glass screen and is reflected upward through a glass prism into the faceted gemstone seen resting upon the prism. Some is reflected downward through an additional optical system, and made to pass through a calibrated screen. Through the eyepiece one can read the refractive index directly on the screen. Note the small bottle of high index fluid in the foreground. Before the specimen is placed on the glass prism of the instrument, a small droplet of this fluid must first be applied to establish optical contact between the specimen and prism.

and finally lose themselves in the specimen. Those which are lower in angle are reflected and reappear on the right side of the glass prism. From this point they pass through a transparent screen marked with refractive index numbers and thence to the eye through the eyepiece. Because the reflection of rays depends on the refractive power of the test specimen, a sharp division between dark and light appears on the screen, one sector marking lost rays (dark), the other marking reflected rays (light). The readings are taken at the point where dark meets light. Illustrations of the Rayner refractometer screen are shown in the lower sketches of Figure 105. The left sketch shows a single reading at 1.717, while the right sketch shows two lines, one fainter than the other, where two sets of rays have emerged as from a doubly refractive mineral. Both scales show a very faint line at 1.81 belonging to the index of the fluid used on the prism.

The sharp readings shown in Figure 105 occur only when monochromatic light is used to illuminate the refractometer, or special filters are used over the eyepiece to screen unwanted portions of ordinary white light. If white light is introduced, it will divide into separate spectral colors as it passes through the instrument and these appear as blurred colored bands at the junction between dark and light. It then becomes a problem as to where to make the reading. For practical purposes, the reading is taken at the point where yellow verges on green because this is close to the yellow spectral band of 5890-5896 A produced by sodium vapor, which is used as the standard wavelength for measurement of refractive indices. To obtain sharp readings most gemologists use a sodium vapor lamp instead of white light, or a special dark yellow filter which fits over the eyepiece and effectively screens out practically all light except a yellow close to the wavelength of sodium vapor. Refractive index values given here and in other books are assumed

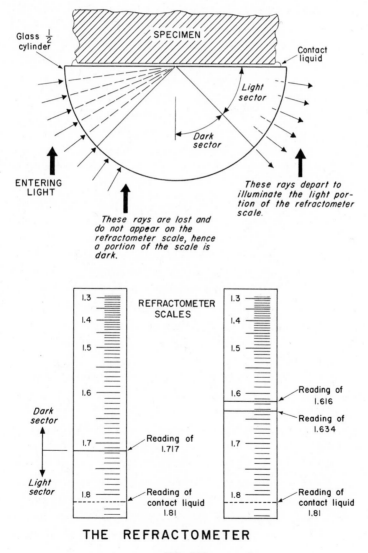

THE REFRACTOMETER

FIG. 105.

to be measured at the sodium band, unless otherwise stated. With some practice, however, the user of a refractometer without these special provisions is able to estimate readings with fair to good accuracy.

Singly refractive minerals as fluorite, and amorphous minerals as opal, produce only one degree of refraction and consequently produce only one reading on the refractometer. The single line shown at the lower left of Figure 105 is opposite 1.717 and may therefore belong to the mineral spinel for which this reading is typical. The double readings, at the right, of 1.616 and 1.634, belong to tourmaline, the difference between readings being the difference between the refractive index of the extraordinary ray and the ordinary ray. Subtracting the smaller value from the larger gives 0.018, known as the *birefringence* of tourmaline. Double readings are always obtained from polished pieces of uniaxial minerals, except if they happen to be cut across the *c* axis. In this case, only one reading, that of the singly refractive ray which travels along this axis will be obtained. For biaxial minerals, two readings are obtained no matter how the specimen is cut.

Because the refractive index varies according to the direction in which the light ray travels in a doubly refractive crystal, as described in a previous section, it is always necessary gently to rotate a polished specimen on the refractometer prism to various positions to be certain that maximum and minimum readings are obtained. Care in turning the specimen is essential to prevent scratching the soft glass prism; usually a gentle nudge with the fingertip is enough to change the position. As the specimen is rotated, it will be seen that the two shadow edges of a uniaxial mineral open to a maximum, and then close to one reading. One of the values remains constant; this is the index of the *ordinary* ray, the other belongs to the *extraordinary* ray. The reading is taken when both are moved as far apart on the scale as possible by turning the test specimen. The difference between them is the birefringence, while the value of the moving or extraordinary ray determines the *optical sign,* that is, if the *extraordinary ray is greater in value, the sign is positive* and vice versa.

For biaxial crystals, the behavior of the shadow edges differs because *both* shadow edges move as the specimen is rotated. Again the object is to obtain the highest reading of the upper edge, and the lowest reading of the lower shadow edge. The difference between them is the birefringence. The maximum reading on the upper edge is the gamma γ index; the lowest reading of the lower edge is alpha α. If the higher edge passes below the halfway position between maximum and minimum values, it means that the crystal is *positive in sign.* Conversely, if the lower reading passes upward across the halfway point, then the mineral is *negative.* For all readings, a piece of paper and a pencil should be handy to insure accurate recording of values.

As mentioned before, singly refractive minerals and substances will produce only one reading which is unchanging as the specimen is rotated.

Limitations of the Refractometer. Aside from the obvious limitation of expense, the refractometer cannot give readings from truly opaque minerals nor from those which refuse to assume a good polish and therefore provide poor optical contact with the glass prism. Nor can readings be obtained from species higher in index than the contact fluid, that is, over $n = 1.81$, the limit of the fluid usually sold for gem refractometers. Thus a considerable number of high refractive index species cannot be tested, as sphalerite $(n = 2.36+)$, cassiterite $(n = 2.0+)$, etc. Opaque minerals include most sulfides, all metals and alloys, and some oxides. Other species, normally transparent, may be rendered so opaque for various reasons that they may also be unsuitable. On the other hand, seemingly opaque minerals, as black tourmaline and pyroxene, etc., are often sufficiently transparent to pass light and allow readings. This is easily determined by holding up a thin splinter to the light and seeing if any light passes through its edge. A light color left on the streak plate is also indicative of enough transparency to enable refractometer readings. Despite the limitations mentioned, even a failure to obtain a reading on the refractometer at least indicates that the specimen's index is higher than 1.81—a fact that is helpful.

POLISHING TEST SPECIMENS

A piece of the unknown mineral, preferably free of cracks, flaws, and inclusions, is selected for polishing. It should be no larger than $\frac{1}{2}''$ in diameter, nor less than this because smaller sizes are awkward to handle. A pinch of 200 carborundum grit is sprinkled on a flat piece of steel and moistened with a few drops of water. The flattest part of the specimen is gently rubbed on the grit to remove high spots, and then rubbed with more pressure when a definite flat area develops. Continue until pits disappear and the surface is uniform. This usually requires several minutes work. Wash off the grit and substitute 400 grit; repeat grinding until all coarse grit marks are removed. Wash off and repeat the process using 1200 grit, rubbing back and forth gently until the surface is perfectly uniform. Carefully clean the specimen using a toothbrush. Polish the specimen on a flat block of tin, lead, or very hard wood, by dabbing on a pinch of tin oxide, Linde A powder, or tripoli. Use very little water to moisten the powder; it must be damp but not wet. Vigorous rubbing, with pressure, is necessary to bring up a polish. When all grinding marks are removed, the specimen is ready to place on the refractometer. *Caution:* all plates and test specimens must be scrubbed clean between steps to avoid carry-over of coarse grit.

After polishing, the test specimen must be carefully scrubbed and dried before placing on the soft and easily scratched prism of the refractometer. To take readings, follow the instructions furnished with the refractometer. The necessary grinding powders and polishing agents can be obtained from amateur lapidary supply houses. If preferred, waterproof sanding papers may be used instead of loose grits.

LUSTER AND OTHER REFLECTIVE EFFECTS

Reflection of light from mineral surfaces, or *luster,* varies according to refractive index in a fairly consistent manner, affording a valuable indication when other evidence is lacking. With some practice in closely observing luster on freshly broken surfaces, it is possible to make reasonably accurate estimates of refractive index. In general, minerals of low refractive index appear "dead," and although one is conscious that the surfaces are clean, the impression is gained that some sort of light-absorptive film has been sprayed on them. Sometimes the surfaces look slightly oily or greasy. As the refractive index rises, more sparkle and flash is evident, like that from freshly broken glass. With still further rise, the flash becomes more intense and takes on a distinctive brilliancy. When indices are particularly high, the luster appears half-silvery and half-glassy, as if an extremely thin coating of silver had been sprayed on the surfaces in a uniform but almost invisible layer. These appearances may be summarized as follows using standard mineralogical terminology:

TABLE OF LUSTERS

Name	Approx. R.I. range	Remarks
Dull, earthy	—	Formed on earthy or porcelainous fracture surfaces
Waxy	—	Formed on minutely granular surfaces; more light reflective than above; typical of chalcedony and colloform minerals
Greasy, oily	1.3 — 1.4	Reminiscent of oily coatings; poorly reflective
Glassy, vitreous	1.5 — 1.8	Bright, like the surface of freshly broken glass
Resinous	1.6 — 1.9	Reminiscent of the luster of freshly broken rosin or shellac; wide variations in range of R.I. in species in which this luster appears
Adamantine	1.9 — 2.5	Named after typical luster of diamond crystals; part silvery in character
Submetallic	2.5 +	Silvery or metallic luster more pronounced but still not fully metallic
Metallic	—	Typical brilliant highly reflective luster of metals

Luster properly refers to reflection of light from surfaces, but it is commonly observed that reflections also arise from inclusions in minerals which range from fully transparent to nearly opaque. Such reflections mingle with surface reflections and impart a characteristic *sheen*. Where beautifully developed and in species of sufficient hardness, valuable gems may be cut from such specimens. In general, reflective effects are easily noted but may require dipping of the specimen in water or some other convenient immersion fluid to make them more conspicuous. In many instances, inclusions which cause sheen effects are oriented in respect to the host crystal or aggregate and require turning the specimen about to bring them into the best viewing position. Lusters and sheens of selected species are listed in Table 1, Appendix A.

TABLE OF SHEENS

Name	Remarks
Pearly	Formed by numerous partly-developed cleavages
Schiller	Similar in appearance but reflections are from many small separations or platey inclusions; often requires magnification to make out details
Aventurescence	*Same* as above but reflections are from many small inclusions giving a *spangled* effect
Silky	Numerous reflections from hair-like inclusions; like the reflection from a spool of silk
Chatoyant	*Same,* but inclusions are oriented in one direction affording a single streak of light when a curved surface is cut and polished, then producing *catseyes*
Asterism	*Same,* but affording several crossing streaks when appropriately cut and polished, then producing *star* stones. Some transparent to translucent minerals exhibit asterism via *transmitted* light
Opalescence	Hazy appearance, like water to which a few drops of milk have been added; commonly observed in *opal,* hence the name

Transparent scums or tarnishes due to chemical alteration on the surfaces of some minerals result in colored *iridescent* films; the effect is described as *iridescence* or *tarnish*. It is actually possible to watch iridescence develop on some sulfides which alter quickly when exposed to the atmosphere. For example, a piece of freshly broken bornite reveals a brownish-golden color and metallic luster, indicative of its true appearance. However, in a matter of several hours, the color alters in hue indicating development of tarnish. In several days, the film turns to dark blue, usually the final color assumed. Interference operates in precious opal to produce extremely vivid or nearly monochromatic colors known en masse as *play of color*. Vivid colors also arise from within labradorite feldspar, the effect being known as *labradorescence*. The strong pearly-to-blue sheen of moonstone feldspars may also be due partly to interference; it is commonly called *adularescence* because it is pronounced in the adularia variety of feldspar. The exact cause of the optical phenomena in these minerals is not clearly understood, but is believed to arise from interference of light as it passes through networks of very small slender crystals contained in the larger host crystal.

9

The Formation
and Association of Minerals

The endless variety presented by minerals is at once the delight and despair of the student of mineralogy. Boredom is never a serious threat for scarcely a month passes before a new species is announced or an older species crops up in novel form. On the other hand, the wild profusion of species within all save the most mineralogically simple deposits poses formidable problems in identification; no less formidable is the problem of unraveling the formational relationships among the various species occurring together. Fortunately all is not as hopeless as it seems because minerals, like all kinds of chemical compounds, obey the same laws of chemistry and physics during formation, resulting in restrictions not only upon the number of species which may form in any given deposit, but also upon which species may form in company, or in *association*. It is the purpose of this chapter to point out important types of mineral deposits to show how characteristic assemblages occur within them. The value of this exposition is two-fold: first, knowing the kind of deposit from which specimens are obtained immediately suggests a restricted list of species within which identification efforts should be confined; second, identifying several species known to be characteristic of a certain association also suggests the kind of deposit being dealt with. This aspect is perhaps the most useful to the active collector or prospector, for he may not immediately realize the nature of any deposit he has discovered until specimens are cleaned and examined at leisure.

The great value of associations is fully appreciated by the experienced mineralogist because the detection of one or two species known to fall within a specific association immediately leads to a search for other members, often with fruitful results. In some instances, associations are so distinctive that a glance is sufficient to tell from what kind of mineral deposit a specimen came, if not the locality itself. As an example, topaz and danburite

236

crystals are nearly identical in color, habit and crystal form, and crystals of each placed side by side might easily be confused. However, they do not occur together in mineral deposits of the same type—topaz is primarily found in granitic rocks, while danburite is usually found in metamorphosed limestones. Therefore it is possible to identify each on the basis of characteristic associated species or the rock matrix, or one can be even surer if both associates and matrix are present. In comparison, some species, as quartz and calcite, are so widespread in occurrence that identifying either upon a specimen may not be very helpful, although even this bit of knowledge is far better than nothing at all.

MINERAL FORMATION

The formation of minerals depends on three factors: (1) the *raw material* available, (2) *pressure,* and (3) *temperature.* All three factors vary considerably, but changes in composition of raw material cause widest variations in the mineralogy of any deposit. Some of the variations possible may be appreciated from considering two kinds of mineral-forming systems, *closed systems* and *open systems,* and how each is capable of producing small to large suites of species.

If any mass of mineral-forming elements is considered by itself, perhaps as if it were confined in a gigantic chemist's retort and subjected to a variety of pressure and temperature changes, it would correspond to the closed system because nothing is permitted to enter and nothing to escape. Such systems occur in nature, as in the center of large masses of molten rock, or in rock openings injected with igneous material. In the simplest case of all, that involving a single element not influenced by neighbors, as masses of sulfur or carbon, the atoms of such native elements promptly adopt a crystal structure most likely to satisfy their bonding requirements under the pressure/temperature conditions prevailing at the time of formation. In the case of carbon, which produces spectacularly different polymorphs, diamond forms in a high region of pressure/temperature but graphite crystallizes in a lower region. The crystal structure of diamond is cubic, and more closely packed than that of graphite which forms a sheeted hexagonal structure. However, if additional elements are present it is very likely that native elements will not result, but that sulfides may form if metals are present in the same system with sulfur, and if oxygen and a metal is present with carbon, the results may well be carbonates. Furthermore, if pressure/temperature varies, additional species may replace those originally present as atoms rearrange themselves to create crystal structures more stable under the new conditions.

In open systems, of which many examples are known, as veins of mineral matter brought up from the depths along narrow fissures, constituents may not be added from beneath, but may be added from above by infiltration of surface water. Many veins show progressive changes in mineralogy due to an initial supply of ingredients from some hot source far below the surface, followed by a supply of differing ingredients during a prolonged cooling period. Frequently upper parts are drastically changed from what they used to be by solution and reaction caused by infiltrating surface water. Many open systems are therefore complex in mineralogy although the most abundant species still tend to be limited in number and characteristically associated.

Restrictions on Mineral Formation. Despite the nearly 80 elements known to participate in formation of minerals, strong preferences of abundant elements to join in only a few crystal structures results in limited numbers of truly *abundant* species which together comprise practically all of the stony matter of the earth's crust. The vast majority of species are really quite rare, but seem abundant only because special pains are taken to discover and mine the deposits in which they are concentrated. Thus our impressions of the relative abundances of species is apt to be grossly distorted by the numerous rarities included in collections. Indeed, if collections contained specimens in numbers proportionate to actual quantities found in nature, feldspar and quartz, as the most abundant of all species, would comprise practically all of such collections, followed by a much lesser number of other silicate minerals, and a vanishingly small suite representative of all the rest. The estimated abundances of elements in the crust, expressed by percentages of atoms, emphasizes these statements:

Distribution of Abundant Elements

Oxygen	62%
Silicon	21

83% (largely in silicate minerals)

Aluminum	6.5%
Sodium	2.6
Calcium	1.9
Iron	1.9
Magnesium	1.8
Potassium	1.4

16.1% (major portions combined in silicate minerals)

As can be seen in the above table, many of the "common" metals as copper, zinc, lead, tin, etc., are not even mentioned because they are exceedingly rare, and actually are abundant only in the small ore bodies in which their minerals are concentrated. Oxygen, silicon and aluminum comprise nearly 90 percent of all atoms in the earth's crust, leading to the intriguing conclusion that almost all of the crust is a more or less continuous network of silicate ions linked together by sodium, calcium, iron, etc.

Another limitation on formation of species is the tendency of some rare elements whose chemical behavior is close to that of more abundant elements to occur in company with the latter and avoid formation of their own species. For example, hafnium chemically behaves like zirconium, and is therefore common in zircon where it substitutes for zirconium, but is rare elsewhere. Because much zirconium is used up in very small crystals sprinkled throughout igneous rocks, neither this element nor its close chemical associate, hafnium, are likely to be well represented by specimens in the average collection. Similarly, scandium rides along with aluminum as it were, and its own compounds are therefore extremely rare; similarly rubidium, occurring with potassium, does not even form a recognized mineral compound. Finally, the rare-earth elements, as cesium, lanthanum, etc., of which there are fifteen, are so close in chemical behavior, that all of them occur in the same minerals instead of forming distinctive compounds.

ROCK ENVIRONMENTS

While the detailed discussion of rock compositions and fabrics is beyond the scope of this book, it is important to summarize briefly the formation and principal features of each of the three great divisions of rocks, namely, *igneous, sedimentary,* and *metamorphic,* because numerous fine mineral specimens originate within the rocks themselves, or owe their origin to material derived from certain types of rocks. Specific deposit types will be discussed later in this chapter, at which time detailed remarks on associations will be made.

The Igneous Environment. Igneous rocks are *melt* rocks, ranging from fluid types issuing from volcanic vents and known as lavas or *extrusive* rocks, to the very large bodies formed at depth known as magmas or *magmatic* rocks. Many variations occur in composition, size, and modes of formation, but within any given mass, the mineralogy tends to be simple toward the core although it may become complex near margins because of intermixtures or reactions with other rocks. Some, as magmas, appear to have formed in place, but others, as volcanic flows, or intrusions into other rocks, have been transported considerable distances. In general, two broad classes of igneous rocks are recognized on the basis of characteristic mineral content, depending on the relative abundance of silica within them. In granites and similar *light-colored* rocks, silica is in excess and is commonly found as grains of free quartz; the pale coloration is due to this mineral and light-colored feldspars. In basalts and other *dark-colored* igneous rocks, silica is less abundant causing free quartz to be absent or nearly so; the dark coloration is due to grayish feldspars plus predominant black to green-gray iron-magnesium silicates. Practically all deep-seated rocks, often forming the highest and most widespread mountain chains, are granitic, while practically all extrusive rocks, as lava flows, "trap rock" sills, etc., are basaltic.

As the variety of constituents in any igneous mass increases, drops in pressure/temperature may cause early formation of certain minerals, usually the darker iron-magnesium silicates, which may sink or otherwise become *segregated*. The remainder of the igneous material changes drastically in character because of this natural filtering of constituents, and if the entire mass cooled, distinctive rocks would be found in different portions. Sometimes pressure/temperature changes bring about formation of high temperature minerals which may persist if the mass is suddenly cooled, providing rocks also of different character. In other instances, reactions take place within the slowly cooling mass resulting in the destruction of species formerly present and creation of new ones. These processes are believed to account for the many variations noted in igneous rocks and in many mineral deposits of complex mineralogy.

Not all igneous rocks form simply from solidification of melts; some are richly mineralized with rare elements trapped in gases composed principally of water vapor and borne along by the latter to be deposited in favored places. Mobility of gases allows them to seek out regions of lower pressure, usually toward the surface, and finally to condense wherever space is found and environments are suitable. Such escaping constituents commonly cross the boundaries of the igneous bodies in which they originated, and invade rocks which may be of vastly different composition and geographically far removed. Furthermore, as the intense heat of deeper regions is left behind, the gases condense into liquids in which much mineral matter remains dissolved until lowering pressures and temperatures nearer the surface encourage formation of minerals from solution.

Thus emanations from deep-seated igneous bodies can transport elements to great distances, and deposit them in a wide variety of circumstances not only within cooler parts of the igneous mass, but in totally unrelated rocks which happen to be porous or to contain fissures through which passage is possible. If the invaded rocks are compositionally similar to the mass from which the gases escaped, chemical reactions may be minor in nature, and the resulting mineral deposits essentially represent the composition of the escaping material. On the other hand, if the invaded rock is easily attacked, as limestone, new constituents are robbed from the host rock and added to the others, resulting in a greater variety of species. Injections of more or less fluid material into openings in rocks immediately surrounding igneous bodies, usually from granitic magmas, results in *pegmatites*. The latter are small in size but coarsely crystallized, frequently containing sufficient concentrations of rare elements to make them especially prospected for and exploited. Cavities within them commonly are lined with exceptional crystals of feldspar, quartz, tourmaline, and other desirable minerals. Escaping gases and fluids, often from igneous bodies so far removed that their presence can only be conjectured, are believed responsible for most ore veins mined for metallic elements as copper, zinc, gold, silver, lead, etc. Some igneous rocks, as basalt and diabase, contain gas cavities filled by interesting though economically unimportant minerals deposited from hot-water solutions at low pressures and temperatures.

The Sedimentary Environment. The next great division of rocks, the *sedimentary,* includes those formed directly or indirectly from the decay of other rocks, as simple aggregates of particles, or as accumulations of mineral and organic materials subjected to a modest degree of chemical change. Sedimentary rocks composed of particles form sandstones, shales, conglomerates, etc., while those formed through chemical change include saline rocks, created by the precipitation of water-soluble minerals from brines, and limestones created by limey material deposited on the floors of ancient bodies of water, largely from the skeletons of various organisms. Still other types are created by the circulation of water which dissolves mineral constituents, transports and deposits them within some specific formation. If chemical reactions are extensive, the final rock may little resemble the starting material. Sometimes the mineralogy of sedimentary rocks is complicated by chemical activity, but mostly it is simple because even in those formed from the decay of mineralogically complex rocks, low surface pressures and temperatures do not permit the variety and intensity of reactions possible in igneous rocks. However, as pointed out before, some sedimentary rocks may play host to very interesting mineral deposits if, like limestone, they are easily subject to chemical attack by invading igneous gases and solutions.

Circulating waters in some sedimentary rocks are responsible for formation of well-crystallized minerals in openings. Limestones commonly contain excellent calcite, barite, celestite, strontianite, millerite and fluorite crystals in cavities. The splendid sulfur-aragonite-celestite associations from Sicily are emplaced in limestone but are believed to derive from igneous emanations. Quartz veins, often containing large cavities lined with splendid crystals, are abundant in sandstones of Arkansas, however, the supply of silica in this instance again appears related to some deep-seated source, probably igneous.

The Metamorphic Environment. The last great division of rocks, the *metamorphic,* includes rocks formed from the more or less complete transformations of previous rocks, ordinarily through the agencies of temperature and pressure, but also through introduction of foreign constituents by gases and solutions, usually accompanied by the influx of considerable heat.

If much heat and pressure are involved, as when a sedimentary rock becomes deeply buried beneath additional sediments, or is buckled by crustal movements, a closed system commonly results, and the minerals formed contain the original constituents as modified by the new pressure/temperature environments. Chemical changes may range from very slight to drastic, sometimes reducing the rock to material indistinguishable from an igneous melt rock. Igneous rocks subjected to metamorphism ordinarily show little change in mineralogy except for characteristic streaking and layering. On the other hand, easily metamorphosed materials, as carbonate rocks, particularly if impure to begin with, readily convert into snow-white marbles spotted with segregations of minerals formed from impurities and the reactions between them and the carbonate ions. The most complex mineralogies occur in metamorphic limestones invaded by igneous emanations, and sometimes further invaded by intrusions of granitic pegmatites or tongues of igneous rock. Sandstones, if pure, may convert into hard compact masses of quartzite, but if impure, silicate minerals as tourmaline, staurolite, kyanite, and others, may be formed.

COMMON ROCK-MINERAL ASSOCIATIONS

Although it is impossible to describe all recorded associations in a brief summary, the following table provides a few of the most important. The table is divided according to the three principal rock divisions, with columns in each for the rock type, and some minerals commonly found in them. In later sections, selected mineral deposits producing specimens of interest to the collector will be described to show additional mineral relationships helpful not only in identification but in gaining an appreciation of the variety of change which occurs in some mineral deposits.

GRANITIC PEGMATITES

Strictly speaking, the term *pegmatite* refers to any exceptionally coarse-grained phase of rock, but the pegmatite bodies of greatest interest are those which occur within granite, or have been injected into rocks surrounding the parent granitic magma. Small bodies enclosed in granite are believed to be places where accumulations of gases prevented normal solidification of the rock, thus affording openings toward which fluids and gases later migrated and induced formation of exceptionally large crystals. The term *miarolitic* is also used to describe pegmatitic granitic rocks more or less honeycombed by gas openings. Many miarolitic cavities do not exceed several inches in diameter, but much larger openings are known. Pegmatites intruded into rocks surrounding granitic magmas, commonly into overlying

SUMMARY TABLE OF ROCK-MINERAL ASSOCIATIONS

IGNEOUS GROUP

Rocks	Deposits and Minerals
Rhyolite	Tin ores in fissures: cassiterite (wood tin), topaz, black tourmaline; cavities and fissures: hyalite opal, common opal, precious opal. *Examples:* in San Luis Potosi, Queretaro and other Mexican States
Trachyte	Fissures: turquois, kaolinite, minor sulfides; cavities: opal. *Examples:* in southwest United States; Mexico
Basalt	*Amygdaloidal:* native copper and silver, quartz, datolite, prehnite, calcite, epidote, chlorite, pumpellyite, adularia, zeolites. *Example:* Keeweenaw Peninsula, Michigan *Pillow basalt:* quartz, calcite, prehnite, pectolite, apophyllite, zeolites, chlorite, babingtonite, minor copper and silver, minor sulfides, etc. *Examples:* basalt sills of Pennsylvania-New Jersey-Connecticut, Oregon, etc.
Diabase	Seams and fissures: silver, copper, cobalt-nickel ores, apophyllite, prehnite, datolite, zeolites. *Examples:* diabase bodies in Ontario, Canada, and east United States
Granite	Veins, seams, and altered zones: cassiterite, topaz, scheelite, wolframite, black tourmaline, quartz, fluorite, arsenopyrite, muscovite, molybdenite, bismuth, etc. *Examples:* Saxony, Germany; Czechoslovakia. Uranium ores in veins: *Example:* Great Bear Lake, Canada Gas cavities lined with pegmatitic material: feldspar, quartz, topaz, beryl, phenakite, fluorite, siderite, goethite, etc. *Examples:* Pikes Peak granite region, Colorado; Conway red granite region, New Hampshire
Granitic Pegmatites	Often emplaced in metamorphic rocks as small to large bodies, usually of vein-like shape: feldspar, quartz, mica, tourmaline, beryl, garnet, columbite, tantalite, amblygonite, pollucite, spodumene, topaz, and many others. *Examples:* pegmatite regions of southern California, Maine, Brazil, Madagascar
Syenite	Veins: quartz, gold, sulfides. *Example:* Ontario
Nepheline Syenite	Distinct crystals (phenocrysts): corundum, zircon; masses: sodalite, cancrinite. *Examples:* bodies of nepheline syenite rock in Ontario, Maine, Arkansas

SUMMARY TABLE OF ROCK-MINERAL ASSOCIATIONS (Continued)

IGNEOUS GROUP

Rocks	Deposits and Minerals
Granodiorite Monzonite	Disseminated: "porphyry" copper ores. *Examples*: in numerous places in southwest United States. Molybdenite. *Example*: Climax, Colorado Altered zones: scheelite. *Example*: Bishop, California
Diorite Gabbro	Veins containing native metals and sulfides in numerous localities. Large masses: magnetite, ilmenite, pyroxenes, copper, nickel, and iron sulfides. *Example*: Lake Superior region
Anorthosite	Large coarsely crystalline masses: labradorite, magnetite, ilmenite. *Examples*: New York; Iron Mountain, Wyoming; Labrador
Peridotite	Masses: chromite, nickel ores with olivine, pyroxene, spinel, platinum, magnetite, pyrrhotite, etc. *Examples*: New Caledonia; eastern Pennsylvania; Siskiyou County, California. Single crystals: diamond, pyrope garnet, enstatite, etc. *Examples*: Murfreesboro, Arkansas; numerous localities in Africa

SEDIMENTARY GROUP

Conglomerate	Copper, silver. *Examples*: Keeweenaw Peninsula, Michigan; Ontario. Diamond. *Example*: Brazil. Gold. *Example*: Witwatersrand, Union of South Africa
Sandstone	Disseminated: uranium-vanadium ores. *Example*: Colorado plateau. Cavities: quartz crystals. *Example*: Hot Springs region, Arkansas. Seams, veins, and disseminated: cinnabar, galena, sphalerite, barite, celestite, strontianite, calcite, gypsum, anhydrite, etc. *Examples*: numerous localities. Fossil replacements: uranium-vanadium ores, as above; quartz after wood, etc. *Example*: Colorado plateau
Shale	Seams and cavities: boron minerals, sulfates and carbonates in playa deposits. *Example*: Searles Lake, California. Strata and seams: anhydrite, gypsum. Concretions and replacements: pyrite, marcasite, goethite; fossils: jet, amber. *Examples*: numerous localities
Limestone	Sulfide veins and disseminated deposits: sulfides of lead, copper, zinc principally. *Examples*: Cananea, Sonora; Mapimi, Durango, Mexico; many others. Seams and geode-like openings: barite, celestite, strontianite, fluorite, quartz, minor sulfides. *Example*: Clay Center, Ohio

Chert	Breccia openings, seams, veins: lead, zinc, and iron sulfides, minor quartz; calcite. *Example:* tri-state district of Oklahoma, Kansas, Missouri
Saline Rocks	Strata: halite, gypsum, anhydrite, sylvite, carnallite, etc. *Examples:* Carlsbad, New Mexico, deposits in New York, Michigan, California, Germany, etc.

METAMORPHIC GROUP

Gneiss	Disseminated pods, single crystals, lenses, stringers: almandite, graphite, cordierite, corundum, mica, andalusite, kyanite, staurolite, ilmenite, etc. *Examples:* Gore Mt., New York; Swiss Alps
Schist	Disseminated pods, single crystals, lenses, stringers: garnet, graphite, talc, pyrophyllite, serpentine, chlorite, magnetite, tremolite, actinolite, epidote, brucite, mica, etc. *Examples:* Wrangell, southeastern Alaska; numerous localities in California, Alpine regions, Appalachian Mountains, etc. Often intruded by pegmatites
Serpentine	Disseminated and vein-like bodies: garnierite, pyrrhotite, chromite, chrysotile asbestos, magnesite, brucite, actinolite, jadeite, cinnabar, etc. *Examples:* southern counties of Oregon and northern counties of California; "barrens" of Maryland, etc.
Quartzite	Cavities: quartz. *Examples:* numerous localities
Slate	Cavities: quartz. Veins: quartz with gold and sulfides. Scattered crystals: pyrite. *Examples:* numerous localities
Phyllite	Scattered crystals and pods: staurolite, andalusite, kyanite
Marble	Scattered crystals, pods, lenses: sulfides, phlogopite mica, graphite, corundum, spinel, chondrodite, grossularite garnet, scapolite, pyroxenes, and amphiboles, etc. *Examples:* northern New Jersey-southern New York marble belt
Marble Skarn	Scattered crystals, pods, lenses: magnetite, spinel, corundum, graphite, andradite and grossularite garnet, wollastonite, scheelite, pyroxenes and amphiboles, scapolite, sulfides, zincite, willemite, many other species. *Examples:* scheelite ore bodies in California; Franklin, New Jersey; Brewster, New York; Riverside County deposits, California, etc.

schists, occur in a wide range of sizes, from bodies only several inches thick to some which may be thousands of feet in length and several hundred feet thick. It is from such larger bodies that commercial feldspar, mica, and a number of rare-element minerals are won, not to mention gemstones such as tourmaline, beryl, quartz, etc., occurring in exceptional crystallizations within cavities.

The *essential* mineralogy of all granitic pegmatites comprises the principal species noted in the parent magma, that is, feldspar is predominant, with quartz abundant, and mica usually occurring in lesser quantity. Matrix specimens commonly show the characteristic feldspar-quartz intergrowth known as *graphic granite,* a certain clue to granite pegmatite origin. Simple mineralogy is the rule in pegmatites injected in one period in the closed system mode. Occasionally, associates in addition to the usual feldspar-quartz-mica are black tourmaline, beryl, and more rarely, columbite-tantalite. Complexly mineralized injections may result in a much wider variety of species, and if the system is open, additional constituents may be introduced from neighboring rocks or from the surface. Complex pegmatites sometimes furnish dozens of species from one body and are eagerly searched for rare species by collectors who have access to them.

The following list includes principal species found in miarolitic cavities. Excellent specimens are obtained from the Pikes Peak red granite region of Colorado and the Conway, New Hampshire, red granite region; also from Mount Antero-White Mountain, Colorado, from Baveno on Lake Maggiore, Italy, the Mourne Mountains of Ireland, Cairngorm Mountains of Scotland, Striegau, Lower Silesia, Poland, etc.

MIAROLITIC CAVITY ASSOCIATES

Abundant: microcline (commonly amazonite) quartz commonly smoky)
Common: albite, beryl, fluorite, muscovite, topaz
Uncommon: apatite, bertrandite, columbite-tantalite, goethite, monazite, phenakite, siderite, zeolites

The following list includes important species occurring in the much larger granitic pegmatite bodies as those of San Diego County, California, also Maine, Connecticut, and abroad in Brazil, Mozambique, Africa, Madagascar, etc.

GRANITIC PEGMATITE ASSOCIATES

Abundant: microline-perthite, quartz, muscovite, biotite
Common: albite, almandite, apatite, beryl, oligoclase, orthoclase, schorl
Uncommon: allanite, beryl (morganite), cassiterite, chysoberyl, columbite-tantalite, cookeite, fluorite,

	goethite, halloysite, hematite, ilmenite, kaolinite, lepidolite, monazite, montmorillonite, rutile, spessartite, spodumene, topaz, tourmaline (colored)
Rare:	amblygonite, andalusite, arsenopyrite, bismuth, bismuthinite, bornite, braunite, calcite, chalcocite, chalcopyrite, corundum, covellite, cyrtolite, dolomite, fergusonite, gadolinite, gahnite, galena, graphite, gummite, herderite, loellingite, magnetite, manganite, molybdenite, psilomelane, pyrite, pyrolusite, pyrrhotite, rhodochrosite, samarskite, scheelite, siderite, sphalerite, sphene, stilbite, tapiolite, thorianite, thorite, triphylite, uraninite, wolframite, zircon
Very rare:	augelite, autunite, bavenite, bastnaesite, bertrandite, beryllonite, beyerite, betafite, brazilianite, cerussite, childrenite, dickinsonite, eosphorite, eucryptite, euxenite, fairfieldite, graftonite, hambergite, helvite, heterosite, hureaulite, laumontite, lazulite, lithiophilite, ludlamite, microlite, palaite, parisite, petalite, phenakite, phosphophyllite, phosphouranylite, pollucite, polycrase, purpurite, pyrochlore, reddingite, rhodizite, salmonsite, scorodite, scorzalite, sicklerite, stannite, stibiotantalite, strengite, torbernite, triplite, vivianite, xenotime, zinnwaldite

ALUMINO-FLUORIDE PEGMATITES

A distinctive class of granitic pegmatites contain unusual concentrations of alumino-fluoride species, particularly cryolite, and from them are obtained most of the specimens of such species. Very few bodies are known; the largest by far is an approximately spherical body, several hundred feet in diameter, on the shore of Arsuk Fjord in southwestern Greenland, from which come all commercial supplies of natural cryolite. Other smaller bodies occur near Miask in the Ilmen Mountains of the U.S.S.R., and at St. Peters Dome in Colorado. The following are associates:

ALUMINO-FLUORIDE PEGMATITE ASSOCIATES

Greenland

Abundant:	cryolite, comprising nearly all of the deposit
Common:	chalcopyrite, fluorite, galena, mica, microcline, quartz, siderite
Uncommon:	arsenopyrite, cassiterite, chiolite, columbite, cryolithionite, gearksuktite, molybdenite, pachnolite, ralstonite, topaz, wolframite, zircon
Rare:	jarlite, prosopite, thomsenolite, weberite

ALUMINO-FLUORIDE PEGMATITE ASSOCIATES (*Continued*)

Ilmen Mountains

cryolite, cryolithionite, chiolite, fluorite, gearksuktite, pachnolite, phenakite, thomsenolite, zinnwaldite

Colorado

cryolite, chiolite, cryolithionite, elpasolite, fluorite, gearksuktite, pachnolite, prosopite, ralstonite, thomsenolite, weberite

TIN-TUNGSTEN FISSURE VEINS IN GRANITE

Formerly, considerable tin and tungsten ores were won from narrow fissure veins in granite through which high temperature aqueous solutions passed, depositing a characteristic suite of minerals, among them cassiterite, the oxide of tin, and wolframite, a tungstate. Alteration of wall rocks commonly results in narrow zones of granular rock known as *greisen,* impregnated with cassiterite, topaz, and quartz. Fissures are sometimes closed by more or less solid masses of similar material, but in a number of deposits, openings are lined with splendid crystallizations affording numerous specimens of high quality. Typical specimens show micaceous coatings on cavity wall linings, usually stained with brown iron oxides, upon which rest splendent crystals of cassiterite and wolframite. The list of associated species is small:

TIN-TUNGSTEN FISSURE VEIN ASSOCIATES

Apatite, arsenopyrite, axinite, bismuth, cassiterite, fluorite, kaolinite, muscovite, molybdenite, prosopite, pyrite, quartz, scheelite, sphalerite, topaz, tourmaline, wolframite, zinnwaldite

Prominent localities formerly furnishing numerous fine specimens are many mines in Saxony, Germany, in Bohemia in Czechoslovakia (Zinnwald, Schlaggenwald), Montebras in France, and at many places in Cornwall, England. The famous cassiterite occurrences in Bolivia are sulfide veins which, as will be described later, differ in character from the deposits above described.

HYDROTHERMAL DEPOSITS

The term *hydrothermal* is used in connection with minerals deposited through the agency of heated water. When sufficiently high in temperature, and also under high pressure, water is easily capable of carrying considerable mineral matter in solution. The rapid laboratory growth of synthetic quartz demonstrates the effectiveness of the hydrothermal process in dissolving such seemingly imperishable minerals as quartz and transporting its ions to be deposited elsewhere. Hydrothermal processes are active during late stages of formation in many pegmatite bodies but principally operate to fill long fissures with mineral matter brought up from deep-seated igneous bodies. Many of the resulting veins are filled with quartz, some with quartz and sulfides, others almost entirely with sulfides. In some instances, porous rocks through which fissures pass become permeated by hydrothermal solutions and mineralization extends considerably beyond the walls of the original openings. Some low grade copper deposits in Arizona and elsewhere, formerly mined by open pit methods to remain economical, are essentially porous igneous rocks impregnated by sulfides, which commonly are later altered by seepage of surface waters. Practically every type of opening in all classes of rocks is susceptible to hydrothermal invasion. As a consequence, there are many mineral deposits, both large and small, that assume a great variety of shapes. Hydrothermal veins have been mined for centuries for the sake of their native metals, metallic sulfides, and the alteration products that result.

The exploration of veins to depth shows mineralization changing considerably from the surface downward, pointing to a parallel change in pressure/temperature conditions, increasing with depth. Thus certain species are characteristically associated with high pressure/temperature regions, and others with moderate conditions, and still others with conditions that are close to the ordinary surface pressure/temperature environment. Many hydrothermal veins consist of solid granular mixtures of various species, but others contain centerline openings or narrow vugs lined with excellent mineral druses, commonly of quartz, on which may be implanted other species.

The following lists of species are divided into regions of high, medium, and low pressure/temperature, conveniently placing certain characteristic associates together and thus aiding in identification as well as appreciation of how such associates form. A number of species, as quartz, sphalerite, pyrite, etc., deposit throughout entire vein systems, others are somewhat less restricted, while still others require relatively narrow ranges of pressure/temperature to form. Not *all* of the associates listed will occur together, and the following lists are not meant to imply that they do.

Common Hydrothermal Associates

HIGH PRESSURE/TEMPERATURE—DEEP VEIN PORTIONS

(Amphibole,) apatite, arsenopyrite, axinite, bismuthinite, breithauptite, cassiterite, chalcopyrite, childrenite, chlorite, cobaltite, cubanite, epidote, fluorite, galena, gold, gummite, hausmannite, hematite, huebnerite, ilmenite, loellingite, magnetite, molybdenite, muscovite, niccolite, orthoclase, pentlandite, pyrite, pyrrhotite, quartz, rutile, scheelite, sphalerite, stannite, tetradymite, topaz, tourmaline, uraninite, wolframite, wurtzite, zinnwaldite

MEDIUM PRESSURE/TEMPERATURE—MID VEIN PORTIONS

Andorite, ankerite, arsenopyrite, axinite, barite, bismuth, bismuthinite, boulangerite, bournonite, calaverite, calcite, chalcopyrite, chloanthite, cobaltite, dolomite, enargite, epidote, fluorite, franckeite, galena, gold, gummite, jamesonite, loellingite, niccolite, polybasite, pyrite, quartz, rammelsbergite, rhodochrosite, scheelite, siderite, silver, skutterudite, sphalerite, tennantite, tetradymite, tetrahedrite, uraninite, wolframite, wurtzite

LOW PRESSURE/TEMPERATURE—UPPER VEIN PORTIONS

Allemontite, anhydrite, antimony, apophyllite, argentite, arsenic, arsenopyrite, augelite, barite, bismuth, bismuthinite, boulangerite, brucite, calaverite, calcite, cervantite, chalcocite, chlorite, cinnabar, dolomite, fluorite, galena, gibbsite, goethite, gold, greenockite, hessite, jamesonite, jarosite, krennerite, ludlamite, magnesite, manganite, marcasite, miargyrite, millerite, opal, orpiment, petzite, phosphophyllite, polybasite, proustite, pyrargyrite, quartz, realgar, rhodochrosite, siderite, silver, sphalerite, stephanite, stibiconite, stibnite, strontianite, sylvanite, tellurium, tetrahedrite, wavellite, witherite, zeolites

Despite the seemingly large number of species listed above, hydrothermal veins, like most pegmatite bodies, consist essentially of several species only; the rest of the vein contains rarer species in much lesser quantities, sometimes only as sprinklings of grains within other minerals, as is commonly the case with sulfide species. A large number of veins are essentially quartz, usually a milky type, unrelieved in monotony of mineralization except by occasional patches of sulfide minerals. Thus most veins are quartz-sulfide veins with the most abundant sulfides being pyrite, galena, chalcopyrite, and sphalerite.

Hydrothermal veins are widely distributed throughout all continents but are exposed and exploited chiefly in mountainous countries. A notable exception to this rule is Switzerland, where very few economic metallic veins of any description occur despite an abundance of mountains. Celebrated vein-mining centers are found in Saxony, Germany, also in Hungary,

Romania, Czechoslovakia, and Cornwall and the Midlands of England, in Australia, South America, particularly along the Andes Cordillera, in Mexico in very many places, in the United States, Canada, etc. American collectors of minerals are predominantly supplied by the Mexican deposits, which currently produce a wide variety of species from numerous small to large mines. The very characteristic galena-sphalerite-dolomite-chalco-pyrite-marcasite association of the tri-state district, comprising numerous mines in chert and limestone beds centered at the junction of Oklahoma, Kansas and Missouri, is well represented in almost every collection. The unique Bolivian tin deposits, essentially sulfide veins rich in cassiterite and rare tin species, still provide unexcelled examples for collectors the world over.

ALTERATION OF SULFIDE VEIN MINERALS

Among the most zealously collected specimens are the numerous colorful or interesting species which form from the chemical alteration of original minerals within hydrothermal veins. Such species are called *secondary*, to distinguish them from the *primary* minerals which first constituted the deposit. As noted before, typical hydrothermal veins are rich in sulfides, and it is these primary sulfides which afford numerous secondary minerals through chemical alteration, usually by the introduction of oxygen and carbon dioxide from surface waters seeping downward through the vein. Two classes of secondary minerals are prominent: *carbonates* and *sulfates*. The important reactions are conversions of sulfides into sulfates, and reactions of sulfates with primary minerals and with limestone, if this is the host rock, to form carbonates.

CONVERSIONS OF SULFIDES TO SULFATES

Sulfide	*to*	*Sulfate*
Pyrite		Melanterite
Chalcopyrite		Chalcanthite
Galena		Anglesite
Sphalerite		Goslarite

The upper leached zone is called the *oxidized zone* inasmuch as it was here that oxygen was introduced into the system. Below this zone occurs an *enriched zone,* in which reactions between ions in downward percolating solutions and existing primary sulfides, caused formation of new sulfide species. The term "enrichment" is applied to copper deposits in particular, because it was found that the new sulfides contained greater proportions of copper, hence became enriched in this element. Below the enriched zone occur the unaltered primary species. The oxidized zone usually provides the widest variety of species, and frequently, because space is created by removal of sulfides, splendid crystallizations occur within the honeycombed masses of brown limonite (gossan) which are found in the oxidized zones as alteration products of iron and copper pyrites. Although copper-lead-zinc deposits are more common than other types of sulfide bodies, and hence secondary species from them are best known among collectors, almost any mineral deposit can be affected by oxidation. Furthermore, if a hydrothermal deposit is oxidized, such alterations as occur take place in the pressure/temperature vein region exposed during the geologic period concerned. Erosion may have removed an upper vein portion, for example, thus leaving a mid-vein portion exposed to oxidation, in which the alteration naturally will affect a somewhat different association of primary minerals than would have been the case if the upper vein portion were exposed. Variations in oxidized zone minerals are therefore considerable from deposit to deposit. The following secondary minerals are observed in oxidized veins.

Oxidized Hydrothermal Vein Associates

Adamite, anglesite, annabergite, antlerite, aragonite, atacamite, aurichalcite, autunite, azurite, bismutite, bornite, brochantite, calcite, caledonite, calomel, cerargyrite, cerussite, chalcanthite, chalcocite, chalcophanite, chalcopyrite, chrysocolla, copiapite, copper, coquimbite, covellite, crocoite, cuprite, cyanotrichite, descloizite, digenite, dioptase, dufrenite, enargite, epsomite, erythrite, fluorite, goethite, goslarite, gypsum, halotrichite, hemimorphite, hydrozoncite, iodyrite, jarosite, kroehnkite, leadhillite, libethenite, linarite, liroconite, malachite, massicot, matlockite, melanterite, mimetite, minium, mottramite, olivenite, phosgenite, plattnerite, pucherite, pyromorphite, roemerite, rosasite, scorodite, senarmontite, shattuckite, silver, smithsonite, spangolite, tarbuttite, tenorite, torbernite, tungstite, vanadinite, vivianite, willemite, wulfenite

HYDROTHERMAL DEPOSITS IN BASALTIC ROCKS

Igneous flow rocks are largely barren of interesting mineralization because movement causes thorough mixing of ingredients while outpourings on the surface or injections into near-surface rocks cause rapid chilling and discourage segregation or growth of large grains. Basalts, diabases, and similar types are therefore fine-grained, hard and dense. However, in some favored places, gas cavities occur of bread-loaf shape or, as some prefer to describe them, *amygdaloidal*—in reference to the almond shape which is sometimes more common. Many cavities remain unfilled, but some, as in the mineralogically prolific basalt flow areas of Rio Grande Do Sul and adjacent areas in Uruguay, contain fillings or linings of chalcedony on which may be further deposited drusy quartz, commonly amethyst. Chalcedony is deposited hydrothermally at low pressure/temperature, often coating or replacing earlier species as aragonite, anhydrite, calcite, glauberite, etc. Bandings concentric to cavity walls are the rule, but straight bands (onyx), botryoidal or stalactitic growths, with or without coatings of drusy quartz, are also common. The variations in mode of formation, inclusions, colors, and patterns are almost endless.

The outstanding specimen occurrences in flow rocks occur in basalts and diabases, more commonly in the former where angular cavities are present between the curious ball-like masses of altered rock which vaguely resemble pillows in shape and give this type its appellation of *pillow basalt*. Formations of this type suggest that lavas flowed over wet ground, possibly shallow salt marshes, the "pillows" forming within the flow during forward movement and becoming permeated with hot gases from moisture trapped beneath, and later, when the flows congealed, by rising hydrothermal mineralization. The unusual number of cavities provided ready access for ascending solutions and space for excellent crystallizations. The borders of pillows are commonly altered, thus soft and brittle, allowing specimens to be removed without much damage. In some flows, the quantity and variety of specimen material is astounding, particularly in the famous Watchung sills of northern New Jersey.

The sequence of mineralization is generally as follows: anhydrite and glauberite, followed by quartz, prehnite, datolite and pectolite, zeolites, and lastly calcite. Not all stages may be present, there being considerable deviation from flow to flow or even within different parts of the same flow. Usual cavity contents consist of altered basalt linings, commonly coated by drusy colorless quartz crystals, over which other species form. Quartz is scarce or absent as in some prehnite-lined cavities. The following are typical species but suites found in various deposits again vary widely both in respect to number and specific association.

PILLOW BASALT CAVITY ASSOCIATIONS

Common: analcime, apophyllite, calcite, chabazite, datolite, heulandite, laumontite, mesolite, natrolite, pectolite, prehnite, quartz, scolecite, stilbite, thaumasite

Uncommon: albite, babingtonite, bornite, chalcopyrite, chlorite, gmelinite, harmotome, hematite, phillipsite, thomsonite

Rare: actinolite, allanite, anhydrite, apatite, arsenopyrite, aurichalcite, axinite, azurite, barite, chrysocolla, copper, cuprite, dolomite, epidote, glauberite, goethite, greenockite, gypsum, magnesite, malachite, opal, orpiment, orthoclase, pumpellyite, pyrolusite, siderite, silver, sphene, stevensite, stilpnomelane, talc, tourmaline, ulexite

Similar associations are noted in diabase, except that minerals occur in fault and fracture fissure openings rather than in the angular cavities typical of pillow basalt or the bread-loaf cavities of amygdaloidal basalt. Considerably less variety is noted. Eminent pillow basalt localities are in the Deccan of India, at several places in Germany, in France, the British Isles, Iceland, Nova Scotia, in the extensive region of sills in the northeastern United States from Pennsylvania to Massachusetts, and in Oregon. Lately, excellent specimens, rivaling those from India which long have been conceded to be the best, have come from amygdaloidal igneous rocks in Rio Grande Do Sul in Brazil.

Because zeolites and species associated with them form at low pressure/temperature, small quantities are commonly found encrusting other minerals in cavities in sulfide veins, pegmatites, and many other occurrences. As examples, stilbite is common as a last mineral to form in the cavities of complex pegmatites; apophyllite is sometimes abundant on drusy quartz in the silver-lead-zinc sulfide veins of Mexico, etc. Although zeolites occur most abundantly in igneous flow rocks, they are by no means confined to them.

FUMAROLIC AND HOT SPRING DEPOSITS

Gaseous emanations and rising waters from deep-seated but still hot igneous masses form important deposits of sulfur and mercury ores. Sometimes excellent specimens are obtained from them. The following associates are commonly noted.

FUMAROLIC AND HOT SPRING ASSOCIATES

Alunite, anhydrite, aragonite, arsenic, barite, calcite, calomel, chalcedony, cinnabar, dolomite, fluorite, gypsum, hematite, magnetite, marcasite, mercury, metacinnabar, opal, pseudobrookite, pyrite, quartz, realgar, sal ammoniac, stibnite, sulfur

SEDIMENTARY DEPOSITS AND ASSOCIATIONS

As mentioned before, much of the bulk of ordinary sedimentary rock is without interest mineralogically, particularly sandstones, conglomerates, and similar rocks formed from accumulations of rock fragments. Quartz is the most resistant common mineral and is therefore preponderant. However, its resistance to chemical change also insures that few changes of interest will occur within siliceous rocks. On the other hand, limestone frequently develops cavities, sometimes of lens-like shape but commonly in spherical or geodal shape. Low pressure/temperature hydrothermal infiltrations introduce quartz and other species, sometimes creating handsome crystallizations. The following are prominent geode species.

Associates in Sedimentary Rock Cavities

Anhydrite, aragonite, barite, calcite, celestite, dolomite, fluorite, glauberite, gypsum, marcasite, millerite, pyrite, quartz, strontianite, sulfur

A far greater variety of species is provided by the fractional precipitation and crystallization of water-soluble species from concentrated brines. Some deposits represent dehydrated sea-water brines, others, brine minerals formed from long-continued influxes of mineralized surface waters to inland lakes with high rates of evaporation and no outlets. The latter are characteristic of desert regions with high mountains and nearly flat valley floors, on which are "dry salt lakes" or playas. The famous Searles Lake in San Bernardino County, California, is an example, while the extensive salt beds of Michigan and the potash beds of New Mexico are examples of sea-brine concentrations. The following form typical associates within such deposits.

Saline Deposit Associates

Anhydrite, aphthitalite, aragonite, bakerite, bloedite, boracite, borax, burkeite, calcite, carnallite, celestite, colemanite, copiapite, coquimbite, epsomite, gaylussite, ginorite, glauberite, gypsum, halite, hanksite, howlite, hydroboracite, inderite, inyoite, kainite, kieserite, kernite, kramerite, krausite, kurnakovite, langbeinite, meyerhofferite, mirabilite, nahcolite, pirssonite, polyhalite, priceite, probertite, sassolite, searlesite, soda niter, sulfohalite, sylvite, thenardite, tincalconite, trona, ulexite, vanthoffite

METAMORPHIC ROCKS AND ASSOCIATIONS

Numerous bodies of weakly-to-strongly-metamorphosed rocks display characteristic associations, created from the application of pressure/temperature due to accumulation of overlying sediments or distortions of the earth's crust, as buckling and folding. Specific species noted of course depend on the raw material originally present in the metamorphosed rocks, particularly if these were sedimentary. As pointed out before, metamorphosis of igneous rocks is not likely to result in much change in mineralogy because the new environment is not greatly different from the one under which such rocks formed to begin with. Despite the wide variety of ingredients which can enter into sedimentary rocks, and later enter into chemical recombinations during the metamorphic process, most metamorphic rocks fall into two great divisions: those derived from sedimentary rocks rich in quartz and silicates, and those derived from carbonate rocks, as limestone. Commonly, smaller metamorphic bodies are found which are largely siliceous but contain some carbonate matter. The first great class, the siliceous metamorphics, characteristically provide the following associates:

SILICEOUS METAMORPHIC ROCK ASSOCIATES

STRONG METAMORPHISM

Almandite, andalusite, biotite, corundum, diaspore, graphite, ilmenite, margarite, plagioclase and potassium feldspars, pyroxenes, quartz, sillimanite, tourmaline

MODERATE METAMORPHISM

Almandite, amphiboles, andalusite, apatite, biotite, cordierite, corundum, dumortierite, epidote, hornblende, kyanite, lazulite, muscovite, plagioclase and potassium feldspars, pyrophyllite, quartz, rutile, staurolite, tourmaline

WEAK METAMORPHISM

Albite, amphiboles, anatase, apatite, braunite, brookite, chlorite, epidote, glaucophane, hausmannite, lawsonite, muscovite, perovskite, pyrite, quartz, rhodonite, spessartite, sphene, talc

Far greater variety of species, often splendidly crystallized, is afforded by the conversion of impure limestone to marble. Where metamorphism occurs without addition of material from sources outside the limestone mass, the result is white marble dotted in places with new species formed from the combination of former impurities. The range of species depends on original impurities and may vary from very few, as in the case of numerous marble bodies of such purity that they are quarried for statuary and building purposes, to some which contain so many segregations of minerals that they become useful only for crushed rock, for cement making or for agricultural lime. The greatest variety in species is caused by the introduction of impurities into marbles on a large scale by influx of hot gases and solutions from adjacent igneous bodies, particularly if these are granitic magmas or their offshoots. Such influxes not only introduce hosts of new ions but induce drastic changes by virtue of the large quantities of heat passing into the marble. This metamorphic process receives the special name *contact metasomatism*, the latter word signifying an exchange of constituents occurring at the same time. Thus, when metasomatic processes occur, the easily attacked calcite of the marble is removed and new minerals introduced. Bodies of ores and minerals noted for their large numbers of species caused by this process include the Franklin-Ogdensburg, New Jersey, zinc deposits, and the complexly mineralized marble bodies quarried in the vicinity of

Riverside, California. A partial list of species often associated in them is furnished below.

Calcareous Metamorphic Rock Associates

Actinolite, afwillite, andradite, anorthite, apatite, aragonite, arsenic, arsenopyrite, axinite, bastnaesite, biotite, bornite, braunite, brucite, calcite, chalcopyrite, chlorite, chondrodite, clinochlore, clinozoisite, corundum, danburite, diaspore, diopside, dolomite, epidote, fluorite, forsterite, franklinite, galena, graphite, grossularite, hausmannite, hedenbergite, hematite, humite, idocrase, ilmenite, ilvaite, lazurite, loelliningite, ludwigite, magnetite, manganosite, molybdenite, norbergite, periclase, phlogopite, pyrite, pyrochlore, pyrrhotite, quartz, rhodochrosite, rhodonite, rutile, scapolite, scheelite, serpentine, siderite, sphalerite, sphene, spinel, tourmaline, tremolite, willemite, wolframite, wollastonite, zincite, zircon, zoisite

10

Identification Procedures and Tests

The easy ability of an expert to identify minerals seems little short of miraculous to the beginner, and perhaps an attainment smacking of occult powers. Actually, proficiency in sight recognition comes from the expert's first-hand familiarity with characteristic features, plus his skill in relating these features to a specific mineral. Everyone has this ability to some degree —the amateur gardener sees nothing remarkable in his ability to recognize many plants, nor does the seamstress consider unusual her skill in accurately sorting out dozens of textiles. Minerals are also identifiable on sight in many instances, but, in fairness, it must be admitted that a large number are nearly impossible to distinguish visually because they occur in exceedingly small crystals, formless masses, or in a variety of confusing disguises. In such examples, sight identification becomes sheer speculation and definitive tests must be used.

Handling a specimen furnishes general impressions of crystal size and shape, color, and approximate specific gravity. Closer study with use of a lens often brings out more details which, together with initial impressions, may be enough for identification, particularly if a similar specimen has been examined before. Actual examinations are far more valuable than most skillfully worded descriptions, and it is for this reason more than any other that it pays to assemble a study collection of important species—they do not have to be large or fine so long as they are typical.

A SUGGESTED IDENTIFICATION SCHEME

The purpose of any identification scheme is to insure that no important evidence is overlooked. Sometimes the evidence is positive, pointing directly to one or two suspects: a heavy malleable mineral of metallic luster certainly points to a native metal. But sometimes the evidence is valid for many species, as so often happens when colorless species are examined, particularly in granular form, and the species may turn out to be one of a large number of silicates, carbonates, or sulfates. It is then that identification steps must be taken, using each shred of new evidence to systematically eliminate suspects until the field is narrowed to a very few. Occasionally, the best effort with means at hand leads to a dead end, and one must seek advice. Nevertheless, it is satisfying to have whittled down a long list of suspects to the point where only several fit. Although sight identification becomes increasingly easy as experience is gained, it still pays to examine specimens thoroughly and avoid leaping to unjustified conclusions as to identity. The prudent scheme is to marshal all possible clues, beginning with those that are obvious to the naked eye; next look for those that can be discovered by magnification, and, in the event an identification is still uncertain or impossible, finally try physical or chemical tests. The identification tables of Appendix A should be consulted to narrow possbilities.

There are many schemes for mineral identification, most depending on discovery of some important and fairly reliable clue as cleavage, hardness, luster, etc., followed by cross-checking with other properties to drastically reduce the size of the possibilities-list. Amateurs who have acquired some experience will resort to more direct methods which may smack of guessing but which actually are backed up by considerable first-hand knowledge. The following scheme is nothing more than a procedure designed to prevent overlooking significant clues. By the time a number of properties have been checked and results jotted down on paper, one or more of them will strongly suggest a certain species or group of species; at this time a direct test to confirm or disprove such suspicion may well be accomplished immediately, eliminating the need for further testing.

THE VISUAL EXAMINATION

Visual examinations should be made of clean specimens under light strong enough to bring out small but important details. The specimen should be turned slowly and each area carefully examined. It is surprising how many unsuspected details emerge during such methodical inspection, such as etch markings, inclusions, parallel growths, twin reentrants, differences in luster on various crystal faces, color zoning, interferences in growth, etc. Even more startling is the unfolding of additional detail when the specimen is examined under stereoscopic binocular magnification. The realistic three-dimensional effect provides the illusion that

General Scheme of Identification

GENERAL EXAMINATION

Color: Classify as black, white, or colored.

Luster: Classify as metallic, submetallic or non-metallic; further classify latter as vitreous, adamantine, etc. Note sheen or optical effects if present.

Diaphaneity: Classify as transparent, translucent or opaque; further classify opaque minerals if actually transparent or translucent on thin edges or slivers.

Specific Gravity: Classify as light, medium, or heavy, after "hefting" specimen in the hands.

Toughness: If material available for test, classify from very tough to very brittle, malleable, sectile, etc.

CLOSE EXAMINATION

Fracture: Classify as fracture, cleavage, or parting; further classify as to exact nature of each surface, using terms as perfect, imperfect, very smooth, etc.

Cleavage: Classify as to quality; check ease of production, observing same under magnification; attempt to correlate cleavage planes to faces, noting number and directions; note if all cleavages are similar in respect to surface character.

Hardness: Check under magnification.

Streak: Crush sample or check streak on porcelain plate; also check under magnification if possible.

Inclusions: Note type, orientation, color, etc. Use magnification.

CLOSE EXAMINATION OF CRYSTALS

Aggregate Habit: Classify as radiated, concentric, drusy, etc.

Specific Habit: Classify as prismatic, blocky, acicular, etc.

Faces: Observe shapes and angles made at corners; observe *like* and *unlike* faces in respect to luster, etch marks, striations, etc.; observe placement of matching faces; note angles between sets of faces; establish symmetry patterns.

Twins: Observe faces for signs of twinning as reentrant angles, striations of two directions on same face, differences in luster, markings, etc.

Cleavage: Note partly developed internal cleavages; orient to faces if possible; note if cleavages are single or multiple; note if belonging to same cleavage form, e.g., cubic, octahedral, etc.

Color: Note if evenly distributed or zonally distributed, or due to inclusions.

Optical Properties: Double refraction and pleochroism often noticeable under simple magnification.

ASSOCIATIONS

Carefully note *all* species, relative quantities, and placement on specimen. If possible, identify matrix.

SPECIFIC GRAVITY

Tests conducted as described in previous chapter.

OPTICAL TESTS

Double Refraction: Observe *through* clear crystals with magnifier focused on opposite face edges; note if doubling or blurring present; if possible, detach crystal or single crystal fragment and test for double refraction in polariscope.

Pleochroism: Often visible with naked eye or under hand magnifier as decided differences in color according to crystal direction; test further using detached crystal or fragment.

Other Optical Tests: As described in previous chapter.

CHEMICAL TESTS (to be described in this chapter)

Acids: Solubility tests.

Flame Color: Note characteristic color obtained.

Bead Test: Note characteristic color of bead.

MAGNETISM

Check using Alnico magnet; also check in conjunction with fusion tests.

FUSION (to be described in this chapter)

one has stepped into the forest of crystals on the specimen itself. For the examination of crystal faces, and their arrangement on crystals, the stereoscopic microscope is unsurpassed.

Magnifiers. In this connection, a few words on magnifiers are in order because some kind is indispensable in the examination of specimens. Magnifiers may be had in all qualities, some so poor that they confuse rather than reveal, and some, at the opposite end of the scale, carefully designed to eliminate as many optical errors as possible. Single-lens magnifiers, as commonly sold in stationery stores for broad examination of printing, stamps, etc., are satisfactory for slight enlargement but completely useless for fine detail. Their size generally prohibits taking them along on field trips although this may not really be a disadvantage because the value of magnification in the field is frequently over-emphasized, and many expert collectors do not bother with hand lenses. In any event, much more is to be learned from specimens in the quiet of the home laboratory, under good lighting, and with plenty of time for unhurried examination. If a specimen in the field seems to offer possibilities for further study, it should be placed in the collecting bag and taken home instead of trying to discover much about it under less than ideal conditions.

The most useful hand magnifiers range in power from five to ten. Because of optical design limitations, higher powers result in narrower fields of view until one gets the impression that he is peeking through a pinhole. Actually, high powers always bring about losses in optical accuracy unless extremely expensive lenses are purchased. As power goes up, the lens must be brought closer to the specimen in direct proportion to the power. Thus a 10x lens usually requires that the front of the lens be one inch from the object being viewed, but a 20x lens requires that it be held *one half inch* away. Needless to say, it is very awkward to examine a specimen with such a lens because fingers and lens practically cut off all light. It is therefore recommended that a hand lens be either 5x or 7.5x, and if one must have higher power that it not be more than 10x.

In respect to lens design, which of course reflects itself in price, the worst lenses are cheapest because they do not correct for *color dispersion* nor for *spherical aberration*. Failure to correct for color results in annoying and confusing spectral haloes appearing around brilliantly reflective objects in the field of vision particularly from crystals of metallic luster. Failure to correct for spherical aberration causes distortion, steadily worsening from the center of the field outward. In very poor high-power hand lenses, practically all that can be seen clearly is a small circular area in the exact center of the field of view; everything else is extremely blurred and colored by dispersion. The method used by lens makers to combat these inherent defects is to sandwich several lens components together, one or more of which are made from different types of glass. Naturally this multiplies expense in proportion to the components used, it being fair to say that a triplet hand magnifier containing three separate lens elements will cost at least three times as much as an equivalent single lens magnifier. Good optical instruments involve considerable hand labor in manufacture, and for this reason prices rise steeply with quality. In view of the great value of a hand magnifier however, the amateur is well advised to stretch his financial resources to get a good one.

Several knacks in using hand magnifiers are worth describing. To steady the instrument and prevent movement of the image, place the magnifier close to the eye, and rest the supporting thumb against the cheekbone and the forefinger against the bony ridge above the eyeball. The other hand, holding the specimen, should then rest against the first hand; in this manner, hands, magnifier and specimen are all braced against the skull. This reduces vibration and allows the sharpest vision possible. Light from above and light from the side are both useful for bringing out details; sometimes one is better than the other, so experiment until the best light is obtained. Avoid dazzling reflections by turning the specimen slightly as necessary, but whenever possible use a very strong light. Keep both eyes open to avoid fatigue; at first this is disconcerting, but with a little practice the image obtained through the lens prevails and the mind learns to ignore the image obtained through the other eye.

It is unfortunate that good binocular microscopes are much more expensive than hand lenses, for they are far superior in performance, and any amateur with the means should certainly buy one. Several standard models are regularly available from American manufacturers, and there are also imports from foreign sources. Stereo microscopes are very popular, in fact it is seldom possible to find one secondhand, while ordinary monocular biological microscopes go begging for buyers. When secondhand stereo microscopes are obtainable, the price is not very far from retail. The chief values of such instruments lie in stereoscopic vision, coupled with the use of *both* eyes, adding that much more vision and avoiding the fatigue so often noted with monocular instruments. The stereoscopic feature is very useful for examining faces on small crystals and obtaining accurate ideas as to general crystal proportions and symmetry. Again, high magnifications are never as useful as low magnifications, a satisfactory range of powers being from 10x or 15x to 40x or 60x. Of these, the range from 10x to 40x is adequate for most purposes. A small light source capable of throwing an intense spot of light on the specimen is quite necessary for best results. The lamp may be purchased with the instrument or improvised—it matters little which—but it must be easily adjustable to point its beam from any angle, because frequently a change of angle is necessary to illuminate all parts of the specimen. A most convenient arrangement is to attach it to the microscope body so that it points to the place where the specimen will be when the latter is in focus.

During visual examination, a pad of paper should be handy to take notes and to make comments on each of the topics listed in the general scheme of identification. Small sketches of crystals are valuable for comparison to textbook crystal drawings; they may be decisive in identification. A stout darning needle mounted in a jeweler's pin vise is excellent to test hardness and develop small cleavages. By wedging its point in crevices and breaking off small bits, some idea of brittleness is obtainable. The needle, or the sharp tip of a broken-off injector razor blade, is also useful for removing surface coatings, as iridescent tarnishes, deposits of clay minerals, etc. With judicious use, a streak test can be made with the needle when the crystals are too small to rub on a porcelain plate.

IMPORTANCE OF ASSOCIATIONS

The great importance of associations has been emphasized in the previous chapter, however a few remarks will be useful on how to employ associations in identification. In general, when two or more species occur on the same specimen, the most prominent is examined to see if it recalls a specific species, and if not, the others are examined toward the same end. It is best to look for species most familiar and most likely to be recognized in order to get them out of the way and allow concentration on those which seem strange. In this process, species likely to be found under many formational circumstances as quartz, calcite, mica, feldspars, garnet, etc., should be recognized and noted, but by themselves should not greatly influence the final decision as to what kind of association is at hand. On the other hand, the discovery of quartz crystals penetrated by needles of black tourmaline immediately eliminates a sedimentary environment but does suggest a high temperature environment which may be in an igneous or metamorphic rock. Calcite by itself is meaningless but if the base of the specimen contains grains of galena, the calcite probably crystallized in an opening in a hydrothermal ore vein. Similarly, a reddish garnet enclosed in a mass of mica scales suggests the well-known association of almandite in metamorphic biotite schist, while discovery of one sulfide in any specimen leads to a search for other sulfides which commonly occur together, and so on.

Perhaps the most skillful use of associations is made by micromounters, who cannot test directly in many cases but must examine minute crystal-lined cavities and determine from recognition of one or two species the most likely association at hand. By consulting a list of associates, as from among those furnished in the previous chapter, he attempts to match observed features on unknown crystals against those described under the species in the descriptive mineralogy portion of any textbook. Usually a match is made and one more species in the association is identified. Where much larger hand specimens are available, a similar process of determining associates and methodically ticking off possibilities from the list is also necessary, but here actual physical and chemical tests can be performed which are not ordinarily possible in microscopic examination. This advantage is somewhat offset by the greater perfection of the small crystals in which the micromounter specializes; they are apt to be more "textbook perfect" than the larger crystals observed on the usual run of hand specimens.

CHEMICAL TESTING

In general, extensive chemical testing is seldom called for if the amateur uses all physical testing means available to him. Physical tests are easier to perform and far less bothersome than chemical tests as a rule. Furthermore, unless chemical testing is a continuing requirement, reagents and laboratory equipment purchased for this purpose tend to gather dust and deteriorate with time. However, several simple chemical tests require only a minimum outlay of funds for chemicals and equipment, and are useful enough to warrant description. Among the simplest of all is the use of hydrochloric acid for dissolving minerals, the reactions during dissolution being indicative of the species. Hydrochloric acid is sold in drug stores in concentrated solutions of various grades of purity as color-less fluids. For laboratory use, the standard dilution is one part acid to one part water by volume. Several fluid ounces will last a long time, unless the acid is to be used in considerable quantity for cleaning goethite (*limonite*) stains from quartz and other minerals, or for removing calcite. Large quantities are much more inexpensively purchased as "commercial" grade acid in quart or gallon containers from paint or hardware stores. They are labeled with the common trade name *muriatic acid*. Commercial acid is pale yellow in color due to slight iron impurities. Removing these in technical or chemically pure grades causes the same acid to cost far more than commercial grades.

Use of Hydrochloric Acid (HCl). A number of species, notably the carbonates, react with gratifying promptness and exhibit unmistakable signs when touched with hydrochloric acid. Other species react less spec-tacularly, sometimes so slowly that several days may pass before a definite effect is noted even when specimens are left submerged in acid. Minerals are usually tested by applying a drop of acid to some surface whose damage will not affect the value of the specimen, or by detaching small pure frag-ments from the specimen and placing them in test tubes, adding acid to cover. The acid is usually used dilute at first, and if no reaction occurs, concentrated acid is substituted. In some instances, reactions proceed quickly at room temperatures, but others do not produce definite results unless the test tube is cautiously warmed over a gas flame. In still other instances, it may be necessary to crush the mineral to a powder in order to expose enough surface area to produce a satisfactory reaction. An excellent and very convenient source of heat for this and other testing purposes is a propane torch, using throwaway metal tanks. The kind using a small can, about five inches tall, is easier to handle than larger models. The blow torch tip should be the type producing a pointed flame. By carefully regu-lating the valve, it is possible to obtain very small flames when these are required.

Hydrochloric Acid Tests. *Dissolution.* Carbonates dissolve with release of large numbers of small carbon dioxide bubbles. Some are attacked by cold acid (calcite), but others only by warm to very warm acid (rhodo-chrosite); see Part II, Descriptive Mineralogy for reactions of specific spe-cies. Some sulfides dissolve slowly, giving off the rotten egg odor of hydrogen sulfide (sphalerite). Manganese oxides dissolve slowly and release acrid "swimming pool odor" of chlorine. Other species dissolve more or less completely, often leaving whitish powdery residues or jelly-like transparent coating (natrolite).

Solution Color. Much iron in a test mineral commonly results in staining solutions yellow to reddish-yellow; copper stains blue or green; cobalt yields a pink solution. Several elements present in the same specimen may confuse coloration, indicating the desirability of pure test samples.

Flame Color Tests. Some elements impart color to torch flames when minerals containing them are heated to incandescence. A useful tool for carrying out this test is a piece of soft iron wire, available in spools from hardware stores, looped at one end to a diameter of about $\frac{1}{16}''$ to $\frac{1}{8}''$, as shown in Figure 106. If a 4″ piece is used, one end may be held in the fingers without fear of burning the skin even when the opposite looped end is extremely hot. The best results are obtained by crushing a pure sample of the mineral, moistening the loop in hydrochloric acid, and touching it to the powder to pick up a few grains. It is then placed in the colorless tip portion of the flame and heated. The light in the home laboratory should be dimmed because flame colors are seldom strong and require darkness to see them well. Usually the wire is contaminated by sodium, as can be quickly ascertained by placing an unused loop in the flame. If so, heat and dip in concentrated acid and reheat. It may be necessary to repeat this several times until the wire glows in the flame without producing the sodium yellow color. When it is clean, it is ready for use. It is very difficult to avoid the bright yellow flame produced by sodium because sodium chloride is very common in minute water-filled inclusions in many minerals. For example, a piece of milky quartz held in the flame will "pop" into many small bursts of yellow as the liquid inclusions explode and release minute quantities of dissolved salt. A cobalt blue glass filter is useful in such cases for absorbing the sodium flame and permitting others to be seen in their true color. The following are typical flame colorations of the elements named:

Flame Colors of Elements

Red	Strontium: vivid and persistent.
	Lithium: usually in streaks, but often obscured by sodium.
Red-orange	Calcium: readily obtained, but often obscured by sodium.
Yellow	Sodium: vivid and extremely persistent.
Yellow-green	Barium
	Boron: in flashes.
	Molybdenum: faint color.
Green	Antimony: pale.
	Bismuth: pale greenish-white.
	Copper: strong.
Blue-green	Phosphorus: pale.
	Zinc: in streaks.
Blue	Lead: pale.
	Arsenic: vivid.
Violet	Potassium: persistent, but usually obscured by sodium.

It is to be noted that the above flame tests tell no more than that one of the elements is present. In some instances, this information may be decisive, as when the relatively rare strontianite, and the relatively common celestite, both of which may be confused with barite, are tested. The char-

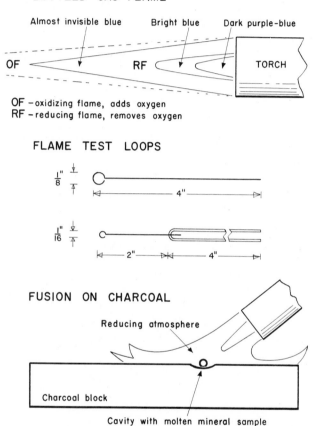

BOTTLED GAS FLAME

Almost invisible blue   Bright blue   Dark purple-blue

OF   RF   TORCH

OF – oxidizing flame, adds oxygen
RF – reducing flame, removes oxygen

FLAME TEST LOOPS

$\frac{1}{8}$"   4"

$\frac{1}{16}$"   2"   4"

FUSION ON CHARCOAL

Reducing atmosphere

Charcoal block

Cavity with molten mineral sample

USING THE TORCH FLAME FOR TESTING MINERALS

FIG. 106.

acteristic flame colors of each often distinguishes them easily. Most elements are so abundant that many species contain them, but some narrowing of suspects is still possible with this test. Because sodium contaminates so easily, care must be used in selecting test samples not to include any sodium compound accidentally and to be sure that the iron wire loop is thoroughly cleaned in acid before it is used again.

Bead Tests. If a suitable flux is melted into a bead in the torch flame, but deliberately made impure beforehand by addition of a very small quantity of an unknown mineral, it frequently assumes a color characteristic of an element within the mineral. This is the principle of the *bead test*. In effect, a miniature sample of stained glass is being made, not nearly as durable as ordinary glass to be sure, but following the same principle involved in the manufacture of the real material. The fluxes used are either powdered borax, available from grocery stores or from druggists, or sodium ammonium phosphate. The latter is obtainable from large mineralogical supply houses or on special order from druggists. Sodium phosphate is also satisfactory, being the product of fusion of sodium ammonium phosphate.

The making of a bead requires a platinum wire support, this metal being the only easily obtainable metal capable of withstanding the intense heat of the torch as well as the chemical attack of the fused minerals, acids, and fluxes. Platinum wire, about 28-gauge, is ordered from a mineralogical supply house in lengths of about 2″. One end is imbedded in a short piece of chemist's glass tubing merely by heating the end of the tube and allowing it to fuse around the wire, as shown in Figure 106. The other end is looped around a pencil tip to make a loop of about $\frac{3}{32}$″ diameter. The loop is prepared for bead testing by first ridding it of impurities. It is heated strongly in the tip portion of the torch flame, then dipped in borax or sodium ammonium phosphate, and reheated to fuse the powder into a clear glassy globule, repeating the process as necessary to build up the bead to approximately spherical form. If any trace of color is observed in the bead, it is flipped from the wire by striking the glass rod against the edge of the hand, first being certain that the molten bead will not fall on one's clothing or on some combustible material. When a bead is made and found to be clear and colorless, it may be touched to the powdered mineral sample, attempting to take up only a speck or two of material, and reheated to dissolve the mineral. Only the slightest trace of mineral is needed to develop color; if too much is taken up, the bead becomes opaque. The bead is then placed in the oxidizing or reducing flame of the torch, as shown in Figure 106, and heated to fluidity. The color is observed when the bead is cold.

Several precautions are needed to obtain good results, the most important being to clean the platinum loop before each new test. This is done using the method described before. Because the bead flux is colored by *oxides* of the elements being tested for, it is often necessary to roast sulfides and sulfosalts on charcoal using the oxidizing flame, in order to drive off sulfur and volatile semi-metals, particularly arsenic. The latter element quickly combines with hot platinum, causing it to become so brittle that the bead loop easily snaps off. If its presence is suspected, the crushed mineral sample should be strongly heated for a considerable time to insure driving off this volatile element.

The procedure in bead testing is to try the bead first in the *oxidizing* flame, noting the color after removal and cooling, followed by inserting a fresh bead in the *reducing* flame, and again noting color. Figure 106 shows the location of oxidizing and reducing flame sections. The oxidizing flame is completely consumed gas, heated to very high temperature. By its motion, it sweeps air against any specimen placed in it, thus supplying oxygen and promoting oxidation. The reducing portion of the flame contains much partly-consumed gas, one of the products of combustion being carbon monoxide, which strongly attracts oxygen to convert itself to carbon dioxide. It does this at the expense of the mineral placed in it, which, if it contains oxygen, suffers its loss and hence is *reduced*. The following table indicates the bead colors which may be expected from the elements listed.

Bead Colors

Color	Element	Borax O.F.	Borax R.F.	Phosphate O.F.	Phosphate R.F.
Red (opaque)	Copper		x		x
Yellow	Iron	x			
	Nickel			x	x
	Tungsten		x		
	Vanadium			x	
	Uranium	x			
(pale)	Iron			x	
Green	Chromium	x	x	x	x
	Molybdenum				x
	Uranium				x
	Iron		x		
	Vanadium		x		x
(pale)	Uranium		x	x	
Blue-green	Copper	x			
Blue	Cobalt	x	x	x	x
	Copper			x	
	Tungsten				x
Violet	Manganese	x		x	
	Titanium		x		x
Brown	Nickel	x			
	Molybdenum		x		
Gray (opaque)	Nickel		x		
Colorless	Iron				x
	Manganese		x		x
	Molybdenum	x		x	
	Titanium	x		x	
	Vanadium	x			
	Tungsten	x		x	

O.F. = Oxidizing Flame, *R.F.* = Reducing Flame.

FUSION TESTS

As a whole, minerals are highly resistant to melting in ordinary flames, as that of the propane torch, but some do melt quickly, others partially, and still others merely incandesce without losing sharpness of edges. Observing such behavior sometimes adds useful information to that already learned by other means. The procedure is to insert a thin sliver or chip, no more than ⅛″ to ¼″ in length, into the flame at the point just beyond the inner colored cone, and note what happens. A pair of sharp-pointed tweezers is needed to hold the sliver; the kind sold for jewelry soldering purposes is satisfactory. During the test, the ease with which the fragment fuses is noted as well as other behavior, as sudden breaking apart (decrepitation), swelling, or production of color in the flame. To avoid needless cracking of the chip, it should be warmed gradually by passing through the outer flame before inserting into the hottest part. If the sample decrepitates, take a fresh fragment, crush to powder, and moisten the powder with a drop of saliva. Place the paste into a small depression on charcoal and heat strongly until the powder partly fuses. Take up the mass with the tweezers and test in the flame as before. The following *scale of fusibility* provides useful measures of the fusibility of unknown minerals:

SCALE OF FUSIBILITY

Number	Mineral	Appr. Temp.	Remarks
1.	Stibnite	525°C(977°F)	Very easily fused, even in candle flame.
2.	Chalcopyrite	800°(1472°)	Fuses easily in propane flame.
3.	Almandite	1050°(1922°)	Fuses slowly in propane flame (taking about one min.); globule becomes magnetic.
4.	Actinolite	1200°(2192°)	Very thin splinter fuses.
5.	Orthoclase	1300°(2372°)	Fuses only on edges with difficulty.
6.	Enstatite	1400°(2552°)	Only finest splinters somewhat rounded.
7.	Quartz	1710°(3110°)	Infusible.

Fusion on Charcoal. It is sometimes possible to obtain beads of more or less pure metal by smelting small crushed samples of suitable minerals mixed with flux. Charcoal blocks are used because they do not burn rapidly and reflect heat into the melt, and they assist greatly in removing oxygen from the mineral. The flux used is sodium carbonate (washing soda) or sodium bicarbonate (baking soda), both of which are likely to be found in the average household or obtained easily from the grocer. Charcoal blocks are less easy to obtain, and it may be necessary to order some from mineralogical supply houses, or from jewelers' supply houses. Sometimes ordinary kindling and broiling charcoal (*not* briquettes) provide pieces large enough to be suitable, however, the necessary flat-bottom slabs must be sawn from such pieces and smoothed with sandpaper. Professional fusion test tablets are several inches long and about one inch wide and slightly less in thickness, but dimensions are not important so long as the block rests firmly on its base. A fireproof sheet of asbestos or sheet metal must be placed beneath blocks to prevent setting off accidental fires or damaging table tops. A block is prepared for use by scooping out a small depression, about the size and shape of half a pea, near one side of the block.

To prepare a mineral sample for smelting, as for example one of the sulfides, crush to fine powder and mix one part of the mineral to three parts of flux. Moisten the mixture to a paste, and place a dab in the charcoal block cavity. First play the torch gently to remove water, then strongly to cause fusion. The position of the flame is shown in the lower part of Figure 106. Note that the hottest part of the flame, the area just ahead of the inner cone, is brought close to the same sample, while the remainder of the flame spreads over the sample, enclosing it in a strong reducing atmosphere. When sufficiently reduced, the sample displays small metal beads of brilliant mirror-like luster quite unlike the luster of the molten carbonate flux. Under favorable conditions, separate beads will flow together to make one bead of considerable size; under poor conditions, only small beads may form and it is necessary to crush and examine the cold mass to discover if any metal beads are actually present. Magnification may be necessary to see them. The following beads are obtained from some of the species containing the elements listed:

Metallic Beads Produced on Charcoal

Antimony	Grayish-white; brittle bead.
Bismuth	Reddish-white; slightly malleable bead.
Cobalt	No metal; dark magnetic bead.
Copper	Black-coated malleable beads; difficult to fuse.
Gold	Bright yellow beads; very malleable; do not tarnish when cold.
Iron	No metal; dark magnetic bead.
Lead	Gray malleable bead.
Nickel	No metal; dark magnetic bead.
Silver	White, malleable bead; stays bright.
Tin	White, malleable bead; assumes white coating when cold.

Because fusion of iron compounds frequently results in beads which are strongly attracted toward the magnet, it pays to test each cooled bead with a small Alnico magnet. Note that cobalt and nickel compounds also produce magnetic beads. During fusion, do not inhale fumes, particularly from sulfides and sulfosalts which contain sulfur and sometimes arsenic. Testing for odor is useful in determining the presence of either element, but is accomplished most safely by wafting some fumes toward the nose rather than leaning over and smelling the fumes arising from the charcoal block. Sulfur forms sulfur dioxide, the familiar acrid penetrating odor emitted by ordinary striking matches, while arsenic announces itself by a garlicky odor.

PART 2

Descriptive Mineralogy

Descriptive Mineralogy

INTRODUCTION

To suit the needs of the amateur, nearly three hundred species are described in this part of the book. The most likely to be encountered during personal collecting activities in the field, or placed in collections through exchanges or purchases, are accorded fullest treatments, with briefer treatments given to species which are rarer or which do not ordinarily appear in nature in good cabinet specimens. For example, the mineralogically, geologically, and economically important *clay minerals* are treated very briefly because they seldom produce remarkable specimens although clays appear nearly everywhere and numerous specimens of other species frequently display one or more clay minerals upon them. Furthermore, it is almost impossible to identify specific clay minerals without the laboratory equipment available to the professional mineralogist, and therefore to describe the properties of such species in detail wastes space which could be more fruitfully devoted to fuller descriptions of species more likely to be collected by the amateur and more readily identified. Conversely, species which are economically unimportant or which may be exceedingly rare in terms of quantities in the earth's crust, sometimes receive full treatment because they happen to occur in fine specimens, eagerly sought for by the amateur.

The selection of specific data within mineral descriptions is also aimed for amateur use, laying greatest emphasis on properties readily determined within the limited means of the average home laboratory. To assist the amateur in becoming a discriminating collector, and to judge the worth of specimens which he collects on field trips or buys from dealers, statements are also made concerning sizes and qualities of crystal and matrix specimens where such information is known.

In a few instances, the maximum known crystal sizes are included. If minerals are suitable for cutting into gems, appropriate statements are included but, of necessity, cannot be detailed.

SCHEME OF PRESENTATION

The presentation of species follows the scheme previously described in Chapter 8, beginning with the Native Elements and ending with the Silicates as *major classes.* Within each there may be *subclasses* as the Metals in the Native Elements, or Tektosilicates in the Silicates. A number of related species may be placed together under *Groups,* or if the relationship is very close, under a *Series.* The names of individual species belonging to a group or series appear in page centers but species not assigned to either

275

of these subdivisions appear to the left. Each species considered worthy of full description is treated as follows:

Name—Chemical Formula

Name: Derivation, alternate names, pronunciation.

Varieties: Names and brief descriptions of varieties based upon significant differences in chemical composition or properties; trivial varieties are omitted.

Crystals: System. Forms, habits, twinning, etc.

Physical Properties: Tenacity; fracture. Cleavage. Hardness. Specific gravity.

Optical Properties: Transparency. Color, color zoning. Streak. Luster and sheen, special optical effects caused by inclusions. Refraction, refractive indices in sodium light, and birefringence. Pleochroism. Fluorescence under ultraviolet light.

Chemistry: Compound name. Variations in composition.

Distinctive Features and Tests: Prominent features likely to lead to identification. Flame and chemical tests.

Occurrences: Formational environments and associates. Localities furnishing good cabinet specimens; special features of specimens; sizes and qualities.

MINERAL NAMES

If mineralogy is rich in anything, it is rich in names. The names themselves are like bookmarks slipped between pages of mineralogical history, marking improvements in man's knowledge of the earth sciences. As civilizations progressed and new uses for a larger variety of mineral substances were discovered, new names had to be coined for such substances. But without sure methods of discriminating between species, mineralogical literature soon burdened itself with hundreds of trivial names, based upon real or fancied differences even within the same species. For example, at least 300 names have been used at one time or another for varieties of just one mineral—quartz, and some authors have gone to the trouble of collecting and listing such names, thereby showing how ridiculous the unbridled license to coin names can become.

Paradoxically, improvement in the knowledge of chemistry brought with it additional nomenclature difficulties, particularly in the period of the seventeenth to nineteenth centuries, when it became possible to identify elements within minerals more certainly, leading to an unjustified overconfidence in the naming of "new" species based on demonstrated differences in chemical composition. It is from the nineteenth century more specifically, that we inherited the abundance of varietal names which one may find listed in *Dana's Textbook of Mineralogy,* or older editions of *Dana's System of Mineralogy.* In the *Textbook,* for example, about 3,000 names are listed in the species index, but only about a third were considered as valid species by its author-reviser, Professor W. E. Ford. It was not until the significance of atomic structure was fully realized that many so-called "species" could be restudied by x-ray and better chemical methods, and placed where they belonged—as varieties of established species or members of series. Thus the seventh and latest edition of *Dana's System of Mineralogy,* Volumes I-III inclusive, drastically reduces the number of varieties, eliminates some species, and includes new species discovered in the interim.

Vigorous steps have been taken by mineralogists to reduce the list of species and varieties, if for no other reason than to add room for valid new species which continue to be discovered and which require space for description. As a substitute for separate varietal names based on chemical grounds, Waldemar T. Schaller proposed in 1930 that modified element names be used to show that a species contained significant amounts of that element. For example, pyrite sometimes contains nickel substituting for iron, and such varieties have been called "blueite," "bravoite," and in German "Kobaltnickelpyrit," but under Schaller's proposed nomenclature, these would be eliminated in favor of *nickelian* pyrite. Thus the use of one term, quite unmistakable as to meaning, replaces three words that were burdening the literature unnecessarily. Where the element has two valences, the last portion of the word is modified to show this, as in *manganoan* (valence of 2) and *manganian* (valence of 3). This was adopted by Palache, Berman, and Frondel in their Seventh Edition of *Dana's System of Mineralogy,* and it is used in this book also. However, objections have been raised in England, and there at least the preference of some authorities is to retain older varietal names.

The right to name a new mineral rests with the first person who adequately describes the species. Despite a rather rigid set of self-imposed requirements which must be met by professional mineralogists before a mineral is considered to be "adequately" described, errors creep in and "new" species sometimes turn out to be older species in confusing disguise, or intimate mixtures of several species difficult to separate in the samples used by workers. Such mistakes happened frequently in the nineteeth century and still occur now, although to a lesser extent. In view of the problems connected with naming of minerals, it is particularly deplorable that a few ill-advised individuals in the amateur earth science fraternity in the United States are coining species names for varieties of quartz and other minerals useful for lapidary work. Some of the names are meant to honor individuals, but others seem no more than proper-name trademarks, designed to publicize saleable rough gemstones. It is vastly preferable to use accepted varietal names as *agate* or *jasper,* suitably modified by geographic terms as "Montana," "Morgan Hill," etc., or descriptive terms as "flowering," "poppy," etc., thus simultaneously identifying the basic material and pointing out that it differs from other subvarieties. Under no circumstances should the endings *-ite* or *-lite* be used because these have been assigned to true mineral species by a custom of several centuries standing.

PRONUNCIATION

To spare the reader the necessity of consulting a dictionary, and it must be a large one to be sure most species names are in it, a simplified pronunciation is furnished for species and some varietal names in the descriptions which follow. The pronunciation marks used are given below, with word examples.

Pronunciation Guide

ā—āle	ē—bē	ī—bīte	ō—ōld	ū—cūbe
ă—ăccount	ĕ—ĕnd	ĭ—ĭll	ŏ—ŏdd	ŭ—ŭp
à—àdd	ẽ—bakẽr		ô—ôrb	
ä—ärm				

ABBREVIATIONS AND SYMBOLS

H.—hardness.
G.—specific gravity.
R.I.—refractive indices.
n—single refractive index, or approximate refractive index in unixial or biaxial minerals.
O—refractive index of ordinary ray in uniaxial minerals.

E—refractive index of extraordinary ray in uniaxial minerals.
α—lowest refractive index in biaxial minerals.
β—intermediate refractive index in biaxial minerals.
γ—highest refractive index in biaxial minerals.

NATIVE ELEMENTS

About twenty elements occur in native form and are placed in groups according to similarities in crystal structure, as in the Gold Group, or for convenience as placing diamond and graphite in the Carbon Group. The *metals,* characterized by malleability and other properties which we ordinarily associate with the term "metal," appear first, followed by the *semi-metals,* and the *non-metals.* The semi-metals *look* like metals but are usually brittle and display properties considerably different from those of the metals. The non-metals do *not* look like metals as a rule, but are still elements in essentially pure forms.

METALS

Gold Group

Gold—Au

Name: Anglo Saxon, of uncertain origin; gōld.
Varieties: Silver-bearing, pale yellow color, *argentian,* or, *electrum.*
Crystals: Isometric; rarely in distinct crystals, then mostly rude, rounded, cavernous, hoppered or in dendritic growths of numerous crystals in parallel position. Octahedrons most common, usually flattened on a face of the octahedron $o[111]$, sometimes as extremely thin plates with equilateral triangle markings (Figure 107); also dodecahedrons and cubes. Rarely over $1/8''$ diameter, but octahedron plates to $12''$ broad have been recorded. Complex twinning a feature of dendritic specimens, especially from Verespatak.
Physical Properties: Malleable, sectile, ductile; hackly fracture. H. $2\frac{1}{2}$-3. G. 19.3 (pure) to as low as 15 in solid solution with silver.
Optical Properties: Opaque. Rich yellow, paling to whitish-yellow with increasing silver. Streak shining yellow; luster metallic.
Chemistry: Ordinarily contains 10%-15% silver; the name *electrum* applied when containing over 30% silver.
Distinctive Features and Tests: Typical color, malleability, specific gravity. Unlike pyrite and chalcopyrite, can be hammered flat without shattering; pyrite and chalcopyrite yield dark powders when crushed and are attacked by nitric acid, giving off rotten egg odor of hydrogen sulfide. Gold insoluble in acid except in aqua regia. Fuses easily at 3, yielding a perfectly spherical shining bead which does not tarnish.
Occurrences: Primarily in hydrothermal quartz veins with pyrite and minor amounts of other sulfides; host rocks commonly schists and slates. Abundant in some gravels where released by erosion of enclosing rocks; nuggets commonly contain small pieces of attached quartz (Figure 108). Uncommonly, as brilliant crystals in vugs. Splendid crystallized specimens from mines in the Mother Lode country of California, particularly from the counties

FIG. 107. Native gold. *Left:* a bright
yellow mass of flattened octahedral
crystals, weighing 55 grams and
measuring 2″ by 1″, from Grass
Valley, Nevada County, California.
Right: a plume-like growth of pale
yellow *electrum* on drusy quartz;
specimen size 2″ by 1¾″. From Rosa
Montana, Verespatak, Bihar Mountains, Transylvania, Romania.

FIG. 108. Stream-worn native gold
nugget with some adhering quartz.
Weight 61.9 troy ounces. From Un-
ion Placer Mine, Greenville, Plumas
County, California. *Photo courtesy
Smithsonian Institution.*

of El Dorado, Nevada, Placer, Siskiyou and Tuolumne; skeletal crystals to 1″ on edge and octahedral plates to 12″ broad, sprinkled with sharp octahedral crystals, have been recorded. Fine crystals from Breckenridge District, Summit County, Colorado, as broad leaves and interlocked wires in sponge-like masses; also beautiful crystals on white quartz from Red Mountain Pass, San Juan County, and fine micro crystals from Dixie Mine, Idaho Springs, Clear Creek County. Gold tellurides from Cripple Creek mines, Teller County, are purposely heated to reduce to globules of gold on rock matrix and sold as "native gold." Splendid crystallized specimens from localities in Mexican states of Sonora, Chihuahua, Hidalgo and Zacatecas. Complexly twinned brilliant pale yellow (*electrum*), in fern-like growths on drusy quartz, and highly prized for collections, from Verespatak, Bihar Mountains, Romania (Figure 107). Coarse subhedral crystals, but also small sharp crystals, from Bendigo and Ballarat, Victoria, Australia; also fine wire gold, sponge gold, and other forms, from mines in Flinders Range, South Australia, and from Coolgardie, Western Australia. Nugget, flake and dust gold from a great many localities.

All specimens are costly, usually commanding several times bullion value if interesting, and sometimes far more if well-crystallized. Finest specimens are bright crystallized flakes, leaves, etc., on white quartz matrix; veinlets in white quartz are next, followed by nuggets, flakes or dust. Good micro specimens are usually available.

Silver—Ag

Name: Anglo Saxon, but origin lost; sĭl′vẽr.
Varieties: Containing gold, *aurian;* also *mercurian, arsenian* and *antimonian.*
Crystals: Isometric; rarely distinct. Commonly in rude masses or wires; also in sheets and plates. Crystals mostly crude cubes, but also octahedrons and dodecahedrons, and in twinned parallel growths forming arborescent "herringbones" and dendrites (Figure 109). Twinning on *o*[111].
Physical Properties: Malleable, ductile, sectile; hackly fracture. H. 2½-3. G. 10.5 (pure), rising with gold to about 11, or falling to about 10 with other metals.
Optical Properties: Opaque. Brilliant white when freshly cut; most specimens tarnished gray to black due to alteration in surface layers. Streak shining white; luster metallic.
Chemistry: Usually contains gold or small amounts of mercury, arsenic, antimony.
Distinctive Features and Tests: Malleability; black tarnish; specific gravity. Fuses at 2, yielding spherical shining bead. Soluble in nitric acid, forming white curdy precipitate when hydrochloric acid is added.
Occurrences: In oxidized zones of hydrothermal sulfide veins; also as a primary mineral in some sulfide deposits; rarely in basalts and diabases, except in the notable occurrence in the Keeweenaw Peninsula, Michigan. In Canada, much silver is obtained in sheets, slabs, and irregular masses many inches across, from silver-nickel veins in diabase at Cobalt, Gowganda and O'Brien, Ontario; silver embedded in dolomite occurs on Silver Islet, Thunder Bay district; dendritic masses in pitchblende deposits in Great Bear Lake district, Northwest Territories; wires, and massive, with calcite, pyrite and sphalerite, from Beaver Dell, British Columbia. Exceptional specimens have been obtained from copper deposits of the Keeweenaw Peninsula, the silver being intimately associated with copper in cavities in amygdaloidal basalt, plus prehnite, calcite, analcime, stilbite, etc. usually the silver occurs as rudely-crystallized arborescent growths (Figure 109), but also as fairly distinct crystals, wires, and dendrites in calcite and dolomite. Commonly, masses show distinct portions of silver and copper, and locally are called "halfbreeds." Small platelets of silver occur within earthy chalcocite at Bisbee, Cochise County, Arizona, also some wire silver in gossans. Fine micro specimens from Creede, Mineral County, Colorado. Excellent

wire and arborescent silver (Figure 109) from Batopilas, Chihuahua, Mexico, and from numerous localities in the states of Sonora, Durango, and Zacatecas. Sheets, wires, and euhedral crystals, sometimes uncommonly large and sharp, occur with calcite, quartz, and sulfides in the famous mines of Kongsberg, Norway; the finest wire silvers originate here, some specimens showing cleavage masses or crystals of calcite from which rise twisted

FIG. 109. *Native silver and copper. Left:* crystals of copper on limonite matrix from Campbell Shaft, Bisbee, Cochise County, Arizona. Size about 2″ by 1½″. *Bottom center:* native copper crystals showing hopper development. Size 1¾″ diameter. New Cornelia Pit, Ajo, Pima County, Arizona. *Top center:* native silver in rude branching crystals from Wolverine Mine, Cheboygan County, Michigan. About 3″ by 2¾″. *Right:* arborescent silver forming sprays of twinned crystals in limonite matrix on limestone. About 2¼″ by 2″. Batopilas, Chihuahua, Mexico.

filaments of silver from several inches to more than 10″ length. Good wire specimens from Saxony and Baden, Germany, and from Broken Hill, New South Wales, Australia. Small sheets and wires with bornite and chalcocite from Tsumeb, South-West Africa.

Distinct crystals are rare and command high prices; the finest wire silvers are conceded to be those from Norway; many outstanding specimens came from Michigan but are now scarce; the "herringbones" of Batopilas are also very desirable.

Copper—Cu

Name: From the Greek, *Kyprios,* the name of the island of Cyprus, once producing this metal; kŏp′ẽr.
Crystals: Isometric; uncommon. Predominately in nodular, sheet-like, or branching masses on which crystals may be rudely developed. The latter are usually dodecahedrons, but cubes and octahedrons also occur. Twinning on *o*[111].

Physical Properties: Malleable, ductile, sectile; hackly fracture. H. $2\frac{1}{2}$-3. G. 8.95.

Optical Properties: Opaque. Pale red when fresh, rapidly tarnishing to brownish shades; sometimes green-stained. Streak shining pale red; luster metallic.

Chemistry: Usually contains small amounts of other metals. Commonly coated with crusts of dark purplish-red cuprite or greenish powdery deposits.

Distinctive Features and Tests: Looks like copper when fresh surface exposed; malleability; high specific gravity. Fuses at 3, producing a bright globule which becomes black-coated on cooling. Dissolves in acid, staining solution green.

Occurrences: In oxidized portions of sulfide ore bodies containing copper; in seams and cavities in sedimentary rocks adjacent to basaltic intrusions; in small quantities in cavities in basalts, associated with zeolites, but exceptional both in respect to quantity and quality from the deposits of Keeweenaw Peninsula, Michigan, where it is deposited in conglomerate near contacts with basaltic rocks. Commonly associated with cuprite and other copper minerals. Crystals to 1″ or better, occurred in the Keeweenaw Peninsula mines, sometimes associated with silver. Splendid specimens, probably the best of their kind, are clear calcites enclosing bright, untarnished copper platelets and wires (Figure 177); calcite crystals reach 4″ in length. Abundant in many mines in New Mexico, Arizona, and elsewhere in the western United States; occasionally in good crystals to $1\frac{1}{4}$″ diameter from the New Cornelia pit, Ajo, Pima County, Arizona (Figure 109); forms are primarily dodecahedrons, tetrahexahedrons, octahedrons; also beautiful fern-like twinned aggregates of great delicacy in calcite. Large rudely-crystallized masses, commonly encrusted with euhedrons of cuprite to $\frac{1}{4}$″, from Chino Pit, Santa Rita, Grant County, New Mexico. Sheets in sandstone in New Jersey, especially adjacent to basaltic sills; a mass of 40 pounds weight has been recorded from within basalt near Somerville. Highly prized because of their rarity are specimens of native copper in franklinite-willemite ore from Franklin, Sussex County, New Jersey. Excellent crystallizations from copper deposits at Bogoslovsk and Nizhni-Tagil, Urals, U.S.S.R. Dendritic and arborescent specimens from Cornwall, England. In quantity, as wires and sponges, from Tsumeb, South-West Africa. Sharp pseudomorphs after aragonite sixling twins, faithful in every detail and most unusual, from sedimentary deposits at Corocoro, Bolivia.

Best specimens show sharp crystals, but these are rare; very desirable are intricately branched dendrites or fern-like twinned aggregates, also the unusual sixlings from Corocoro. Very highly prized are Michigan clear calcites with copper enclosures. Masses of native copper reach many pounds weight, and it is ordinarily possible to obtain specimens from several inches to a foot or better.

Lead—Pb

An extremely rare mineral, seldom represented in average collections; lĕd. Dull gray masses associated with hausmannite and pyrochroite occur in quantity only in mines near Pajsberg and Langban, Vermland, Sweden; sizes from one inch to several inches; weights to over a pound.

Mercury—Hg

Relatively rare, only occasionally being found as small shining globules in crevices in ordinary mercury ores; měr′kū rī. Localities are New Almaden, Santa Clara County, and mercury deposits in Sonoma, Napa, and Lake counties, California; also at Idria, Gorizia,

Italy, and Almadén, Ciudad Real, Spain. Native mercury looks like mercury, the small droplets being easily moved about by the point of a pin.

Platinum Group

Platinum—Pt

Name: From the Spanish, *plata,* "silver," in allusion to its color; plăt′ĭ nŭm.
Varieties: Up to 28% iron, *ferrian;* up to 37% palladium, *palladian;* also *cuprian, rhodian, iridian.*
Crystals: Isometric; very rare. Predominately in nuggets or grains.
Physical Properties: Malleable, ductile; hackly fracture; very tough. H. 4-4½. G. 21.46 (pure); usually 14-19.
Optical Properties: Opaque. Slightly bluish-gray, like steel; iron-rich varieties dark gray to nearly black. Streak shining grayish-white; luster metallic.
Chemistry: Always contains iron, and commonly, palladium in substantial quantities.
Distinctive Features and Tests: Color and metallic properties; high specific gravity. Does not melt in ordinary torch flames; insoluble except in hot aqua regia.
Occurrences: Associated with chromite and magnetite in olivine-rich rocks as olivine-gabbro, dunite, peridotite, and found in alluvium resulting from their decay. As nuggets, and rarely, as very small cubes and octahedrons, from alluvial deposits in the Urals of the U.S.S.R., particularly from headwaters of the Tura River, Province of Perm. From the department of Cauca, Colombia; also from various places in New South Wales and New Zealand, Australia. In very small grains from Alaskan alluvial deposits and from the sands of the Trinity River, Trinity County, California.

Costs at least three times as much per ounce as gold; large nuggets rare, and very few collections have specimens exceeding size of pea.

Iron Group

Iron—Fe

A very rare mineral, found only in basaltic rocks as nodular masses, or in meteorites; ī′ẽrn. The prime locality is at Ovifak, Disko Island, Greenland, where many tons of specimens have been quarried from basalt; a single mass of 19 tons is preserved in Denmark. Usual specimen size, 1″ to 3″, commonly with adhering basalt. In smaller masses to 12 pounds, but very good, from Buhl, near Weimar, Germany. Specimens are cut through, lapped flat, and varnished to prevent rusting.

Nickel-Iron—Ni,Fe

Very rare, but found in considerable abundance as rolled masses in stream gravels in Josephine (*josephinite*) and Jackson Counties, Oregon. The latter have been found to several pounds weight, but usual specimens are not over 1″ to 3″ diameter. Iron and nickel-iron are the principal metallic constituents of meteorites, and are known respectively as *kamacite* and *taenite.*

SEMI-METALS

Arsenic Group

Arsenic—As

Name: From the Greek, *arsenikon,* a name originally applied to the mineral orpiment; är′sĕ nĭk.

Crystals: Hexagonal; very rarely, as rhombohedrons resembling cubes. Usually nodular or reniform masses, the latter forming crusts; also lamellar and then easily cleaved.

Physical Properties: Brittle; uneven fracture. Perfect cleavage parallel to basal pinacoid $c[0001]$, commonly with curved surfaces. H. $3\frac{1}{2}$. G. 5.7.

Optical Properties: Opaque. Tin-white color when fresh; tarnishes quickly to black. Streak shining tin-white; luster metallic.

Chemistry: Commonly contains small amounts of antimony, iron, nickel, silver or sulfur.

Distinctive Features and Tests: Crusted masses of granular texture; metallic luster; specific gravity. Can be confused with antimony and allemontite. Strongly heated on charcoal, volatilizes into dense white fumes, giving off garlic odor; antimony melts to metallic globule; allemontite also melts to metallic globule which takes fire.

Occurrences: With silver, nickel and cobalt ores in hydrothermal veins. In good specimens, but usually tarnished, from Freiberg, Schneeberg, Marienberg, and Annaberg in Saxony, Germany, associated with limonite and sometimes proustite. From Saltash, Cornwall, England. Spherical masses to several pounds weight from Washington Camp, Santa Cruz County, Arizona. Masses to 5″ across from Atlin, British Columbia. Specimens are usually badly discolored and unattractive, regardless of source.

Allemontite—AsSb

Name: After the type locality, Allemont, Isère, France; àl′ĕ mŏn′tīt.

Crystals: Hexagonal, but very rarely distinct. Usually in reniform masses or thick crusts with granular to fibrous aggregate structure.

Physical Properties: Brittle. One perfect cleavage. H. 3-4. G. 5.8-6.2.

Optical Properties: Opaque. Tin-white to pinkish-gray; gray to brown-gray when tarnished. Streak tin-white; luster metallic.

Distinctive Features and Tests: Brilliant luster; tin-white color; bright cleavages. Can be confused with arsenic and antimony. Fuses at 1, emitting fumes of arsenic and antimony, and producing a metallic globule which burns.

Occurrences: In hydrothermal sulfide veins; sometimes in granitic pegmatites. Usually associated with arsenic, antimony, stibnite, quartz, calcite and sphalerite. From Mine des Chalanches, near Allemont, France. Fine specimens to 3″ diameter, with quartz, from Atlin, British Columbia.

Antimony—Sb

Name: From Arabic, *al-uthmud,* to Medieval Latin, *antimonium;* orginally applied to stibnite, the sulfide of antimony; àn′tĭ mŏ′nĭ.

Crystals: Hexagonal; rare. Usually massive-lamellar, distinctly cleavable, forming brilliant reflective surfaces; also as thick crusts.

Physical Properties: Brittle; tough in finely granular masses. Perfect cleavage parallel to basal pinacoid $c[0001]$. H. 3-3½. G. 6.61-6.72.

Optical Properties: Opaque. Brilliant tin-white on fresh surfaces, usually coated by films of valentinite, an alteration product. Streak gray; luster metallic.

Chemistry: Sometimes contains small amounts of arsenic, iron or silver. Alters to valentinite and stibiconite.

Distinctive Features and Tests: Similar in appearance to allemontite and arsenic, with which it is easily confused. Fusibility 1; difficult to distinguish from arsenic and allemontite but melted globule on charcoal develops small protruding acicular crystals of white valentinite.

Occurrences: In hydrothermal veins, with arsenic, allemontite, stibnite and sulfides. Fine large pure masses formerly obtained from mines in Havilah and Kernville areas, Kern County, California; reported in masses to 300 pounds from Erskine Creek, near Hot Springs. From Arechuybo, Chihuahua, Mexico, in nodules several inches in diameter coated with pale yellow translucent valentinite; similarly from mines in Sonora. Coarsely cleavable plates in matrix, to several inches diameter, from Los Animos, Bolivia. Fine specimens from Sarawak Province, Borneo, and from localities in Germany, Czechoslovakia, and Sweden. Interesting specimens but seldom seen in average collections.

Bismuth—Bi

Name: From the German, *Wismuth,* of unknown origin; bĭs'mŭth.

Crystals: Hexagonal; natural crystals rudely formed, usually as platey aggregates in subparallel position. Often polysynthetically twinned on $e[10\bar{1}4]$.

Physical Properties: Brittle, somewhat sectile. One perfect easy cleavage parallel to basal pinacoid $c[0001]$, usually conspicuous on fracture surfaces. H. 2-2½. G. 9.7-9.8.

Optical Properties: Opaque. Distinctive pinkish silvery-white hue on fresh fracture surfaces, soon tarnishing to grayish iridescent film. Streak silver-white; luster metallic.

Chemistry: Commonly contains traces of sulfur, tellurium, arsenic, antimony.

Distinctive Features and Tests: Broader cleavage surfaces than other semi-metals; pinkish silver-white hue; polysynthetic twinning striations. On charcoal, fuses readily at 1, giving orange-yellow oxide coating on block, turning to lemon-yellow when cool. Soluble in nitric acid.

Occurrences: With other semi-metals and sulfides in hydrothermal silver, cobalt and nickel ore veins; rarely in small masses in granitic pegmatites. Massive in pitchblende at Great Bear Lake, Northwest territories, Canada, and in silver-cobalt-nickel ores of Ontario at O'Brien and Cobalt; the latter locality furnished specimens to 6″ broad, showing brilliant cleavage plates on diabase with calcite and quartz. Fine brilliant gray coarsely crystalline masses with bismuth and bismutite, from El Carmen Mine, Durango, Mexico. Coarsely cleavable, from Sorata, La Paz, Bolivia, sometimes in individual plates to 2″ diameter. Abundant in Saxony, Germany; Cornwall, England; Sweden, etc. With molybdenite from pegmatites at Kingsgate, New England Range, New South Wales, Australia.

Tellurium Group

Tellurium—Te

A rare mineral, seldom seen in collections due to scarcity and unattractiveness of the very small crystals; tĕl loo'rĭ ŭm. Fine specimens from Cripple Creek, Teller County, Colorado, as acicular crystals in parallel growths in rock fissure openings; more recently, from an unstated locality in New Mexico with emmonsite, pyrite and tellurite. The largest known crystal, about ¾″ by 1″, was reported from Balia, Turkey. Often in fine crystals from the Kawazu Mine near Simoda, Izu, Japan.

Selenium—Se

Excessively rare; sĕl ēn′ĭ ŭm. Recently, as red acicular crystals in felted masses or as coatings on sandstone from Homestake Mine, Ambrosia Lakes district, near Grants, Valencia County, New Mexico.

NON-METALS

Sulfur Group

Sulfur—S

Name: Of uncertain origin; also *sulphur,* but modern usage favors the simplified spelling given; sŭl′fĕr.
Varieties: Containing over 1% selenium, *selenian,* the color then inclining to orange; rare.
Crystals: Orthorhombic; commonly in fine crystals. Bipyramidal, also tabular on *c*[001] (Figure 110); faces usually developed unequally, leading to difficulty in orientation. Also

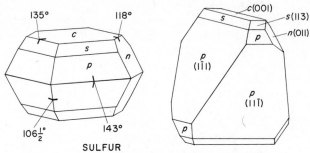

SULFUR
FORMS—pinacoid *c/*001/, prism *n/*011/, dipyramids *p/*111/, *s/*113/.

FIG. 110.

granular massive in spherical or encrusting shapes. Rarely twinned on *n*[011]; also *e*[101] and *m*[110]. Crystals usually sharp, smooth or even glassy. Observed along the *c* axis, the crystal cross-section is diamond-shaped.
Physical Properties: Very brittle, weak, slightly sectile; large crystals audibly crack from the heat of the fingers when held close to the ear. Fracture conchoidal. Imperfect cleavages parallel to *c*[001], *m*[110], and *p*[111]. H. 1½-2½. G. 2.07.
Optical Properties: Transparent to translucent. Vivid pale to deep yellow; also inclining toward orange; sometimes greenish, grayish, if inclusions present. Streak colorless. Luster resinous to greasy in larger crystals, but vitreous in small, perfectly-formed crystals. Biaxial positive. R.I. $\alpha = 1.958$, $\beta = 2.038$, $\gamma = 2.245$. Birefringence 0.29. Slightly pleochroic, pale yellow, deeper yellow.

Chemistry: Sometimes contains selenium and tellurium. Often contaminated with clay and bitumen.

Distinctive Features and Tests: Unmistakable transparent yellow crystals; softness; low melting point; acrid sulfur dioxide fumes when burnt. Fuses at 1; burns at 270°C with almost invisible flame giving off intensely acrid fumes (striking match odor).

Occurrences: Common as thin crusts, sometimes coated with small sparkling crystals in volcanic fumaroles but the finest specimens are from cavities in limestones and other sedimentary rocks, associated with celestite, calcite, aragonite and gypsum. The best specimens, displaying sharp transparent crystals of wonderful perfection and size, commonly 2″ to 3″ in diameter, are obtained from Agrigento (formerly Girgenti), Cianciana (crystals to 5″), Racalmuto, and Cattolica on the Island of Sicily, Italy; also fine from Perticara in Romagna with asphalt. Small crystal druses common in volcanic regions. Lately, druses of brilliant smooth small crystals, seldom over ¼″, from near San Felipe, Baja California, Mexico.

Classic specimens from Sicily commonly display large crystals sprinkled in singles and groups on limestone matrix with white bladed celestite crystals and large sixling twins of white to pale green aragonite, the latter to 2″ in diameter. Matrices may reach a foot or more across. Good average specimens are several inches in diameter sprinkled with sulfur crystals of about ½″; also desirable are specimens of globular crystal aggregates of sulfur alone, or large, nearly complete, single crystals.

Carbon Group

Diamond—C

Name: From the Greek, *adamas,* meaning "invincible" or "hardest"; dī′ mŭnd.

Varieties: Granular to cryptocrystalline, grayish and translucent *bort;* massive finely granular and black opaque, sometimes slightly porous, *carbonado.*

Crystals: Isometric; predominantly octahedrons of greater or lesser perfection, the finest appearing as if cut and polished to geometrical perfection ("glassies"). Less common are dodecahedrons, cubes, and tetrahedral crystals, the latter usually smooth, but dodecahedrons and cubes mostly rough, pitted, striated, or displaying several crystal forms. Commonly twinned on *o*[111] (Figure 111), sometimes forming tabular aggregates flattened on *o*[111]; also rarely twinned on *c*[001]. Growth hillocks abundant on crystal faces, less commonly, solution pits. Very small crystals vary greatly in form and are sought by micromounters for purposes of study.

Physical Properties: Brittle; fracture conchoidal. *Bort* and *carbonado* extremely tough; twins less so, single crystals still less. Perfect cleavage parallel to *o*[111]. H. 10. G. 3.50-3.53.

Optical Properties: Transparent to opaque. Usually colorless or nearly so; rarely, distinct yellow, brown, green, pink, blue; very rarely deep hued. *Bort* medium to pale gray; *carbonado* black. Adamantine luster, reminiscent of half-silvered glass. R.I., n = 2.4175. Under ultraviolet, some stones fluoresce pale blue, green, yellow, and, rarely, red. Irradiated stones may turn green or blue; the green can be heat-treated to impart brownish or yellowish hues.

Chemistry: Nearly pure carbon.

Distinctive Features and Tests: Supreme hardness; luster; single crystals commonly octahedrons. Infusible, insoluble.

Occurrences: In kimberlite pipes and dikes associated with pyrope, enstatite, phlogopite, pyrite, zircon, augite, perovskite, chromite and magnetite; also in conglomerates and

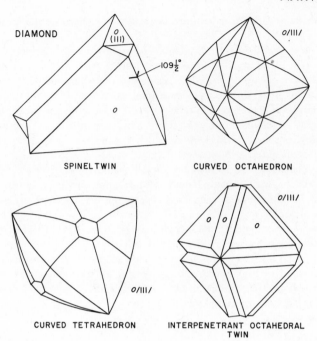

DIAMOND

$109\frac{1}{2}°$

o (111)

o

o/111/

SPINEL TWIN

CURVED OCTAHEDRON

CURVED TETRAHEDRON

o/111/

INTERPENETRANT OCTAHEDRAL TWIN

o/111/

o o o

FIG. 111.

gravels. Alluvial diamond has been found at various places in the Appalachian Mountains, especially in North Carolina; also in a circular belt of glacial moraines around the Great Lakes. The only diamond-bearing kimberlites in the United States occur near Murfreesboro, Pike County, Arkansas, where crystals to about 40 carats weight have been recovered (Figure 112); smaller crystals are still being found. Numerous crystals were found in gold placers in California, particularly in the Mother Lode country. Crystals in conglomerate and alluvium derived from its erosion, occur in a number of Brazilian states; rarely, unusual matrix specimens are obtained, displaying lustrous irregular crystals imbedded in goethite- and hematite-stained conglomerate, with particles of sharp sand; fine micromount crystals also come from Brazil. Practically all commercial diamond is supplied by kimberlite mines and alluvial workings in Africa, from countries on or near the Equator to the Union of South Africa. Also in minor amounts from India, Borneo, and elsewhere.

Because of expense, crystals over 1/8″ diameter are seldom seen in average collections; especially prized specimens are crystals in matrix, now mainly from Brazil, but rarely obtainable, and formerly, from African kimberlites but now unobtainable because of the mechanization of mining.

Graphite—C

Name: From the Greek, *graphein*, "to write"; gràf′ĭt.
Crystals: Hexagonal; usually thin tabular, compressed along the *c* axis, making *c*[0001] the broadest faces (Figure 112); the latter may show triangular markings. Most commonly in coarse to fine foliated masses; also scaly, columnar, earthy, and in radiate aggregates.

FIG. 112. Diamond crystals and graphite. The diamond crystals at the left are from the Murfreesboro, Arkansas, mines, and show the rude forms which these crystals assume. Many are yellow, brown, greenish, etc. and only a few are colorless. The largest crystal in this group, the distorted octahedral specimen to the left of and slightly below center, weighs 17.85 carats. *Right:* tabular crystals of graphite in matrix of fine-granular metamorphosed limestone, from near Warwick, Orange County, New York. Size of specimen 3″ by 2½″. *Diamond photo courtesy Smithsonian Institution.*

Physical Properties: Flexible but not elastic; sectile. Perfect cleavage parallel to pinacoid $c[0001]$. Greasy feel. H. 1-2. G. 2.1-2.2.

Optical Properties: Opaque. Black; rarely, steel-gray. Streak shining black; luster submetallic on cleavage plates; also dull, earthy.

Chemistry: Carbon, but commonly impure with clays, iron oxides, etc.

Distinctive Features and Tests: Color; softness and streak; greasy feel. Can be confused with molybdenite but latter decidedly bluish in color and much more metallic in luster. Manganese minerals are not flaky nor as soft as graphite. Insoluble; infusible.

Occurrences: In metamorphic rocks as marbles, schists, gneisses; in fissure veins with quartz, orthoclase, biotite; sometimes in basaltic rocks. Abundant in small quantities but good crystals or attractive masses scarce. Massive, from Ticonderoga, Essex County, New York; good flakes, and occasionally fine small crystals, from marbles in Orange County, New York, and Sussex County, New Jersey. Large masses of interesting structure, from Ragadera Mine, near Galle, Ceylon. Graphite is seldom represented in collections because of the scarcity of attractive specimens.

SULFIDES

The sulfides include about eighty species, but of this number, only those of iron, copper, and zinc are abundant. The remaining sulfides are uncommon or rare, particularly in specimens suitable for collections. Sulfides concentrate in hydrothermal ore veins, tending to form small to large masses of several species together and, less commonly, substantial masses in which one sulfide predominates. Mostly, veins are filled wall-to-wall and euhedral crystals are difficult to obtain despite the abundance of the sulfides concerned. As a class, sulfides are characterized by dark colors, metallic lusters, high specific gravities, and low tenacities. Most are therefore easily broken, and readily produce dark streaks on a porcelain test plate. Additionally, many fuse easily on charcoal and otherwise react distinctively to various tests. Practically all are opaque or at best translucent.

Copper Arsenide Group

Algodonite—Cu₆As

Domeykite—Cu₃As

Rare copper arsenides, occurring in abundance only in Michigan; ăl gŏd'ō nīt; dō mā'kīt. Algodonite, a steel-gray mineral of bright metallic luster when fresh, but dulling on exposure, is found with copper and domeykite in quartz at the Mohawk Mine and others in Keweenaw and Baraga counties in Michigan. Domeykite, a tin-white mineral, tarnishing to brown on exposure, occurs in the Sheldon-Columbia Mine on Portage Lake in Houghton County, with niccolite, rammelsbergite and algodonite in quartz; also from the mines previously mentioned for algodonite. Masses up to 6″ across have been recovered.

Argentite Group

Argentite—Ag₂S

Name: After *argent,* "silver"; är'jĕn tīt.
Variety: About 14% copper, *cuprian.*
Crystals: Isometric; rare. Cubic and octahedral; also dodecahedral; seldom sharp or well-formed. Predominately in parallel growths, skeletal crystals, and subhedral aggregates. Twinning on $o[111]$.
Physical Properties: Sectile, almost like lead. Fracture subconchoidal. Poor cleavages on $c[001]$ and $d[110]$. H. 2-2½ G. 7.2-7.4.
Optical Properties: Opaque. Blackish lead-gray; bright metallic on fresh surfaces, tarnishing quickly. Streak shining metallic black; luster metallic, but dull on most crystals.
Chemistry: Silver sulfide. Below 179°C, the structure loses isometric symmetry and becomes orthorhombic acanthite; thus specimens labeled "argentite" are really the latter species. Alters to native silver, some specimens being part silver, part argentite.

Distinctive Features and Tests: Skeletal crystals in parallel growths; resembles galena but more sectile and lacking the perfect cleavage of that species. Fuses at 1½, yielding malleable silver bead.

Occurrences: A primary, low-temperature hydrothermal vein mineral, commonly associated with silver, proustite, pyrargyrite, stephanite, polybasite and cerargyrite. Occurs massive in many mines of California. Large masses were obtained from the Comstock Lode of Virginia City, Nevada. Sometimes well-crystallized from mines in Guanajuato and Zacatecas, Mexico; recently in skeletal crystals from El Rosario, Urique district, Chihuahua. Especially fine and large crystals from Chanarcillo, Chile. Exceptional crystals, mainly octahedrons modified by cubes, to 1½″ diameter, from the silver mines of Schneeberg, Annaberg, Marienberg, and others in the Freiberg district and at Andreasberg, Harz, Germany. Fine and large from Pribram, Czechoslovakia.

Good crystals are rare and very expensive; average crystals are ¼″ to ⅜″; over ½″ exceptional. Usually offered in small groups not over 2″ across.

Chalcocite—Cu₂S

Name: From the Greek, *chalkos,* "copper"; kăl′kō sĭt.

Crystals: Orthorhombic; uncommon. Quite sharp tabular pseudo-hexagonal prisms formed by sixling twinning (Figures 113, 114). Twinning plane is prism *m*[110]; broad face of sixlings is basal pinacoid *c*[001]; latter striated. Faces dull because of alteration.

Physical Properties: Brittle, somewhat sectile. Fracture conchoidal. Cleavage indistinct parallel to *m*[110]. H. 2½-3. G. 5.5-5.8.

Optical Properties: Opaque. Dull gray-black to dead black. Streak dark gray; luster metallic on fresh surfaces, otherwise dull.

Chemistry: Cuprous sulfide. Alters to copper or covellite.

Distinctive Features and Tests: Dead black color; pseudo-hexagonal twins; softness; sectility. Crystals jumbled on matrix, usually on edge. Fuses at 2-2½ on charcoal, yielding copper globule in reducing flame. Soluble in nitric acid.

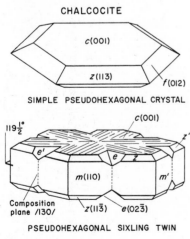

CHALCOCITE

c(001)

z(113)   *f*(012)

SIMPLE PSEUDOHEXAGONAL CRYSTAL

119½°   *c*(001)   *z*′

e′   *e*   *z*

m(110)   *m*′

Composition plane /130/   *z*(11̄3)   *e*(02̄3)

PSEUDOHEXAGONAL SIXLING TWIN

FIG. 113.

Fig. 114. Chalcocite sixling twins on matrix of massive chalcocite, from Messina, North Transvaal, South Africa. Size of specimen 1½″. Note striations on the c(0001) plane of the crystal at the top.

Occurrences: Usually in formless masses, sometimes of considerable size, with covellite, bornite, and other sulfides, in the enriched zones of hydrothermal sulfide veins; crystals from openings in oxidized zones. Massive chalcocite abundant in many western United States copper deposits, e.g., Butte, Montana; Miami, Morenci, Bisbee, and Ray, Arizona; also at Kennecott in the Copper River district of Alaska. In Mexico, Chile and Peru. In large amounts at Tsumeb, and in excellent crystals, very similar to the Cornish, from Messina, South-West Africa (Figure 114). Crystals are rare and highly prized despite the dull black coating which soon covers them after removal from the ground. Thick tabular to sharp-edged tabular individuals up to about ¾" have been found in a sulfide deposit near Bristol, Hartford County, Connecticut. The best crystals came from Redruth, St. Just, St. Ives, and Cambourne in Cornwall, England. These were found sprinkled on matrix, usually standing on edge, with individuals about ½" diameter; also as star-like twins.

Bornite—Cu_5FeS_4

Name: After Austrian mineralogist Ignatius von Born (1742-1791); bōr'nĭt.
Crystals: Isometric; very rare. Rough-surfaced curved cubes and dodecahedrons. Predominately massive (Figure 115).
Physical Properties: Brittle, somewhat sectile. Conchoidal fracture. Indistinct cleavage parallel to $o[111]$. H. 3. G. 5.07.
Optical Properties: Opaque. Coppery or bronzy on fresh fracture surfaces, very quickly tarnishing to darker hues, and ultimately to intense purplish hues, hence the common

FIG. 115. Bornite cavity filling, studded with pyrite crystals and lightly coated in places with small patches of wire silver. The impressions upon the mass suggest that the bornite occupied a space between calcite crystals with the latter subsequently dissolving. About 4" by 2½". From lower oxidized zone, Tsumeb, South-West Africa.

name "peacock ore." Streak pale gray-black; luster metallic when fresh, submetallic when deeply tarnished.

Chemistry: Copper iron sulfide.

Distinctive Features and Tests: Purplish tarnish; bronzy color on fresh fractures; specific gravity. On charcoal, fuses at 2, yielding brittle black magnetic globule. Soluble in nitric acid, staining the solution blue.

Occurrences: Most abundant in massive intergrowths with other sulfides in hydrothermal veins; also in contact metasomatic deposits, and in small nodules in basalt cavities associated with prehnite and zeolites. Numerous localities but small crystals in druses are known only from very few localities, e.g., Bristol, Hartford County, Connecticut; Butte, Silver Bow County, Montana; Redruth, Cornwall, England.

Galena Group

Galena—PbS

Name: From the Greek, *galene,* "lead ore"; gă lē′nă.

Crystals: Isometric; very common. Predominately cubes, or cubes with octahedral modification (Figure 116); uncommonly in octahedrons. Sometimes repeatedly twinned on

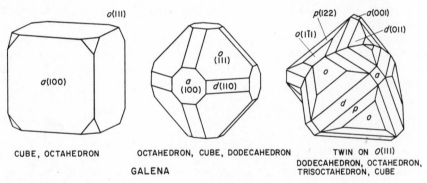

CUBE, OCTAHEDRON   OCTAHEDRON, CUBE, DODECAHEDRON   TWIN ON *O*(1Ī1)
                                                                 DODECAHEDRON, OCTAHEDRON,
                         GALENA   TRISOCTAHEDRON, CUBE

FIG. 116.

o[111] forming complex, nearly spherical crystals; also twinned on μ[114], then forming lamellar twins with crystals encircled by numerous narrow bands marking twin junctions. Sharp crystals are common but faces usually interrupted by numerous offsets; skeletal and hopper crystals rare. Face quality ranges from dull to brilliantly smooth and brightly metallic. Cubes commonly studded with "hobnail" overgrowths of octahedrons, also oriented overgrowths of marcasite or chalcopyrite.

Physical Properties: Brittle; subconchoidal or stepped fracture, but the very easy and perfect cleavage parallel to cube faces *a*[100] is invariably evident in all fractured specimens and provides the most reliable recognition clue. H. 2½. G. 7.58.

Optical Properties: Opaque. Blue-gray metallic color when fresh, but usually some shade of pale to dark gray because of surface alteration. Streak lead-gray. Metallic luster when freshly fractured; some crystals also display bright metallic luster but most are dull.

Chemistry: Lead sulfide. Silver-bearing galenas commonly contain argentite and tetrahedrite in solid solution; there are no outward signs of this intergrowth.

Distinctive Features and Tests: Perfect cleavage and color; luster; crystals. On charcoal, fuses at 2, coating block yellow near flame, emitting sulfurous fumes, and yielding malleable lead globule.

Occurrences: Present in practically all hydrothermal sulfide ore bodies, associated with pyrite, chalcopyrite and sphalerite. In large deposits throughout the Mississippi River Valley, especially in the Oklahoma-Kansas-Missouri area, where chert-limestones are replaced and cavities filled by sulfides, presumably from rising hydrothermal mineralization. Occurs in small nodules in some granitic pegmatites; also in contact metasomatic deposits. A very widespread mineral; localities are so numerous that it is impossible to list more than a few. Splendid crystals from mines in the tri-state district mentioned above, especially near Joplin, Missouri (Figure 117); cubes of 6″ or better reported but commonly offered in specimens covered by cubes ranging in size from ½″ to 2″; octahedrons rare, seldom over ¾″. Associates in the tri-state deposits are very typical: yellow scalenohedral calcite crystals, sometimes to 24″ in length, but usually far less, small sharp tetrahedral chalcopyrite crystals, cockscombs of marcasite, black to dark red sphalerite crystals, curved pink dolomite crystals, drusy quartz, and, rarely, enargite. Fine specimens come from similar deposits in Wisconsin, Illinois, and Iowa. Splendid lustrous cubo-octahedrons from Leadville district, Lake County, Colorado. Fine crystals in the silver deposits of Freiberg, Saxony, and elsewhere in Germany. Exceptional specimens, often brilliant and complexly twinned, from Truro, Liskeard, etc., in Cornwall, and from the lead-mining districts at Alston Moor and Weardale in England; similarly from Wanlockhead, Scotland. An interesting occurrence is at Broken Hill, New South Wales, Australia where massive galena encloses deep red-brown translucent to transparent crystals of rhodonite.

Galena occurs in such variety that a large single-species collection could be made from it alone. Excellent specimens of tri-state material are relatively inexpensive and can be obtained in all sizes from single crystals to large slabs of matrix studded with crystals. Specimens from abroad are costlier, usually because localities are no longer productive.

Sphalerite Group

Sphalerite—(Zn,Fe)S

Name: From the Greek, *sphaleros,* "treacherous," in allusion to its similarity to more easily smelted ores. The alternate name, blende, is derived from the German *blenden,* "to deceive." sfă′lĕr ĭt.

Varieties: Containing up to 26% iron, *ferroan* or *marmatite,* very dark brown to black. Nearly iron-free and very pale in color, *cleiophane;* klē′ō făn.

Crystals: Isometric; common. Predominately tetrahedral (Figure 118), often developed into forms resembling octahedrons. Seldom sharp or with smooth faces; usually curved, twinned, striated, etc. (Figure 119). Commonly in fine-grained to coarse-grained masses; rarely cryptocrystalline. Twinning very common on o[111], and repeated, manifested by narrow striations and bands of slightly differing luster on crystal surfaces and cleavage planes.

Physical Properties: Brittle; uneven fracture. Perfect cleavage, easily developed, parallel to dodecahedron d[110]; with care, a large fragment can be cleaved to this form. H. 3½-4. G. 3.9-4.1, lowering with iron.

Optical Properties: Transparent to translucent; dark varieties nearly opaque. Exceptional pale colored specimens transparent through 2″ of material. Usually some shade of brown, reddish-brown, red or green; also pale yellow, very pale green, and nearly colorless. Iron-rich varieties nearly black. Color zoning common as red-brown streaks in green or yellow material. Streak pale brown to colorless; luster resinous to adamantine. R.I., $n =$ 2.369. Transparent material is suitable for gems.

FIG. 117. Galena crystals on chert matrix from the Grace Walker Mine, Picher, Ottawa County, Oklahoma. *Top:* Cube-octahedrons sprinkled on breccia fragment of chert. Size of specimen 9″ by 5″. *Bottom:* striated cubes coated with minute cube-octahedrons of galena and some marcasite and chalcopyrite. Size 5½″ by 4¼″.

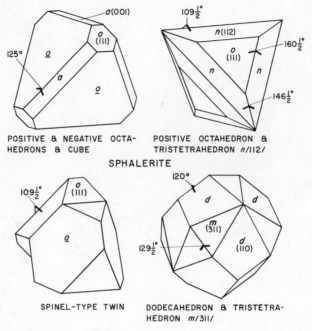

POSITIVE & NEGATIVE OCTA-   POSITIVE OCTAHEDRON &
HEDRONS & CUBE   TRISTETRAHEDRON *n*/112/

SPHALERITE

SPINEL-TYPE TWIN   DODECAHEDRON & TRISTETRA-
                          HEDRON *m*/311/

FIG. 118.

Chemistry: Zinc sulfide, invariably with more or less iron, sometimes a little manganese or cadmium.

Distinctive Features and Tests: Typical dark red-brown or green color; luster; cleavage. Practically infusible; when roasted, emits sulfurous fumes. Slowly soluble in hydrochloric acid, emitting rotten egg odor of hydrogen sulfide.

Occurrences: Usually with other sulfides in hydrothermal ore veins and in disseminated deposits; also in veins in igneous rocks and in contact metasomatic deposits. Good crystals, some to 1½″ (Figure 119), from numerous lead deposits in the Mississippi Valley region; fine examples have been obtained from the tri-state region, centered about Joplin, Missouri, ranging from nearly black crystals to transparent individuals of deep red color. Fine dark olive-green crystals from the Breckenridge district, Summit County, Colorado. From Butte, Silver Bow County, Montana in exceptional specimens. Green to very pale green (*cleiophane*) crystals to 1″, and massive, from Franklin, Sussex County, New Jersey, occasionally cut into gems of splendid luster and color. Fine pale to deep green, and green and red banded gemmy material from Cananea, Sonora, Mexico, sometimes in complex twinned crystals to 4″ in diameter with calcite and drusy quartz. Probably the sharpest, most nearly textbook-perfect crystals, to ¾″, occur in small openings in dolomitic marble at the famed Lengenbach quarry, Binnatal, Switzerland; crystals are yellow-brown, exceptionally smooth-faced, and principally single or twinned tetrahedrons. The silver-lead-zinc ore veins of Czechoslovakia, Germany and Romania, yield fine specimens, displaying sharp crystals associated with galena, pyrite, etc. Similar specimens of high quality from Cumberland, Durham, Derbyshire, Cornwall, etc., in England. Splendid gem quality, pale yellow to orange cleavage masses, also brown and green, from Picos de Europa, Santander, Spain; also excellent crystal druses to 6″ across, covered by brilliant irregular twinned crystals.

FIG. 119. Sphalerite. *Left:* Rude twinned black crystals on chert from southwest Missouri. Size 5½″ by 3½″. *Top right:* Transparent greenish-yellow curved and twinned crystals from Picos de Europa, Santander, Spain. Size 1¾″ by 1¾″. *Bottom right:* pale green cluster of transparent crystals from Franklin, Sussex County, New Jersey. Size 1½″ by 1⅝″.

Specimens are abundant, especially from the tri-state district. Smooth-faced crystals with easily recognizable forms are scarce. Lengenbach specimens are difficult to obtain and expensive.

Chalcopyrite Group

Chalcopyrite—CuFeS$_2$

Name: From the Greek, *chalkos*, "copper," hence "copper pyrite"; kàl′kō pī rīt.

Crystals: Tetragonal; commonly resembling tetrahedrons, sharp and distinct when small, but becoming curved, striated and otherwise irregular with increasing size (Figure 120). Broad faces marked by triangular striations. Usually as isolated individuals on matrix, but also drusy, in parallel growths, and massive. Seldom very bright because of tarnishing, the latter often irridescent. Twinning on $p[112]$.

Physical Properties: Brittle; more so than pyrite. Fracture uneven to conchoidal. Poor cleavage parallel to $z[011]$. H. 3½-4. G. 4.1-4.3.

Optical Properties: Opaque. Bright brass-yellow when fresh, the color being more intense than pyrite. Streak greenish-black; luster metallic to dull.

Chemistry: Copper and iron sulfide.

Distinctive Features and Tests: Sharp tetrahedral crystals; rich brass-yellow color; streak; brittleness. May be confused with pyrite and marcasite, but pyrite paler, and crystal forms are different; marcasite forms are also different as well as color. Fuses at 2, giving off sulfurous fumes and yielding black magnetic globule. Soluble in nitric acid, tinging the solution green.

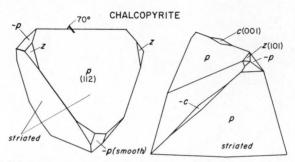

POSITIVE DISPHENOID *p* /112/, NEGATIVE DISPHENOID −*p* /112/,
DIPYRAMID *z* /011/, AND PINACOIDS *c*/001/ AND −*c*/001/

FIG. 120.

Occurrences: The most abundant copper mineral, very commonly formed in medium to high temperature hydrothermal ore veins associated with pyrite and other sulfides. Abundant in contact metasomatic deposits, particularly in marble skarn. Also with pyrrhotite and pentlandite in sulfide nickel deposits. Sharp small crystals, about $\frac{1}{4}''$, occur sprinkled on pink dolomite in the tri-state district. In exceptional crystals from the French Creek Mines, Chester County, Pennsylvania, to $1\frac{1}{4}''$, but usually tarnished. Large tarnished crystals from various mines in Clear Creek County, Colorado. Lately, fine iridescent-tarnished tetrahedral crystals to $\frac{1}{4}''$, with quartz crystals, from La Bufa, Chihuahua, Mexico. From the Ground Hog Mine, Vanadium, near Silver City, New Mexico, as large crude interpenetrant crystals to $1''$. From numerous localities in Germany, France, Italy, Romania, etc. Forms unusual colloform material, the so-called "blister copper," in the Carn Brea Mine, Illogan, Cornwall, England; in fine crystals from Greenside Mine, Westmoreland. In Scotland, from the Wanlockhead Mine in Dumfriesshire. Fine groups from Ani and Arakawa, Ugo, Japan. Also fine groups, with quartz crystals, from Wallaroo, South Australia.

Good specimens are abundant, the best showing lustrous sharp crystals, usually on matrix of white drusy quartz; crystals in excess of $\frac{1}{2}''$ are uncommon.

Stannite—Cu_2FeSnS_4

A rare copper, iron and tin sulfide, seldom represented in collections; stăn′ĭt. Massive material occurs at some European localities, principally in Cornwall, England, but the finest specimens are those from tin deposits in Bolivia. At Llallagua, complexly twinned crystals were found to $1\frac{3}{4}''$ diameter, although the usual run of specimens from this locality exhibit crystals to about $\frac{1}{2}''$; also from Uncia. Crystals are black and unattractive, but when associated with bright pyrite, as is commonly the case, excellent showy specimens result.

Wurtzite Group

Wurtzite—(Zn,Fe)S

Name: After French chemist, Adolphe Wurtz; wĕrtz′ĭt.

Crystals: Hexagonal; tapering hexagonal pyramids, blunt or pointed, with basal face $c[0001]$ prominent. Pyramid faces usually striated parallel to base; hemimorphic, thus not *di*pyramidal. Usually in small isolated crystals on matrix.

Physical Properties: Brittle; fracture even to conchoidal. Cleavage parallel to $a[11\bar{2}0]$ easy; $c[0001]$ difficult. H. $3\frac{1}{2}$-4. G. 3.98.

Optical Properties: Opaque to translucent. Brownish black. Streak brown; resinous luster.

Chemistry: Zinc and iron sulfide.

Distinctive Features and Tests: Small tapering hexagonal pyramids; luster.

Occurrences: A very rare mineral. Fine crystals have been found at Oruro and Llallagua in Bolivia, reportedly to 1″. Also fine micro crystals in vugs in massive wurtzite from the Original Mine, Butte, Silver Bow County, Montana. Micro crystals may be more common than supposed but are usually overlooked.

Greenockite—CdS

Cadmium sulfide; similar in crystal structure and habit to wurtzite, and noted for the smallness and excessive rarity of crystals; grēn′ŭk ĭt. The record crystal appears to be only a $\frac{1}{2}$″ individual, found in Scotland in 1810 and originally thought to be sphalerite! Very small elegant crystals occur in cavities in the tin veins of Llallagua, Bolivia. Some have been found with prehnite in cavities in the pillow basalts of New Jersey, mainly around Paterson, Passaic County. On prehnite, with natrolite and calcite in cavities in porphyry at Bishopton, Renfrew, Scotland. Elsewhere, greenockite occurs mostly as earthy films resulting from alteration of zinc minerals, especially sphalerite. Almost unobtainable and a prize worth seeking by micromounters.

Niccolite Group

Pyrrhotite—FeS

Name: From the Greek, *pyrrhotes,* "redness," in allusion to color; pĭr′ō tĭt.

Variety: Certain massive types are called *troilite.*

Crystals: Hexagonal; uncommon. Usually thin tabular (Figure 121) flattened along $c[0001]$, in subparallel rosettes of hexagonal outline. Usually tarnished, sometimes coated with small pyrite crystals, or partly altered to whitish coating. Far more abundant massive.

Physical Properties. Brittle. No cleavage but parting sometimes observed on $c[0001]$. Fracture uneven to subconchoidal. H. $3\frac{1}{2}$-$4\frac{1}{2}$. G. 4.6-4.65. Magnetic.

Optical Properties: Opaque. Bronze-yellow to brownish; tarnishing to dark brown, often with iridescence. Streak gray-black; luster metallic.

Chemistry: Iron sulfide. Ordinary pyrrhotite is deficient in iron, a number of sites in the crystal structure, normally containing iron, being unfilled; the *troilite* variety is close to theoretical composition. The absence of iron contributes to easy alteration, some specimens quickly developing a powdery coating of iron sulfate and crumbling.

Distinctive Features and Tests: Peculiar bronze-yellow color; distinctive flat crystal rosettes; basal parting. Particles and powder unmistakably attracted to magnet in contrast to behavior of similar yellow metallic-lustered species as pyrite and chalcopyrite. Fuses at 3 to black magnetic globule; dissolves in hydrochloric acid, giving off rotten egg odor of hydrogen sulfide.

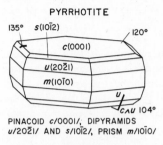

PYRRHOTITE

PINACOID *c*/0001/, DIPYRAMIDS
u/20$\bar{2}$1/ AND *s*/10$\bar{1}$2/, PRISM *m*/10$\bar{1}$0/

FIG. 121.

Occurrences: With sulfides of nickel and chalcopyrite in the large ore bodies in basic igneous rocks at Sudbury, Ontario. In high temperature hydrothermal sulfide veins with other sulfides and oxides; in contact metasomatic deposits; also in very small masses or wafer-like crystals in granitic pegmatites. Common as *troilite* nodules in meteorites. Good hexagonal crystals to 1″ occur in the Bluebell Mine, British Columbia. Some fine specimen crystals were obtained in pegmatite at Standish, Cumberland County, Maine. The best crystals, up to 3″ to 4″ in diameter, were obtained from Trepča, Yugoslavia, and from Kisbanya, Transylvania, Romania. Extremely large though poorly formed crystals, replaced by marcasite and covered by wavellite with quartz crystals, were recovered from tin mines of Llallagua, Bolivia; crystals to 3″ by 4″ by 6″ were recorded; also massive, with basal parting surfaces continuous for 10″. Fine specimens from the Morro Velho Mine, near Ouro Preto, Minas Gerais, Brazil.

Niccolite—NiAs

Name: From the Latin, *niccolum*, referring to composition; nĭk′ō līt.

Crystals: Hexagonal; very rare, tabular on *c*[0001], or pyramidal. Mostly massive and featureless.

Physical Properties: Brittle; uneven fracture. No cleavage. H. 5-5½. G. 7.78.

Optical Properties: Opaque. Pale coppery red, tarnishing to gray. Streak pale brownish-black. Luster metallic.

Chemistry: Nickel arsenide. Contains iron, cobalt and sulfur in small amounts. Alters on surface to pale green annabergite.

Distinctive Features and Tests: Distinctive pink metallic hue. Fuses at 2, giving off arsenical garlic odor fumes, and producing bronze-colored metallic globules. Dissolves in nitric acid, staining solution green.

Occurrences: With other nickel sulfides and arsenides, pyrrhotite and chalcopyrite, in sulfide ore bodies; also in hydrothermal veins. Relatively rare, found only in quantity in the Cobalt, Gowanda, and Sudbury districts of Ontario, sometimes in specimen masses to 6″ or more across. Occasionally, attractive intergrowths with associated sulfide minerals are polished in slab form.

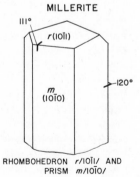

MILLERITE

RHOMBOHEDRON *r*/10$\bar{1}$1/ AND
PRISM *m*/10$\bar{1}$0/

FIG. 122.

Millerite—NiS

Name: After English mineralogist, W. H. Miller (1801-1880); mĭl′ẽr ĭt.

Crystals: Hexagonal; predominately as very slender filaments, greatly elongated in the direction of the *c* axis, and terminated by a low rhombohedron (Figure 122). In radiate sprays and matted fibrous masses. Rarely, granular to massive.

Physical Properties: Brittle; hairlike crystals elastic. Perfect cleavage parallel to $r[10\bar{1}1]$ and $e[01\bar{1}2]$. H. 3-3½. G. 5.5.

Optical Properties: Opaque. Brass-yellow, bronze-yellow, tarnishing to gray-green. Streak greenish-black. Luster metallic.

Chemistry: Nickel sulfide.

Distinctive Features and Tests: Fibrous crystals of metallic luster and yellow color cannot be mistaken for other species. Fuses readily at 1½-2, yielding black magnetic globule.

Occurrences: A low temperature hydrothermal mineral in carbonate veins associated with other nickel minerals. Fine specimens occur in quartz-lined geodes of the Keokuk area near the junction of Iowa, Illinois and Missouri, and in cavities in limestones in other places in the Mississippi River valley. Occasionally enclosed in clear calcite crystals from these localities. Excellent specimens from Siegen, Westphal, Germany; at Kladno-Rodna, Czechoslovakia, and elsewhere in Europe. Unusual cleavable masses to several inches across, occur with gersdorffite, quartz, bravoite, and chalcopyrite at Timagami, Ontario. From the collector's viewpoint, prize specimens are those within geodes, showing sprays of millerite to 2″ in length, and contrasting well against white drusy quartz.

Pentlandite—(Fe,Ni)$_9$S$_8$

An iron-nickel sulfide which occurs only granular massive; pĕnt′lånd ĭt. Resembles pyrrhotite but is distinguished from the latter mineral by lack of magnetism. Associated with pyrrhotite in the nickel ores of Ontario; also from various places in southeastern Alaska, and massive, in single-crystal grains to several inches across, from the Yale district, Emory Creek, British Columbia. Seldom placed in amateur collections.

Covellite Group

Covellite—CuS

Name: After N. Covelli (1790-1829), an Italian mineralogist; kō′vĕl ĭt.

Crystals: Hexagonal; rare. Thin tabular hexagonal plates, compressed along the *c* axis making the faces of basal pinacoid *c*[0001] the broadest; these faces may show hexagonal striations (Figure 123). Aggregates consist of many subparallel plates standing on edge. Also foliated massive.

Physical Properties: Brittle to somewhat sectile; thin plates somewhat flexible. Uneven fracture. Cleavage easy and perfect parallel to *c*[0001]. H. 1½-2. G. 4.6-4.76.

Optical Properties: Opaque. Deep indigo-blue, often with strong purplish iridescence. Streak shining gray-black; luster submetallic to dull.

Chemistry: Copper sulfide.

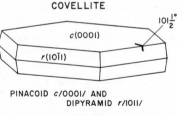

COVELLITE

$c(0001)$

$r(10\bar{1}1)$

$101\frac{1}{2}°$

PINACOID *c*/0001/ AND
DIPYRAMID *r*/1011/

FIG. 123.

Distinctive Features and Tests: Color and cleavage prevent confusion with bornite or chalcocite; crystals unmistakable, commonly being paper-thin. Fuses at 2, thin flakes catching fire and burning with blue flame, yielding a dark globule.

Occurrences: Principally in enriched zones of copper sulfide deposits with chalcopyrite, chalcocite, pyrite, enargite and bornite, usually intergrown with pyrite or chalcopyrite. Large masses to 6″ across, virtually pure, from Kennecott, Copper River district, Alaska. Splendid shining paper-thin crystals, often to 1″ or more in diameter, forming distinctive boxworks on massive covellite, from the Leonard and other mines at Butte, Silver Bow County; similarly from the Calabona Mine, Alghero, Sardinia, Italy. Crystallized specimens are highly prized and expensive; matrices to 6″ to 8″ across displaying numbers of 3/4″ to 1″ crystals thickly intermeshed have been obtained from Montana. Massive material sometimes cut into cabochons or polished in flats.

Cinnabar—HgS

Name: Of uncertain original meaning, from the Latin, *cinnabaris;* sĭn′ ă bär.

Crystals: Hexagonal; rare. Rhombohedral to thick tabular, compressed along the *c* axis; seldom over 1/8″ (Figure 124). Usually in druses or drusy coatings on other minerals. Faces sharp but commonly tarnished to nearly silvery luster. Twinning on *c*[0001].

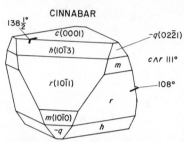

CINNABAR

138½°

c(0001)

h(10⊺3)

m

r(10⊺1)

m(10⊺0)

−*q*

h

−*q*(022⊺1)

c∧*r* 111°

108°

r

PINACOID *c*/0001/, PRISM *m*/10⊺0/, POSITIVE RHOMBOHEDRONS *r*/10⊺1/ AND *h*/10⊺3/, AND NEGATIVE RHOMBOHEDRON *q*/022⊺1/

FIG. 124.

Physical Properties: Brittle, somewhat sectile: fracture conchoidal to uneven. Perfect cleavage parallel to *m*[10⊺0]. H. 2-2½. G. 8.1.

Optical Properties: Transparent to translucent. Intense red, verging on slightly brownish-red. Streak bright red; luster adamantine to submetallic silvery. Uniaxial positive; R.I., 0 = 2.905, E = 3.256. Birefringence 0.351.

Chemistry: Mercuric sulfide.

Distinctive Features and Tests: Bright red color; softness; cleavage; in massive material, the unusual heaviness. Crystals may be mistaken for realgar, but the latter lacks the strong silvery adamantine luster observed on cinnabar crystals, and its cleavage is not as perfect. Volatilizes completely in the flame, but fumes are *dangerously poisonous* and must not be inhaled! Mercury condenses on cold surfaces from powder heated in an open tube.

Occurrences: In low-temperature deposits, usually near recent volcanoes or hot springs; associated with pyrite, stibnite, marcasite, metacinnabar, opal, quartz, chalcedony and calcite. Fine crystals in calcite, sometimes to 3/4″, occur at the Cahill Mine, Humboldt County, Nevada, and perhaps provide the best United States cinnabar specimens. Lately, considerable brilliant red cinnabar in veinlets and drusy crystals to 1/2″, on white granular calcite, has appeared from Charcas, San Luis Potosi, Mexico. Almadén, Ciudad Real, Spain, a locality known since ancient times for its cinnabar, produces sharp crystals up to 1/2″ sprinkled on matrix rock. Fine crystals also from Mount Avala, near Belgrade, Yugoslavia. The world's finest crystals are the 1/2″ to 1″, sharp, deep red, transparent twinned crystals on drusy quartz from mines in Kweichow and Hunan Provinces, China. The general scarcity of crystals leads to high prices being asked for even small specimens; many sources of large crystals are no longer productive.

Realgar—AsS

Name: From Arabic, *rahj al ghar,* "powder of the mine"; rē ăl′gẽr.

Crystals: Monoclinic. Short prisms, diamond-shaped in cross-section with wedge-or chisel-shaped terminations (Figure 125) ; often with fine flat faces when fresh but dulling on exposure to light and becoming coated with an orange powder. Also granular massive. Rarely, twinned on *a*[100].

Physical Properties: Very brittle; sectile. Conchoidal fracture. Good cleavage parallel to $b[010]$. H. $1\frac{1}{2}$-2. G. 3.56.

Optical Properties: Transparent. Dark red, inclining toward orange-red. Streak orange-red. Luster resinous. Biaxial negative. R.I., $\alpha = 2.538$, $\beta = 2.684$, $\gamma = 2.704$. Birefringence 0.166. Distinctly pleochroic: dark red, orange-red.

Chemistry: Arsenic monosulfide. When exposed to light, alters readily to yellow orpiment and arsenolite, requiring specimens to be preserved in darkness.

Distinctive Features and Tests: Color; softness; prismatic crystals. Fuses very easily at 1, yielding garlic odor and white coating on charcoal block; powder heated in closed tube yields red sublimate.

Occurrence: In low-temperature veins with antimony sulfides. Practically all good specimens presently come from the Getchell Mine, Humboldt County, Nevada, where prismatic crystals to $\frac{1}{4}''$ by $\frac{3}{4}''$ occur sprinkled on dark gray clayey matrix; also at Manhattan, Nye County, in spectacular large masses with foliated vivid yellow-orange orpiment, the latter commonly crystallized. Small brilliant crystals in ulexite occur in the open pit borate mine at Boron, Kern County, California. In fine crystals in cavities in calcite at Mercur, Tooele County, Utah. Excellent prismatic crystals to $\frac{1}{2}''$ on matrix from Nagyag and Felsobanya, Romania. In exceptionally sharp crystals from cavities in dolomite, Lengenbach quarry, Binnatal, Switzerland. Crystals to $\frac{1}{4}''$ to $\frac{1}{2}''$ are considered good; matrix specimens over $2''$ are exceptional. A popular thumbnail mineral, particularly from the Getchell Mine, where crumbly matrix allows judicious trimming.

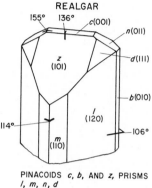

REALGAR

PINACOIDS *c, b,* AND *z,* PRISMS *l, m, n, d*

FIG. 125.

Orpiment—As$_2$S$_3$

Name: From the Latin, *auripigmentum,* in allusion to vivid gold hue; ôr′pĭ mĕnt.

Crystals: Monoclinic. Poorly formed, thick wedge-shaped crystals with curved faces of dull luster; seldom over $\frac{3}{8}''$.

Physical Properties: Sectile, yielding like wax along the very prominent cleavage plane parallel to $b[010]$. In broad crystals, cleavage resembles mica in its perfection, but surfaces are commonly curved, bent or rippled; cleavage laminae flexible but inelastic. H. $1\frac{1}{2}$-2. G. 3.49.

Optical Properties: Transparent to translucent. Vivid yellow-orange in compact masses and crystals, but orange-yellow on cleavage surfaces. Streak pale yellow. Luster resinous to pearly on cleavages. Biaxial negative. R. I., $\alpha = 2.4$, $\beta = 2.81$, $\gamma = 3.02$. Birefringence 0.62.

Chemistry: Arsenic trisulfide. Commonly alters from realgar.

Distinctive Features and Tests: Mica-like cleavage with pearly luster; vivid color. Fuses at 1, volatilizing and forming yellow coating on charcoal block.

Occurrences: In low temperature veins and hot spring deposits, nearly always with realgar in characteristic orange masses mottled by red veinings. In fine large crystals at Mercur, Tooele County, Utah; in Nevada at the Getchell Mine, Humboldt County, sometimes in druses of $\frac{1}{2}''$ crystals covering rock areas of $3''$ by $4''$, and as splendid large foliate masses, with individual folia occasionally to several inches diameter; also in very beautiful masses from Manhattan, Nye County. Foliated masses from Morococha and Acobambilla, Peru. In crystals from clay at Tajowa, Neusohl, Hungary; at Kresevo,

Bosnia, Yugoslavia, and Allchar, Macedonia, Greece. Fine crystals from Balin, Turkey. Large masses of compact realgar, up to 6″, perhaps consolidated from smaller particles, are occasionally carved into statuettes by the Chinese.

Stibnite Group

Stibnite—Sb₂S₃

Name: From the Greek, *stimmi* or *stibi*, "antimony," thence to the Latin, *stibium;* stĭb′nĭt.

Crystals: Orthorhombic. Usually in slender prismatic striated individuals, with pyramidal terminations (Figure 126). Some crystals bent, twisted, occasionally in "s" fashion. In

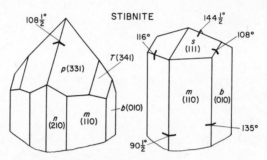

PINACOID *b*, PRISMS *m* AND *n*, DIPYRAMIDS *p*, *s* AND *T*

FIG. 126.

jackstraw aggregates or radiated groups. Prisms usually display $m[110]$ and $b[010]$, also $n[210]$; terminations predominately low dipyramid $s[111]$ and steep dipyramid $p[331]$. Many other forms are known, some crystals becoming very complex on terminations. Faces range in smoothness from dull, like galena, to brilliant, like polished steel. Also granular to cleavable massive.

Physical Properties: Brittle; somewhat sectile and flexible, but inelastic. Fracture sub-conchoidal to uneven, interrupted by highly perfect and easily developed cleavage parallel to $b[010]$. H. 2. G. 4.63.

Optical Properties: Opaque. Pale to dark lead-gray, inclining toward blue; also nearly silver white, brilliant. Sometimes tarnished iridescent blue-purple. Streak gray to dark gray. Luster metallic to exceptionally brilliant on some specimens.

Chemistry: Antimony trisulfide. Alters to white senarmontite, or completely to pale grayish-yellow stibiconite, which retains faithfully the shapes and markings of original crystals.

Distinctive Features and Tests: Prismatic gray, striated crystals unmistakable. In massive form, the perfect cleavage is useful recognition feature. Fuses at 1, coloring flame green-blue, and melting to a puddle which then completely volatilizes. Slowly soluble in hydrochloric acid.

Occurrences: Predominately in low-temperature hydrothermal veins associated with realgar, orpiment, galena, marcasite, pyrite, cinnabar, calcite, and quartz. Very fine matrix specimens formerly from the Darwin district, Inyo County, California, in blades to 4″, some partly altered; from Antimony Peak, near Hollister, San Benito County, and in large crystals from the Rand district, San Bernardino County. Masses of jackstraw crystals, some to 14″ diameter, studded with singles to 4″ in length, occurred in the

Manhattan Mines of Nye County, Nevada. These are the best North American specimens. Excellent matrix specimens, displaying needle-like prisms (Figure 127) on quartz
and barite, from Felsobanya and Kapnik, Romania. Fine crystals with realgar and cinnabar from Pereta, Tuscany, Italy; also from the departments of Mayenne, Haute-Loire,

FIG. 127. Slender crystals of stibnite on quartzite matrix. From Felsobanya, Romania. Size 5″
by 2½″.

and Cantal, in France. Excellent acicular crystals from Allchar, Macedonia, Greece. The
supreme stibnite locality is the Ichinokowa deposits at Saijo, Iyo Province, on the Island
of Shikoku, Japan, where groups of magnificent prismatic crystals to 20″ length were
found grown freely in cavities. Crystals usually range from ½″ diameter and 4″ to 6″
in length, to 3″ diameter and well over a foot in length. Very few large groups are now
available and only rarely are smaller groups offered. Single crystals are often obtainable.
Probably the acquisition of a fine group of Japanese stibnite crystals is to an amateur
mineralogist an attainment equivalent to an art patron's acquisition of an old master.
Needless to say, fine large groups are extremely expensive, but smaller crystals, and
acicular crystal groups from Manhattan and European localities are very fine and less
costly.

Pyrite Group

Pyrite—FeS$_2$

Name: From the Greek, *pyrites lithos*, "stone which strikes fire," in allusion to the
sparking produced when iron is struck by a lump of pyrite; pī′rīt.

Varieties: When nickel or cobalt substitute for iron in substantial amounts, *nickelian* and *cobaltian*. When more nickel than iron is present, the species bravoite is formed, i.e., $(Ni,Fe) S_2$.

Crystals: Isometric; abundant and fine. Predominately in cubes, pyritohedrons, or modifications thereof; sometimes octahedral, and more rarely, distorted octahedral; also other forms, but appearing mostly as modifications on cubes, pyritohedrons and octahedrons (Figure 128). Crystals often complex and rendered nearly spherical by an abundance

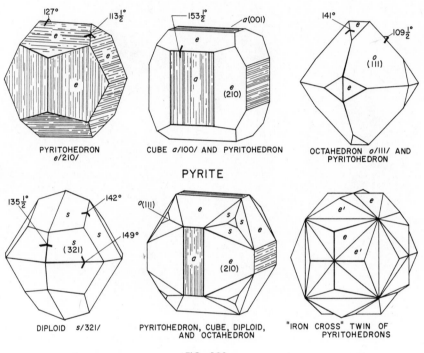

PYRITOHEDRON
e/210/

CUBE *a*/100/ AND PYRITOHEDRON

OCTAHEDRON *o*/111/ AND PYRITOHEDRON

PYRITE

DIPLOID *s*/321/

PYRITOHEDRON, CUBE, DIPLOID, AND OCTAHEDRON

"IRON CROSS" TWIN OF PYRITOHEDRONS

FIG. 128.

of forms (Figure 129). Principally in druses, but also in clusters and single crystals; very abundant in massive types when it may assume fibrous, radiate, mamillary, stalactitic, and other aggregate architectures. Parallel striations common on cubes and pyritohedrons (Figure 128). Pyritohedrons sometimes found in interpenetrating twins, such that a pair of edges cross each other like the face of a star drill; to them is given the name *iron cross* twins (Figure 128).

Physical Properties: Brittle; uneven to conchoidal fracture. No cleavage. H. 6-6½, considerably higher than sulfides of similar appearance. G. 5.02.

Optical Properties: Opaque. Pale brass-yellow color, often lightly coated by brown iron oxide film, or somewhat iridescent. Streak greenish- to brownish-black. Luster metallic.

Chemistry: Iron disulfide; *nickelian* up to 17% Ni, *cobaltian* to 14% Co. Commonly altered into sulfates or limonite in upper or oxidized portions of ore veins; faithful pseudomorphs of goethite after pyrite abundant.

Distinctive Features and Tests: Striations on crystals very distinctive; harder than other similar-appearing sulfides as marcasite and chalcopyrite. Fuses at 2½-3. Insoluble in hydrochloric acid.

Occurrences: Very widespread and most abundant sulfide; common in hydrothermal ore veins, in contact metasomatic deposits, and as an accessory in many kinds of rocks, including sedimentary. Numerous localities yield fine or interesting crystallizations. Favorite collectors items are the large crystals to 5″ diameter from the American Mine, Bingham, Tooele County, Utah. From mines at Park City, Wasatch County, Nevada, in brilliant, sharp crystals to 1½″ forming broad sheets and groups. Splendid pyritohedrons from Colorado as at Rico and Gilman, also in cubes to 3″ diameter; reported in cubes to 6″

FIG. 129. Pyrite. *Top left:* modified pyritohedrons on matrix of specular hematite. Size 3½″ by 3″. Rio Marina, Elba, Italy. *Top right:* sharp octahedral crystals on massive pyrite from Morococha, Peru. Size 3½″ by 2½″. *Bottom left:* highly modified crystals in a cluster from Elba. Forms include cube, pyritohedron, octahedron and diploid. Size 1¾″. *Bottom center:* simple cubes from Elba.

on edge at Leadville, Colorado. Beautiful disk-like concretions, "pyrite suns," from coal shales near Sparta, Randolph County, and pyritized snail fossils in open pit coal mines near Middlegrove, Knox County, Illinois. The French Creek Mines in Chester County, Pennsylvania are noted for fine octahedral pyrite crystals to 1¾″, elongated along one axis and with curved faces. Splendid crystals occur in many places in Mexico; also in South America, particularly in the tin veins of Bolivia; very large octahedrons are reported from Ubina, Llallagua, Bolivia, also from Morococha, Peru (Figure 129). Consistently fine and large crystals occur with hematite and quartz in the famed iron deposits of Rio Marina, Island of Elba, Italy (Figure 129). Octahedral crystals to 3″ on edge and cubes to 5″ on edge have been found, although recent specimens exhibit mainly pyritohedral crystals from about 1″ to 3″ diameter. Elba crystals occur in splendid sharp single crystals, clusters of two or three crystals, and larger groups; nearly every conceivable form and combination is found. Fine crystals also occur in other parts of Italy, as at Brossa and in the Piedmont; in France, England, Germany, Scandinavia, and elsewhere. The variety in pyrite affords the single-species specialist a wonderful selection. For the amateur collector, specimens are abundant, and considering size and beauty, are good bargains. Perfect single crystals are much desired as are also sharp octahedrons, a rarer form than cubes or pyritohedrons.

Bravoite—(Ni,Fe)S$_2$

Nickel iron sulfide; brä'vō īt. A bright steel-gray mineral, related to pyrite, usually found coating pyrite crystals, as on the pyritohedrons lately coming from the Rico Argentine Mine at Rico, Dolores County, Colorado. Rare.

Hauerite—MnS$_2$

Manganese sulfide; how'ẽr īt. A rare mineral, prized by single crystal collectors. Black octahedrons, usually ⅜″ to ¾″ across, from Destricello and Raddusa, Sicily; the record from the latter place is an octahedron 2″ on edge.

Cobaltite Group

Cobaltite—CoAsS

Name: From the German, *Kobold,* "underground spirit" or "goblin," in allusion to the refusal of cobaltiferous ores to smelt properly, hence "bewitched"; kŏ'bôl tīt.
Crystals: Isometric; crystals rare. Usually in cubes and pyritohedrons; also octahedrons. Sometimes striated similar to pyrite. Predominately massive granular. Occasionally in isolated crystals, but more commonly as outgrowths from massive material. Faces lustrous and smooth, but seldom as sharp as pyrite.
Physical Properties: Brittle; fracture uneven. Good cubic cleavage. H. 5½. G. 6.33.
Optical Properties: Opaque. Tin-white. Streak gray-black. Luster metallic.
Chemistry: Cobalt sulfarsenide.
Distinctive Features and Tests: Crystals resemble pyrite in forms but tin-white hue is distinctive; softer than pyrite; slight pink cast distinguishes from smaltite in massive material. Fuses at 2-3, giving off sulfurous and garlicky fumes. Fused globules magnetic, and when taken up in borax bead, impart a deep blue color to the oxidizing flame. Powdered mineral imparts pink color to warm nitric acid.
Occurrences: In high-temperature deposits in metamorphic rocks; also in hydrothermal veins with other cobalt and nickel sulfides. Fine octahedral and cubic crystals occur in the Columbus Mine, Coleman Township, near Cobalt, Ontario, on matrices to 3″ by 4″. Finest crystals from Tunaberg, Riddarhytten, and several other localities in Sweden; Tunaberg crystals often occur as isolated euhedrons to ⅝″ diameter. Good crystals occur in calc-silicate skarn near Bimbowrie, South Australia.

Gersdorffite—NiAsS
Ullmannite—NiSbS

Both species are found associated with other nickel minerals, but are comparatively rare; gẽrz dôrf'īt; ŭl'mȧn īt. Tin-white gersdorffite crystals in quartz, with chalcopyrite, have been found at the Falconbridge Mine, Sudbury, Ontario, and as masses in ore from other localities, principally in central Europe. Ullmannite is usually massive but fine cubic crystals occur at Montenarba, Sarrabus, Sardinia. Also in large gray masses with massive millerite and siderite, Cochabamba, Bolivia. Seldom represented in collections.

Loellingite Group

Loellingite—FeAs$_2$

Name: After the locality at Lölling, Carinthia, Austria; lĕr'lĭng ĭt.

Varieties: Containing sulfur, cobalt or antimony: *sulfurian, cobaltian,* and *antimonian.*

Crystals: Orthorhombic. Predominantly prismatic, elongated in the *c* axis direction, and striated parallel to this direction on prism faces (Figure 130); terminations blunt, wedge-shaped. Twinning on *l*[011], sometimes producing sixlings of distinct star shape. Often in small isolated crystals; also massive.

Physical Properties: Brittle; uneven fracture. Cleavage parallel to *b*[010] sometimes distinct. H. 5-5½. G. 7.40.

Optical Properties: Opaque. Silver-white to slightly grayish. Streak grayish-black. Luster metallic.

Chemistry: Iron diarsenide; sometimes containing sulfur, cobalt or antimony, to approximately 6%.

Distinctive Features and Tests: Similar in appearance to arsenopyrite but crystals usually slender prismatic instead of stubby; difficult to distinguish in massive form. Fuses at 2, giving garlic odor of arsenic and yielding magnetic globule. Dissolves in nitric acid, imparting yellow hue to solution.

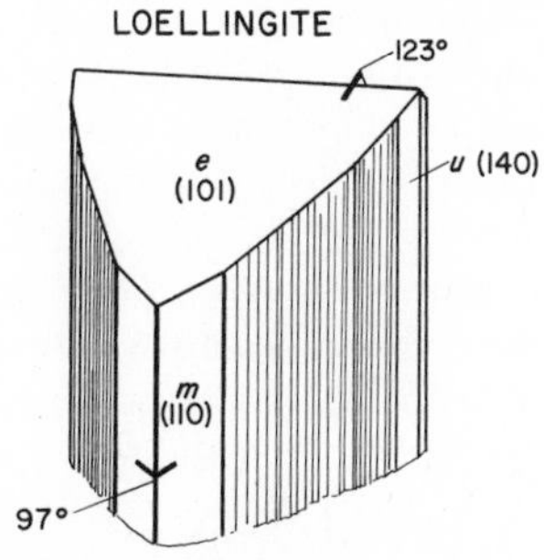

LOELLINGITE

123°

e (101)

u (140)

m (110)

97°

PRISMS *m*, *u*, AND *e*

FIG. 130.

Occurrences: In medium- to high-temperature hydrothermal sulfide veins, commonly with calcite; also sparingly as small prismatic crystals in some granitic pegmatites; sometimes in good euhedral crystals in marble as in the Franklin metamorphic limestone of northern New Jersey and southern New York. In good crystals from augite-syenite, Langesund Fjord district of southern Norway. A rare mineral seldom represented in amateur collections.

Marcasite—FeS$_2$

Name: Arabic or Moorish, of uncertain origin; mär'kă sīt.

Crystals: Orthorhombic. Predominately tabular, compressed on *b*[010]; sometimes prismatic (Figure 131). Fine sharp crystals are rare, the preference being to form aggregates, often in concretionary masses, stalagtites, disks, balls, spear heads, etc. An exceptionally common aggregate type is the so-called "cockscomb," formed by fan-like sprays of stubby crystals, oriented with *c* axes pointing outward, and twinned along the plane *e*[101]; the broad lateral faces in such aggregates are *b*[010], the wedge faces are *e*[101] (Figure 132). Faces are seldom smooth; often stepped, notched, grooved, or curved. Another twinning habit produces fivelings, particularly when crystals are flattened on *b*[010]. The spear-shaped aggregates found in many localities are twinned in this manner.

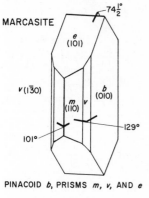

MARCASITE

74½°

e (101)

v (130)

b (010)

m (110)

v

101°

129°

PINACOID *b*, PRISMS *m*, *v*, AND *e*

FIG. 131.

Physical Properties: Brittle; uneven fracture. Cleavage on *e*[101] sometimes distinct. H. 6-6½. G. 4.89.

Optical Properties: Opaque. Pale brass-yellow, inclining toward a peculiar greenish hue, and tarnishing to dingy gray-green. Streak grayish- to brownish-black. Luster metallic but seldom splendent.

Chemistry: Iron disulfide.

Distinctive Features and Tests: Confused with pyrite only in massive types, inasmuch as crystals are distinctive; also paler in hue than pyrite, but tarnishes more quickly and assumes the dingy greenish hue mentioned above. Many specimens readily decompose, becoming coated with a white efflorescence of iron sulfate. Careful soaking in pure water to remove solubles, followed by prolonged drying and spray-coating of colorless lacquer is often used to preserve specimens.

Occurrences: Primarily a low-temperature mineral, created under surface conditions in sedimentary formations, clays, shales, etc., and commonly as concretions, encrustations around twigs, in spear-head shapes, etc. Abundant in some sulfide veins, and exceptionally abundant in the lead-zinc deposits of the tri-state district, and other lead-zinc deposits in the Mississippi Valley. Splendid cockscombs are obtained from numerous mines in the tri-state district, sometimes in sheets several feet across, or as large rosetted areas on chert (Figure 132), or smaller rosettes on galena crystals; also common as small sharp crystals sprinkled on galena, sphalerite, etc. The sharpest and most brilliant cockscombs, though of small size, occur in cavities in dolomite at Mineral Point and Racine, Racine County, Wisconsin. Beautiful spear heads are dug from brick clays near Red Bank, New Jersey; in the chalk of the Dover Cliffs, between Folkestone and Dover, England; in France, and other European sites. Fine large marcasite replacements of pyrrhotite occur in the tin veins of Llallagua, Bolivia.

FIG. 132. Marcasite in cockscomb growths upon chert breccia fragment. Size 3½″ by 3″. From Joplin, Missouri. The sharp wedge-shaped terminations consist of a pair of e(101) faces, many of which are exposed because of repeated twinning which forms the fan-like sprays. The sides of the sprays are b(010) faces.

Arsenopyrite Group

Arsenopyrite—FeAsS

Name: After chemical composition; är′sē nō pī′rīt.

Varieties: With cobalt replacing iron, *cobaltian;* also *bismuthian.*

Crystals: Monoclinic. Fine crystals common, usually as stubby sharp prisms somewhat elongated parallel to the *c* axis or, less commonly, parallel to the *b* axis (Figure 133). Striated on *m*[110] parallel to the *c* axis. Crystals from ¼″ upward are usually aggregates of crystals in approximately parallel position, causing curved or stepped faces and are very distinctive of this species (Figure 134). Twinning common on *m*[110], sometimes repeated

FIG. 134. *Right:* Arsenopyrite. Aggregates of numerous sub-parallel crystals resembling rhombohedral forms. From Parral, Chihuahua, Mexico. Size 2″ by 2″.

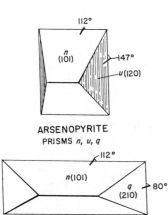

ARSENOPYRITE
PRISMS *n, u, q*

FIG. 133.

as in marcasite; also on *a*[100] and *b*[010]; uncommonly, twinning on *e*[012] produces cross-shaped twins.

Physical Properties: Brittle; uneven fracture. Cleavage distinct on *m*[110]. H. 5½-6. G. 6.07.

Optical Properties: Opaque. Silver-white, somewhat grayish; often tarnished to dull pale gray. Streak dark gray-black. Luster metallic.

Chemistry: Iron arsenide-sulfide; cobalt is usually present, and rarely, bismuth.

Distinctive Features and Tests: Sharp wedge-shaped crystals characteristically striated, curved or offset. Fuses at 2, emitting white fumes, garlic odor, and eventually, yielding a small magnetic globule. Decomposed by nitric acid, yielding insoluble sulfur.

Occurrences: The most common arsenic mineral, occurring in a wide variety of deposits as in sulfide veins, contact metasomatic deposits, and in metamorphic rocks; also sparsely in some granitic pegmatites. An early-forming mineral with crystals commonly enclosed by other species; also in druses. Fine crystals, sometimes as cyclic twins in quartz, from Deloro, Hastings County, Ontario. Rarely in good crystals from the United States, but fine specimens have come from gneiss at Franconia, New Hampshire, and from the Franklin marble of southern New York and northern New Jersey. Recently, superior specimens, consisting of druses to 12″ across of sharp, bright individuals to as much as 2″ in length, have come from Hidalgo de Parral, Chihuahua, Mexico (Figure 134). Single crystals to 1½″ by 1½″ have been recorded from Llallagua, Bolivia. From numerous localities abroad, particularly fine from Sulitjelma, Norway; Tavistock, Devonshire, England; Freiberg, Germany. Sharp single crystals have been obtained from schists at the Mitterberg, near Muhlbach, Austria, and in the Trentino district of Italy. Fine Mexican specimens are readily available but those from European sources turn up only occasionally; crystals to ½″ are average, while those over ¾″ are exceptional.

Molybdenite Group

Molybdenite—MoS₂

Name: After molybdenum content; mō lĭb′dĭ nīt.

Crystals: Hexagonal. Predominantly thin to thick tabular, usually with very poor faces and rough hexagonal profiles with faces of basal pinacoid $c[0001]$ prominent (Figure 135). Also in clusters, foliate or radiate masses (Figure 136). Good crystals are rare.

FIG. 136. *Right:* Molybdenite. Mass of radiating tabular crystals of rude form, showing the perfect cleavage parallel to $c(0001)$ and bright metallic luster. Taehwa Mine, Chung Wun Gun, Korea. Size 2″ by 1¾″.

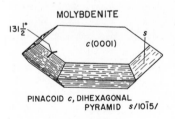

MOLYBDENITE

$131\frac{1}{2}°$

$c(0001)$

s

PINACOID c, DIHEXAGONAL
PYRAMID $s/10\bar{1}5/$

FIG. 135.

Physical Properties: Flexible but not elastic; sectile. Very easily cleaved along $c[0001]$, the pressure of the fingers being sufficient to cause slipping of plates. H. 1-1½. G. 4.6-4.7.

Optical Properties: Opaque. Blue-gray in hue, like freshly cut lead. Streak blue-gray. Luster metallic, splendant.

Chemistry: Molybdenum sulfide. Sometimes coated with alteration products in the form of dustings or very small grains of ferrimolybdenite or powellite (bright yellow).

Distinctive Features and Tests: Can only be mistaken for graphite, but distinguished by bluish hue, much finer luster, and better-formed hexagonal crystals. Infusible; on charcoal, yields characteristic red-yellow-white coatings, progressing outward from the sample being heated. Decomposed by nitric acid; soluble in aqua regia.

Occurrences: Widespread in small amounts in pegmatites, calc-silicate skarns, in granites and aplites, and sometimes in high-temperature veins with wolframite, scheelite and cassiterite. Splendid crystals, to 3″ diameter, often quite sharp, occur in the calc-silicate skarns and associated pegmatites in the Grenville marbles of Ontario and Quebec; a notable locality is Aldfield, Pontiac County, Quebec, where numerous showy crystals occur in pyroxene matrix. Small but good crystals occur in the Franklin marbles of southern New York and northern New Jersey. Large folia to 5″ across from Urad Mine, Empire, Clear Creek County, Colorado, associated with ferrimolybdite, quartz and beryl. Excellent crystals in quartz near Lake Chelan, Okanogan and Chelan counties, Washington; crystals reach 2″ diameter and about $\frac{1}{4}$″ thickness; matrix specimens are rare. Fine sharp crystals from Hirase Mine, Honshu, Japan. Very large crystals, among the best, from Kingsgate and Deepwater districts, New South Wales, Australia. Also fine specimens from various localities in Korea. Indifferent specimens of molybdenite are not difficult to obtain, but the very easy cleavage prevents most crystals from surviving the shock of removal from matrix; this accounts for the high esteem in which reasonably perfect single crystals or crystals in matrix are held.

SKUTTERUDITE SERIES

Skutterudite—(Co,Ni)As$_3$ (Cobalt-rich end member)

Smaltite—(Co,Ni)As$_3$ (Intermediate member)

Chloanthite—(Ni,Co)As$_3$ (Nickel-rich end member)

Names: After Skutterud, Norway, skŭt′ĕr ŭd īt. In reference to *smalt,* a deep blue glaze in which this mineral was used to furnish the necessary cobalt, smôl′tīt. After the Greek *chloe* and *anthos,* "green flower," in allusion to green efflorescent coating (annabergite) frequently observed on this mineral; klō ăn′thīt.

Crystals: Isometric. Mostly massive but good crystals are common in some deposits: usually as cubes, cubo-octahedrons, or octahedrons; also dodecahedral and pyritohedral modifications. Crystals sometimes complex and then resemble faceted balls. Faces seldom sharp but often lustrous. Rarely, sixling twins on plane *n*[112]. Also finely granular; colloform.

Physical Properties: Brittle; fracture conchoidal to uneven. Cleavage sometimes distinct parallel to *a*[100] and *o*[111]. H. $5\frac{1}{2}$-6. G. 6.1-6.9.

Optical Properties: Opaque. Tin-white, occasionally obscured by iridescent tarnish. Streak black. Luster metallic.

Chemistry: This series represents variations in cobalt/nickel content; there are no arbitrary divisions between them and only chemical analyses can determine the position of any given specimen. Iron, to a maximum of 12%, is so consistent in analyses that some prefer to consider iron as essential as cobalt and nickel. Nickel-rich members alter to green annabergite, cobalt-rich members to magenta erythrite.

Distinctive Features and Tests: Similar in color to arsenopyrite but crystals distinctive; can be confused with cobaltite and then distinguished only by advanced tests. Fuses at 2-$2\frac{1}{2}$, emitting garlic odor of arsenic; white coatings on charcoal; melt globules magnetic.

Occurrences: In medium-temperature hydrothermal veins with other nickel and cobalt minerals and sulfides. Found in large masses in the silver veins of Cobalt, Ontario, associated with niccolite and native silver in calcite; niccolite and smaltite sometimes form interesting coarse plumose intergrowths which are quite attractive when massive specimens are sawed and polished. Skutterudite in crystals occurs at Skutterud near

Modum, Norway. Occasionally very fine crystals of chloanthite occur at Andreasberg in the Harz, Germany; similarly at Annaberg in Saxony. From the latter locality, masses nearly 6″ in diameter have been obtained, covered by brilliant crystals to ¾″ diameter. Fine crystals of chloanthite occur at Schneeberg, Saxony, primarily in cubes and pyritohedrons up to 1¼″ on edge. Massive specimens are commonly represented in collections but crystals rarely; the latter are seldom available and command good prices even for specimens an inch or two across and large specimens, from 4″ to 6″, are very expensive.

SULFOSALTS

There are nearly 85 species in the sulfosalts but by far the greater number are quite rare and seldom available in more than small crystals or inconspicuous masses. The semi-metals antimony, arsenic, and bismuth, combined characteristically with sulfur as anions, favor low-temperature environments, thus making the sulfosalts among the last minerals to form in various types of deposits, or in the case of hydrothermal veins, more likely to be found in near-surface portions.

Polybasite Group

Polybasite—$(Ag,Cu)_{16}Sb_2S_{11}$

Silver-copper antimony sulfide; pŏl′ĭ băs′ĭt. An unattractive black species, sometimes crystallized in pseudohexagonal tabular crystals. Rarely offered from various localities in Colorado, Idaho, Montana, and Nevada. Good crystals from Andreasberg, the Harz, Germany. In fine crystals from Las Chiapas and Arizpe, Sonora, Mexico. Specimens invariably expensive; seldom over ¾″.

Stephanite—Ag_5SbS_4

Silver antimony sulfide; stĕf′ăn īt. Another iron-black mineral of considerable rarity, especially in crystals, and highly prized on this account. Excellent crystals from Arizpe, Sonora, Mexico; from mines in Cornwall, England; from silver mines in the Harz and Saxony, Germany, and from Czechoslovakia. Specimens expensive; crystals seldom over 1″ in length.

Ruby Silver Group

Pyrargyrite—Ag_3SbS_3

Name: From the Greek, *pyr* and *argyros,* "fire-silver" in allusion to color and silver content; pī rär′jĭ rīt.
Crystals: Hexagonal. Usually in rude hexagonal prisms with blunt pyramidal terminations; distinct sharp crystals rare; commonly in aggregates of interlocked prismatic crystals. Also massive. Hemimorphic. Sometimes twinned on $u[10\bar{1}4]$, forming divergent "swallowtail" sprays, also twinned on $r[10\bar{1}1]$.
Physical Properties: Very brittle; fracture conchoidal to uneven, granular. Poor cleavage on $r[10\bar{1}1]$. H. 2½. G. 5.85.

Optical Properties: Translucent, showing deep red color through sections up to ¼″ thick when held before a strong light. Apparent color is black with indistinct red cast which is more evident on bruised places. Streak dark red. Adamantine to submetallic luster. Uniaxial negative; R.I., O = 3.084, E = 2.881. Birefringence 0.203.

Chemistry: Silver antimony sulfide, commonly with small amounts of arsenic.

Distinctive Features and Tests: Much darker than proustite with which it is most easily confused; crystal forms distinguish from cuprite which has nearly the same color and luster. Fuses at 1 to bright black globule which shatters readily when tapped; this further distinguishes it from cuprite which fuses at 3, eventually yielding a globule of malleable copper.

Occurrences: One of the last minerals to form in low-temperature environments in sulfide veins. Usually associated with proustite, argentite, tetrahedrite, silver, calcite, dolomite and quartz. Splendid specimens from Colquechaca in Bolivia, and with proustite at the Delores Mine, Chanarcillo, Atacama, Chile. Exceptional crystals to 1⅛″ from Andreasberg, in the Harz, and Annaberg and Freiberg, Saxony, Germany. Rude prismatic crystals formerly from silver mines near Guanajuato, Mexico, some to 1¼″ by ½″. Single crystals, or small groups with good crystals, are much in demand by collectors but quite expensive. Crystals on matrix are very rare.

Proustite—Ag₃AsS₃

Name: After French chemist, J. L. Proust (1754-1826); proos′tīt.

Crystals: Hexagonal. Usually as rude prismatic crystals of approximate hexagonal cross-section; sometimes triangular in cross-section and tapered; seldom well-terminated. Prism faces grooved and ridged in larger crystals, causing crystals to assume approximately circular cross-sections. Also massive. Twinning similar to pyrargyrite.

Physical Properties: Very brittle; slightly sectile. Cleavage similar to pyrargyrite. H. 2-2½. G. 5.55.

Optical Properties: Transparent to translucent; a bright red color is readily seen through ½″ to ⅝″ clear sections held before a strong light. The hue is deep rich red, far less blackish than pyrargyrite. Streak bright red. Adamantine to submetallic luster. Exposure to light causes both pyrargyrite and proustite to alter slightly on surfaces, intensifying the submetallic luster. Uniaxial negative. R.I. O = 3.087, E = 2.792. Birefringence, 0.295. In sections about ⅛″ thick, weak dichroism in shades of red.

Chemistry: Silver arsenic sulfide, containing some antimony.

Distinctive Features and Tests: Rich deep red prismatic crystals of characteristic silvery luster. Fuses at 1, giving off sulfurous and arsenical fumes; after prolonged heating the resulting gray globule of silver can be hammered flat, thus distinguishing it from pyrargyrite which yields a brittle globule.

Occurrences: Similar to pyrargyrite and from the same localities. The finest crystal specimens are from the Dolores Mine, Chanarcillo, Chile, reaching lengths to 3″ and thicknesses to 1″. Much is so transparent that pieces could be used for faceting into gems if the difficulties of cutting are overcome. Also very fine crystals from Joachimsthal, Bohemia, Czechoslovakia, and from various silver deposits in Saxony. Proustite specimens are more prized than pyrargyrite by virtue of better color and the tendency to form better crystals. Occasionally, small fine crystals on limonite matrix, with arsenic encrustations, are offered from Saxony sources. Such make very fine small specimens as "thumbnails" and "miniatures."

TETRAHEDRITE SERIES

Tetrahedrite—$(Cu,Fe)_{12}Sb_4S_{13}$(Antimony-rich end member)

Tennantite—$(Cu,Fe)_{12}As_4S_{13}$(Arsenic-rich end member)

Names: Tetrahedrite after the predominant crystal form; tĕt'ră hē drīt. Tennantite after English chemist Smithson Tennant (1761-1815); tĕn'ănt ĭt.

Varieties: Containing silver, *argentian,* or *freibergite.* See below under Chemistry.

Crystals: Isometric. Tetrahedral crystals are most common but are often modified by other forms. Larger and sharper tetrahedral crystals are characteristic of tetrahedrite but much smaller crystals, usually modified cubes, are more characteristic of tennantite (Figure 137).

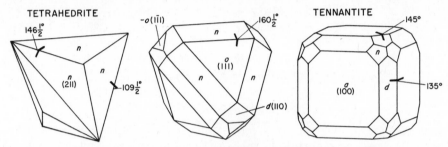

POSITIVE TRISTETRAHEDRON *n*, POSITIVE TETRAHEDRON *o*, DODECAHEDRON *d*, CUBE *a*

FIG. 137.

Crystals commonly form druses. Faces often smooth, lustrous, but inclined to be dull on tennantite. Also massive, granular. Twinning parallel to *o*[111]; also interpenetrant along an axis perpendicular to an octahedral face, usually showing the sharp corners of one crystal arising from the faces of the other. Minute oriented chalcopyrite crystals sometimes completely cover tetrahedrite crystals and disguise the ordinary hue.

Physical Properties: Brittle; subconchoidal to uneven fracture. No cleavage. H. 3-4½, tennantite the harder. G. 4.6-5.1, rising with silver and antimony.

Optical Properties: Opaque. Bright gray, like freshly cut lead, to dark gray, to dead black. Streak black to brown; rarely, dark red. Metallic luster, sometimes splendent.

Chemistry: Essentially copper antimony-arsenic sulfides with iron always present to about 13% substituting for copper; sometimes zinc to 8% and silver to about 5%. Antimony and arsenic substitute for each other through a complete range from one end member to the other. To complicate matters, the copper is also commonly replaced by iron and zinc, and less often by silver and mercury; bismuth sometimes takes the place of arsenic.

Distinctive Features and Tests: Distinctive crystals, but massive material resembles a number of other sulfides and is distinguished with difficulty. Lack of cleavage is helpful in eliminating possibilities. Fuses easily at 1, emitting arsenical, sulfurous, and antimonial fumes. Powder of fused globule stains nitric acid blue.

Occurrences: Common in low- to medium-temperature hydrothermal ore veins, associated with sulfides, calcite, dolomite, siderite, barite, fluorite and quartz. Sometimes in contact metasomatic deposits. Very fine crystals as beautiful lustrous sharp gray tetrahedrons to ½″, occur at Bingham, Salt Lake County, Utah, with pyrite and chalcopyrite. Smaller crystals occur with siderite in the Sunshine Mine at Kellogg, Shoshone Country, Idaho. Micro crystals of tennantite occur with enargite at Butte, Silver Bow County, Montana.

Lately, large black tetrahedrite crystals to ¾″, but usually coated and indistinct, and associated with pyrite, sphalerite, minor quartz and very small bournonite crystals, have come from the Mina Bonanza, Concepcion del Oro, Zacatecas, Mexico. Very fine crystals came from Cerro de Pasco and Morococha, Peru. The finest crystals are undoubtedly those which came from European localities, particularly from Germany, as from Clausthal in the Harz, to 1″ on edge, coated with chalcopyrite; also at Horhausen in Rhenish Prussia as sharp tetrahedrons to ¾″ on edge, and at Dillenburg, Hesse-Nassau. Also exceptional crystals on matrix from Kapnik and Botes, Romania. Small, very complex, lustrous crystals of nearly spherical shape (tennantite) from cavities in dolomite from the Lengenbach quarry, Binnatal, Switzerland. Fine chalcopyrite-coated tetrahedrite crystals to ⅝″ on drusy quartz, with galena, from Herodsfoot Mine, Liskeard, Cornwall, England, and elsewhere in this mining district; also good micro specimens of tennantite from Trevisans. Small but fine micro tennantite crystals with sphalerite, germanite, galena, etc., occur at Tsumeb, South-West Africa. Many other localities are known, but many are also presently unproductive, and the "classical" specimens from Europe must be purchased from old collections as opportunity offers.

Enargite Group

Enargite—Cu_3AsS_4

Name: From the Greek, *enargos,* "visible," in allusion to the distinct cleavage; ĕn är′jĭt.
Crystals: Orthorhombic. Commonly prismatic along *c* axis, with broad faces striated parallel to this axis. Terminated by flat basal planes *c*[001]. From the front (*a* axis), crystals appear square to rectangular in outline, but observed from the top, that is, looking down the *c* axis, the cross section resembles a lens (Figure 138). Crystals also

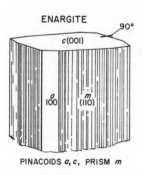

ENARGITE

PINACOIDS *a, c*, PRISM *m*

FIG. 138.

FIG. 139. Enargite crystals, showing striations parallel to the c axis caused by repeated combination of the faces a(100) and m(110). From Butte, Silver Bow County, Montana. The largest crystal is about 1″ tall.

tabular, compressed along the *c* axis (Peru), and sometimes twinned cyclically as sixlings on plane *x*[320]. Faces lustrous to somewhat dull; sometimes coated by thin films of chalcocite. Also granular massive.

Physical Properties: Brittle; fracture uneven, interrupted by cleavages. Perfect cleavage, seldom absent, parallel to $m[110]$; distinct cleavages on $a[100]$ and $b[101]$. H. 3. G. 4.4-4.5.

Optical Properties: Opaque. Gray-black to sooty-black. Streak grayish black. Luster sub-metallic to dull.

Chemistry: Copper arsenic sulfide. Antimony may substitute for arsenic to 6% by weight; iron is commonly present to 3% by weight.

Distinctive Features and Tests: Tabular striated crystals unmistakable. In masses, the twinkling reflections from the ever-present perfect cleavage on $m[110]$ easily distinguishes this species from other sulfides which resemble it in color. In the flame, it promptly decrepitates into powder, but may be fused very easily at 1, forming a brittle, black globule.

Occurrences: In hydrothermal ore veins, forming at medium temperatures and some-times at low temperatures, usually associated with sulfides and quartz. Fine specimens occur in the Leonard Mine, Butte, Silver Bow County, Montana, usually as $\frac{1}{2}''$ crystals studding matrices of massive enargite (Figure 139); crystals to $1''$ are far less common, but ex-ceptional crystals have been obtained to $2''$ length, and nearly as wide. Formerly, very fine from Bingham, Tooele County, Utah. Small crystals, not over $\frac{1}{8}''$ to $\frac{1}{4}''$, occur on matrices of massive enargite, galena, sphalerite, tetrahedrite and calcite in the Argentine Tunnel, Ouray, Ouray County, Colorado. Rarely, as micro crystals and massive in the assemblage of sulfides of the tri-state district. Brilliant black crystals at Cerro de Pasco, Peru, and large cleavages to $2''$ by $3''$ at La Paz, Bolivia.

Bournonite Group

Bournonite—PbCuSbS₃

Name: After French mineralogist Count J. L. Bournon; boor'nŏn ĭt.

Varieties: Arsenic partly substituting for antimony, *arsenian*. The varietal name *endel-lionite* is often applied to the exceptional specimens obtained from Wheal Boys Mine, Endellion, Cornwall, England (Figure 141).

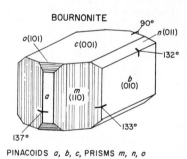

PINACOIDS *a, b, c*, PRISMS *m, n, o*

FIG. 140.

FIG. 141. Bournonite crystals in "cogwheel" twins from Wheal Boys, Endellion, Cornwall, England. Size $1\frac{1}{2}''$ by $1\frac{1}{4}''$.

Crystals: Orthorhombic. Simple crystals are stubby prismatic, sometimes tabular, com-pressed on $c[001]$, as in Figure 140, or terminated by prism $e[210]$, resulting in two low faces atop the c axis giving crystals a blunt wedge-shaped appearance, but very com-monly, bournonite twins on prism plane $m[110]$ in fourling fashion, forming the so-called "cogwheel" crystals, much prized by collectors. Commonly striated with twin reentrants parallel to the *c* axis. The general aspect of twinned crystals is therefore flattened and rough on the *c*[001] faces, forming round cogwheel or gear-like tablets with brilliant faces on the edges. Also massive.

Physical Properties: Brittle; fracture subconchoidal to uneven. Imperfect cleavage parallel to *b*[010]; poor on *a*[100]. H. 2½-3. G. 5.83.

Optical Properties: Opaque. Steel-gray to black, the steel-gray hue often slightly bluish-gray. Streak gray to black. Luster metallic, often splendent.

Chemistry: Lead copper antimony sulfide. Arsenic substitutes for antimony to 3%.

Distinctive Features and Tests: Cogwheel crystals; also the pattern of brilliant faces on edges of the cogwheels and dull, rough surfaces on the broad *c*[001] faces. Cogwheels usually edge-up on matrices. Fuses readily at 1 to a brittle silvery globule; white and yellow sublimates on charcoal next to the sample. Decomposed by nitric acid, staining solution pale blue-green color.

Occurrences: In medium-temperature hydrothermal ore veins with other sulfides, mainly galena, tetrahedrite, sphalerite, chalcopyrite and pyrite. Widespread but seldom in good crystals. Large crystals with siderite and sphalerite from Park City, Summit County, Utah. Very small brilliant untwinned crystals coating tetrahedrite crystals, with galena, from Mina Bonanza, Concepcion del Oro, Zacatecas, Mexico. Magnificent crystals to 4″ diameter, from Vibora Mine, Machacamarca, Bolivia. In England, in exceptional cogwheel specimens to 2″ diameter, from Wheal Boys Mine, Endellion (Figure 141), and from Herodsfoot Mine, Liskeard, Cornwall, the latter on white quartz, pale steel-gray and very brilliant. Crystals from ½″ to 2″, from Willroth, near Horhausen, Westerwald, Germany, and from Andreasberg, Harz, as splendent tabular crystals on matrix. Very brilliant sharp crystals of ⅜″ to ½″ on matrix from Nakaze Mine, Hyogo Prefecture, Honshu, Japan. Because good crystals are not easy to obtain, bournonite is prized by collectors. Small specimens in the range of from 1″ to 2″ across, with crystals on matrix are usual; larger matrix specimens and crystals are difficult to obtain.

Boulangerite—$Pb_5Sb_4S_{11}$

Lead antimony sulfide; boo lån'jĕr ĭt. A gray mineral occurring mainly in fibrous radiate masses of little attractiveness, as at the Cleveland Mine, Stevens County, Washington, and from many other deposits in the world. Occasionally, very slender acicular crystals are found in cavities as felted masses. Very similar to jamesonite and easily confused with this species (see below).

Jamesonite—$Pb_4FeSb_6S_{14}$

Lead iron antimony sulfide; jăm'sŭn ĭt. Also a gray mineral occurring in fibrous masses or felted aggregates of exceedingly thin acicular crystals. Recently, some good specimens have been obtained from Mexico as felted masses on brilliant cockscomb pyrite from the Mina Noche Buena near Mazapil, Zacatecas. The following tests may help in distinguishing boulangerite and jamesonite. In the flame, boulangerite decrepitates and forms a bubbly mass; jamesonite fuses more easily and dissolves in hot hydrochloric acid, giving off the rotten egg odor of hydrogen sulfide.

OXIDES AND HYDROXIDES

About 150 oxides are known, including simple oxides, oxides with water or hydroxyl and multiple oxides, but relatively few are abundant and of these, still fewer provide attractive specimens. Oxides occur in a variety of environments, many being found in metamorphic or igneous rocks, some in ore veins, a considerable number in granitic pegmatites, and others in near-surface deposits, formed under conditions of low pressure and temperature. Because of the strength of bonds between the oxygen anions and metal cations, oxides are resistant to further chemical attack and generally provide hard, durable minerals, often evident as crystals in sands, gravels, etc., and testifying to their permanence as compared to other minerals in the rocks formerly enclosing them. A few oxides, as hematite and magnetite, occur in enormous quantities providing economic ore deposits. The hydrated iron oxide, goethite, the principal constituent of *limonite,* is present practically everywhere, sometimes in quantities sufficient to create economic ore deposits. A number of exceptionally strong oxides, as corundum, spinel and chryso-beryl sometimes occur in clear gem-quality specimens of attractive coloration.

Cuprite—Cu_2O

Name: From the Greek, *cuprum,* for "copper"; kŭ′prīt.
Varieties: Elongated crystals, *chalcotrichite* (see below).
Crystals: Isometric. Commonly, very small, highly modified crystals, displaying the octahedron, dodecahedron and cube; less commonly as simple octahedrons, dodecahedrons or cubes (Figure 142). Crystals seldom over ¼″. Faces range from very smooth to dull.

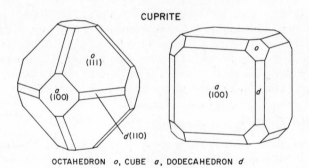

CUPRITE

OCTAHEDRON *o*, CUBE *a*, DODECAHEDRON *d*

FIG. 142.

Acicular micro crystals are termed *chalcotrichite,* and occur as wads or mats, sometimes enclosed in calcite and coloring the latter a vivid orange-red. Also granular to earthy massive.
Physical Properties: Brittle; fracture conchoidal to uneven. Indistinct and interrupted cleavage parallel to the faces of *o*[111]. H. 3½-4. G. 6.14.
Optical Properties: Translucent to transparent; rarely transparent through as much as ¼″ of material. Brownish-red, purplish-red, a peculiar purplish cast being typical. Streak brownish-red, shining. Adamantine luster, verging on submetallic. R.I., $n = 2.85$ (red light).
Chemistry: Cuprous oxide.
Distinctive Features and Tests: Strong luster; usually closely associated with native copper. Hair-like *chalcotrichite* is unmistakable. On charcoal, fuses at 3, yielding a malleable copper bead after prolonged heating in reducing flame. Dissolves in hydro-chloric acid, staining the solution blue.
Occurrences: In oxidized portions of copper sulfide veins, usually coating native copper

from which it forms. Specimens are commonly held together by cores of copper, which fact may be discovered in attempts to trim them. Fine specimens from Bisbee, Cochise County, Arizona, as cubes to ½″ on native copper, sometimes very lustrous and transparent; skeletal crystals, called "plush copper" in allusion to the surface texture resembling plush or velvet, commonly line cavities in brown limonite; also masses of *chalcotrichite* fibers, and as great prizes, euhedral crystals of calcite rendered brilliantly orange by *chalcotrichite* inclusions. Complex, nearly spherical crystals to ¼″ occur on native copper at Ray, Pinal County, Arizona and at the Chino Pit, Santa Rita, Grant County, New Mexico (Figure 143). Formerly in fine large crystals from Bingham, Tooele

FIG. 143. Highly modified cuprite crystals growing upon mass of native copper with adhering fragments of altered rock. From Chino Pit, Santa Rita, Grant County, New Mexico. Size exposed in picture about 3″ by 2″.

County, Utah. Exceptional sharp complex crystals obtained many years ago from copper mines near Bogoslovsk, Ural Mountains, U.S.S.R. At Chessy, near Lyon, France, in outstanding crystals to 1¼″ diameter, as dodecahedrons and octahedrons coated with bright green earthy malachite altering from cuprite cores; many single crystals occur loose in clay and are much prized for single-crystal collections. Also fine crystals from sulfide veins in Cornwall, England, particularly from mines near Redruth and Liskeard. Tsumeb, South-West Africa, provided crystals to ⅜″ on edge, but present offerings are seldom over ⅛″. Fine *chalcotrichite* also known from Tsumeb, and from Mt. Isa in Queensland, Australia.

Zincite Group

Zincite—ZnO

Zinc oxide; zĭnk′ĭt. A very rare mineral except at the unique deposits in the Franklin marble at Franklin and Ogdensburg, Sussex County, New Jersey. Here it occurs in veinlets to about 4″ thick, or in peculiar rounded blebs enclosed in calcite, and abundantly, in granular masses of franklinite, green to brown willemite, and white calcite. The color is always some shade of orange-red, ranging from extremely dark to vivid deep orange red. Most is only translucent but some masses to about 3/4″ diameter are clear and provide the raw material for fine transparent red gems. Though crude, crystals are exceedingly rare, and are counted as prizes by collectors (Figure 144). Some reach about 1¼″ in length, consisting of simple hexagonal pyramids coming to a point at one end of the c axis, and terminated at the other by the pedion, c[0001]; the record crystal is about 2″ in diameter. Crystal faces are rough but a few sharp micro crystals have been found. The usual collection specimens are more or less pure masses of zincite to several inches diameter, gaining more favor as the color brightens; also desirable are colorful mixtures of blebby zincite in white calcite with crystals of franklinite and willemite. The latter association is characteristic and serves to identify zincite in hand specimens. Furthermore, the associates calcite and willemite, usually fluoresce vividly under ultraviolet, calcite red, and willemite green.

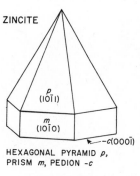

ZINCITE

p (10$\bar{1}$1)

m (10$\bar{1}$0)

c(000$\bar{1}$)

HEXAGONAL PYRAMID p,
PRISM m, PEDION $-c$

FIG. 144.

Tenorite—CuO

Copper oxide; tĕn ôr ĭt. Mentioned here only because it is common in copper deposits of the southwestern United States as dead black earthy to pitchy masses displaying very few features of any sort but asociated with more deirable copper minerals. The varietal name *melaconite* is sometimes applied to massive material. Occasionally, a colloform variety is found, especially at Bisbee, Arizona, much resembling ordinary asphalt (*copper pitch ore*).

Litharge—PbO

Massicot—PbO

Minium—Pb$_3$O$_4$

Oxides of lead occurring as powdery coatings and earthy masses in oxidized portions of lead deposits. Litharge (lĭth′ärj) is red, massicot (măs′ĭ cŏt) vivid orange-yellow, and minium (mĭn′ĭ ŭm) brownish-red to scarlet. Litharge and massicot are very seldom offered in specimens, which, truth to tell, would not be very attractive in view of the form these species assume. However, masses of minium, sometimes to several inches diameter, are occasionally offered as rarities from various localities, among such may be mentioned: Broken Hill, New South Wales, Australia; Castelberg, Moselle, Germany; and from several localities in Arizona, Colorado, and Montana.

Hematite Group

Corundum—Al₂O₃

Name: Probably derived from the Sanskrit, *kuruvinda*, "ruby"; kô rŭn′dŭm.

Varieties: A very impure material, known as *emery,* is an intimate mixture of granular corundum, magnetite, hematite, and spinel. Color phases in gem material are known as *ruby* when intense red; other colors are called *sapphire.*

Crystals: Hexagonal. Red corundum characteristically in tabular crystals, usually displaying the hexagonal prism $a[11\bar{2}0]$, broad basal planes of $c[0001]$, sometimes beveled by $n[22\bar{4}3]$ and truncated on corners by $r[10\bar{1}1]$; tablets sometimes compressed on $c[0001]$ to very thin wafers (Figure 145). Basal plane markings are usually distinctive in red corundum, consisting of sharp striations meeting at 60° interior angles, often as equilateral triangles. Other color phases of corundum, including the large crude grayish crystals sometimes mined for abrasive purposes, incline toward stubby to elongated hexagonal prismatic habits, usually tapering like cigars toward the ends of c axis (Figures 145, 146). Few crystals display smooth faces; those along the sides of prisms are very

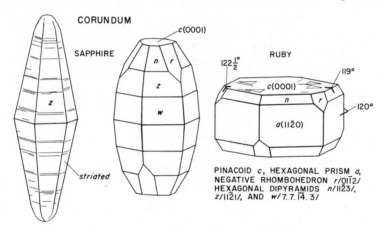

CORUNDUM

SAPPHIRE

striated

RUBY

$c(0001)$

$122\frac{1}{2}°$

$119°$

$120°$

$a(11\bar{2}0)$

PINACOID *c*, HEXAGONAL PRISM *a*, NEGATIVE RHOMBOHEDRON *r/0112/* HEXAGONAL DIPYRAMIDS *n/1123/,* *z/1121/,* AND *w/7.7.14. 3/*

FIG. 145.

commonly striated horizontally, thus affording an excellent clue to identity along with the typical cigar-shape. Lamellar twinning on $r[10\bar{1}1]$, of which there are three planes on each end of each crystal, sometimes causes easy partings of considerable flatness and smoothness as shown in the righthand crystal of Figure 146.

Physical Properties: Tough to very tough, except in lamellar twins which tend to crumble, or in those with basal inclusions, leading to ready parting on $c[0001]$. No cleavage. H. 9, the hardest mineral next to diamond. G. 4.0-4.1; transparent gem material usually 3.997-4.00.

Optical Properties: Transparent to translucent. Large variety of hues from perfectly colorless to nearly black when filled with inclusions; commonly greenish-gray, blue-gray to brown in coarse translucent varieties, but in gem varieties, fine red through pink, yellow, orange, green, blue, and violet. Often parti-colored within same crystal, showing zones of color in layers prallel to the basal plane $c[0001]$, as thin overlays of color on pyramidal faces, or as indefinite clouds, bands, etc. Fragments observed through the

FIG. 146. Corundum crystals. *Top* and *bottom left:* large water-worn crystals from the gem gravels of Ceylon; the larger dipyramidal crystal at upper left is yellow in hue and measures about 3″ tall. *Top center:* tabular hexagonal prism of ruby corundum from Matabatu Mountains, Tanganyika. *Top right:* dark blue parting fragment from Mozambique showing parting parallel to c(0001) at the top of the specimen, and partings on the sides parallel to the rhombohedron r(10$\bar{1}$0). *Lower right:* an irregular dipyramidal crystal from Ceylon and two ruby crystals from Mogok, Burma.

basal plane usually show color zones arranged in hexagonal patterns around the c axis. Minute platelets, possibly of hematite, occur oriented parallel to c[0001] in some of the brown corundum of India and the "black star" corundum of Thailand, and elsewhere, producing a broad schiller and promoting parting on the basal plane, and when polished, sometimes providing asteriated gems. The finest asterias result from the silky reflections of light from sets of inclusions, presumably rutile but possibly also tubular voids, oriented along basal planes and parallel to prism faces. The usual asterias are six-rayed, but single-ray (catseye) and twelve-rayed gems are known. Luster vitreous inclining toward adamantine; silky chatoyant in star stones on c[0001]; sometimes bronzy-submetallic on this plane when schiller-type inclusions present. Transparent gem-quality specimens usually display one- or two-phase inclusions in veils and clouds. Uni-axial negative R.I., O = 1.7687, E = 1.7606; sometimes reaching: O = 1.78, E = 1.77. Bire-fringence 0.008. Dichroism pronounced except in yellow varieties; in ruby commonly intense purplish-red when observed down the c axis (ordinary ray), and a slightly orange red when observed through the sides of crystal prisms (extraordinary ray). For blue sapphires in corresponding directions, it is deep pure blue, greenish-blue, etc.

Chemistry: Essentially pure aluminum oxide. Red coloration is imparted by chromium substituting for aluminum; ferric iron imparts a slight brownish cast to red. Blue is caused by iron and titanium but opinions as to color-causing mechanism differ, the latest investigations suggesting that both elements are present as colloidal ilmenite, $FeTiO_3$.

Distinctive Features and Tests: Hardness; crystal forms. Ruby crystals identified by color, shape, and triangular markings on basal faces. Crystals of other varieties characteristically cigar-shaped. Hexagonal color-banding and inclusion banding across basal plane are also distinctive. Infusible. Insoluble.

Occurrences: Widespread in a variety of deposits, but most abundant in metamorphic rocks; also in pegmatites, and in considerable quantity in some nepheline syenites; less commonly, in basalt. Corundum is highly resistant to alteration and abrasion, being found in stream gravels in Burma, Ceylon, Montana, Australia, etc. Associated minerals are zircon and spinel in metamorphosed limestones; margarite and kyanite in basic rocks and feldspars in pegmatites. Fine crystals occur in syenitic rocks of Bancroft, Hastings County, and Craigmont, Renfrew County, Ontario, usually as stubby cigar-shaped crystals with rounded terminations; single crystals to 60 pounds weight are recorded. Blue and pink crystals usually rough and poorly formed, to 1″, occur in the Franklin marble belt of southern New York and northern New Jersey. From metamorphic rocks in the Appalachian Mountains of North Carolina, South Carolina, Georgia, and northern Alabama, as small rude crystals to large masses, sometimes in pseudo-cubic parting blocks of exceptional flatness and sharpness. In Montana, as gem quality tabular blue crystals seldom over ½″ diameter from an igneous dike at Yogo Gulch, Judith Basin County; crude tabular to stubby prismatic crystals from alluvial deposits at Rock Creek, Granite County, and elsewhere in this state. The finest rubies occur in alluvium in and around Mogok, Burma, in crystals to 2″ in length and are derived from weathering of marble; also sapphires in crystals to 4″ in length. Sapphires and rubies occur in Cambodia and Chantabun, Thailand. Fine crystals occur in the gravels in the Ratnapura and Rakwana districts of Ceylon, often reaching lengths of 3″. In India, as large coarse crystals, and recently, as dark red asteriated corundum in crude barrel-shaped individuals. The finest blue gem sapphires occurred in feldspar pegmatite in Kashmir, northern India. Large crystals from a number of localities in Madagascar; from Cape Province and Transvaal, South Africa, the latter once yielded a crystal of 335 pounds!; also from Mozambique. Colorful specimens of bright red corundum in green chromiferous zoisite occur in Tanganyika. Good specimens from near Anakie, Queensland, Australia, also several places in Europe, lately, in fine red tabular crystals in syenite, from Froland, Norway; from the Urals in the U.S.S.R.

It is difficult to obtain sharp crystals of colored corundum because these are likely to be gem material and cut before leaving the country of origin. From among the usually small fragments, pebbles, and crystals in discarded gem gravel, some crystals displaying good forms and sharp faces are obtained and sold to collectors. Crude crystals of no value for gems are usually available.

Hematite—Fe_2O_3

Name: From the Greek, *haimatites,* "bloodlike," in allusion to vivid red color of the powder; hĕm′ ă tīt.

Varieties: Small mica-like tabular crystals of steel-gray color and very brilliant metallic luster are called *specularite;* reniform masses are sometimes called *kidney ore.* Hematite pseudormorphic after magnetite is called *martite* (Figure 148).

Crystals: Hexagonal. Predominately thin to thick tabular; less commonly, as complex crystals approaching spherical shape (Figure 147). Tabular crystals often beautifully

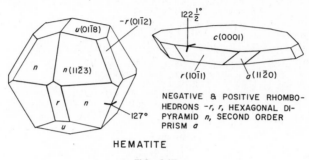

NEGATIVE & POSITIVE RHOMBO-
HEDRONS $-r, r,$ HEXAGONAL DI-
PYRAMID $n,$ SECOND ORDER
PRISM a

HEMATITE

FIG. 147.

arranged in flower-like fashion in "iron roses" (Figure 148), with depressed centers surrounded by smaller and thinner crystals, and with thicker and broader crystals in additional rows outward; in some examples, the aggregate outline is nearly perfectly circular. Edges of single crystals usually small and flat, as if artificially faceted; those near the basal plane $c[0001]$ are commonly ridged, rippled, striated, or covered by shallow growth hillocks. Twinning on $c[0001]$ as penetration twins; lamellar twinning on $r[10\bar{1}1]$ causing striations on basal planes. Rutile often grows in oriented position on the basal plane, forming bundles of acicular crystals radiating outward at 60° apart.

Physical Properties: Brittle; uneven to perfect conchoidal fracture on large single crystals. No cleavage; parting commonly prominent on $r[10\bar{1}1]$ and on $c[0001]$ due to twinning; some pseudo-cubic parting masses of hematite several inches across have been developed from large twinned crystals. H. 6.5 (crystals). G. 5.3 (crystals).

Optical Properties: Opaque black or steel-gray except through thinnest splinter edges or through some minute platey inclusions in calcite, quartz, and feldspars, when the color is seen to be deep blood red. Streak, deep red. Luster earthy to dull on massive varieties; crystals submetallic to metallic. Often tarnished iridescent blue, purple, red.

Chemistry: Ferric oxide.

Distinctive Features and Tests: Color; hardness; deep red streak. "Iron roses" unmistakable. Infusible; slowly soluble in hot concentrated hydrochloric acid, staining the solution intense reddish-yellow. Fragments strongly heated in reducing flame become magnetic. Latter test with streak test distinguishes from ilmenite and magnetite, the last being magnetic without heating. Black sulfides rarely produce a deep red streak.

Occurrences: In large beds of sedimentary origin; in some metamorphic rocks; in tabular crystals in Swiss-Alpine-type cavities; in quartz veins; in contact metasomatic deposits; abundant and widespread in small quantities nearly everywhere. Bright drusy crystals with quartz crystals from mines near Edwards, Gouverneur, and Antwerp, St. Lawrence County, New York. Unique rhombohedral partings, to 4″ on edge, from Franklin, Sussex County, New Jersey. Excellent reniform hematite from iron mines in Michigan and Minnesota; similar material recently discovered on Aztec Peak, Gila County, Arizona. Hematite pseudos after sharp siderite rhombs in cavities in amazonite-quartz pegmatites of the Crystal Peak area, Teller County, Colorado; some rhombs reach

FIG. 148. Hematite. *Top left:* highly modified and curved crystals implanted on quartz matrix from Rio Marina, Elba, Italy. Size 3″ by 2½″. *Top right:* tabular crystal displaying broad c (0001) faces and narrow faces of $a(11\bar{2}0)$ and $r(10\bar{1}0)$. Some golden rutile crystals are attached in oriented positions. From Itibiara, Bahia, Brazil. Size 2¼″ by 2″. *Top center:* An "iron rose" from Burnier, Minas Gerais, Brazil. Size 2¼″ by 1¼″. *Bottom:* Hematite replacing large magnetite octahedrons in the variety known as *martite*. From Twin Peaks, Millard County, Utah. Size 3″ by 3″.

several inches length. Hematite (*martite*) replacing octahedrons of magnetite to 5″ on edge (Figure 148), occur in clusters and coarse druses at Twin Peaks, Millard County, Utah. Sharp brilliant black complex crystals occur with quartz at Swansea near Bouse, and in small cavities as isolated plates to an inch across, near Quartzite, Yuma County, Arizona. Interesting replacements after pyrite cubes to 2″, occur in the garnet-idocrase deposit near Lake Jaco, Chihuahua, Mexico. Octahedrons of *martite* to 1″, occur with the yellow apatite of Durango, Durango, Mexico. Consistently fine crystallized specimens from Rio Marina, Elba, where single crystals, clusters, and lustrous black crystals on matrix, with quartz crystals, have been obtained for many years. Elba crystals are seldom over 1″ in diameter, but are arranged nicely and often iridescent; matrix specimens to about 9″ by 7″ have been recorded (Figure 148). Outstanding iron roses from a number of Swiss localities, to 2″ diameter, but usually not over ½″ to ¾″; many are unbelievably delicate and wonderfully symmetrical in their arrangement of platey crystals. Splendid striated rhombohedral crystals to 1¼″ from Achmatovsk, Ural Mountains, U.S.S.R. Fine crystals to ½″, in druses and clusters, from Dognaska, Banat, Romania. Tabular crystals also occur in Switzerland, sometimes with partly-imbedded slender prisms of blood-red rutile in oriented position, a notable locality being Cavradi, Tavetschtal. Small but very fine crystals occur in crevices in volcanic rocks and ash at Vesuvius and Etna, and elsewhere in Italy; also very fine from the Madeira Islands, Portugal. Fine crystals from France, as at Framont, Alsace, and in the Puy-de-Dôme district. Lustrous sharp crystals from Cleator Moor, Cumberland, England; also hematite "kidney ore" to 50 pounds weight from Ulverstone, Lancashire. Large iron roses, commonly 2″ in diameter, and occasionally to 6″, occur in hematite ore beds in Minas Gerais, Brazil as at the Burnier Mine, Itabira, and at Lago de Netto near Ouro Preto (Figure 148). Recently, exceptional tabular single crystals to 3″ or more in diameter, from cavities in a large quartz vein near Itibiara, Bahia, Brazil (Figure 148). Choice, nearly complete tabular crystals to 2″ diameter, occur near Bimbowrie, South Australia.

Despite the prejudice to all-black or all-white minerals on the part of many collectors, the numerous varieties of hematite offer a fine field for specialization. Perhaps of all varieties, "kidney ore" from England, and Elba crystals are most commonly represented, and less commonly, iron roses. Of all types, iron roses, especially from Switzerland, are most expensive.

ILMENITE SERIES

Ilmenite—FeTiO$_3$ (Iron end member)

Geikielite—MgTiO$_3$ (Magnesium end member)

Pyrophanite—MnTiO$_3$ (Manganese end member)

Name: After the locality in the Ilmen Mountains of the U.S.S.R.; ĭl′mĕn īt. After Sir Archibald Geikie (1835-1924), English geologist, gē′kĭ līt. After the Greek, *pyro* and *phane,* "fire" and "appear," in allusion to red color, pī rŏf′ă nīt.

Varieties: A variety of ilmenite containing about 6% ferric oxide, *menaccanite; ferroan* geikielite and *ferroan* pyrophanite have been recognized.

Crystals: Hexagonal. Commonly thick tabular on c[0001], with periphery modified by prism and rhombohedron faces (Figure 149). Rarely in rhombohedral crystals. Also lamellar, massive. Crystals seldom as sharp or glossy as hematite (Figure 150). Twinning on c[0001] and lamellar twinning on r[10$\bar{1}$1].

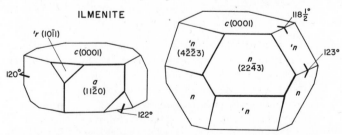

ILMENITE

PINACOID c,SECOND ORDER PRISM a ,NEGATIVE RHOMBOHEDRONS
'r, 'n, POSITIVE RHOMBOHEDRON n

FIG. 149.

Physical Properties: Brittle; conchoidal to uneven fracture. Ilmenite: no cleavage; geikielite, on $r[10\bar{1}1]$; pyrophanite, on $s[02\bar{2}1]$, perfect. Ilmenite parts on basal plane $c[0001]$ and rhombohedron $r[10\bar{1}1]$. H. 5-6. G., ilmenite, 4.72; geikielite, 4.05; pyrophanite, 4.54.

Optical Properties: Opaque. Ilmenite, iron-black; geikielite, brownish-black; pyrophanite, very dark red. Streaks: ilmenite, black; geikielite, brownish to dark yellow-brown; pyrophanite, brownish-red. Luster submetallic to dull.

Chemistry: Iron-magnesium-titanium oxides. Magnesium replaces iron in the series ilmenite-geikielite, and iron partly replaces manganese in the series ilmenite-pyrophanite.

Distinctive Features and Tests: Dull black to steely-gray thin to thick tabular crystals. Although resembling hematite in form, they are seldom as lustrous or as deep black in hue. Streak test further distinguishes from hematite; lack of magnetism from magnetite. Infusible; thin splinters very slightly rounded in hot flame. Slowly soluble in hot hydrochloric acid. Ilmenite powder sometimes slightly magnetic.

Occurrences: For the sake of completeness, data on all series species has been given above, but only ilmenite is common, occurring principally in gabbros, diorites, and anorthosites as grains, masses, or veins; common as an accessory mineral in some granitic rocks. Crystals have been found in the extinct emery mines near Chester, Massachusetts. In serpentine and marble outcrops in Orange County, New York, associated with spinel, chondrodite and rutile. Lately, as abraded tabular crystals to 2″ diameter scattered on the surface over a wide area near El Alamo, southeast of Ensenada, Baja California, Mexico. Very large fine crystals occur in diorite veins with hornblende at Kragero, Froland, Norway; crystals are thick tabular to equant, sometimes skeletal, in singles and clusters; some singles reaching 5″ diameter; commonly

FIG. 150. Ilmenite. Stubby crystal showing a brightly reflective $n(22\bar{4}3)$ face. From Froland, near Arendal, Norway. Size 2″ by 2″.

rounded on edges but many are sharp and lustrous (Figure 150). Also as crystals in the Binnatal, Switzerland; at St. Cristophe, Bourg D'Oisans, France; and in thick tabular crystals to 1″ diameter near Lake Ilmen, Miask, Urals, U.S.S.R. Ilmenite is abundant in massive material but is remarkably rare in crystals and is therefore seldom represented in collections. The finest specimens are from Norway, and have recently reappeared on the market.

Arsenolite Group

Senarmontite—Sb_2O_3

Antimony trioxide; sĕn′är mŏn tīt. A rare mineral providing worthy crystals only from Hamimat, southwest of Guelma, Constantine, Algeria. Crystals occur loose in a grayish marl, associated with valentinite, cerussite, hemimorphite, cinnabar, calcite, and barite.

Crystals translucent grayish-white, simple octahedrons with dull but flat faces, ranging from $\frac{1}{4}''$ to $\frac{3}{8}''$; crystals to $1\frac{1}{4}''$ diameter have been recorded Prized by single crystal collectors.

Bixbyite—$(Mn,Fe)_2O_3$

Manganese iron oxide; bĭx′bē ĭt. Generally rare, but relatively common in small, nearly perfect cubic crystals associated with topaz, hematite, garnet, and a curious deep red beryl, in cavities in rhyolite, Thomas Mountains, Juab County, Utah. Very rarely to $\frac{1}{2}''$ on edge; the usual run of specimens display small lustrous black cubes not over $\frac{1}{4}''$ on edge.

Braunite—$(Mn,Si)_2O_3$

Manganese silicon oxide; brown′ĭt. Common in manganese deposits, usually in massive or earthy form, but seldom represented by specimens in collections. Fine black crystals have been obtained from several localities and sometimes afford interesting if not colorful specimens. From Jacobsberg, Nordmark, Sweden, with calcite, richterite and schefferite; from Limenau, Thuringia, Germany; crystals to $2''$ from St. Marcel, Piedmont, Italy; also fine from Kacharwaki, Nagpur, India.

Rutile Group

Rutile—TiO_2

Name: From the Latin, *rutilus,* "red, golden-red," in allusion to color; roo′tĕl.
Varieties: With iron, *ferrian;* also *tantalian* and *niobian.*
Crystals: Tetragonal. Stubby prismatic crystals, usually striated parallel to c axis obliterating prism faces (Figure 151); rather rare in simple smooth-faced crystals; also slender

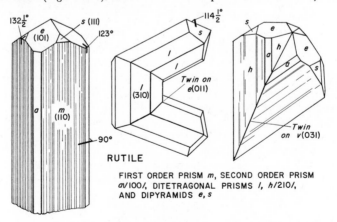

RUTILE

FIRST ORDER PRISM *m*, SECOND ORDER PRISM
a/100/, DITETRAGONAL PRISMS /, *h*/210/,
AND DIPYRAMIDS *e, s*

FIG. 151.

prismatic to acicular. Commonly twinned on $e[011]$, forming "bent" crystals, which if cyclically twinned, may close completely in rings. Figure 152 shows a cyclic twin in which one ring penetrates another. The criss-cross pattern of many slender crystals, known as *sagenite* and often seen enclosed in *rutilated quartz,* is also due to this twinning. Also granular massive.

Physical Properties: Brittle; conchoidal to uneven fracture. Distinct cleavage parallel to prism faces $m[110]$; poor on $a[100]$ and $s[111]$. H. 6-6½. G. 4.2-4.25.

Optical Properties: Usually translucent, but sometimes transparent through as much as ¼″ of material. Varieties containing iron, niobium or tantalum are nearly black and opaque. Usually rich red-brown, dark brown-red, gray-black and black; rarely red. Acicular types usually coppery, red, yellow, orange-yellow. Streak pale brown in dark types to nearly colorless in paler types. Splendent adamantine luster; also submetallic. Uniaxial positive. R.I., O = 2.6124, E = 2.8993. Birefringence 0.287. Transparent red specimens dichroic, dark red and very dark red, nearly black. Rarely, in crystals clear enough to facet very small gems.

Chemistry: Titanium dioxide. Substantial amounts of iron, niobium and tantalum may be present.

Distinctive Features and Tests: Striated crystals; strong luster; high specific gravity; cyclic twins. Sagenitic growths and acicular prisms in clear quartz are unmistakable. Infusible. Insoluble.

Occurrences: Widespread in small amounts in many kinds of rocks and mineral deposits; perhaps most common in strongly metamorphosed rocks as quartzites, schists, and gneisses; also in cavities in quartz veins. A common alluvial mineral. Magnificent crystals in quartzite with lazulite, kyanite and pyrophyllite on Graves Mountain, Lincoln County, Georgia; crystals usually simple in form, often with smooth shining flat faces, but also striated or twinned, to 5″ by 4″ by 3″. Fine black striated singles and cyclic twins to 1″, from Magnet Cove, Hot Springs County, Arkansas, commonly as closed rings of eight cyclically twinned crystals; also as sharp pseudomorphs after brookite. Fine sagenitic crystals from open fissures and cavities in pegmatites at various places in North Carolina, particularly from near Hiddenite, Alexander County. Excellent crystals resembling the Georgia crystals except in respect to size, occur in soft pale brown pyrophyllite in the

FIG. 152. Rutile. *Left:* a large striated cyclic twin penetrated by another twin. Minas Gerais, Brazil. Size 2¼″ by 1¾″. *Right:* small cyclic twins in quartz from Graves Mountain, Lincoln County, Georgia.

Champion Mine near Laws, Inyo County, California; crystals seldom exceed 1½" in length, but are often very sharp and perfectly smooth.

Fine simple prisms to 1" from Lofthus, Snarum, Norway. Small but exceptionally sharp and brilliant crystals occur in numerous Alpine localities, sometimes as crystals lining vugs, sometimes as oriented *sagenite* on hematite; noteworthy localities are Tavetschtal and Binnatal. Splendid "butterfly" twins of rutile occur at Cerrado Frio, Minas Gerais, Brazil, and consist of two flattened crystals, twinned on v[031], and furnishing unique specimens to 2" in length and about 1" in width (Figure 151). Lately, considerable quantities of large dark red twinned crystals to 3" in length and nearly the same in thickness (Figure 152), from Minas Gerais. The finest specimens of rutilated quartz occur at Itibiara, Bahia, Brazil; clear quartz crystals to 18" length, mostly smoky, but also colorless, enclose coarse sprays of brilliant golden rutile, the latter oriented on tabular crystals of hematite. In some examples, flat sprays of rutile, 60° apart, radiate outward from original cavity walls and are enclosed in later quartz. Some sprays are 4" in length and ½" wide, consisting of almost solid ribbons of brilliant yellow rutile.

Good rutile crystals are in demand for collections, particularly those displaying obvious red color and shining faces. Graves Mountain crystals are "classics" and very much desired.

Pyrolusite—MnO₂

Name: From the Greek, *pyro* and *louein,* "fire" and "to wash," because used to remove or "wash out" the greenish color imparted to glass by iron compounds; pī′rō loo′sīt.

Crystals: Tetragonal. Rarely in prismatic crystals elongated parallel to *c* axis; by far most common as reniform sooty masses (Figure 153). The delicate dendritic traceries found in narrow seams in all kinds of rocks and commonly in chalcedony, are usually pyrolusite. Also earthy.

Physical Properties: Brittle; in some massive forms, tough. Cleavage perfect on m[110]. Masses fracture with dull earthy surfaces, sometimes displaying fibrous structure. H. 6-6½, for crystals, to much lower for massive material. G. 5.06 for crystals; 4.4-5.0, massive.

Optical Properties: Opaque. Iron-black; dark steel-gray in crystals. Streak black. Luster submetallic to metallic in crystals, otherwise dull.

Chemistry: Manganese dioxide.

Distinctive Features and Tests: Usually massive, reniform; soils the fingers during handling. Infusible. Dissolves in hydrochloric acid, giving off choking fumes of chlorine. Crushed material placed in test tube and heated, gives off oxygen, causing glowing wood splinter to burst into flame when placed inside. Borax bead colored amethyst in oxidizing flame. *Note:* It is not practicable to distinguish pyrolusite from the very similar species psilomelane and wad, except that crystals can be assigned to pyrolusite inasmuch as psilomelane and wad occur only massive.

Occurrences: Pyrolusite, psilomelane, and wad occur in oxidizing conditions near the surface as coatings or crusts on other manganese minerals, or as linings and fillings in soils and rocks. Associates commonly are hausmannite, goethite, limonite and braunite. Characteristically coating or enclosing weathered rhodonite. Specific localities for crystals are few, but such are likely to be found lining cavities wherever thick masses of pyrolusite occur. Thus crystal-lined geodes, displaying small prismatic crystals, occur in manganese deposits in the Cartersville district, Bartow County, Georgia; similarly in the Maroco Mine, Ironton, Crow Wing County, Minnesota in the United States; also near Belem, Brazil and other places in that country. Large crystals occur at Platten in Bo-

FIG. 153. Pyrolusite. Black reniform mass from surface workings in a manganese deposit in Sonora, Mexico. Size 4″ by 4″.

hemia, Czechoslovakia. The usual run of crystallized specimens show earthy dead-black material becoming more dense and visibly crystalline toward an opening, and assuming a steel-gray color and metallic luster, and finally lining the cavity with brilliant drusy crystals, seldom over ⅛″ length. Interesting concretionary and reniform masses are often obtainable.

Cassiterite—SnO$_2$

Name: After the Greek, *kassiteros,* "tin"; kă sĭt′ĕr ĭt.

Varieties: Iron, tantalum and niobium are sometimes present, in *ferrian, tantalian,* and *niobian* varieties. Colloform material is called *wood-tin,* in allusion to internal concentric structure.

Crystals: Tetragonal. Usually in stubby prisms, sometimes bipyramidal; commonly twinned on *e*[101], resulting in complex crystals whose faces may be difficult to decipher (Figure 154). Faces often brilliant, flat, and meeting in sharp junctions. In druses; rarely, as single crystals. Also granular and colloform (*wood-tin*) as in Figure 155.

Physical Properties: Brittle; colloform masses very tough. Fractures subconchoidal to un-

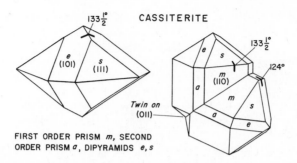

CASSITERITE

FIRST ORDER PRISM *m*, SECOND
ORDER PRISM *a*, DIPYRAMIDS *e*, *s*

even; fibrous in *wood-tin*. Good cleavage parallel to *a*[100] but usually interrupted; indistinct on *m*[110]. H. 6-7. G. 6.8-7.1; lower for *wood-tin*.

Optical Properties: Crystals opaque to transparent but many seemingly black crystals, as those from Bolivia, are actually transparent in outer zones to $\frac{1}{8}''$ depth, the blackness being imparted by central zones of intensely dark brown material. Black, very dark brown, but also pale brown and colorless. *Wood-tin* displays wide range of dull hues, usually tans, browns, reds and brown-greens. Streak colorless; rarely brownish. Luster greasy on broken crystal surfaces; brilliant vitreous on crystals, but adamantine on rubbed surfaces as on stream worn pebbles or damaged edges of crystals. Uniaxial positive. R.I., O = 2.003, E = 2.101. Birefringence 0.098. Dichroism weak.

Chemistry: Tin oxide. Iron substitutes for tin to 3% by weight; niobium and tantalum may also be present in small amounts. *Wood-tin* contains inclusions of hematite and quartz.

Distinctive Features and Tests: Brilliant black crystals; high specific gravity. Pale streak, hardness and transparency in outer crystal layers distinguishes from other heavy black minerals. Infusible. Insoluble. Pieces of zinc and cassiterite placed in dilute hydrochloric acid result in the cassiterite becoming coated with gray metallic tin which is brightened by rubbing.

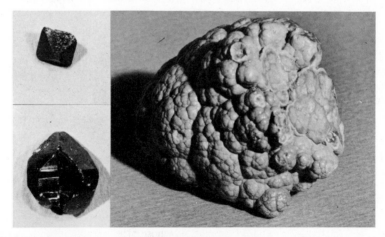

FIG. 155. Cassiterite specimens. Small crystals showing typical dipyramidal form at *top left* and twin at *bottom left*. *Right:* colloidal cassiterite or "wood tin," from the Arroyo Carrizal, Guanajuato, Mexico. Size 1½" diameter.

Occurrences: A high-temperature hydrothermal vein mineral in fissure veins in granitic rocks. Also as small masses and crystals in granitic pegmatites and as *wood-tin* concretions in openings in rhyolite or in oxidized portions of tin-bearing veins. Abundant in some alluvial deposits. Associates in primary deposits are wolframite, tourmaline, topaz, quartz, fluoride, muscovite, lepidolite, bismuthinite, bismuth, and, in Bolivian tin veins, sulfides.

Good specimens are seldom found in North America, except for fine botryoidal masses of *wood-tin* to several inches diameter, from Mexico (Figure 155), and less commonly, from rhyolite, in Lander County, Nevada, and from Taylor Creek, Sierra County, New Mexico. Fine sharp bipyramidal crystals, rarely over ⅝″, occur in granitic pegmatites in the New England states, at Amelia, Virginia, and in San Diego County, California. Occasionally, as well-formed crystals in the tin deposits of Irish Creek, Rockbridge County, Virginia, and in pegmatite at Silver Hill, Spokane County, Washington. Rarely as fine crystals at Lost River, Seward Peninsula, Alaska. The finest crystals, sometimes to 3″ diameter, occur in the deposits of the Llallagua, Araca and Oruro districts of Bolivia. A large unbroken druse measuring 24″ by 12″ by 8″, and weighing 350 pounds, coated with crystals averaging about 1″, was recovered from Araca; it was unsaleable in one piece and was eventually broken up. Some Oruro crystals are acicular, forming the so-called "needle tin." Bolivian specimens characteristically display very lustrous crystals, apparently black but often transparent in outer zones, and commonly twinned; druses are abundant but single crystals rare. Splendid crystals formerly from numerous tin veins in Cornwall, England, but now seldom available. Highly prized large lustrous black crystals formerly occurred in narrow fissure cavities in altered granitic rocks (greisen), at Schlaggenwald and Zinnwald, Bohemia, Czechoslovakia, and at Ehrenfriedersdorf, Saxony, Germany. Schlaggenwald crystals reached 3″ in diameter, but averaged ¾″; specimens commonly display small clusters and singles on matrix (Figure 156). Small but fine single prisms and twinned groups in crumbling granular quartz, from La Villeder, Morbihan, France. Fine crystals, rarely available on the specimen market, from the Malayan Peninsula, Borneo, etc. Choice druse specimens from Australia, particularly from Emmaville, New South Wales, and the Benlomond District, and as pseudos

FIG. 156. Sketch of cassiterite crystals on matrix of quartz, muscovite and topaz from Schlaggenwald, Bohemia, Czechoslovakia. Size about 3″ by 4″.

after orthoclase as at Rex Hill, Tasmania, Australia. Also from Arandis, South-West Africa; Nigeria, etc.

Cassiterite specimens are prized because of their sharp brilliant crystals. Most available specimens are Bolivian, but English and Bohemian specimens occasionally turn up. Schlaggenwald specimens have much to commend them, perhaps more than Bolivian because of better color contrast between dark crystals and paler matrix. Interesting rarities are flattish orthoclase crystals to 2" by 1" by $\frac{1}{4}$" from Bottalack, Cornwall, England, completely replaced by granular quartz and cassiterite; surfaces are dull but crystal forms are unmistakable; many specimens show Carlsbad twinning (Figure 34).

Anatase—TiO₂

Titanium dioxide; ȧn' ă̇ tăs. A rare mineral, except in the crystal-lined cavities of the Alpine region, where it is commonly associated with quartz, brookite, rutile, adularia, hematite, and chlorite. Crystals are sharp bipyramids, usually striated at right angles to the c axis, and seldom exceeding $\frac{1}{4}$" although specimens to 1" have been recorded from Binnatal, Switzerland. Swiss material varies from pale brown or yellowish to deep blue; other hues are known. Small crystals are prized for micromounts. Notable localities include Tavetschtal, Binnatal, Maderanertal, and the St. Gotthard, Switzerland; Bourg d'Oisons district, France; at various places near Salzburg, Austria, in the Italian Alps, etc. Transparent gemmy crystals occur in diamond-bearing gravels, Minas Gerais, Brazil. Fine blue crystals along joint planes in diorite, on Beaver Creek, Gunnison County, Colorado.

Brookite—TiO₂

Titanium dioxide; brook'īt. Uncommon except in the Alpine vugs mentioned under Anatase, and in a few other places in the world where notable crystals occur. Fine tabular brown crystals to 2" from Maderanertal, Switzerland, and from Nillgrabental near Pragatten, Austrian Tyrol. Black to dark brown crystals from cavernous gray quartz in Magnet Cove, Hot Springs County, Arkansas; crystals reach 1" across and assume a pseudohexagonal-pyramidal habit; they are implanted on dull gray rude quartz prisms, several inches long.

Stibiconite—Sb₃(OH)O₆

Hydroxyl oxide of antimony; stĭb' ĭ kō nīt. A secondary mineral produced by alteration of stibnite and commonly forming faithful pseudomorphs of its crystals. The color is usually yellowish or grayish-yellow. Interesting and sometimes spectacular single crystals and groups of stibiconite after stibnite come from Charcas, San Luis Potosi, where single crystals (Figure 157) to 12" and entire groups to nearly 20" across have been recovered. Some Charcas crystals are very sharp and fairly smooth though not at all lustrous; others are crude but recognizable. Smaller pseudos also occur in the Sacramento Mine, Mercur, Tooele County, Utah.

Uraninite Group

Uraninite—UO₂

Name: After its composition: ū rȧn' ĭ nīt.
Varieties: At least six names have been given to minerals later proved to be uraninite, but only several varietal names are now used. Cryptocrystalline or colloform material is

FIG. 157. Stibiconite replacing large stibnite crystal. Note how faithfully the typical striations of stibnite crystals are preserved. From Catorce, San Luis Potosi, Mexico. Size 6¼".

known as *pitchblende*. On the basis of composition, *thorian, cerian* and *yttrian* varieties are also recognized.

Crystals: Isometric; rare. Crude cubes, sometimes with slight octahedral or dodecahedral modifications; also octahedrons or dodecahedrons. Much uraninite is massive to earthy.

Physical Properties: Brittle; fracture conchoidal to uneven. No cleavage. H. 5-6. G. 7.5-9.7 (crystals); 6.5-9.0 (colloform).

Optical Properties: Opaque. Black to brownish-black. Streak brownish-black. Luster submetallic; also pitchy or greasy.

Chemistry: Ideally, uranium dioxide, but usually containing thorium substituting for uranium in substantial amounts; cerium and yttrium also present but in smaller amounts. Easily alters, then commonly surrounded by vivid haloes of yellow, orange, and green secondary uranium minerals. *Gummite* has been applied to such alteration products, but is now known to contain, among other species, clarkeite, fourmarierite, vandendriesscheite, uranophane, curite, becquerelite, schoepite and kasolite.

Distinctive Features and Tests: Color; luster; vividly colored alteration haloes; high specific gravity. Cube crystals distinctive. High radioactivity. Infusible. Borax bead fluoresces brilliant pale green in ultraviolet light. Dissolves easily in nitric acid and in aqua regia.

Occurrences: Chiefly in metalliferous hydrothermal veins, but also in pegmatites. Vein deposits usually contain *pitchblende,* but sometimes crystals as in the Shinkolobwe deposit in Katanga, Congo, where a few well-formed crystals to 1½″ on edge have been found. The largest crystals known came from pegmatite at Wilberforce, Renfrew County, Ontario, where cubes and modified dodecahedrons have been found several inches in diameter, while one broken crystal weighed five pounds; associates in these pegmatites are salmon-colored calcite, very dark purple fluorite, mica and apatite. Masses of uraninite, to a number of pounds weight, occur in the pegmatites of New England and North Carolina, and in pegmatites in a number of western states. Sharp black cubes to 1″ on edge have been found in the Rock Landing Quarry, near Portland, Connecticut. The Ruggles Mine at Grafton, Grafton County, New Hampshire is noted for colorful specimens of dendritic uraninite in feldspar (Figure 158), surrounded by orange, brown and

FIG. 158. Uraninite in pegmatite. The uraninite forms plumose and dendritic black growths surrounded by vividly-colored haloes of alteration products, mainly orange fourmarierite and vandendriesschite, and yellow becquerelite and schoepite. Sawed and lacquered section from the Ruggles Mine pegmatite, Grafton Center, Grafton County, New Hampshire. Size 3″ by 3½″.

green alteration minerals (*gummite*). Fine crystals from pegmatites at Elvestad, Annerod Peninsula, Norway; in crystals to ¾″ from Morogoro, Uluguru Mountains, Tanganyika. Few collections contain crystals owing to their considerable rarity; masses of pitchblende to 6″ across, are usually available but are far from attractive.

Brucite Group

Brucite—Mg(OH)$_2$

Name: After A. Bruce (1777-1818), an early New York mineralogist; broos'īt.

Varieties: Iron or manganese substituting for magnesium, *manganoan* and *ferroan*.

Crystals: Hexagonal; rare. Thin tabular hexagonal plates with broad $c[0001]$ faces. Usually foliated massive, resembling talc or deformed mica; sometimes fibrous.

Physical Properties: Can be split easily into flexible but inelastic sheets along the perfect $c[0001]$ cleavage. H. $2\frac{1}{2}$. G. 2.39.

Optical Properties: Transparent to translucent. White, pale greenish, grayish or very pale blue. Manganoan varieties: yellow, brown, deep brownish-red. Streak white. Pearly luster on cleavage surfaces; otherwise waxy. Uniaxial positive. R.I., O = 1.56-1.59, E = 1.58-1.60. Birefringence, 0.02.

Chemistry: Manganese hydroxide.

Distinctive Features and Tests: Pearly luster; foliated masses; harder than talc. Folia often "rippled" in contrast to ordinary flat cleavage of gypsum which it resembles. Easily soluble in acids where gypsum is not. Infusible, but glows brightly in the flame.

Occurrences: A low-temperature hydrothermal mineral typically found in veins in serpentine associated with calcite, aragonite, hydromagnesite, artinite, talc, magnesite and deweylite; also in chloritic or dolomitic schists, and in marble. The world's finest specimens came from the Wood's and Low's chromite mines near the Pennsylvania-Maryland border, Lancaster County, Pennsylvania; crystals to 7″ across have been recorded as well as broad cleavages of snow-white color and fine pearly luster over 8″ across (Figure 159). Large pale blue crystals also occurred in the magnetite ore body of the Tilly Foster Iron Mine, near Brewster, Putnam County, New York. Fine white cleavage plates to

FIG. 159. Foliated mass of pure white color from the Low Chrome Mine, near Rock Springs, Lancaster County, Pennsylvania. Size 7½″ by 3¼″. The pearly luster upon the excellent cleavage plane is easily apparent.

4″ x 7″ in marble from Wakefield, and pale blue fibrous or platey from asbestos mines at Asbestos and Black Lake, Quebec. Unusual in most collections, crystals now being rare because of exhaustion of the Tilly Foster and Pennsylvania deposits.

Manganite—Mn(OH)O

Name: From the manganese content; măng′ gă nĭt.

Crystals: Monoclinic. Small striated prismatic crystals with wedge-shaped or flat terminations (Figure 160), often grouped in short bundles (Figure 161); also fibrous massive. Faces sharp and brilliant. Twinning on plane $e[011]$, as contact and penetration twins; uncommon. Some specimens are covered by uniform bristly crystals; others, as in the Ilfeld specimens, display stout circular rods composed of numerous parallel crystals, the tops of which are ridged or grooved by individual wedge-shaped terminations (Figure 161).

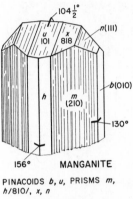

MANGANITE

PINACOIDS *b, u,* PRISMS *m,*
h/810/, *x, n*

FIG. 160.

Physical Properties: Brittle; uneven fracture. Perfect cleavage on $b[010]$; less perfect on $m[110]$; also basal cleavage on $c[001]$. H. 4. G. 4.3.

Optical Properties: Opaque. Dark steel-gray to black. Streak dark red-brown. Luster submetallic.

Chemistry: Hydroxyl oxide of manganese.

Distinctive Features and Tests: Shining black crystals; distinctive crystal aggregates; streak. Infusible. Dissolves in concentrated hydrochloric acid, emitting choking fumes of chlorine.

Occurrences: A low-temperature hydrothermal vein mineral associated with braunite, hausmannite, barite, calcite, siderite; also formed in near-surface deposits by the action of surface waters and then associated with pyrolusite, goethite, barite. Relatively rare in good crystallized specimens. Formerly in ½″ crystals from Picton County, Nova Scotia. Good to fine druses in massive hematite ore at Jackson Mine, Negaunee, Marquette County, Michigan. Good crystals, altered to pyrolusite, at Powells Fort, near Woodstock, Shenandoah County, Virginia; also at the Vesuvius Mine, Augusta County, the Midvale Mine, Rockbridge County, and in Smyth and Bland counties. The finest specimens known, and zealously sought by knowing collectors, are the Ilfeld, Harz, Germany, groups of large and fine aggregate crystals heavily coating massive quartz and associated with barite and calcite (Figure 161). Commonly the crystal bundles reach diameters of ½″ and lengths of 1″. Some bundles are square in cross-section, with flat tops, others are rudely square to nearly round and striated on the sides where individual crystals meet. Ilfeld crystals alter slightly, assuming very thin earthy coatings which, however, can be removed by brushing in soap and water. Of the Ilfeld specimens, matrix groups to about 4″ by 4″ are considered excellent.

Goethite Group

Diaspore—HAlO₂

Name: From the Greek, *diaspora,* "scattering," in allusion to the decrepitation of crystals when heated in the flame; dī′ ă spôr.

Varieties: With iron or manganese, *ferrian* and *manganian.*

Crystals: Orthorhombic; good crystals rare. Thin platey, compressed on $b[010]$; also

FIG. 161. Manganite specimen showing numerous aggregate crystals forming parallel bundles. From the famous Ilfeld, Harz, Germany, locality. Size 3¾" by 2¾".

prismatic elongated parallel to the *c* axis. Commonly in lamellar aggregates, foliated, scaley.

Physical Properties: Very brittle; conchoidal fracture. Perfect cleavage on b[010]; less perfect on m[110]. H. 6½-7. G. 3.3-3.5.

Optical Properties: Transparent to translucent. Colorless, white, greenish, brown, pale violet, pink. Streak colorless. Luster pearly on cleavage plane b[010], elsewhere bright vitreous. Biaxial positive. R.I., $\alpha = 1.702$, $\beta = 1.722$, $\gamma = 1.750$. Birefringence 0.048. Colored crystals strongly pleochroic.

Chemistry: Brown material usually contains small amount of iron substituting for aluminum; deep red or rose material contains small amounts of manganese.

Distinctive Features and Tests: Coarsely crystalline lamellar masses resemble axinite but distinguished by superior hardness and cleavage. Infusible; decrepitates violently.

Occurrences: Primarily with corundum in emery deposits, sometimes with corundum in marble; commonly in fine-granular aggregates in bauxite. Clear material has been faceted into small gems. Fine bladed crystals to 2" from cavities in emery at Chester, Hampden County, Massachusetts. Also fine from Corundum Hill, Newlin, Chester County, Pennsylvania, where crystals to 2″ length were found. From Culsagee Mine, near Franklin, Macon County, New Hampshire. Bladed aggregates from the Champion Mine, near Laws, Inyo County, California. Good small crystals occur near Dilln, Schemnitz, Hungary. Associated with corundum as small transparent colorless to slightly yellow blades in cavities in sugary dolomite at Campolungo, Lake Tremorgio, Switzerland.

Goethite—HFeO$_2$

Name: After German author-scientist, Johann Wolfgang von Goethe (1749-1832); gō′thīt, gĕr′thīt.

Crystals: Orthorhombic; rare. Crystals prismatic along *c* axis (Figure 162), sometimes as thin blades flattened on *b*[010] and elongated along the *c* axis. In tufts, druses, or radiating clusters (Figure 163). Usually in fibrous colloform or earthy material (Figure 163).

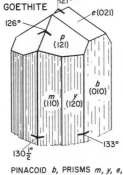

PINACOID *b*, PRISMS *m, y, e,* DIPYRAMID *p*

FIG. 162.

Physical Properties: Tough; splintery fracture in colloform masses; otherwise brittle. Perfect cleavage parallel to *b*[010] visible as very narrow shining streaks along fracture surfaces. H. 5-5½. G. 3.3-4.3 in massive material; crystals 4.28.

Optical Properties: Opaque but slightly translucent on thin crystal edges. Black, brownish-black; various shades of brown-yellow in earthy or colloform types. Streak brownish-yellow. Silky luster on fibrous fracture surfaces; crystals submetallic to dull.

Chemistry: Hydrogen iron oxide. Sometimes contains manganese. Slowly soluble in oxalic and hydrochloric acids.

Distinctive Features and Tests: Streak distinguishes from colloform hematite. Becomes magnetic after strong heating. In closed tube, yields water and converts to hematite. Fuses with difficulty at 5-5½. Soluble in hydrochloric acid, staining solution yellow.

Occurrences: Very widespread and abundant as an alteration product from other iron minerals, especially sulfides as pyrite and chalcopyrite, or siderite and magnetite. Abundant in masses in surface gravels forming colloform concretions, often with openings lined by very compact material with shining black surfaces (Figure 163). Forms the gossan cappings of sulfide veins, and is very commonly the earthy porous material on which oxidized zone minerals form. Practically all dark brown to yellow-brown massive material, usually called "limonite" is largely composed of goethite. Replaces pyrite to form faithful pseudomorphs as in Figure 164. Fine stalactitic growths with shining black surfaces occur in many places as at Horhausen and in Westerwald, Rhine district, Germany. As fine small crystals forming attractive druses from Pribram, Czechoslovakia. Shining stalactitic material and small sharp crystals to ⅛″, from Restormel Mine, Lostwithiel, and mines near Bottalack and St. Just, in Cornwall, England. Early described in brilliant yellow acicular growths within amethyst from Lake Onega, near the Finno-Russian border, and called *onegite;* lately similar material in large amethyst-citrine-smoky quartz crystals to 6″ diameter, with the vivid orange-yellow tufts of goethite arranged normal to rhombohedral faces and forming sharp phantoms, from Minas Gerais, Brazil. Fine stalactitic material from a number of places in the United States, as concretions in alluvium from Nova Scotia southward along the Atlantic seaboard into Alabama; in the Lake Superior hematite districts, often in large compact masses, attractively colored and banded and suitable for lapidary treatment. Good druses occur in the Jackson and Superior Mines, near Marquette, Marquette County, Michigan. Fine

FIG. 163. *Above:* Goethite. *Left:* botryoidal mass on quartz, from Cornwall, England. Size 4″ by 2¼″. *Right:* bladed crystals forming rosettes on quartz crystal from within a pegmatite cavity. Crystal Peak area, north of Lake George, Teller County, Colorado. Size 3¼″ by 3″.

FIG. 164. *Right:* Goethite replacing striated cubes of pyrite from Pelican Point, Utah Lake, Utah County, Utah. About 3″ diameter. *Photo courtesy Schortmann's Minerals, Easthampton, Massachusetts.*

large bladed crystals to 2″ length, forming attractive tufts and rosettes on quartz and feldspar from cavities in small pegmatites in the Crystal Peak area, Teller County, Colorado (Figure 163); crystal-covered masses to 12″ across have been recovered. Goethite pseudomorphs after pyrite occur in many places; a noteworthy deposit at Pelican Point, on the west shore of Utah Lake, Utah, furnishes very sharp examples of cubes to 3½″ on edge. Goethite also replaces siderite, barite and calcite. Distinct crystals of goethite are desired for collections and can be attractive though black in color. Fine lustrous stalactite growths are also desirable and easier to obtain than crystals; the best display long stalactites with rounded, undamaged tips.

Spinel Group

The spinels are primarily oxides of aluminum with magnesium, iron, zinc, and manganese or chromium. The various species fall into three series: *Spinel Series,* of which spinel, $MgAl_2O_4$, and gahnite, $ZnAl_2O_4$, are best known to collectors; the *Magnetite Series,* of which magnetite, Fe_2O_3, and Franklinite, $(Zn,Mn,Fe)(Fe,Mn)_2O_4$, are best known; and the *Chromite Series,* of which only chromite, $(Mg,Fe)Cr_2O_4$, is commonly represented in collections. All form similar crystals but properties vary considerably. Only spinel, magnetite, franklinite, and chromite will be treated here.

SPINEL SERIES

Spinel—MgAl₂O₄

Name: Of uncertain origin, possibly derived from Latin, *spina,* for "thorn," in allusion to sharply-pointed crystals; spĭ nĕl′.

Varieties: The Spinel Series consists of the ideal end members spinel, a magnesium-aluminum oxide, hercynite (hĕr′ sĭ nīt), an iron-aluminum oxide, gahnite (gä′ nīt), zinc-aluminum oxide, and galaxite (gă lăks′ĭt) the manganese-aluminum oxide. Substitutions involving the metals named, and also chromium, take place extensively and numerous varieties have been named on chemical grounds but lack of space prevents full discussion.

Crystals: Isometric; common. Simple octahedrons predominate but sometimes are modified by dodecahedron faces (Figure 165). Crystals range from sharp, glassy, textbook-perfect individuals to those which are rounded, as if partly melted on edges, or etched or skeletal. Twinning common on the octahedral plane, $o[111]$ (Spinel Law), twins often flattened into wafers of hexagonal outline (Figure 165). Clear crystals commonly contain veins of small octahedral inclusions. Also massive, granular, or in rounded grains.

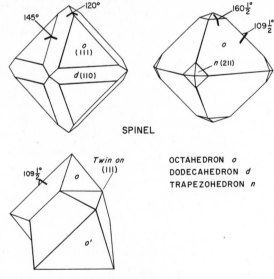

FIG. 165.

Physical Properties: Brittle but tough; uneven to conchoidal fracture, the latter often perfectly smooth. No cleavage but larger crystals commonly contain networks of fairly flat fractures aligned with octahedral planes. H. 8. G. 3.58-3.98; usually 3.58-3.61 in gem material.

Optical Properties: Transparent to opaque. Wide color range: black, pink, red, orange-red, dark green, blue, purple, purplish-gray, yellow, brown. Colorless specimens unknown. Gahnite is usually dark blue-green; hercynite and galaxite black. Streak usually color-less; also pale gray to green or brown. Vitreous luster; silky in some translucent crystals due to oriented acicular inclusions, sometimes capable of producing four-ray star gems when cut. R.I., $n = 1.712$-1.80; hercynites $n = 177$-1.80; gahnites $n = 1.715$-1.746 to 1.792; galaxite $n = 1.923$.

Chemistry: Oxides of aluminum and magnesium, iron, zinc or manganese.

Distinctive Features and Tests: Isolated octahedral crystals; hardness. Infusible. Insoluble.

Occurrences: Spinel forms at high temperatures in various environments, the finest crystals coming from metamorphosed impure limestones, particularly those affected by contact metasomatism. Also as an accessory mineral in some igneous rocks, and in granite pegmatite (gahnite). Crystals of 1″ to 2″ diameter, some fine and sharp, have been found in salmon-colored calcite veins in the Grenville marbles of Ross Township, Renfrew County, and South Burgess, Leeds County, Ontario; also fine at Wakefield, Ottawa County, Quebec. Similar Grenville rocks yield spinels at several localities in St. Lawrence County, New York, while the narrow belt of Franklin marble, extending southward from Orange County into Sussex County, New Jersey, provided what are undoubtedly the world's largest single crystals of this species. An early nineteenth century account talks of an imperfect crystal found ½ mile south of Amity in Orange County, which weighed 29 pounds! Numerous simple octahedrons were recovered from many points in Orange County where low rounded hills of marble rise above marshland. The locality producing the largest spinels is now "lost," and this is not surprising in view of the dense, jungle-like underbrush characteristic of this region, which makes it easily possible to walk within feet of an old open pit and never notice it. Farther south, large sharp octahedrons have been found in the Franklin marble of New Jersey along contacts with the zinc orebodies of Franklin and in small lenses in impure marbles at Ball's Hill west of Ogdensburg. The Sterling Hill Mine in the latter town produced a clean spinel octahedron measuring 5″ on edge. Elsewhere in the United States, dark blue-green gahnite crystals, occasionally to ½″, occur with muscovite mica in granitic pegmatites in New England and North Carolina. A famous locality at Rowe, Franklin County, Massachusetts, provides dark blue-green spinel to ½″ diameter within a granular quartz-pyrite lens in schist; the crystals are rude octahedrons with each face raised by triangular growth plateaus. In Europe, excellent crystals to about 1″ come from Aker, Helsingland, Sweden. From Slatoust in the Urals of the U.S.S.R. Small sharp crystals in ejected marble blacks at Monte Somma, Vesuvius, Naples, Italy. Splendid sharp crystals, of perfect octahedral form, occur abundantly in Ceylon and Burma gem gravels; the finest specimens are usually the clearest and are cut into gems prior to leaving their homeland. Occasionally, large non-gem crystals to 2″ diameter, some red, red-brown, and blue-gray, reach the mineral specimen market. Burmese spinels are commonly slightly rounded on edges or depressed on faces, as if the crystal had been partly fused in a flame. Splendid black modified crystals to 1″ occur at Antanimora and Ambatomainty, Madagascar, associated with diopside, schorl, phlogopite and tremolite.

From the collector's viewpoint, sharp spinel crystals are desirable for single crystal collections but are not easy to obtain, while matrix specimens are even more difficult to acquire.

Magnetite—FeFe$_2$O$_4$

Name: From being attracted to a magnet; màg′nĕ tīt.

Varieties: The magnetite series consists ideally of magnesioferrite (màg nē′sē ō fĕr′ĭt), MgFe$_2$O$_4$, magnetite, FeFe$_2$O$_4$, franklinite, ZnFe$_2$O$_4$, jacobsite, MnFe$_2$O$_4$, and trevorite (trĕ′vĕr ĭt), NiFe$_2$O$_4$, Of these, only magnetite is abundant, while the enormous quantity of franklinite emanating from the unique zinc deposits of Franklin, New Jersey, has made examples of that species nearly as commonly represented in collections as its more abundant relative magnetite. As in the spinel series, where aluminum and oxygen are present in all formulas, iron and oxygen are represented in the magnetite series formulas. Substitutions are common and several varieties of magnetite occur.

Crystals: Isometric. Good crystals uncommon. Usually in octahedrons, from simple to complexly modified; also dodecahedrons (Figure 166). Magnesioferrite, jacobsite and trevorite are rarely found in crystals. Twinning common on *o*[111], (Spinel Law). Often massive, granular.

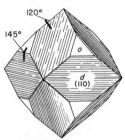

MAGNETITE
STRIATED DODECAHEDRON *d*
OCTAHEDRON *o/III/*

FIG. 166.

Physical Properties: Brittle; fracture subconchoidal to uneven. Good parting on *o*[111], sometimes producing cleavage-like surfaces. H. 6. G. 5.18.

Optical Properties: Opaque. Iron-black, gray-black. Streak black. Luster splendent submetallic to dull.

Chemisty: Ferrous-ferric iron oxide. Magnesium, zinc, manganese, and rarely, nickel, substitute for ferrous iron (Fe^{+2}); aluminum, chromium, manganic manganese and vanadium substitute for ferric iron (Fe^{+3}).

Distinctive Features and Tests: Attraction to magnet, not approached in strength by any other mineral. Naturally magnetized pieces (*lodestones*) attract bits of iron or steel. Infusible. Dissolves very slowly in hot hydrochloric acid.

Occurrences: Abundant and widespread, forming under high temperature conditions as segregations from molten igneous rocks; in marbles, especially in contact metasomatic deposits; in replacement bodies; in sulfide veins, etc. In large coarse crystals from Faraday, Hastings County, Ontario; a single crystal was recorded of over 400 pounds weight! Small, beautifully crystallized, in the Triassic trap of Kings and Annapolis counties, Nova Scotia. Excellent 1″ octahedral crystals occur in the iron deposits of Mineville and Moriah, Essex County, New York; also fine lustrous sharp ½″ dodecahedral crystals on matrix, enclosed in calcite, in the Tilly Foster Iron Mine, Brewster, Putnam County. Sharp single octahedrons to ½″ occur enclosed in dark green prochlorite schist with pyrite cubes near Chester, Windsor County, Vermont. Brilliant sharp octahedral crystals on matrix from French Creek mines, Chester County, Pennsylvania. The Twin Peaks area of Millard County, and also Iron County, Utah, have produced large rude octahedrons of magnetite to 3½″ on edge, but much is converted to hematite (*martite*) as in Figure 148. Sharp dodecahedral crystals of magnetite to ½″ in calcite with tremolite and serpentine on Asbestos Peak, near Globe, Gila County, Arizona. Sharp crystals also from many European localities, as at Nordmark, Vermland and Gammalkroppa, Sweden. In rare cubic crystals in serpentine at Gulsen, Styria, and sharp octahedrons in schist in Pfischtal and Zillertal, Austria. Exceptional crystals from the Traversella region, Piedmont, Italy, etc.

Franklinite—(Zn,Mn,Fe)(Fe,Mn)$_2$O$_4$

Basically, a zinc-iron oxide; frànk′lĭn ĭt. Abundant only in Franklin and Ogdensburg, (Sterling Hill) deposits of Sussex County, New Jersey, as crude, nearly round, black blebs, to sharp octahedral crystals enclosed in white calcite and commonly associated with willemite, zincite, garnet, etc. (Figure 167). The record crystal is a simple octahe-

FIG. 167. Franklinite octahedrons in calcite matrix from Franklin, Sussex County, New Jersey. Size 3″ by 2″.

dron, 7″ on edge, now in the Canfield Collection, Smithsonian Institution, Washington, D.C. Good specimens, worthy of private collections, show rude to sharp octahedrons, from ½″ to 2″ across, perched on white calcite matrix, the latter chipped away to expose the crystals. In contrast to magnetite, franklinite is weakly magnetic and because of this property, and its characteristic associates, it is readily distinguished.

Chromite—(Mg,Fe)Cr$_2$O$_4$

A magnesium-iron, chromium oxide; abundant and widespread in granular masses but remarkably rare in crystals; krōm′ĭt. Practically all chromite deposits are irregular lens-like masses in peridotites and serpentinites, and only rarely are single octahedral crystals found of a size worthy of collections. Recently, sharp octahedrons from ¼″ to ⅝″, priced astonishingly high, have been offered from a source in West Africa. Chromite is weakly magnetic, and produces a decided brown streak which serves to distinguish it from other species in the spinel group.

Chrysoberyl—BeAl$_2$O$_4$

Name: From the Greek, *chrysos* and *beryllos,* in allusion to golden yellow color and beryl, which mineral it was once thought to be a variety of; krĭs′ō bĕr′ĭl.
Varieties: On the basis of color: *alexandrite,* showing a distinctive color change from green to bright red, when observed in daylight and tungsten light respectively. On the basis of inclusions: *catseye,* showing a broad to narrow brilliant bluish to white silky line when cut in cabochon form.

Crystals: Orthorhombic; usually in crystals. Commonly tabular on $c[001]$, which face often striated parallel to a axis. Twinning on plane [130], forming pseudo-hexagonal sixlings which are more common than untwinned crystals (Figure 168). Some twins, as from Brazil, are much extended along the c axis, then forming groups of tabular individuals radiating outward, the twinned cluster approximately ball-like (Figure 169). Faces often brilliantly smooth, but also striated or rough. Also granular massive.

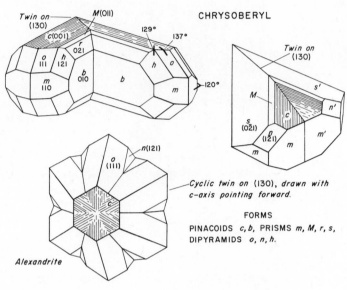

CHRYSOBERYL

Cyclic twin on (130), drawn with c-axis pointing forward.

FORMS
PINACOIDS c, b, PRISMS m, M, r, s,
DIPYRAMIDS o, n, h.

FIG. 168.

Physical Properties: Brittle, tough; fracture conchoidal to uneven. Cleavage sometimes evident on $m[110]$ and $b[010]$. H. $8\frac{1}{2}$, a valuable distinctive property. G. 3.71-3.72 in gem material, to 3.75 in ordinary material.

Optical Properties: Transparent to translucent. Inclusions causing bluish opalescence and chatoyancy are oriented parallel to the c axis, with strongest effects on $b[010]$ faces; in ordinary flattened crystals, the opalescence usually appears on a narrow edge. Commonly yellow, brownish-yellow, greenish-yellow, green, brown. *Alexandrite* from the Urals is blue-green in daylight, bright purplish-red under tungsten light; similar material from Ceylon, is olive-green to brown in daylight, while the red color under artificial light is also less vivid. Biaxial positive. R.I., variable, generally in the range: $\alpha = 1.75 - \gamma = 1.76$. Birefringence 0.008-0.010. Trichroism strong in *alexandrite*: red, orange-yellow and green; weak in ordinary material.

Chemistry: Beryllium aluminum oxide. Slight amounts of iron and chromium have been detected, the latter assumed to be responsible for the green-red color shift in *alexandrite*.

Distinctive Features and Tests: Superior hardness; crystals; chatoyancy. Infusible. Insoluble.

Occurrences: In granite pegmatites; in mica schists; rarely, in dolomite marble; in alluvium. Small flattened twins, some shaped like spear points (Figure 168), and occasionally with bluish chatoyancy, occur in granitic pegmatites in the New England states, and at Greenfield, near Saratoga, Saratoga County, New York. Large crude crystals to $5\frac{1}{2}''$ were found in a pegmatite near Golden, Jefferson County, Colorado. Splendid rich yellow to greenish-yellow sixling twins, largely transparent, occur near Collatina, Espirito Santo, Brazil; most are not over ¾″ diameter, but a considerable number reach as much as 3″. The finest *alexandrites*, invariably as sixlings of tabular form (Figure 169), and ranging from ½″ to 3″ diameter, occur imbedded in biotite schist with phenakite and emerald on Takowaya River, northeast of Sverdlovsk, Urals, U.S.S.R. Small Uralian crystals are usually sharp but larger specimens cruder with faces seldom as smooth and lustrous; twins are sometimes preserved in matrix, affording exceptionally

FIG. 169. Chrysoberyl. *Top left:* yellow twinned crystals from Collatina, Rio Doce, Espirito Santo, Brazil; 1⅜″ by 1″. *Top right:* alexandrite sixling twin from Krasnobolskaia Mine, Takowaya River, Ural Mountains, U.S.S.R. Size 1¾″ by 1½″. *Bottom left:* a smaller alexandrite sixling from the same locality. *Bottom right:* pale yellow transparent sixling twin striated upon the c(0001) faces, from Lac Alaotra, Madagascar.

valuable specimens eagerly sought for by every collector. Fine gem specimens, but seldom showing faces, occur abundantly in Ceylon gem gravels, and occasionally in those of Mogok, Burma. Ceylon *alexandrites* are sometimes fine but never as vividly colored as Urals specimens; fine catseyes come mainly from Ceylon but also from Brazil. Excellent flat crystals (Figure 169), came lately from Lake Alaotra, Madagascar; gravels in various places on this island also provide gem material; also, large tabular single crystals to 5″ and much smaller vee-shaped twins, from Miakanjovato. Lately, very dark *alexandrite* crystals to ¾″ from Southern Rhodesia.

Most mineralogical specimens offered are from Brazil, none being cheap except in very small sizes. Very rarely, Uralian *alexandrite* is offered but specimens appearing on the market were mined during pre-Soviet eras. Occasionally, good vee-shaped twins in white feldspar matrix are offered from Maine localities.

Perovskite—CaTiO$_3$

Calcium titanium oxide; pĕ rŏf′skĭt. A rare mineral, occurring in black, simple cubic crystals, rarely exceeding $3/8''$ diameter. Small black crystals occur in marble in the Oka District, Montreal, Quebec. Octahedrons and cubo-octahedrons to $1/2''$ of the *columbian* variety, *dysanalyte,* occur in some abundance with monticellite, thomsonite and calcite at Magnet Cove, Garland County, Arkansas. Fine sharp gray-black cubes to $3/8''$ sprinkled on gray altered serpentine in narrow crevices, occur within a mile of the benitoite locality, San Benito County, California. Splendid black to brown cubes to $3/4''$ on edge, occur in chlorite schist near Slatoust, and at Achmatowsk, Urals, U.S.S.R. Similar crystals imbedded in talc schist occur near Findeln glacier, Zermatt, Switzerland.

PYROCHLORE-MICROLITE SERIES

Pyrochlore—NaCaNb$_2$O$_6$F

Microlite—(Na,Ca)$_2$Ta$_2$O$_6$(O,OH,F)

Pyrochlore commonly occurs in nepheline syenite and associated pegmatites, microlite in granitic pegmatites; well-formed crystals of either are decidedly uncommon and seldom represented in collections. Pyrochlore, pī′rō klôr, is an oxide of alkali metals with niobium and fluorine; microlite, mī′krō līt, is similar, but contains tantalum in place of niobium, and additionally, hydroxyl. Because of compositional complexity, and substitutions by other elements other than those named, varieties of pyrochlore have formerly been given a considerable number of names such as *hatchettolite, ellsworthite, koppite, pyrrhite,* etc. Because uranium and thorium occur in substitution, both species yield metamict crystals. Hues are brown to black. Crystals are usually cubes or octahedrons. Uranian pyrochlore (*ellsworthite*) in salmon-colored calcite, Monteagle Township, Hastings County, Ontario. From Mbeya, Tanganyika, in sharp octahedrons to $3/8''$. From near Larvik, Norway, in syenite. The finest microlite crystals, some completely transparent, but usually opaque to translucent greenish-brown, brown or brownish-yellow octahedrons, often greatly modified and then ball-like but with sharp faces, occur in albite in the Rutherford Mines, near Amelia, Amelia County, Virginia; crystals range from $1/8''$ to $3/4''$ diameter; sizes over $1/2''$ are decidedly rare; the largest crystal of record is an octahedron measuring $2 1/2''$ on edge. Numerous small transparent glassy grains of brownish-yellow microlite occur in scaley lepidolite at the Harding Mine, near Dixon, Taos County, New Mexico.

Fergusonite—(Y,Er,Ce,Fe)(Nb,Ta,Ti)O$_4$

A rare earth, iron, and niobium-tantalum-titanium oxide; fĕr′gă sŭn ĭt. Occurs in some pegmatites, usually as crude tapering square prismatic crystals, dark brown and pitchy in luster on fresh fracture surfaces, but pale red-brown opaque on exteriors. Sometimes offered in specimens, but the best are unprepossessing objects. Large masses and rude crystals occur in a number of pegmatites in the Grenville marble of Ontario.

Formerly from Baringer Hill, Llano County, Texas, in 8″ crystals about 1″ thick. From the Rutherford Mines near Amelia, Virginia. From several localities in Madagascar in ¾″ crystals, also from many places in southern Norway, etc.

Stibiotantalite—SbTaO₄

A very rare oxide of antimony and tantalum, occurring in very few localities and seldom offered in specimens; stĭb′yō tǎn′tǎ līt. The finest crystals are thin tubular striated individuals of rectangular profile, commonly transparent to translucent, dark brown to brownish-yellow in color, from pockets in the gem tourmaline pegmatite of the Himalaya and San Diego mines, Mesa Grande, San Diego County, California. Crystals are implanted on quartz, tourmaline or feldspar, but sometimes are recovered loose; sizes range from microscopic to nearly an inch in length. Rarely, clear material has been faceted into small gems.

COLUMBITE-TANTALITE SERIES

Columbite—(Fe,Mn)(Nb,Ta)₂O₆

Tantalite—(Fe,Mn)(Ta,Nb)₂O₆

Name: Columbite from columbium content, but this element is now called niobium; tantalite from tantalum content; kŏ lŭm′ bĭt; tǎn′tǎ līt.

Varieties: Ferrocolumbite, ferrotantalite (ferroan); manganocolumbite and manganotantalite (manganoan).

Crystals: Orthorhombic; common. Predominantly in thick to thin tabular crystals compressed on b[010], and of approximately square to rectangular profile, sometimes appearing rounded when terminations are modified by numerous faces (Figure 170). Also as stubby prisms of nearly square cross-section elongated parallel to the c axis. Crystals commonly rough, with curved faces. Seldom sharp and brilliant. Usually striated parallel to c axis on b[010]. Twinning on plane e[201], sometimes forming twins of arrowhead profile. Also granular massive.

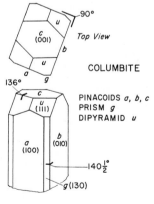

FIG. 170.

FIG. 171. Columbite. Several lustrous black crystals growing in sub-parallel position. The broad faces are b(010). Minas Gerais, Brazil. Size 1½″ by 1″.

Physical Properties: Brittle to very brittle; because of numerous internal fractures, whole crystals are removed with much difficulty from enclosing feldspar. Fracture uneven to subconchoidal; sometimes conchoidal in *manganoan tantalite*. Cleavage distinct on $b[010]$, sometimes perfect in *manganoan tantalite* (Amelia); on $a[100]$ poor. In general, the smoothness of fracture and cleavage surfaces improves toward *manganoan tantalite*. H. columbite 6, to tantalite $6\frac{1}{2}$. G. 5.2-8.0, increasing from columbite to tantalite.

Optical Properties: Columbite opaque; dead gray-black, black, to shining black. Tantalite similar but usually transparent deep red on splinter edges, the color sometimes being obvious through partly developed cleavages on $b[010]$. Streak black to dark red. Luster dull to splendent adamantine, sometimes submetallic. Commonly iridescent.

Chemistry: Iron-manganese, niobium-tantalum oxides in an unbroken series. Sometimes tin substitutes for iron-manganese, or tungsten for niobium-tantalum.

Distinctive Features and Tests: Tabular black crystals associated with typical granitic pegmatite species as feldspar, mica, garnet, etc. High specific gravities, weak magnetism, lack of radioactivity, distinguish them from similar-appearing heavy black minerals. Infusible. Insoluble.

Occurrences: Confined to granitic pegmatites as single crystals or clusters, usually growing away from the walls toward the center of the pegmatite body. Commonly associated with albite, perthite, muscovite, beryl, schorl, apatite; less commonly with fergusonite, gadolinite, microlite, zircon. In some pegmatites (Amelia), columbite forms near walls, but tantalite forms nearer the cores. Crystals of columbite occur in almost every granitic pegmatite district, but not in every pegmatite. In small crystals, seldom over an inch, in pegmatites of Renfrew County, Ontario; also in the New England and Appalachian pegmatite districts. Fine manganotantalite crystals to 3″, as sub-parallel aggregates of thin blades, occur in the Rutherford mines, near Amelia, Virginia. From many localities in North Carolina. In the Bridger Mountains of Wyoming; in Colorado; in San Diego County, California, etc. Also from abroad, from gem-bearing pegmatites in Minas Gerais, Brazil (Figure 171), in crystals to 6″ by 4″ by 2″; from Madagascar, Mozambique, etc.

The desirability of crystals increases with sharpness and luster; undamaged crystals are difficult to obtain. Matrix specimens are very rare. Crystals range from $\frac{1}{2}$″ to as much as 6″ and even larger.

Euxenite–(Y,Ca,Ce,U,Th)(Nb,Ta,Ti)$_2$O$_6$

A rare, complex oxide, usually radioactive, forming crude crystals seldom over 2″ in length, and ordinarily coated with pale yellow, earthy, alteration products. From granitic pegmatites; ŭk′sĕn īt. Crude crystals but sometimes showing distinct faces, are offered from localities in Brazil and Madagascar. Expensive.

Samarskite–(Y,Er,Ce,U,Ca,Fe,Pb,Th)(Nb,Ta,Ti,Sn)$_2$O$_6$

A black radioactive mineral, occurring exclusively in granitic pegmatites; sǎ mär′ skĭt. Sharp crystals very rare; usually in rude rectangular crystals solidly enclosed in feldspar and very difficult to remove without breaking. Good crystals have come from southern Norway pegmatite regions. Commonly as pitchy masses in the pegmatites of North Carolina. Fair crystals to $1\frac{1}{2}$″ from Minas Gerais, Brazil.

Betafite—(U,Ca)(Nb,Ta,Ti)$_3$O$_9$·nH$_2$O

A radioactive granitic pegmatite mineral sometimes forming fairly sharp dark brown dull cubo-octahedral crystals; bĕ' tă fīt. Prized by single crystal collectors. Crystals to 3″ diameter occur in Madagascar; some are reported as weighing several pounds. Fine sharp cubo-octahedrons to 2½″ diameter from Silver Crater Mine, Bancroft, Ontario.

HALIDES

The list of halides—that is, compounds in which a halogen element is an essential anion—is a long one, but very few species are abundant. Many occur in oxidized portions of ore veins, generally as small inconspicuous crystals, chiefly of interest to the collector of micromounts. The best known halide is salt or halite; others also well known include fluorite, sylvite, and carnallite.

Halite Group

Halite—NaCl

Name: From the Greek *hals*, "salt"; hā' līt.

Crystals: Isometric. Usually simple cubes (Figure 172), but sometimes modified by octahedrons; rarely as octahedrons. Faces commonly cavernous, developed into "hopper" crystals. Faces fairly smooth but seldom lustrous. Also coarse to fine granular massive.

Physical Properties: Very brittle; fracture conchoidal but rarely observed. Perfect easy cleavage parallel to cube faces of a[100]. H. 2½. G. 2.17.

Optical Properties: Transparent. Colorless; also pink to medium red, pale green, pale yellow, blue or purple. Blue and purple are caused by free sodium ions. Streak colorless. Luster greasy to vitreous. R.I., $n = 1.5446$.

Chemistry: Sodium chloride. Commonly contaminated by impurities as clay, sand, iron oxide, and inclusions of associated species.

Distinctive Features and Tests: Pale lusterless cubic crystals; salty taste. Readily dissolves in water; surfaces become wet and stay wet for a considerable period of time merely through handling with the bare fingers. Fuses at 1½, coloring flame bright yellow.

Occurrences: Widespread in large saline sedimentary deposits. In saline playa deposits as at Searles Lake, California and elsewhere. Recovered from sea water by evaporative processes. The famous and infamous salt mines of Siberia at Iletskaya Zashchita in the Chkalov region are a source of excellent crystallized specimens as well as jokes. In Galicia, Poland, extensive artificial caverns have been created underground at Wieliczka, complete with chapels, halls, stables, etc. Beautiful crystals and magnificent large cleave plates come from this deposit. Fine blue halite from Stassfurt, Germany. Excellent crystallizations occur at many places in the United States.

Halite readily absorbs water from the atmosphere and must be protected in moist climates. The usual method is to spray-coat crystals with clear lacquer, being certain no spots are missed; specimens can also be placed in airtight containers. Crystals range in size from ½″ to 3″; individuals to 10″ on edge are known. Inexpensive.

FIG. 172. Halite crystals of simple cubic form. From saline lake bed deposits of Searles Lake, San Bernardino County, California. Size 3″ by 2″.

Sylvite—KCl

Potassium chloride; sĭl′vīt. Far less common than halite and crystals are seldom included in collections. Properties, color, etc., similar to halite but crystals usually display octahedral faces developed equally with cube faces. Crystals to 2″ diameter from Stassfurt, Germany; also good specimens from the deposits near Carlsbad, New Mexico.

CERARGYRITE-BROMYRITE SERIES

Cerargyrite—AgCl (Chlorine end-member)

Bromyrite—AgBr (Bromine end-member)

Cerargyrite is silver chloride; sē̆ rär′jĭ rīt, bromyrite is silver bromide; brō′mĭ rīt. Rarely represented in collections not only because of natural scarcity but because most specimens are unattractive. Silver halides are light-sensitive and it is necessary to shield specimens from light lest they turn from their fresh color of white to violet-brown or purple. Specimens in this series occur in small masses, and rarely, in cubic crystals, in oxidized portions of silver-bearing sulfide deposits. Good specimens are available primarily from Broken Hill, New South Wales.

Calomel—HgCl

Mercurous chloride; kăl′ō mĕl. A secondary mercury mineral occurring massive and in good small crystals in oxidized portions of ore bodies containing mercury. Colors are pale and unattractive; good specimens are scarce.

Fluorite—CaF₂

Name: From the Latin, *fluere,* to "flow," in allusion to easy melting when used as a flux in the smelting and refining of metals; floo′ô rīt.

Varieties: With small amounts of yttrium and cerium substituting for calcium, *yttrian.* *Antozonite* is an old varietal name applied to dark purple fluorite, structurally damaged by radioactivity, and giving off free fluorine when abraded or crushed.

Crystals: Isometric; very common. Prevalently in simple cubes, sometimes modified (Figure 173); less commonly in octahedrons and rarely, in dodecahedrons. Usually in

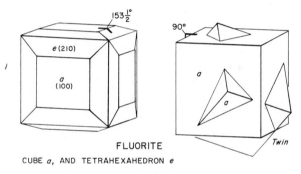

FLUORITE
CUBE *a,* AND TETRAHEXAHEDRON *e*

FIG. 173.

clusters or druses, often as interpenetrant twins on a cube-diagonal twinning axis, as in Figure 173. Faces smooth to rough or etched; octahedrons usually rougher than cubes. Also fine to coarse granular; botryoidal.

Physical Properties: Brittle; fracture subconchoidal to splintery. Easy perfect octahedral cleavage on *o*[111]. Cleavage faces usually mottled in texture because of exposing offsets in growth. H. 4. G. 3.180.

Optical Properties: Transparent to translucent. Colorless, purple, violet, blue, green, yellow, orange, brown, and rarely, red. Commonly varicolored, showing zones of color parallel to cube or octahedron faces. Fluorescent blue, white, pale violet, or reddish, under ultraviolet; in some varieties, thermoluminescent (*chlorophane*). R.I., $n = 1.434$. Transparent material is cut into gems which, however, are not very brilliant due to low refraction; also used as lenses and prisms in various scientific instruments. Small ribbon-like marcasite crystals are frequently included in fluorite, forming distinct phantoms.

Chemistry: Calcium fluoride.

Distinctive Features and Tests: Crystal shape; prominent octahedral cleavage. In massive forms, the greater hardness and inertness in acid distinguish from calcite. Violently decrepitates in flame. Crushed to powder, fuses at 3. Decomposed by hot concentrated sulfuric acid, giving off *extremely dangerous* fumes of hydrofluoric acid.

Occurrences: A very common mineral found in many kinds of mineral deposits. Abundant as masses or crystals in hydrothermal ore veins, often associated with barite, quartz, calcite and dolomite. In geodal cavities in sedimentary rocks, mainly in limestones but also in sandstones, associated with celestite, gypsum, dolomite, calcite. Sparsely in pegmatites. Pale-colored octahedral crystals form at higher temperatures; darker colored cubic crystals at lower temperatures. Excellent specimens, commonly associated with amethyst, occur in Canada in lead-silver veins in the Thunder Bay district. Superb sea-green cubes, often perfectly flawless, sharp, and with exceptionally smooth faces, occur enclosed in barite, or in drusy groups, in Madoc, Marmora and Huntingdon, Hastings County, Ontario; those from Madoc are truly fine, some to 5″ on edge have been reported. Fine green crystals from Derry, Ottawa County, Quebec. Deep purple granular *antozonite* (ȧnt ō′zŭn īt) is abundant in irregular masses and veinlets with salmon calcite in pegmatites and adjacent rocks of the Grenville marbles in the Halliburton-Bancroft area of southern Ontario. Good to fine purple cubes occur at many places in the limestone-sandstone Niagara formation in Ontario. In New Hampshire, lovely bright green material in rude octahedrons and etched cleavage masses, some to 8½″ on edge, often clear and suitable for gems, from mines near Westmoreland, Cheshire County. Large sea-green cubes formerly abundant from the shore of Muscalonge Lake, Jefferson County, New York; old accounts state that cubes over 12″ on edge were recovered. In Ohio, sedimentary limestones yield handsome groups of cubic crystals, distinctive yellow-brown in color, with individuals to 3″ on edge, at Rimer, Putnam County, and Clay Center, Ottawa County; the latter locality is well-known for splendid specimens of celestite as well as fluorite. Fluorite veins in the southern tip of Illinois are systematically mined for steel-making flux and chemical fluorite, and yield magnificent crystal groups to several hundred pounds weight. The most productive deposits are near Rosiclare in Hardin County (Figure 174); also in Pope County, and in adjacent areas in Kentucky, and in central Kentucky in Mercer and Woodford counties. Associated species

FIG. 174. Cube crystals of fluorite from the Ozark-Mahoning Mine, Cave In Rock, Hardin County, Illinois. The color is basically rich yellow but masked by outer zones of purple. Size 4½″ by 3½″.

are drusy quartz, calcite, dolomite, barite, and small amounts of sulfides. Cleavage octahedrons are cobbed from Rosiclare material and are sold everywhere by dealers (Figure 78). Handsome pale green octahedral crystals with pink rhodochrosite occur in the American Tunnel, Sunnyside Mine, near Silverton, San Juan County, Colorado; crystals seldom exceed 1″. Large blue etched cubes to several inches on edge, usually as singles, are found in pegmatite cavities of the Crystal Peak area, north of Florissant, Teller County, Colorado; also octahedrons of pale-green color from the Barstow Mine, Ouray County. The Hansonburg District of Socorro County, New Mexico, has furnished good matrix specimens covered by beautiful blue cubes about ½″ on edge. Fine specimens from the Felix Mine north of Azuza in Los Angeles County, and optical quality crystals from the Floyd Brown Mine near Blythe in Riverside County, California. Good colorless cubes on galena occur in mines near Mapimi, Durango, Mexico.

In Europe, superb examples are furnished by English lead mines. Weardale, Durham, provides fine green and purple transparent interpenetration twins and excellent druses many inches across; cubes to 4″ on edge have been recorded but average specimens display cubes not over ¾″ to 1″ on edge. Alston Moor and Cleator Moor in Cumberland furnish beautiful crystals, some recorded as singles to 9¼″ on edge; most examples consist of purple cubes but those from Weardale are often green; associates are calcite, barite, quartz, galena, sphalerite, and siderite. The Wheal Mary Mine at Menheniot in Cornwall, furnished deep blue cubes to 12″ on edge. Fine specimens also came from Beer Alston near Travistock, Devonshire. Massive, coarse to fine granular deep purple, blue and colorless fluorite, sometimes cut and polished into vases, cups and other ornamental objects, and known popularly as "Blue John," occurs as nodular masses or crusts to 8″ thick in clay at the Blue John Mine, Tray Cliff, near Castleton, Derbyshire. Elsewhere in Europe, fine specimens occur in the mining districts of Saxony, Germany, in the Harz, and in Bavaria; especially popular are the bright yellow cubes from Gersdorf in Saxony. Rose-colored octahedrons to 1½″ on edge occur at various places in the Alps, and command very high prices; these occur as singles and small clusters on quartz crystals, often in interesting arrangements and striking color contrasts. Specimens are not large as a rule and matrixes over several inches are uncommon. Val Sugano, Trentino, Italy, provides fine green cubic crystals.

Excellent fluorites are always obtainable from American deposits, principally from the Illinois-Kentucky area. American specimens can be had in literally any size to several hundred pounds in weight; but the usual groups run from several inches to about 10″ across. Purple is the commonest fluorite hue, followed by green; colorless crystals are rarer. The rose octahedrons from Switzerland are very highly prized. In respect to crystal forms, cubes are most common; cubes modified on edges by the tetrahexahedron are prized more highly than simple cubes; octahedrons are very highly prized, but as remarked before, are seldom smooth-faced.

Laurionite—Pb(OH)Cl

A hydroxyl-chloride of lead; law′rĭ ŭn īt. Occurs in smelting slags from ancient lead-silver mining operations at Laurion, Greece. Laurionite and its associates paralaurionite, penfieldite, fiedlerite, phosgenite, cerussite and anglesite, provide a wealth of material for micromounting.

Atacamite—Cu₂(OH)₃Cl

A hydroxyl-chloride of copper; ă tăk′ă mīt. Another micromounting species occurring as small acicular crystals of vivid dark green color in oxidized zones of copper deposits.

Good specimens occur at San Manuel, Pinal County, Arizona; also from Chuquicamata, Chile, and from Ravensthorpe, West Australia.

Boleite—$Pb_9Cu_8Ag_3(OH)_{16}Cl_{21} \cdot H_2O$

Cumengite—$Pb_4Cu_4(OH)_8Cl_8 \cdot H_2O$

Pseudoboleite—$Pb_5Cu_4(OH)_8Cl_{10} \cdot H_2O$

All three species are excessively rare, complex hydroxyl-chlorides with water, occurring together in the unique copper deposit at Boleo, Baja California Sud, Mexico. Boleite, bō′lē ĭt, is tetragonal but its crystals are nearly perfect cubes reaching ½″ on edge, and composed of three interpenetrant individuals; the color is an unmistakable deep blackish-blue. Cumengite, koo měn′jĭt, is also dark blue, but occurs as octahedral overgrowths on boleite and pseudoboleite and rarely, as small trilling twins not over ¼″ in diameter. Pseudoboleite is also blue, but appears as stepped parallel overgrowths on boleite crystals, forming square short prisms emerging from each of the faces. All three species are highly prized for single crystal-collections. Expensive.

Carnallite—$KMgCl_3 \cdot 6H_2O$

Potassium-magnesium chloride with water; kär′năl līt; mined from large deposits in Germany and in southern New Mexico in Eddy County. Occasionally in good transparent colorless crystals to 1″ by 1½″ from U.S. Potash Company's mine in Eddy County, but seldom included in collections because of its tendency to absorb water in moist atmospheres.

ALUMINO-FLUORIDES

Cryolite—Na_3AlF_6

A rare mineral, sodium aluminum fluoride; krī′ō līt. Found abundantly only in a large pegmatite enclosed in granite on the shore of Arsuk Fjord, Ivigtut, southwestern Greenland; also in minor amounts in deposits in Colorado and near Miask, U.S.S.R. Urals. Usually in white translucent to transparent masses; rarely, in cube-like crystals in parallel position coating massive cryolite. Easily recognized because of extremely low refractive index, about $n = 1.34$, such that small slivers introduced into water disappear from view because the index of the water is so close, i.e., $n = 1.33$. Massive material usually available from dealers, but crystals are rare; the latter occur to ½″ at Ivigtut.

CARBONATES

Although about 70 carbonate species are recognized, practically the whole of carbonate material in the earth's crust is taken up by just three species: calcite, dolomite and siderite. Limestones and marbles are principally calcite and dolomite, while siderite is abundant and widespread in sedimentary deposits where it sometimes forms iron ore beds of economic value. Calcite is noted for the enormous variety in which its crystals and massive forms occur, and needless to say, it is as commonly represented in collections as quartz, and in most instances, by greater variety in respect to crystals and associates. Other carbonates are relatively rare and some, as the hydrated normal carbonates, are water-soluble and therefore seldom found in good specimens, nor are they particularly desirable even when crystallized because of the difficulties in preserving them. Some carbonates containing copper or zinc, as malachite, rosasite, etc., are extremely colorful and especially prized on this account.

Calcite Group

Calcite—CaCO$_3$

Name: From Latin *calx, calcis,* "lime," originally from the Greek *chalx* "burnt lime"; kăl′sīt.

Varieties: Substitutions for calcium provide *manganoan, ferroan, cobaltian, plumbian, barian, strontian,* and *magnesian* varieties. *Satin spar* is a chatoyant, fibrous type.

Crystals: Hexagonal. Two major types of crystals occur: scalenohedral-prismatic, in blunt- to sharp-pointed tapered individuals; and rhombohedral, in which rhombohedral forms are predominant and furnish blocky crystals (Figure 175). Less commonly, in

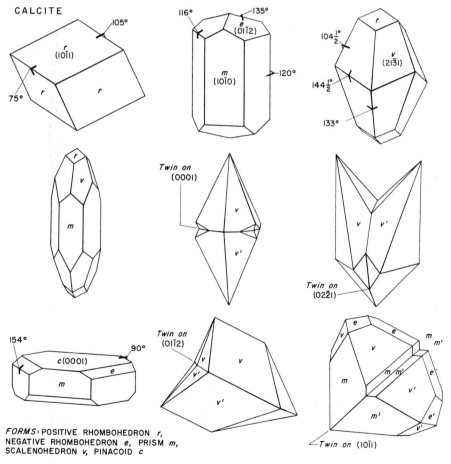

FORMS: POSITIVE RHOMBOHEDRON *r*,
NEGATIVE RHOMBOHEDRON *e*, PRISM *m*,
SCALENOHEDRON *v*, PINACOID *c*

FIG. 175.

tabular to disk-like crystals, compressed along the c axis. Tremendous variety in forms, over 700 being known. Most crystals occur implanted on matrix, thus giving the appearance of a forest of points, more or less at right angles to the plane of the matrix (Figure 176). Doubly-terminated crystals and parallel growths are not common. Quality of faces

FIG. 176. Calcite crystals of scalenohedral habit, essentially colorless but partly coated by vivid red hematite. From Frizington, Cumberland, England. Size 5½″ by 3½″.

varies from brilliantly smooth to rough, but bright crystals are relatively rare because of the easy solubility of calcite in water. Twinning is common according to four laws: twin and composition plane on $c[0001]$, forming reentrant vees around the girdle of crystals, or sometimes a curious aggregate of tabular crystals, resembling a stack of pancakes; twin and composition plane $e[01\bar{1}2]$, forming a pair of individuals diverging from each other at an angle not far from 120°; twin and composition plane, $m[10\bar{1}1]$ commonly forming heart-shaped twins (Figure 177); also on $\phi[02\bar{2}1]$ in which the crystals diverge at an angle about 45° (Figure 178). Cleavage rhombs very commonly striated on surfaces because of repeated twinning on $e[01\bar{1}2]$; when clear examples are held to the light, brilliant flashes of color are seen in certain positions where white light is divided into colored components by differences in refraction along twin boundaries. Often massive fibrous (*satin spar*), fine to coarse granular, and stalactitic ("cave onyx"); also as the cement in sandstone and other porous rocks, sometimes concentrating in spherical concretions or forming rude but distinct crystals enclosing as much as 65% sand by volume ("sand calcites").

Physical Properties: Brittle. Easily and perfectly cleaved on $r[10\bar{1}1]$ (Figure 78), yielding typical sharp-edged, nearly perfect, rhombohedral fragments. Fracture conchoidal but difficult to create. H. about $2\frac{3}{4}$ on $c[0001]$; about $3\frac{1}{4}$ on prism faces in the direction parallel to the c axis. G. 2.71; varying according to composition.

Optical Properties: Transparent to translucent. Usually colorless, or pale to distinct yellow; also brown, pink, violet, blue, greenish, and sometimes strongly colored because of inclusions. Color zoning common. Streak colorless. Luster vitreous to somewhat greasy. Uniaxial negative. R.I., $O = 1.658$, $E = 1.486$. Birefringence large, 0.172, and extremely useful in identifying transparent calcite and other rhombohedral carbonates. Ultraviolet fluorescence marked in some varieties in which manganese is the activator, e.g., brilliant red in calcite from Franklin, New Jersey.

Chemistry: Calcium carbonate. Transparent colorless or faintly colored varieties are quite pure, but substitutions for calcium are common, leading to variations in properties. A complete series exists from calcite to rhodochrosite, and partly toward siderite and smithsonite.

Distinctive Features and Tests: Because of the tremendous and sometimes confusing variety in which it occurs, calcite is most certainly recognized by its perfect rhombohedral cleavage which is always evident on broken places (Figure 78). Judicious use of a steel needle checks hardness and cleavage. Prompt and vigorous bubbling beneath a drop of cold hydrochloric acid. Reaction to acid distinguishes from dolomite; cleavage from aragonite. Infusible; decrepitates violently.

Occurrences: Very common and abundant in all classes of rocks except granitic types and pegmatites; less abundant in diabases, basalts, and related flow rocks, except in cavities and fissures. As a gangue mineral in many kinds of hydrothermal veins. In cavern in limestone formations, or deposited from warm or cold springs. It is virtually impossible to list even the most important localities and only some exceptional occurrences can be mentioned. Fine crystals, to 2″ with fluorite, from Rossie, St. Lawrence County; a cave in limestone one mile east of Sterlingbush in Lewis County provided exceptionally large calcite crystals, one individual measuring $3\frac{1}{2}$ by 3 feet and weighing about 1,000 pounds! Almost six tons of crystals were removed from this cave. Fine scalenohedral or "dogtooth" crystals, colorless and golden, occur in cavities in basalt in quarries in the states of Massachusetts, Connecticut, and New Jersey; especially fine from the latter state in the vicinity of Paterson, where crystals to several inches often occur on bright green prehnite. In splendid colorless to faintly yellow crystals, often clear, enclosing brilliant native copper in mines on Keeweenaw Peninsula, Michigan (Figure 177); undoubtedly these are among the finest specimens of any class, some crystals reaching 4″ length. Throughout the geode region centered about the junction of Iowa-Illinois-Missouri, as rhombohedrons implanted on drusy quartz (Figure 243), sometimes with inclusions of millerite. Fine pale brown curved crystals of low hexagonal prismatic form, terminated by low rhombohedrons and forming nearly spherical crystals, from Jennings County, Indiana (Figure 177). In the lead-zinc regions of southern Wisconsin and the tri-state district of Oklahoma-Kansas-Missouri, in exceptional crystallizations and great abundance, often in sharp yellow scalenohedrons reaching several feet in length (Figure 178). Unusual pale violet cleavages from the Pelican Mine, near Picher, Ottawa County, Oklahoma. "Sand calcites" in singles to 7″ and in large groups to many pounds weight, at Rattlesnake Butte in the badlands of Washington County, South Dakota. White to green-blue stalactites lining cavities in limestone, adjacent to copper deposits

FIG. 177. Calcite specimens. *Top left:* Colorless heart-shaped twin crystals as depicted in the lower right hand drawing of Figure 175. Cumberland, England. Size 3″ by 2″. *Top right:* Large white tabular crystal on matrix from the Mina San Sebastian, Charcas, San Luis Potosi, Mexico. The broad face is c(0001) but is not perfectly flat; the edges of the crystal are bounded by faces of the scalenohedron $v(21\bar{3}1)$ and prism $m(10\bar{1}0)$. Size 3½″ by 2½″. *Bottom left:* Transparent colorless scalenohedrons enclosing bright flakes and platelets of native copper. From Quincy Mine, Houghton County, Michigan. Size 3″ by 2½″. *Bottom right:* pale brown crystal on calcite-impregnated sandstone from North Vernon, Jennings County, Indiana. The c axis is vertical, and the crystal is capped by three faces of rhombohedron $r(10\bar{1}1)$; the sides are highly irregular. Size 3″ by 2″.

FIG. 178. Calcite specimens. *Top left:* "fish-tail" twin of golden calcite from West Side Mine, Treece, Cherokee County, Kansas. The twin is similar to that depicted at the bottom center of Figure 175. *Top right:* scalenohedral crystals from the same mine; size 5″ by 4″. Both specimens have been dipped in acid to make the faces lustrous. *Bottom:* coralloidal calcite from Ground Hog Mine, Vanadium, Grant County, New Mexico. Size 3″ by 3″. Much calcite of this sort is mislabeled "aragonite."

of Bisbee, Cochise County, Arizona; also rarely, in crystals filled with vivid red-orange cuprite (*chalcotrichite*) inclusions, creating spectacular specimens. Optical calcite from large flat crystals, compressed along the *c* axis, from Santa Rosa Mountains, San Diego County, California, and from a number of localities in the states of Durango, Sonora, Chihuahua and Sinaloa, Mexico. Specimens from the La Fe Mine, near Rodeo, Durango, yielded large boxwork clusters composed of thin snow-white plates to 12″ across in openings in calcite veins in recent volcanic rocks. Exceptional tabular crystals (Figure 177) from Mina San Sebastian, Charcas, San Luis Potosi, Mexico, forming groups to several feet across, with crystals to 5″ diameter.

Perhaps the world's most famous source of calcite is the large mass of coarse clear grains, each several feet across, and large single crystals—one measured 20 by 6½ feet!—from within basalt at Helgustadir, on Eskifjord, Iceland ("Iceland spar"). However, specimen crystals from the Iceland deposit are not exceptional. Beautiful specimens, comprising some of the finest cabinet material in the world, come from English sources as follows. In Derbyshire, fine small "butterfly" twins of calcite from near Eyam; Ecton provided calcites enclosing bright chalcopyrite crystals. In Cumberland, the localities near Frizington and Egremont formerly provided splendid, perfectly smooth, colorless hexagonal prisms topped by scalenohedrons with individual crystals to 3″. Egremont also provided fine large colorless butterfly twins, as in Figure 177. The Stank Mine in Lancashire is noted for scalenohedral crystals of calcite in miniature forests of pointed crystals as in Figure 176, often coated with bright red hematite. Low rhombohedral crystals, commonly called "nail head" spar, come from Alston Moor in Cumberland. The original "sand calcite" locality is at Fontainebleau near Paris, France. Fine sharp colorless rhombs of calcite handsomely setting off native silver wires and crystals, occur in silver mines in Kongsberg, Norway. Silver-lead districts in Saxony, Germany, and in Czechoslovakia, also provide fine crystals. Lately, excellent specimens, sometimes in aggregates of curiously warped and twisted colorless or white crystals as in Figure 179, came from oxidized zones in large copper mines at Tsumeb and Grootfontein, South-West Africa; some specimens display perfectly smooth transparent rhombs, exactly like fresh cleavage rhombs.

Calcite occurs in a bewildering variety, with no limit to the number of distinctly different specimens which one may collect and continue to collect—one never seems to run out of possibilities. Indeed, it is possible to assemble a collection of several hundred specimens of English calcites, for example, without fear of repetition. Perhaps the most prized calcite specimens, if one were to rank them, would be the copper-bearing crystals from Michigan, the silver-calcite specimens from Kongsberg, Norway, and, last but not least, the splendid crystals of England.

Magnesite—MgCO$_3$

Magnesium carbonate; măg′ně sīt. Rarely in distinct crystals and seldom occurring in suitable cabinet specimens. The finest specimens are from deposits in the Serra Das Eguas, Bahia, Brazil, near the town of Brumado, formerly Bom Jesus dos Meiras; euhedral crystals to 2″ on edge have been found, many colorless and transparent; massive material has yielded cleavage rhombs to 3″ across and nearly indistinguishable from ordinary calcite. Large gray-white crystals also come from Oberdorf in Austria. Distinguished from calcite by greater hardness, density, and failure to bubble in cold hydrochloric acid.

Siderite—FeCO$_3$

Name: From the Greek *sideros* for "iron," in reference to composition; sĭd′ĕr īt.
Varieties: According to substitutions for iron: *manganoan, magnesian, calcian,* and *cobaltian.*
Crystals: Hexagonal; uncommon in good examples. Usually simple rhombohedrons with curved or warped faces because of aggregation of many offset crystals (Figure 180). Crystals commonly form druses with all crystals about the same height. Faces fairly smooth but seldom brilliant. Also massive, coarse and fine granular, botryoidal.

FIG. 179. Calcite specimen from Tsumeb, South-West Africa. Size 5½″ by 4″. The crystals are simple rhombohedrons $r(10\bar{1}1)$, white in color, and slightly curved. Sulfide and calcite matrix.

Physical Properties: Brittle; uneven to conchoidal fracture but difficult to develop. Perfect rhombohedral cleavage on $r[10\bar{1}1]$. H. 3½-4. G. 3.8-3.9; decreasing when substitution for iron takes place.

Optical Properties: Translucent to subtranslucent; rarely transparent. Predominately some shade of dull greenish or reddish brown. Also grayish, greenish, and colorless. Commonly coated by iridescent film. Streak white; luster vitreous to slightly pearly. Uniaxial negative. R.I., O = 1.873, E = 1.633; indices decrease as substitutions for iron occur. Birefringence large, 0.240.

Chemistry: Iron carbonate. A complete series exists between siderite and rhodochrosite, with manganese substituting for iron; also to magnesite with magnesium substituting for iron. Partial replacement occurs involving calcium or cobalt for iron. Alters to goethite.

Distinctive Features and Tests: Brown color; *rhombohedral* crystals with curved faces. Cleavage and chemical tests distinguish from sphalerite, which it resembles. Slowly soluble in cold hydrochloric acid, but rapidly in hot. Fuses at 4½, forming a black mass.

Occurrences: In large deposits in sedimentary rocks, but good crystallized specimens are obtained mainly from sulfide ore veins where it forms hydrothermally; also in some pegmatites and in metamorphosed sedimentary rocks. Large brown cleavages occur at Mine Hill, Roxbury, Litchfield County, Connecticut, and, rarely, crystals. Light gray to

FIG. 180. Siderite crystals from Erzberg, Styria, Austria. Size 3¼″ by 2½″. The brownish crystals are actually curved aggregates of numerous smaller rhombohedral crystals.

brown crystals are found lining cavities in massive siderite in mines of the Gilman District, Eagle County, Colorado; also as large rhombs to 2″, altered to goethite or hematite, in pegmatite cavities in the Crystal Peak area, Teller County; a rhomb of 5″ on edge has been recorded from the Pikes Peak region. Fine granular dark brown stalactitic siderite coated by iridescent small crystals occurs in the Campbell Shaft, Bisbee, Cochise County, Arizona. Large brown rhombohedral crystals with pyrite from Colavi, Bolivia. Beautiful sharp transparent yellow-brown crystals to 1″, on matrix with quartz crystals, occur in the Morro Velho gold mine, Minas Gerais, Brazil. Fine cleavage rhombs to 12″ from the cryolite pegmatite at Ivigtut, Arsuk Fjord, southwestern Greenland. Beautifully crystallized at many places in Cornwall, as at Camborne, Redruth, St. Austell, etc. Exceptional crystals to 1″ on edge, often in large druses, from Lintorf, Hannover and from Neudorf, the Harz, and mines in Saxony. Exceptional crystals from Erzberg, near Eisenerz, Styria, Austria. Fine crystals at Brosso, Traversella, Piedmont, Italy, and from Allevard, Isère, France, in transparent 1″ crystals.

Rhodochrosite—$MnCo_3$

Name: From the Greek, *rhodo,* for "rose," and *chrosis,* "a coloring," in allusion to the usual rose-red hue; rō′dō krō′sīt.

Varieties: According to substitutions for manganese: *calcian, ferroan, magnesian,* and *zincian.*

Crystals: Hexagonal; rare. Usually as simple rhombohedrons, or, more rarely, as sharp scalenohedrons. Massive types range from fine to coarse granular, often handsomely banded as in botryoidal cavity linings or in stalactitic formations (Figure 181), and then much desired for cutting and polishing into ornamental objects.

Physical Properties: Brittle; fracture uneven to conchoidal but difficult to develop. Perfect rhombohedral cleavage on $r[10\bar{1}1]$. H. $3\frac{1}{2}$-4. G. 3.4-3.6, variable according to composition.

Optical Properties: Translucent to transparent. Usually some shade of red, from pale pink to deep pink, sometimes almost red; also grayish, brownish, greenish. Streak white. Luster vitreous to somewhat pearly on cleavage planes. Uniaxial negative. R.I., usually near O $= 1.816$, E $= 1.597$, but variable according to composition, rising with iron and falling with calcium and magnesium. Birefringence large, 0.219. Distinct to strong dichroism, pale straw yellow, dark pink. Clear material has been faceted into attractive gems.

Chemistry: Manganese carbonate. Complete series extend to siderite and to calcite.

Distinctive Features and Tests: Characteristic pink hue and cleavage distinguish from rhodonite. Dissolves in warm hydrochloric acid with bubbling. Borax bead assumes violet color in oxidizing flame.

Occurrences: Common in hydrothermal silver, lead, zinc and copper ore veins, usually associated with other carbonates and with fluorite, barite, quartz and sulfides. Also in high-temperature contact metasomatic depos-

FIG. 181. Stalactitic rhodochrosite from the famous locality in Catamarca Province, Argentina. Size $5\frac{1}{2}''$ by $3''$. The surface is covered with rude crystal aggregates which form ridges.

its and as a secondary mineral in manganese deposits. Sparingly, as a late species in pegmatites. The finest crystals occur in ore veins in Colorado, forming vivid red simple rhombohedrons sprinkled on vuggy matrix or party imbedded; some crystals are perfectly transparent and furnish faceted gems to 8 carats weight; crystals range in size from about $\frac{1}{2}''$ to as much as $2''$ on edge. Localities yielding especially fine specimens are St. Elmo and Mary Murphy mines, Chaffee County; Moose Mine, Gilpin County; John Reed Mine, Lake County; Sweet Home Mine, formerly a famous source, Park County; and the Eagle and Rawley mines in Saguache County, Colorado. Rhombohedrons of light greenish ferroan rhodochrosite occur in the White Raven Mine, Boulder County; mines in the Silverton District, San Juan County, Colorado, furnish pale pink translucent crystals. Pale pink opaque crystals, sometimes forming druses $6''$ or more across, occur in abundance in the Alice and other mines in Butte, Silver Bow County, Montana. Small sharp scalenohedrons, but seldom over $\frac{1}{4}''$, occur on limonitic matrix in a number of deposits in Europe, particularly at Beiersdorf, Germany, at Kapnik and other places in Romania and in Yugoslavia. Large stalactitic and thick banded crusts occur in galena veins in the Catamarca Province, Argentina, and have been worked for many years for the sake of ornamental rhodochrosite (Figure 181). The best type is deep pink, distinctly banded, and forms complete circular "eyes" in stalactite cross-sections.

Crystals of transparent rhodochrosite from Colorado are difficult to acquire inasmuch as most mines are inactive. Good specimens consist of $\frac{1}{2}''$ to $1''$ rhombs sprinkled on matrix; matrices may reach $6''$ or more across but such specimens are very expensive. Pale pink druses from Butte, Montana, are commonly available and are much cheaper than good Colorado material. Massive banded material from Argentina is usually sliced into slabs and polished.

Smithsonite—ZnCO₃

Name: After James Smithson (1754-1829), founder of the Smithsonian Institution, Washington, D.C.; smĭth'sŭn ĭt.

Varieties: According to substitutions for zinc: *ferroan, cuprian,* and *manganoan;* less commonly, *calcian, cobaltian, cadmian, magnesian,* and *plumbian.*

Crystals: Hexagonal; distinct crystals rare. Usually small sub-parallel aggregates of rhombohedrons, requiring magnification to see well; large rhombohedral crystals only from South-West Africa and Broken Hill, New South Wales. Predominately earthy to fibrous-granular massive, forming cavity fillings or reniform encrustations, stalactites, etc. Minute bright crystal faces impart a twinkling appearance to surfaces of material lining cavities (Figure 182).

Physical Properties: Brittle in single crystals but fibrous aggregates quite tough. Perfect rhombohedral cleavage on $r[10\bar{1}1]$ pronounced in single crystals, but not readily apparent on fracture surfaces of massive material. H. 4-4½. G. 4.0-4.45, varying according to composition and lower in porous massive material.

Optical Properties: Translucent; crystals sometimes transparent. Commonly grayish, colorless, pale to deep yellow, yellow-brown, brown; also greenish to vivid apple-green, blue-green, pale blue; rarely purplish, pink. Massive stalactitic or botryoidal specimens usually display concentric color banding. Streak white. Luster greasy to vitreous in crystals; greasy or oily on fresh fracture surfaces of fibrous material; also earthy dull. Uniaxial negative. R.I., O = 1.848, E = 1.621. Birefringence large, 0.227. Translucent massive material gives vague refractometer readings near $n = 1.74$. Dichroism not apparent.

Chemistry: Zinc carbonate. Only partial substitution takes place between zinc and the other elements previously mentioned.

Distinctive Features and Tests: Specific gravity and hardness higher than other carbonates. Effervescence in hot hydrochloric acid distinguishes from wavellite, prehnite, etc., which form similar massive material. Infusible; in reducing flame, coats charcoal block yellow when hot, turning white when cold.

Occurrences: Predominately a secondary mineral in oxidized zones of sulfide veins, derived from alteration of sphalerite in the presence of limestone; associated with hemimorphite, cerussite, malachite, azurite, anglesite, pyromorphite, etc. Most smithsonite is unattractive, but a few localities furnish fine cabinet specimens or gem material. Intense yellow banded crusts ("turkey fat ore") occur on brecciated limestone in lead mines at Yellville, Marion County, Arkansas; some is capable of being polished into handsome cross-section specimens. Exceptional blue-green material from mines at Kelly, near Magdalena, Socorro County, New Mexico, providing broad botryoidal encrustations to $1''$ to $2''$ thick on limestone breccia and lining pipes and cavities in this rock (Figure 182). Specimens many feet across were encountered in the Kelly workings, but are now scarce and command high prices; much solid material was cut into cabochons and small ornamental objects. Pale greenish-blue or yellow crusts to $1''$ thick, formerly from Cerro Gordo, Inyo County, California. Formerly abundant as splendid thick polishable stalactites and crusts in lead-zinc deposits, Iglesias district, Sardinia, Italy; many fine large stalactites, to $4''$ diameter, were recovered and polished to expose the vivid yellow color

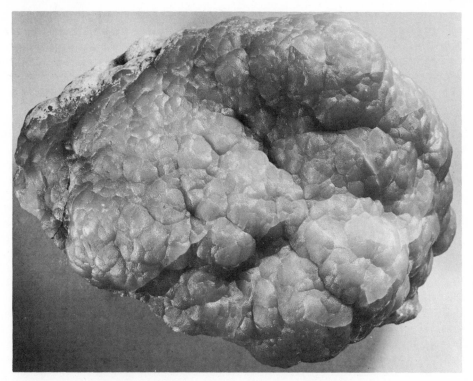

FIG. 182. Large mass of blue-green smithsonite forming a botryoidal cavity lining. From the Kelly Mine, Magdalena, Socorro County, New Mexico. Size about 10″ by 7″. The sparkling reflections from the surface arise from numerous rhombohedral faces. *Photo courtesy Smithsonian Institution.*

and distinct banding. Large quantities of similar material were found in lead-zinc mines at Laurion, southeast of Athens, Greece, as crusts, often very thick, and as stalactites, in almost every known color including fine blue, green, and yellow. The best massive translucent smithsonite ever known came from copper deposits in Tsumeb and Grootfontein, South-West Africa, where crusts to 3″ thick, colored deep vivid green by copper compounds, were found in slabs a foot or better across; also found were colorless and yellow-brown masses, and, rarely, distinct crystals to ¾″ length. The largest crystals known come from Broken Hill, New South Wales, Australia, as acute rhombohedral individuals of pale straw yellow color, largely transparent, and from ¾″ to 1½″ length; the faces are always somewhat curved.

Smithsonite from Kelly, New Mexico, is frequently included in collections because of its vivid color but the vivid yellow-ringed stalactites from Sardinia and Laurion are easily as good. Distinct crystals are rare and much desired.

Dolomite Group

Dolomite—$CaMg(CO_3)_2$

Ankerite—$Ca(Fe,Mg)(CO_3)_2$

Names: After Déodat de Dolomieu (1750-1801), French mineralogist; dŏl′ō mīt. After M. J. Anker (1772-1843), Austrian mineralogist; ăn′kĕr ĭt.

Varieties: Of dolomite, according to substitution for magnesium by iron, manganese, or calcium: *manganoan, ferroan, calcian.* A *manganoan* variety of ankerite is recognized.

Crystals: Hexagonal. Commonly, simple rhombohedrons of $r[10\bar{1}1]$ with curved faces and forming aggregates of many offset crystals. Repeated offsets create exaggerated curved "saddle" crystals which little resemble the basic rhombohedron. Also prismatic in steep rhombohedrons; rarely, tabular on $c[0001]$; scalenohedrons are unknown. Twinning on $c[0001]$. In massive forms, like calcite, but not stalactitic.

Physical Properites: Brittle; fracture subconchoidal. Perfect easy cleavage on $r[10\bar{1}1]$. H. 3½-4. G. 2.8-3.0, increasing with iron.

Optical Properties: Transparent to translucent. Colorless, very pale brown, grayish, greenish; sometimes pink. Brownish alteration stains on crystal surfaces when weathered. Streak white. Luster vitreous to slightly pearly ("pearl spar"). Uniaxial negative. R.I., dolomite, O = 1.680, E = 1.501, rising with increase of iron to ankerite, O = 1.764, E = 1.574. Birefringence large, about 0.179, increasing with iron.

Chemistry: Ideally calcium-magnesium carbonate (dolomite) forming a continuous series by substitution of iron for magnesium, when it becomes calcium-iron-magnesium carbonate (ankerite). Manganese commonly substitutes for magnesium in conjunction with iron; cobalt and zinc also substitute for magnesium but in very small amounts.

Distinctive Features and Tests: Curved "saddle" crystals; pearly luster. Far less vigorous effervescence in cold hydrochloric acid but dissolves readily in warm acid. Infusible, glowing brightly. Ankerite and iron-rich dolomite darken and become magnetic after heating.

Occurrences: Large dolomitic rock masses are found in many places, but euhedral crystals are neither as abundant nor so varied as calcite. Dolomite also occurs in hydrothermal veins, associated with quartz, calcite, barite, fluorite, etc.; in cavities in limestones or dolomite rocks, in sedimentary rock cavities, in veins in serpentines, and in altered igneous rocks containing considerable magnesium. Small pink crystals occur in cavities in Niagara limestone in the Ontario peninsula between lakes Erie, Ontario, and Huron, and in the same formation in adjacent areas of New York and Ohio, with gypsum, celestite, calcite and fluorite. Fine crystals have been obtained from cavities in pegmatites with rutile and muscovite in the vicinity of Stony Point, Alexander County, North Carolina; a rhombohedron of 4″ on edge has been recorded from this county. Abundant and fine, but in small pink "saddle crystals, seldom over ⅜″, in cavities in limestone in the lead-zinc deposits of the Mississippi Valley, commonly with small brassy chalcopyrite crystals; similarly in the Marquette area of Michigan. Simple rhombohedral ankerite crystals, pale brown in color and to 1½″ on edge, occur with galena, pyrite and sphalerite in the Empire Zinc Mine, Gilman, Eagle County, and in the Weston Pass district, Park and Lake counties, Colorado. Small sharp crystals of dolomite and ankerite occur in vugs in quartz in a number of gold mines of the Mother Lode country of California. Formerly in abundance from the silver veins of Guanajuato, Mexico. Excellent crystallized specimens from Morro Velho gold mine, Minas Gerais, and large clear crystals from near Brumado, Bahia, Brazil. In Europe, outstanding simple colorless rhombohedrons, commonly to 2″ on edge, occur at Traversella and Brosso, Piedmont, also in the Pfischtal, Trentino, Italy. From numerous localities in Switzerland, as in Lengenbach quarry in the Binnatal, in perfectly clear, beautifully developed complex crystals from ½″ to 1¼″, in openings in sugary dolomite; similar crystals to ½″ at Campolungo. Curved white dolomite crystals forming druses similar to the Mississippi Valley specimens, occur at Raibl, Carinthia, Austria. Choice crystallized specimens, with hematite and quartz, on kidney ore, from Cumberland, England; also at Frizington, and less commonly, in ore veins of Cornwall. Small crystals, not over $1/4''$ occur in druses with chalcopyrite and quartz on Isle of Man, England. Small black rhombohedral crystals to $3/4''$, discolored by carbonaceous inclusions (*teruelite*), occur in the Barranca del Salobral, Teruel, Spain.

Aragonite Group

Aragonite—CaCO₃

Name: From the locality in Aragon province, Spain; ă răg'ŏ nīt.

Varieties: Containing lead, *plumbian* (*tarnowitzite*); also *strontian* and *zincian*. *Flos-ferri*, literally, "flowers of iron," is applied to coralloidal growths of snow-white color, fantastically intertwined in slender curved branches.

Crystals: Orthorhombic, uncommon. Short to long prismatic along the *c*-axis, in six-sided prisms nearly hexagonal in cross-section and terminated by a pair of faces making a wedge; rarely, in pseudo-hexagonal bipyramids drawn out to very slender points (Figure 183). Practically all crystals are twinned along the plane *m*[110], forming sixling prisms, nearly hexagonal in cross-section (Figures 183, 184). Commonly striated or rough. Also

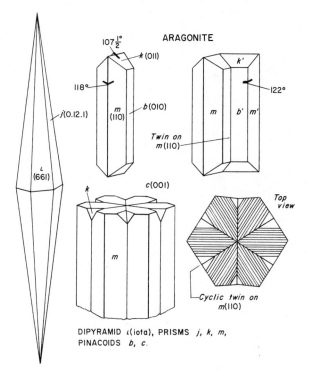

DIPYRAMID *ι*(iota), PRISMS *j, k, m,*
PINACOIDS *b, c.*

FIG. 183.

columnar, fibrous or coarsely crystalline to microcrystalline, like calcite "cave onyx," but much so called stalactitic aragonite is really calcite. Also coralloidal as in *flos-ferri*.

Physical Properties: Brittle to fairly tough in fine granular or fibrous masses. Uneven to subconchoidal fracture. Good cleavage, easily developed, parallel to prism faces

m[110], and certainly distinguishing this species from rhombohedral carbonates as calcite. H. 4. G. 2.95-3.0, increasing in plumbian material.

Optical Properties: Transparent to translucent. Commonly white to very pale yellow; also grayish, very pale blue, green, pink, etc. Streak white. Luster vitreous to greasy; dull to somewhat silky on massive material. Biaxial negative. R.I., $\alpha = 1.530$, $\beta = 1.681$, $\gamma = 1.685$. Birefringence 0.155. Clear crystals have been cut as faceted gems. Fluorescent under ultraviolet: pale red, yellow, rarely, blue.

Chemistry: Calcium carbonate. Lead, strontium, and possibly, zinc, substitute for calcium in small amounts.

FIG. 184. Aragonite sixling twin from Herrengrund, Hungary. Size 2¼″ by 1¾″.

Distinctive Features and Tests: Good prismatic cleavage distinguishes from calcite, dolomite, etc. Faces of c[100] display radial striations forming vee-shaped sectors, each corresponding to one of the sixling twins. Infusible. Dissolves easily with effervescence in dilute hydrochloric acid.

Occurrences: A low-temperature mineral, usually forming near the surface in many types of deposits. Good crystals come from numerous localities abroad but fine specimens are uncommon from North American localities. Fine crystals, with calcite, in mines of the Magdalena district, Socorro County, New Mexico. Large twinned crystals replaced by calcite in clay beds at Las Animas, Bent County, Colorado (Figure 34). In delicate sprays of acicular pyramidal crystals from Frizington, Cleator Moor, and other places in Cumberland, England, and in the Leadhills of Scotland. Abundant and fine *flos-ferri* from Carinthia, Austria, particularly from Erzberg, near Eisenerz, and at Hüttenberg-Lölling; also in fine crystals at Werfen and Leogang, Salzburg, Austria. Good crystals of the plumbian variety (*tarnowitzite*) from Tarnowitz, Silesia, Poland. Splendid transparent yellow crystals to 3″ length, suitable for gems, from veins traversing basalt on Spitzberg, near Horschenz (Bilin), Czechoslovakia. Large fine white to colorless or pale yellow transparent sixlings from Dognacska and Herrengrund, Hungary (Figure 184). Handsome groupings of large white or greenish-white flat pseudo-hexagonal twins, with celestite and sulfur, from Racalmuto, Cianciana and Agrigento, Sicily; a record crystal from Cianciana is 3¼″ in diameter; Agrigento, one of 2″ by 2″. The classical locality for sixling twins is Molina de Aragon, Guadalajara, Spain, where many thousands of crystals occur in reddish gypsum-rich clay; similar crystals from Bastennes and elsewhere in the Landes department, France.

Sixling twins from Spain are readily available and cost little for fine examples. Clear yellow crystals from Horschenz or the large white sixlings from Herrengrund are rarely obtainable and are expensive. Flos-ferri specimens are also commonly available and make interesting specimens in any collection; the average size is about 4″ across, while the best specimens show branches to 3″ to 4″ length. Sicilian specimens are excellent and showy, and usually available.

Witherite—BaCO₃

Name: After English mineralogist W. Withering (1741-1799); wĭth′ẽr ĭt.

Crystals: Orthorhombic. Always twinned on $m[110]$, forming pseudohexagonal dipyramids, or commonly, many such dipyramids grown together in near-parallel position (Figure 185), then forming stubby pseudo-hexagonal prisms with stepped sides and curved

FIG. 185. Witherite aggregate forming a pseudohexagonal crystal. From Cave In Rock, Hardin County, Illinois. Size 3″ by 2″.

tops. Crystal faces usually dull because of numerous striations parallel to the horizontal edges of the dipyramids. Also massive in crusts.

Physical Properties: Brittle; fracture uneven. Fair cleavage on $b[010]$. H. 3-3½. G. 4.29.

Optical Properties: Always translucent. Grayish, slightly yellow; sometimes almost white. Streak white. Luster weakly vitreous; greasy on fracture surfaces. Biaxial negative. Refractometer reading indistinct, about $n = 1.60$.

Chemistry: Barium carbonate. Small amounts of strontium or calcium substitute for barium.

Distinctive Features and Tests: Pseudohexagonal dull grayish dipyramids and low hexagonal aggregates; high specific gravity; easy solubility in cold dilute hydrochloric acid with effervescence. Fuses at 3, coloring the flame greenish-yellow.

Occurrences: A relatively rare low-temperature mineral in hydrothermal sulfide veins, usually associated with barite and galena, but also with fluorite at Rosiclare, Illinois. The finest pseudohexagonal dipyramids occur in lead mines of England, as at Fallowfield, Hexham, in Northumberland, where individuals to 3″ diameter have been found as well as smaller crystals forming druses or stepped growths resembling a series of cones stacked one on top of the other; also fine from Alston Moor, Cumberland, and in the Settlingstones Mine, near Fourstones, Northumberland, and in the Morrison Mine, Durham. The only worthy locality in the United States provides splendid large aggregates, as shown in Figure 185, in sizes to 1″ to 3″ diameter, scattered on grayish shale in the fluorite mines near Rosiclare, Hardin County, Illinois. Illinois specimens are usually available at reasonable prices from dealers, but English specimens are no longer abundant.

Strontianite—SrCO₃

Name: After the locality at Strontian, Argyll, Scotland; strŏn′ shĭ ăn ĭt.
Crystals: Orthorhombic; rare. Usually long sharply-pointed individuals, like aragonite, or bladed and elongated along the c axis; also twinned in sixlings like aragonite but less tabular. Commonly bladed to fibrous massive.
Physical Properties: Brittle; fracture uneven to subconchoidal. Excellent cleavage on $m[110]$, clearly apparent on bladed massive material and on broken crystals. H. 3½. G. 3.72-3.76.
Optical Properties: Transparent to translucent. Colorless, very pale gray, yellowish, greenish. Streak colorless. Luster vitreous to somewhat greasy on fracture surfaces. Biaxial negative. R.I., $\alpha = 1.52$, $\beta = 1.66$, $\gamma = 1.67$. Birefringence 0.15.
Chemistry: Strontium carbonate. Usually contains some calcium substituting for strontium.
Distinctive Features and Tests: Infusible, but strongly colors the flame red (strontium). Soluble in cold dilute hydrochloric acid.
Occurrences: A low-temperature mineral found with barite, celestite and calcite in limestones, or in near-surface portions of lead ore veins; also in geodes in sedimentary rocks. Good crystals are uncommon, and suitable cabinet specimens difficult to obtain. Fine sub-parallel bladed crystals to 2″ with clear terminations, occur at Drensteinfurt, Ascheberg, and Ahlen in Westphalia, Germany; also fine from Leogang near Salzburg, and Brixlegg in the Tyrol of Austria. Seldom available, and then in small crystals, from the type locality in Scotland. Small, sharp, pseudohexagonal crystals occur on massive dolomite at Woodville, Sandusky County, Ohio.

Cerussite—PbCO₃

Name: From the Latin *cerussa*, for *"ceruse,"* a white lead pigment; sĕr′oo sĭt.
Crystals: Orthorhombic; common. Greatly varied in habit; usually twinned. Single crystals mostly elongated along the a axis, and approximately square in cross-section (Figure 186); also tabular, compressed along the b axis. Stubby to long-prismatic. Twinning on $m[110]$, forming sixlings, sometimes almost perfectly developed (Figures 187, 188), in flower-like or star-like groups; also in reticulated networks reminiscent of snowflakes. In sixling twins, the c axis is the "hub," the a axes form "spokes." Silky inclusions parallel to the a axis sometimes cause strong chatoyancy. Also massive, granular to earthy; rarely, fibrous.
Physical Properties: Very brittle; fracture conchoidal. Distinct cleavage parallel to $m[110]$ and $i[021]$. H. 3. G. 6.55.
Optical Properties: Transparent to opaque. Predominately colorless to white, often with streaks or zones of gray, black or brown. Streak white. Luster strongly adamantine, particularly on fresh fracture surfaces; also silky chatoyant with inclusions. Biaxial negative. R.I., $\alpha = 1.804$, $\beta = 2.077$, $\gamma = 2.079$. Birefringence 0.275. Clear material has with

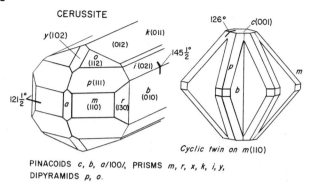

CERUSSITE

PINACOIDS *c, b, a/100/,* PRISMS *m, r, x, k, i, y,*
DIPYRAMIDS *p, o.*

FIG. 186.

difficulty been faceted into gems of considerable beauty with strong dispersion; chatoyant material provides fine catseyes.

Chemistry: Lead carbonate.

Distinctive Features and Tests: Twinned crystals or reticulated aggregates; strong luster; brittleness; high specific gravity; chatoyant inclusions. Reacts distinctively in the flame: decrepitates violently, then fuses at $1\frac{1}{2}$, becoming strongly colored yellow-brown or red-

FIG. 187. Cerussite sixling twins from Tsumeb, South-West Africa. The large crystal to the right measures about $2\frac{1}{2}''$ in diameter. Looking down upon c axes.

brown, finally reducing to a malleable globule of lead. A yellow opaque sublimate sur-
rounds the specimen. Dissolves with effervescence in dilute nitric acid; slowly soluble
in hydrochloric acid.

Occurrences: A secondary low-temperature mineral in oxidized portions of ore bodies
containing galena. Usually in limonite gossan associated with anglesite, phosgenite, pyromorphite, malachite and smithsonite. Fine specimens formerly from the Wheatley mines,
near Phoenixville, Chester County, Pennsylvania, as twins to 2″ diameter and singles to
¾″ to 1¼″. From the Organ District, Dona Ana County, New Mexico, in fine large
crystals, some as heart-shaped twins. From the Mammoth - St. Anthony mine, Tiger, Pinal
County, Arizona, in choice vee-twins and reticulated groups; also from the Flux Mine,
Santa Cruz County, in large slender acicular crystal aggregates of pure white color,
jumbled in jackstraw fashion in gossan openings (Figure 188); sharp, pale gray nearly

FIG. 188. Cerussite specimens. *Top:*
small gray sixling twins on botryoidal
psilomelane from the Campbell Shaft,
Bisbee, Cochise County, Arizona. Size
2¾″ by 1¾″. *Bottom:* "jackstraw" aggregate of acicular white cerussite on
limonite gossan matrix. Flux Mine, Santa
Cruz County, Arizona. Size 3½″ by 3″.

perfect tabular sixlings to ⅝″ diameter, on psilomelane, from Campbell Shaft, Bisbee,
Cochise County. In fine twinned transparent crystals to several inches diameter, of pale
brown color, from Friedrichssegen, Ems, Germany; also in Saxony, and elsewhere in
central European mining districts. From the Leadhills, Lanarkshire, Scotland. Fine specimens from Sardinian lead mines at Monte Poni and Monte Vecchio near Iglesias as
singles to 2¼″ long and 1″ diameter. Especially large and fine from Tsumeb, South-

West Africa, in great variety, ranging from approximately cylindrical single crystals to thick sixlings (Figure 187), and snowflake-like reticulated masses of twins, the latter sometimes to 6″ diameter; also in black sixling twins, elongated along the c axis, and forming stout hexagonal prisms with rounded terminations. At Mibladen, Morocco in ½″ sixlings. Very fine from Broken Hill, New South Wales, as large reticulated masses to 8″ by 5″; also at Dundas, Tasmania, in single crystals to 4¾″ length and 1″ diameter.

Cerussite furnishes fine specimens in almost any form, but the large sixlings, similar to those in Figure 187, are perhaps prized above others. Large reticulated twinned groups, especially when approaching the symmetry and perfection of snowflakes, are especially desirable. Matrix specimens are difficult to obtain because crystals are brittle and lightly attached, the inevitable jarring during extraction easily knocking them off. Expensive in large, fine specimens.

Aurichalcite—$(Zn,Cu)_5(OH)_6(CO_3)_2$

Zinc-copper-hydroxyl carbonate; ô′rĭ kȧl sīt. As lustrous delicate pale blue acicular crystals, forming tufts and ragged carpets on limonite gossan; prized by micromounters, but also commonly included in cabinet collections when areas covered by crystals exceed several inches, as is commonly the case with specimens from the Mina Ojuela, Mapimi, Durango, Mexico. At this locality, it is often associated with calcite, rosasite and hemimorphite. Good specimens also come from various mines in Arizona; at Mindouli, Congo; at Tsumeb, South-West Africa, and elsewhere.

Rosasite—$(Cu,Zn)_2(OH)_2CO_3$

Similar to aurichalcite in composition and commonly associated; rō′ ză sīt. In colorful incrustations of microcrystals on limonite from Mina Ojuela, often covering areas to about 3″ by 3″. Distinguished from aurichalcite by forming dense carpets, similar to velvet in the uniformity of "nap," and bluish-green in hue.

Malachite—$Cu_2(OH)_2CO_3$

Name: From Greek, *malache*, "mallow," in reference to green leaf color; mȧl′ă kīt.
Crystals: Monoclinic; only in microscopic acicular crystals, displaying wedge-shaped terminations atop eight-sided prisms. Crystals are invariably twinned on a[100]. By far the greatest quantity of malachite occurs massive. Some crusts are coated with minute crystals resembling velvet (Figure 189). Very compact types are dense enough to cut and polish to a high luster (Figure 190).
Physical Properties: Brittle, but quite tough in massive types; fracture uneven to splintery. H. 3½-4. G. 3.6-4.05, lowering in massive types.
Optical Properties: Transparent in microcrystals; otherwise opaque to translucent in thin splinters. Bright green, ranging from intense emerald-green in crystals, to very dark to pale green in compact massive material. Distinct color banding common (Figure 190). Streak pale green. Luster sub-adamantine in crystals; otherwise silky to dull. Biaxial negative. Massive material on the refractometer yields an indistinct reading near $n = 1.77$.
Chemistry: Copper hydroxyl carbonate.
Distinctive Features and Tests: Vivid color usually identifies. In masses, concentric color banding is distinctive. Readily dissolves with effervescence in cold dilute hydrochloric

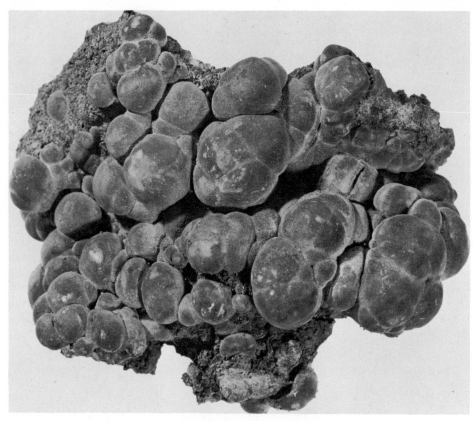

FIG. 189. Botryoidal malachite with velvety surface from Bisbee, Cochise County, Arizona. Size 4″ by 3″. Photo courtesy Smithsonian Institution.

acid. Fusible at 3; fluxed with borax on charcoal, blackens, eventually yielding malleable bead of copper. May be confused with rosasite but latter is distinctly bluish in hue.

Occurrences: A secondary mineral in oxidized portions of deposits in which copper sulfides are available for alteration. Abundant in sulfide vein deposits but also in disseminated deposits, as in the so-called "porphyry coppers" of southwestern United States. Forms crusts and stainings, usually with azurite, in practically all classes of rocks at or near the surface, providing some copper-bearing species, or native copper, was originally present. Very commonly associated with azurite, but more abundant than that mineral; also associated with cuprite, tenorite, limonite, calcite, and other secondary copper species. Commonly alters from azurite crystals (Figure 192), partly or wholly replacing same.

Fine gem quality masses formerly abundant in mines of Bisbee, Cochise County, Arizona; also as pseudomorphs after azurite crystals, to several inches length. Other mines in this state produce fine specimens as at Morenci, Greenlee County, particularly in vividly-colored banded crusts and stalactites, showing alternating layers of blue azurite and green malachite; also in the Globe district, Pima County, etc. In fine specimens from various localities in New Mexico, Utah, and Nevada. Beautiful geodes, lined with velvety malachite, sometimes dotted with rosettes of azurite, provide exceptionally handsome specimens from the areas mentioned, perhaps the best coming from the Copper

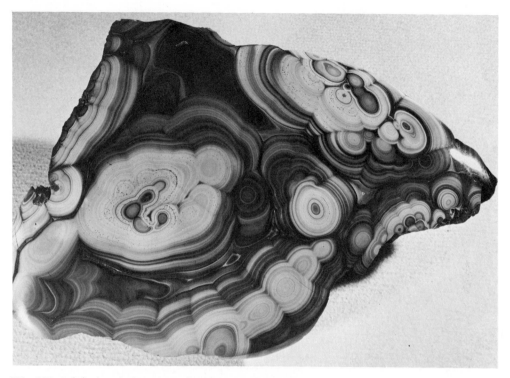

FIG. 190. Polished malachite mass showing complex banding. From Ruwe, near Kolwezi, Haut Katanga, Congo. Size 7″ by 5″.

Queen Mine at Bisbee. Similar material occurs in smaller quantities in various mines of Mexico, but not as fine. In Europe, outstanding localities include several in the Urals, near Sverdlovsk, notably the celebrated Gumeschevsk deposit at Sissersk, and the Mednorudiansk deposit near Nizhni-Tagil, where enormous masses to many feet across were recovered. In France, malachite occurs at Chessy, near Lyon, as an alteration product of single crystals of cuprite, providing specimens much in demand although the deposit is exhausted and specimens must be obtained from old collections. Fine massive material occurs in great quantity at the Etoile du Congo Mine, in the Katanga of Congo, also at Bwana Mkubwa in northern Rhodesia, and at Tsumeb, South-West Africa. Fine polishable masses come from Burra Burra, South Australia.

Because of vivid coloration and handsome patterns, massive malachite is fairly expensive, but few collections are without a representative polished specimen. Prized highly are good examples of "velvet" malachite, particularly if the velvet is uncrushed; also stalactitic forms of compact malachite. Most expensive, and rarely offered for sale, are the Chessy cuprite pseudomorphs and handsome specimens of alternately banded azurite-malachite.

Azurite—$Cu_3(OH)_2(CO_3)_2$

Name: In reference to the color; àzh′ ū rīt.
Crystals: Monoclinic. Commonly in sharp lustrous crystals in a considerable variety of habits, isolated or in druses. In tabular, sharp-edged individuals compressed along the

c axis with *c*[001] dominant, but also prismatic along this axis. Many crystals are wedge-shaped and complex, making it difficult to decipher and identify faces (Figure 191). Striations occur parallel to the *a* axis on *c*[001], or parallel to the *b* axis on *a*[100]. Wafer-like crystals of nearly circular outline usually display an enlarged pair of *c*[001] faces. Also earthy to granular massive.

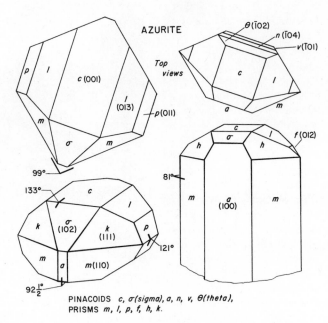

FIG. 191.

Physical Properties: Brittle; conchoidal fracture. Imperfect cleavage parallel to *p*[011]; fair cleavage on *a* [100]. H. 3½-4. G. 3.77.

Optical Properties: Translucent to transparent. Intense blue, in large crystals nearly black; pale blue in earthy types. Streak pale blue. Luster vitreous; slightly greasy on fracture surfaces. Biaxial positive. R.I., $\alpha = 1.730$, $\beta = 1.758$, $\gamma = 1.838$. Birefringence 0.108. Strong pleochroism in shades of blue, but difficult to observe except in very thin splinters. Some transparent crystals have been cut as gems but appear black even in small sizes.

Chemistry: Copper hydroxyl carbonate.

Distinctive Features and Tests: Color, crystals, and intimate association with malachite. Tests as for malachite.

Occurrences: Under the same circumstances as malachite but considerably less abundant. Commonly forms large crystals; massive types are less compact than malachite. Localities for good crystals are numerous and only a few can be mentioned. At Bisbee, Cochise County, Arizona, in exceptional specimens from minute drusy crystals lining cavities within limonite gossan, to clusters of large crystals of several inches length. Especially prized by copper species collectors are the Bisbee spherical aggregates or rosettes to 2″ diameter, consisting of numerous subparallel sharp-edged crystals of deep blue color; similar rosetted azurite occurs at Chessy in France. Nodular concretions of earthy azurite, sometimes with central cavities lined with euhedral crystals of small size, occur in the Blue Ball Mine, near Globe, Gila County, Arizona; nodules reach 3″ diameter. Fine crystals in the Apex Mine, St. George, Washington County, Utah, and other places in this state. From Mexico in Mina San Carlos, Mazapil, Zecatecas, and elsewhere. Consistently the largest and best crystals occur at Tsumeb, South-West Africa, where individuals to 10″ length have been recorded although most are 2″ or less; magnificent specimens have been obtained of large perfect prismatic crystals, some simple in habit, perched in clusters on matrix with separate crystals several inches long (Figure 192). Fine crystals occur in the Urals and Altai Mountains of the U.S.S.R.; also in Sardinia, Italy; at Laurion, Greece; at Chessy near Lyon, France, and at Wallaroo, South Australia, and Broken Hill, New South Wales.

Azurite in almost any form is a prized collector's item because of its beautiful color; earthy types are often vividly pale blue, and desirable on that account, while mixtures with malachite in solid massive types provide spectacular polished specimens. Crystals are available in drusy types consisting of numerous small sparkling individuals or in larger crystals, which though nearly black in hue, display sharp and smooth faces, and commonly, excellent terminations. Interesting replacements by malachite are also prized. The price of specimens rises sharply with increasing crystal size; top-grade examples from Tsumeb are not ordinarily within reach of the average amateur.

FIG. 192. Azurite crystals partly altering to malachite along the bases of the prisms. From Tsumeb, South-West Africa. Size 3″ by 2½″.

Phosgenite—$Pb_2Cl_2(CO_3)$

Lead chloro-carbonate; fŏs′jĕn nīt. A rare mineral, occurring in good crystals only in oxidized portions of a few lead deposits. Crystals usually transparent, of adamantine luster, sharp, and in most respects resembling cerussite crystals of similar size although phosgenite crystals are commonly blocky prisms, displaying four large prism faces terminated by a broad $c[100]$ face. Furthermore, they do not form the twins so commonly observed in cerussite. Prime localities are Monte Poni and Monte Vecchio, near Iglesias, and Gibbas on the Island of Sardinia, Italy, where crystals to 5″ across have been found. Most are transparent in large areas, colorless, white, or somewhat brownish. They are very expensive in large sizes. Good to fine crystals of much smaller size occur at Tarnowitz, Upper Silesia, Poland, and with matlockite at Matlock, and Cromford, Derbyshire, England.

BORATES

About 45 borate species are recognized, of which the *hydrated borates,* formed in sedimentary deposits in the floors of desert playas, are of greatest interest to collectors. Some borates containing halogen or hydroxyl are of interest to micromounters; others occasionally furnish good single-crystal specimens (boracite, hambergite). Hydrated borates require care in preservation to prevent loss or gain of water. For this reason, they are not as commonly represented in collections as may be expected from the relative abundance of fine specimens which presently come from large-scale mining operations in California.

HYDRATED BORATES

Kernite—$Na_2B_4O_7 \cdot 4H_2O$

Name: After the California county in which it occurs; kĕr'nĭt.

Crystals: Monoclinic. Euhedrons rare; usually in large crystal grains commonly reaching diameters of 24″ to 36″ (Boron). Twinning on $m[110]$. Also massive.

Physical Properties: Very brittle; splintery. Easy, perfect cleavages on $a[100]$ and $c[001]$ divide crystals into laths of rectangular cross-section elongated parallel to the b axis; also another cleavage across the two previously mentioned, parallel to $d[201]$. H. 2½-3. G. 1.91.

Optical Properties: Transparent, colorless when fresh, but usually opaque because of alteration coating of tincalconite. Streak white. Luster dull to vitreous to somewhat silky or pearly on cleavage surfaces. Biaxial negative. R.I., $\alpha = 1.454$, $\beta = 1.472$, $\gamma = 1.488$. Birefringence 0.034.

Chemistry: Hydrous sodium borate. Alters readily to tincalconite.

Distinctive Features and Tests: Lath-like cleavage fragments, easily cleaved further. Swells, then fuses at 1½ in flame to opaque mass which becomes clear with further heating; flame colored intense yellow. Slowly soluble in cold water and easily in hot water and acids.

Occurrences: In the playa deposits of the Kramer District, Kern County, specifically in the large open pit mine near Boron. The largest single crystal grain of record measured 8′ by 3′. Commonly available in small to large cleavage fragments.

Tincalconite—$Na_2B_4O_7 \cdot 5H_2O$

Hydrous sodium borate; tĭn căl'cŏn nĭt. A white opaque powdery coating on other borate minerals; crystals unknown in nature.

Borax—$Na_2B_4O_7 \cdot 10H_2O$

Name: From Arabic *buraq,* for "white"; bô'răks.

Crystals: Monoclinic. Short prismatic, tabular, compressed along a axis, commonly studding matrix of massive borax with c axes upright (Figure 193). Faces sharp and lustrous when fresh, but coating quickly with tincalconite. Broadest faces usually $a[100]$. Also massive granular.

Physical Properties: Very brittle; fracture conchoidal. Cleavage plane parallel to $a[100]$ perfect, parallel to $m[110]$ less so. H. 2-2½. G. 1.71.

Optical Properties: Transparent to translucent. Colorless; also grayish, bluish or greenish. Streak white. Luster vitreous. Biaxial negative. R.I., $\alpha = 1.447$, $\beta = 1.469$, $\gamma = 1.472$. Birefringence 0.025.

Chemistry: Hydrous sodium borate. Alters readily to white tincalconite.

Distinctive Features and Tests: Sharp crystals; alteration coating; brittleness. Soluble in water. Fuses at 1½, coloring flame yellow, frothing, then subsiding into clear glass.

Occurences: With other borates in playa deposits, associated with halite, trona, ulexite, gypsum, hanksite, etc. Fine large crystals from mud of Borax Lake, near Clear Lake, Lake County, and splendid large crystals, some to 6″ diameter, but mainly in groups of 1″ crystals on matrix, the latter often better than 12″ across, from Searles Lake, San Bernardino County, California. Also from Boron open pit, Kramer District, Kern County,

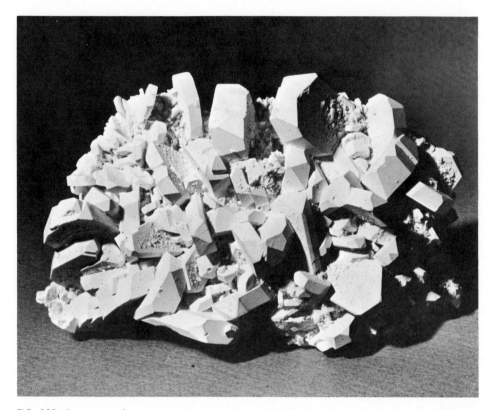

FIG. 193. Borax crystals on matrix of massive borax and clay from Boron, Kern County, California. Size 5″ by 3″. The white coating is tincalconite, to which mineral the originally transparent borax alters with exposure.

California, where masses of borax are found many feet across. Splendid groups are presently available at low cost; preservation requires spraying thoroughly with plastic spray, but specimens may also be allowed to alter to tincalconite which remains reasonably firm.

Ulexite—NaCaB$_5$O$_9$·8H$_2$O

Name: After German chemist George Ludwig Ulex (1811-1883); ū′lĕk sīt.

Crystals: Triclinic; distinct crystals rare. Usually acicular, elongated along the c axis. In tufted cottony masses, in radiate masses of free-standing acicular crystals, or as parallel crystals forming compact cross-fiber seams in clays and shales (Figure 194). Also in dense cauliflower-like growths.

Physical Properties: Very brittle. Fibrous masses readily part parallel to fibers, along perfect cleavages $b[010]$ and $m[110]$. H. $2\frac{1}{2}$. G. 1.955.

Optical Properties: Transparent to translucent. Remarkably transparent along the c axis in cross-fiber masses in which distinct patterns can be transmitted by individual crystals through at least 6″ of fiber length, providing the ends of the masses are first flattened and polished. White; sometimes greenish or grayish from clay inclusions. Streak white. Luster vitreous to strong silky chatoyant on fibrous types (Figure 194). Biaxial positive. R.I. $\alpha = 1.49$, $\beta = 1.51$, $\gamma = 1.52$. Birefringence, 0.03. Suitable cross-fiber specimens provide fascinating optical toys when opposite faces are polished across fibers; when placed on newsprint or lettering, the patterns are totally reflected within each fiber, finally appearing on top. Such specimens are popularly called "television stone." Fine catseyes and spheres have been prepared from chatoyant material but unless coated with plastic, are soon covered by a powdery efflorescence.

Chemistry: Hydrous sodium-calcium borate.

Distinctive Features and Tests: Fibrous masses readily distinguished from gypsum and calcite by very easy splitting and softness; cottony material distinguished by associates. Fuses at 1 into clear glass after preliminary frothing; flame colored yellow. Slightly decomposed in water; soluble in acids.

Occurrences: With other borates in playa deposits. Large masses of brilliantly chatoyant material are abundant in the open pit at Boron, Kern County, California; also from nearby in the Kramer District, San Bernardino County. As tufted "cotton ball" aggregates, especially fine at Mt. Blanco and Corkscrew Canyon, Death Valley, California. Fine fibrous and cotton specimens are readily available at the present time.

Colemanite—$Ca_2B_6O_{11} \cdot 5H_2O$

Name: After William T. Coleman, owner of the California mine where this species was first found; kōl′măn ĭt.

Crystals: Monoclinic. Commonly blocky or pseudo-rhombohedral, but also tapering to sharp points reminiscent of acute scalenohedrons of calcite (Figure 195). Also granular to sugary massive.

Physical Properties: Brittle; subconchoidal to uneven fracture. Perfect and easily developed cleavage parallel to $b[010]$; also distinct parallel to $c[001]$. H. $4\frac{1}{2}$. G. 2.42.

Optical Properties: Transparent to translucent. Colorless, white; also faintly straw-yellow, grayish. Streak white. Luster vitreous; on some specimens, slightly pearly on cleavage plane $b[010]$. Biaxial positive. R.I., $\alpha = 1.586$, $\beta = 1.592$, $\gamma = 1.614$. Birefringence 0.028. Clear specimens have been faceted into collector's gems.

Chemistry: Hydrous calcium borate.

Distinctive Features and Tests: Crystals and borate associates. Hardness and one perfect cleavage distinguish from calcite. Decrepitates in flame, then puffs up, finally fusing at $1\frac{1}{2}$ into an opaque white bead rather than colorless as with other borates. Soluble in hot hydrochloric acid.

Occurrences: With other borates in playa deposits, commonly in druses lining geodes or cavities. In Gower Gulch in the Death Valley region of Inyo County, California, as small to large white blocky crystals $\frac{3}{4}$″ to 3″ diameter; also at other points in Death Valley in druses and single crystals. Fine sharp-pointed colorless or very pale yellow crystals occur abundantly in openings in massive borate ore at the Boron open pit,

FIG. 194. Fibrous ulexite formed in seams in clay in the Kramer District, Kern County, California. Size 6½″ by 4″. The ulexite is pure white while the clay, some of which is still adhering to the top and bottom of the specimen, is pale bluish-gray. The chatoyancy of this material is very apparent in the photograph. *Specimen courtesy Schneider's Minerals, Poway, California.*

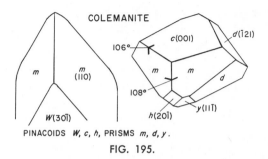

PINACOIDS *W, c, h*, PRISMS *m, d, y*.

FIG. 195.

Kern County, California; some specimens exhibit crystals to 1″ diameter, others display radiating crystals of colemanite, or crystals lining spherical geodes (Figure 196).

Inderite—$Mg_2B_6O_{11} \cdot 13H_2O$

Kurnakovite—$Mg_2B_6O_{11} \cdot 15H_2O$

The species formerly known as "lesserite" has now been discredited and found to be inderite (ĭn′dĕr ĭt). That which was formerly called "inderite" is now identified as kurnakovite; kĕr′nă kō vīt. Both species are magnesium borates and occur in abundance in the open pit at Boron, Kern County, California.

FIG. 196. Transparent straw yellow colemanite crystals of sharp rhombohedral habit lining a geode in massive material. From Boron, Kern County, California. Size 4″ by 3″.

Inderite forms slender prismatic crystals of rectangular cross-section in clayey shale. Specimens are recovered by soaking in water to remove the clay. Colorless to white; brittle, with one good and one poor prismatic cleavage: a very easy fracture across the crystals; commonly striated parallel to the prism. Kurnakovite occurs in large masses and crystals, some crystals to 24″ across; cleavage parallel to $b[010]$ perfect, to $m[1\bar{1}0]$ good. Twinning common. Crystals blocky to approximately cubic; often in groups. Pearly luster on perfect cleavage surface.

Howlite—$Ca_2SiB_5O_9(OH)_5$

Calcium-silicon borate with hydroxyl; how′līt. Occurs abundantly in cauliflower-like masses at the Sterling Borax Mine, Tick Canyon, Los Angeles County, California, usually in masses of 3″ to 4″ diameter (Figure 197), but also commonly to 10″ to 12″ diameter.

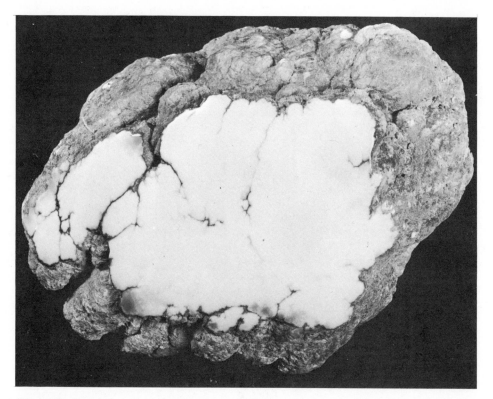

FIG. 197. Nodular mass of fine-grained howlite from the Sterling Borax Mine, Tick Canyon, Los Angeles County, California. Size 4″ by 5″. One portion of the nodule has been polished to expose the howlite.

Specimens to 100 pounds have been recorded. Also at Boron, Kern County, and elsewhere in the playa borate deposits of southern California. At Windsor, Brookville and Wentworth, Newport Station and elsewhere in Hants County, Nova Scotia. Prized for lapidary work in massive form; distinct crystals are exceedingly rare.

BORATES WITH HYDROXYL OR HALOGEN

Hambergite—Be₂(OH,F)BO₃

A rare beryllium hydroxyl-fluorine borate; hăm'bĕrg ĭt. Found exclusively in pegmatites. Crystals are lath-like individuals, colorless to white, sometimes twinned (Figure 198). The finest specimens came from pegmatites at Mt. Bity, Madagascar, in colorless crystals reaching $4\frac{1}{2}''$ in length and $1\frac{1}{4}''$ in width. Clear specimens have been cut into gems. Recently, good sharp crystals, largely opaque white, were found in the Little Three pegmatite near Ramona, and in the Himalaya Mine, near Mesa Grande, San Diego County, California (Figure 198). The best crystals reached $2\frac{1}{4}''$ length and $1\frac{1}{4}''$ width; twinned groups are common. Hambergite is desirable for single-crystal collections.

Boracite—Mg₃ClB₇O₁₃

An uncommon magnesium chloro-borate found in good crystals only in Germany; bŏ'ră sĭt. Crystals occur singly in granular anhydrite, gypsum and halite, but also in carnallite and kainite, forming rough-surfaced cube-octahedrons or dodecahedrons, seldom over $\frac{1}{2}''$. The finest examples are pale green-blue, like pale aquamarine, transparent and fairly sharp. Crystals in matrix are very rare. Some crystals are flawless and have been cut into collector's gems. Principal localities are the salt beds of the Hannover district, Germany.

SULFATES

Of the nearly 160 sulfate species, a mere handful provide good cabinet specimens; many others occur in small inconspicuous crystals which are collected and examined by micromounters. The considerable number of species attests to the easy ability of the sulfate radical to enter into many types of chemical compounds, but also to form species easily soluble in water. The affinity of many sulfates for water makes it difficult to obtain good specimens except in desert region deposits, and imposes upon the collector the burden of taking special pains to preserve such specimens. By far the most popular representatives of the sulfates are the insoluble *anhydrous sulfates,* among them barite, celestite, anglesite, etc.

ANHYDROUS SULFATES

Barite Group

Barite—BaSO₄

Name: From the Greek *barys,* "heavy," in allusion to the high specific gravity; bàr'ĭt.
Varieties: With appropriate substitutions for barium: *strontian, calcian* and *plumbian.*
Crystals: Orthorhombic; common, often large and fine. Usually thin to thick tabular, compressed along the c axis with faces of $c[001]$ the largest of all (Figure 199); also com-

FIG. 198. Hambergite crystals from the Little Three Mine, Ramona, San Diego County, California. Striations parallel to the c axes. A number of crystals are twinned, some in reticulated growths. Largest crystal about 1¼" long.

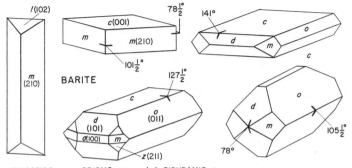

BARITE

PINACOIDS *a, c,* PRISMS *m, o, d, l,* DIPYRAMID *z.*

FIG. 199.

monly elongated along the *a* axis (Figure 200), and less commonly along the *b* axis. Crystals occur as square tablets, beveled-edged tablets, and chisel-tipped prisms. Occasionally, prisms are tapered because of gently-curved faces in the zone around the *a* axis (Figure 200). Aggregates common as rosettes of plates, or as interpenetrant rude crystals with sand or earth inclusions forming the so-called "desert roses." Also concretionary, then strongly resembling stalactitic calcite.

Physical Properties: Very brittle; massive forms fairly tough. Fracture uneven. Perfect

FIG. 200. Barite. *Left:* brownish-yellow transparent crystals on yellow calcite coating dark brown massive barite. From Elk Creek, Meade County, South Dakota. Size 4″ by 3″. Crystals occur in openings in large barite concretions. Crystals elongated along *a* axes, with faces in the zone of this axis curved. *Right:* colorless tabular to prismatic barite crystals from Knoch, Westmoreland, England. Crystals elongated along *a* axes; broad top faces are c(001), sloping side faces are prisms o(011) and d(101).

and easily developed cleavages parallel to $c[001]$, and others parallel to $m[210]$ and $b[010]$. In many crystals tabular along the c axis, the $m[210]$ cleavages appear as criss-cross networks of partly-developed, brightly reflective fissures at right angles to the broadest faces $c[001]$. H. 3-3½. G. 4.3-4.6.

Optical Properties: Transparent to translucent. Usually colorless, but also pale blue, yellow, brown, grayish; rarely, reddish or greenish. Color zoning common. Streak white. Luster vitreous to somewhat greasy on fracture surfaces. Biaxial positive. R.I., $\alpha = 1.636$, $\beta = 1.637$, $\gamma = 1.648$. Birefringence 0.012. Weakly pleochroic. Some varieties fluoresce under ultraviolet; also thermoluminescent and phosphorescent.

Chemistry: Barium sulfate. *Strontian barite* forms a complete series to celestite, and *plumbian barite* to anglesite.

Distinctive Features and Tests: The unusually heavy weight in hand specimens is quite remarkable, and in conjunction with the pale color of crystals, is perhaps the best field clue to identity. Tabular crystals distinctive; also partly developed cleavages which are seldom absent. Flies apart in the flame, fusing with difficulty at 3. Insoluble in acids.

Occurrences: A moderate to low temperature hydrothermal vein mineral, commonly forming the gangue for sulfide ores; also in many sedimentary rocks as veins or lenses, or lining cavities; rarely, in veins in basalt. A great many localities for fine crystallized specimens are known and only a few can be listed. Fine yellow tabular crystals to 1″ on pale green corroded fluorite octahedrons from Grand Forks, British Columbia. Good to fine druses at Five Islands, Cumberland County, also in the Londonderry Mines and along the Bass and East rivers, Colchester County, Nova Scotia. Fine granular massive, suitable for lapidary work, in veins on the shore of Lake Ontario, Pillar Point, Jefferson County; also large tabular crystals at De Kalb and other places in St. Lawrence County, New York. Formerly in splendid museum specimens displaying large, perfect and clear crystals, from Cheshire, New Haven County, Connecticut, in veins in sandstone where it was mined commercially. Splendid tabular blue crystals in clay and on matrix near Sterling, Weld County, Colorado; crystals reach 2″ to 4″ length and may be doubly-terminated. Also in Colorado in fine bright transparent yellow drusy crystals in the Eagle Mine, Gilman, Eagle County; fine tabular from a number of places in the Silver Cliff and Rosita areas of Custer County; sharp yellow-brown crystals to 2″, with calcite, from Clear Creek quarry, Jefferson County; abundant in alluvium at Wagon Wheel Gap, Mineral County; perfect blue single crystals to 3″ in a barite mine four miles southwest of Hartsel, Park County; similarly in the northeastern part of the Apishapa quadrangle of Pueblo County. Barite roses to 4″ diameter abundant in topsoil around Norman, Oklahoma and other places in this state; rosettes are crudely formed tabular interpenetrant crystals including considerable red-stained sand. Splendid, perfectly transparent, largely flawless, rich yellow-brown crystals in cavities in barite concretions in Pierre shales along Elk Creek, Mead County, South Dakota; singles to 1¼″ diameter and 6″ length have been recorded (Figure 200). Good tabular crystals to 1″ in druses within barite seams exposed along sea cliffs of Palos Verdes hills, Los Angeles County, California.

Splendid crystals, undoubtedly among the best from anywhere, occur in a number of places in England, as at Dufton, Cumberland, and Knoch, Westmoreland, as simple prismatic colorless crystals (Figure 200), often many inches long and many pounds in weight; the world's record for a single barite crystal is a 100 pound individual from Dufton. Also in this country, the fine 4″ to 6″, sharp-pointed, tabular blue crystals from Frizington in Cumberland are always highly prized as specimens, particularly those which appear tastefully arranged on matrix; large brown crystals to 8″ from Mowbray, and very long colorless prisms from Parkside; opaque white "cockscomb" groups, about the same size and configuration as cockscomb marcasite from the United States tri-state district, occur in the Settlingstones Mine in Northumberland. Fine stalactitic barite, of rich brown color and handsomely patterned, occurs in surface deposits near Newhaven, Youlgreave, Derbyshire, England; sections at least 3″ across, resembling petrified wood, have been cut and polished. Fine large yellow crystals from Bamle, Norway. Colorless tabular crystals, in groups to 8″ by 6″, from Ober-Ostern in Odenwald, Germany; also fine from Ilfeld, Harz. A unique yellow-tipped variety occurs in splendid groups displaying long prismatic crystals to 2″ at Pribram, Czechoslovakia. Tabular barite crystals about 1″, containing vivid red dots of cinnabar, occur in the mercury mines of Almaden, Spain.

Barite is a favorite collection species but undamaged specimens are sometimes difficult to obtain because of the characteristic softness and easy cleavage which makes packaging and transport serious problems. Specimens from England and other European localities are becoming increasingly scarce, particularly the large blue Frizington crystals. Matrix specimens from Sterling, Colorado, are not easy to obtain, although single crystals are very commonly available.

Celestite—SrSO$_4$

Name: From the Latin, *caelestis,* "of the sky," in reference to the typical blue color of some specimens; sĕ lĕs′tīt.

Varieties: According to substitution for strontium, *barian* and *calcian.*

Crystals: Orthorhombic; similar to barite crystals. Usually thin to thick tabular, compressed along the *c* axis, forming tablets with broad *c*[001] faces (Figure 201); also

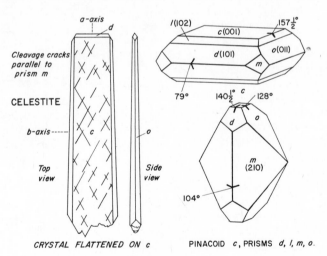

CRYSTAL FLATTENED ON *c*   PINACOID *c*, PRISMS *d, l, m, o*.

FIG. 201.

elongated along the *a* axis or along the *b* axis (Figure 202). Less commonly, blocky. Faces seldom as smooth as barite. Usually in druses and less inclined to form parallel groups than barite (Figure 203). Also massive in cross-fiber veins; in coarsely-crystalline masses in geodes, sometimes forming terminated crystals within them.

Physical Properties: Very brittle; uneven fracture. Perfect cleavage easily developed parallel to *c*[001]; good on *m*[210], poor on *b*[010]. H. 3-3½. G. 3.96-3.98.

Optical Properties: Transparent to translucent. Commonly colorless or pale blue, often zoned; also reddish, greenish or brownish. Streak white. Luster vitreous to somewhat greasy on fresh fracture surfaces. Biaxial positive. R.I., $\alpha = 1.622$, $\beta = 1.624$, $\gamma = 1.631$. Birefringence 0.093. Distinctly pleochroic in dark blue crystals (Austin, Texas); purple-blue, pale bluish-green, colorless. Orange material (Ontario), shows no pleochroism.

Chemistry: Strontium sulfate. Barium substitutes extensively for strontium and a complete series extends from barite to celestite.

Distinctive Features and Tests: Heavy, tabular crystals similar to barite but predominantly blue and zoned in color. Fuses at $3\frac{1}{2}$-4; distinguished by vivid red flame color of strontium, but such appears only when a small sliver held in forceps is heated to brilliant incandescence. Only slowly soluble in concentrated acids.

Occurrences: Mainly in sedimentary rocks of marine origin, and particularly in limestones where it commonly fills veins and other openings; also as a primary hydrothermal min-

FIG. 202. Divergent spray of large, pale blue, terminated celestite crystals, the largest individual about $3\frac{1}{2}''$ tall. From Matehuala, San Luis Potosi, Mexico. The c axis in each crystal is horizontal in this photograph, the a axes are vertical, and the b axes point toward the eye of the reader. The wedge-shaped terminations are pairs of $d[102]$ faces; the faces toward the eye of the reader are pairs of $o[011]$ faces.

eral; rarely, in igneous rocks. Associated with calcite, fluorite, barite, sulfur, aragonite, etc. A unique, vivid orange-red bladed celestite occurs in geodal masses in limestone at Forks of Credit, Caledon Township, Peel County, Ontario. Blue crystals of prismatic habit, usually about $\frac{1}{2}''$ length, but known to reach $1\frac{1}{2}''$, often perfectly transparent and with sharp glossy faces, occur in geodes in the Lockport dolomite in western New York State, as around Syracuse and Rochester. It was first discovered in Niagara County during excavations for the Erie Canal, and lately, fresh supplies have come from Chittenango Falls. An enormous deposit of celestite with tabular pale blue crystals to 18″ length was recovered many years ago from a cave in dolomite at Put-In-Bay, South Bass Island, in Lake Erie; also very fine crystals from 6″ to 8″ were found on Kelleys Island, Ohio. Fine thick tabular blue crystals, reaching $1\frac{1}{4}''$, often curiously etched on edges, occur in fine groups on cavity walls in limestone at Clay Center, Ottawa County, also very thin colorless blades of the type shown in Figure 201. Blue crystals to $1\frac{1}{2}''$ on edge, from Portage, Wood County, Ohio. Large crystals occur at Lampasas, Lampasas County, and in the Mount Bonnell area near Austin, Travis County, Texas; the last locality provides square prismatic crystals in geodes, colored deep purplish blue on the tips with white bases. Light blue to colorless crystals, to $1\frac{1}{2}''$, occur with strontianite in geodes in colemanite ores of Borate in the Calico Hills district, San Bernardino County, California. Recently, fine $\frac{3}{8}''$ sky-blue crystals on matrix from near Emery, Emery County, Utah. Very pale blue crystals, often etched, in singles and groups come from Matehuala, San Luis Potosi, Mexico; a divergent group is shown in Figure 202; the largest crystals are about $3\frac{1}{2}''$ long. Splendid crystals, commonly clear and colorless, to 2″, from Yate, Clifton, and other places around Bristol, Gloucestershire, England (Figure 203). Excellent blue groups

FIG. 203. Tabular celestite crystals from Yate, Glocestershire, England. Size about 3″ by 4″. *Photo courtesy Schortmann's Minerals, Easthampton, Massachusetts.*

from Gembock, and other places in Westphalia, Germany. Fine groups of white crystals, with sulfur, aragonite and gypsum, from the sulfur deposits of Agrigento, Cianciana, etc., Sicily. Fine white crystals to ¾", in singles and groups, and of unusual blocky habit, from Mokattam, southeast of Cairo, Egypt.

A good collector's mineral, usually available in fine specimens with prominent crystals of easily discerned habit. The best specimens come from the United States, especially from the Ohio localities; the clearest and sharpest crystals are those from the Lockport dolomite of New York. Not expensive.

Anglesite—PbSO₄

Name: From the Isle of Anglesea (Anglesey), Wales, where it was first found in Pary's Mine; ăng′glĕ sīt.

Crystals: Orthorhombic. Wide variety of habits; may be elongated on any of the three axes; also tabular, compressed along the c axis (Figure 204). Equant to prismatic. Faces

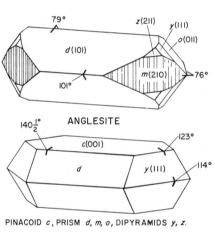

PINACOID *c*, PRISM *d, m, o*, DIPYRAMIDS *y, z*.

FIG. 204.

FIG. 205. Tapered crystals of white anglesite on matrix of tetrahedrite from Tsumeb, South-West Africa. Size 2¼" by 2".

a[100] and m[210] usually striated parallel to the c axis. Many crystals display curved brilliant faces which may be difficult to identify. Crystals form singly or in small groups (Figure 205). Also massive, granular to fine granular; nodular, stalactitic; sometimes as concentric alteration layers around galena cores.

Physical Properties: Very brittle; conchoidal fracture. Good cleavage parallel to c[001]; m[210] distinct. H. 2½-3. G. 6.36-6.39.

Optical Properties: Transparent to translucent. Ordinarily white, colorless, or pale straw yellow; also grayish, greenish, bluish. Streak white. Luster adamantine; sometimes vitreous or resinous. Fluoresces yellow under ultraviolet. Biaxial positive. R.I., $\alpha = 1.877$, $\beta = 1.883$, $\gamma = 1.894$. Birefringence 0.017.

Chemistry: Lead sulfate.

Distinctive Features and Tests: Crystals resemble barite and celestite but are usually white and adamantine in luster. Lack of twinning distinguishes from cerussite. Very heavy even in small hand specimens. Flies apart in flame, but as powder, fuses readily at

1½, eventually yielding on charcoal, a globule of malleable lead. Not soluble in acids except difficultly in nitric acid.

Occurrences: Predominantly in oxidized veins associated directly with galena as its alteration product, either surrounding unaltered galena cores, or occupying cavities within massive galena. Also associated with cerussite, pyromorphite, mimetite, wulfenite, gypsum, etc. Fine crystals were formerly obtained from many localities, of which, unfortunately, very few are now productive and good crystallized specimens are not now easy to obtain. Formerly from the Wheatley Mines, near Phoenixville, Chester County, Pennsylvania, in fine colorless crystals to ½ pound weight, measuring to 5½″ by 1½″! The Coeur d'Alene District, Shoshone County, Idaho, also produced exceptional crystals of unequaled sharpness and transparency from the Last Chance Mine, but best of all, from the Hypotheek Mine, near Kingston, from which came perfect doubly-terminated crystals within cavities in massive galena, an inch or better in size, and sometimes "as large as a hen's egg." Also in excellent crystals from Yellow Pine Mine, Goodsprings, Nevada. Matlock and Cromford, Derbyshire, England, furnished good crystallized specimens from time to time; the crystals from the type locality in Wales are small; good dagger-shaped crystals occurred in the mines of the Leadhills district of Scotland. Splendid crystals to 1½″ by 1″ from Monte Poni, Sardinia, often as acute prisms, but also in equant crystals, in clusters on matrix. More recently, in exceptional, water-clear, brilliant colorless crystals from Sidi-Amor-ben-Salem in Tunis, some to ¾″ diameter. Good crystals to 1½″ by 1½″ from Tsumeb, South-West Africa (Figure 205). The Broken Hill mines in New South Wales, Australia, are also known for production of numerous fine anglesite specimens, usually in yellowish transparent crystals to about ½″, on matrix.

Anhydrite—CaSO₄

Uncommon in collections; calcium sulfate; ản hī′drĭt. Crystals are rare and most specimens are simply cleavage masses ranging in hue from colorless to pale violet or very pale blue. Anhydrite masses are easily recognized by three cleavages intersecting each other at right angles and furnishing fragments of blocky shape. Cleavages occur parallel to b[010] perfect, a[100] nearly perfect, and c[001] good, and with the pearly luster which commonly develops on the best cleavage, serve to distinguish this sulfate from gypsum and other associates. Fine crystals to ⅛″ occur in kieserite in the saline beds of Stassfurt, Germany; also at Douglashall, Hallein, and Leopoldshall. Fine violet cleavages in the St. Gotthard region of Switzerland. Also as cleavage masses in North Burgess at McLaren's Mine, Lanark County, Ontario, associated with gypsum in an apatite deposit. Large lilac cleavages to 1½″ across in the New Cornelia pit, Ajo, Pima County, Arizona. Rarely, anhydrite is found with zeolite species in cavities in basalt of northern New Jersey; most of it has disappeared through solution, leaving rectangular cavities preserved in prehnite (Figure 323), or in chalcedony, opal, or other species formed after anhydrite.

Glauberite—Na₂Ca(SO₄)₂

Sodium-calcium sulfate; glô′bĕr ĭt. Not commonly repersented in collections because of lackluster crystals and easy alteration. In water or exposed to a moist atmosphere, crystals alter on surfaces and turn white. Crystals usually grayish, dull in luster, but are of interest for single-crystal collections. Crystals extremely flat with sharp bevels on edges. Singles and small clusters of 1″ to 2″ from saline deposits in the Salton Sea area, Imperial County, California (Figure 206); also from Camp Verde, Yavapai County, Arizona. Cal-

FIG. 206. Glauberite crystals showing diamond-shaped profiles and alteration coating caused by dehydration. From Salton Sea, Imperial County, California. Largest crystal about 1½".

cite and gypsum replace glauberite (Figure 34). The diamond-shaped cavities in prehnite and other minerals in the cavities in basalt of Northern New Jersey were caused by early glauberite which dissolved after being enclosed.

HYDRATED SULFATES

Gypsum—CaSO$_4$·2H$_2$O

Name: Probably from the Arabic *jibs,* for "plaster," thence to the Greek *gypsos* for "chalk"; jĭp′sŭm.

Varieties: On the basis of habit: *selenite,* in distinct crystals, clear plates, etc.; *satin spar,* numerous parallel acicular crystals in compact masses, often polishable into ornamental objects; *alabaster,* very fine-grained types, much used for carving and polishing into ornamental objects.

Crystals: Monoclinic; abundant, often large and fine, sometimes reaching extraordinary sizes as lath-like or blocky individuals to several feet in length and diameter. Untwinned crystals usually simple in habit, diamond-shaped in profile with more or less evenly beveled edges. Largest faces of diamond shape are ordinarily side pinacoids *b*[010]; beveling faces are prisms *f*[120] with *l*[Ī11] (Figure 207). Crystals commonly elongated along the c axis, or curving, especially when composed of numerous individuals in parallel position (Figure 208). Many crystals are rounded, striated, warped, or bent; also fibrous, granular, etc. The diversity of habit is great and one may devote an entire collection to gypsum without running out of varieties. Twinning common on composition plane *a*[100] forming "swallow tail" twins or interpenetrant twins; also on *d*[101] as butterfly-shaped or heart-shaped twins.

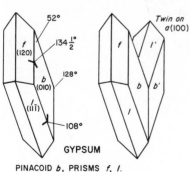

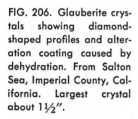

GYPSUM

PINACOID *b*, PRISMS *f*, *l*.

FIG. 207.

Physical Properties: Not brittle as a rule, tending to bend first, especially in thin sections or slender crystals, finally breaking with a splintery fracture. Cleavage parallel to plane *b*[010] is very prominent and perfect, and almost inescapable in development whenever a crystal is broken; also distinct cleavages on *a*[100] and *m*[011], the latter causing the division of broken crystals into thin splinters parallel to the *c* axis. Cleavage laminations bend easily but are not elastic. H. only 2, an extremely valuable diagnostic property; easily scratched by the fingernail. G. 2.32.

Optical Properties: Transparent to translucent. Predominantly colorless or white; also grayish, yellowish, or brownish, when colored by inclusions of clay, iron oxides, etc. Streak white. Luster vitreous to somewhat resinous, and very commonly, slightly to strongly pearly on cleavage planes. Fibrous varieties are strongly chatoyant, showing a fine silky luster on fresh fracture surfaces. Biaxial positive. R.I., $\alpha = 1.521$, $\beta = 1.523$, $\gamma = 1.530$. Birefringence 0.009.

Chemistry: Calcium sulfate with water. When heated strongly, gypsum loses water, converting to anhydrite, but in the manufacture of plaster of Paris, heat is adjusted to drive off only part of the water, resulting in a hemihydrate. When water is added later, the hemihydrate recrystallizes as gypsum in the well-known process of "setting-up" or solidification of plaster.

Distinctive Features and Tests: Fingernail scratch test; crystal habits; very common occurrence in clay; pearly luster. Fuses at 2½-3. Heated in closed tube, loses water which appears as droplets on cool walls. Soluble in hot dilute hydrochloric acid.

Occurrences: Very widespread and abundant wherever extensive sedimentary beds occur, particularly those rich in saline precipitates. Common as single crystals, twins, and crystal

FIG. 208. Gypsum. *Left:* "ram's horn" *selenite,* formed by numerous acicular curving crystals growing from matrix of gypsum and clay. From near Ciudad Chihuahua, Chihuahua, Mexico. Size 3½". *Right:* colorless and transparent *selenite* crystal from Naica, Chihuahua, Mexico. Crystal length 3½".

groups in clay beds; also in salt lakes and playas, and in volcanic fumeroles and deposits associated with fumarolic activity. Sometimes abundant in oxidized portions of ore veins. Only localities producing exceptional specimens can be mentioned. Fine crystals in cavities in dolomite at Lockport and elsewhere in Niagara County, New York. Clear "textbook" crystals to 2½" from Ellsworth, Mahoning County, Ohio. Bladed brown crystals in singles to 6" by 2", and radiate clusters from within the first five feet of saline material in the Great Salt Plain, near Jet, Alfalfa County, Oklahoma (Figure 209), in translucent specimens display striking "hourglass" inclusions of sand. Enormous transparent colorless

FIG. 209. Gypsum crystals of bladed habit forming clusters imbedded in the sand of the saline lake of the Great Salt Plain, near Jet, Alfalfa County, Oklahoma. Much brownish sand is included in the crystals. Size 5" tall.

crystals have been found in Fremont River Canyon, Wayne County, Utah; a prismatic crystal of 42″ by 9″ by 5½″ is recorded, also large cleavages, often showing transparent areas of a square foot or better. Small to very large sharp prismatic crystals, as in Figure 208, but mostly in striated laths with rounded edges and tips, occur in cavities in limestone in the Mina Maravilla, Naica, Chihuahua, Mexico. At this locality is the Cave of Swords which is preserved for sightseeing. It is a cavern large enough to enter and from its floor protrude crystals of several feet to as much as five feet in length. This locality and others in the same limestone formation have provided many fine groups of crystals, in the cabinet range of 6″ by 6″, or better, showing slender transparent *selenites* emerging from the surface in striated prisms to 1″ by 12″. In England, "textbook" single crystals occur at Trowbridge, Wiltshire, to 1″ by 1″. The premier locality for jewelry and ornamental *satin spar,* of pure white color but sometimes faintly pink, and strongly chatoyant, is at East Bridgford, Nottinghamshire. Fine single crystals and clusters occur in various localities in Germany, Austria, and in the Sicilian sulfur mines. The marls and clays of Montmartre, Paris, France, provide rich yellow swallowtail twins of 6″ to 9″ length. Perhaps the record gypsum crystal of all time was from the Braden Mine in Chile—a single crystal 10′ in length and 3″ in diameter!

Gypsum is a fine species to collect because of the great variety of crystals and aggregates. Excellent examples are abundant and prices are modest. Its softness poses field collecting problems if specimens are to be extracted without damage, and to a lesser extent, this problem also exists at home. Perhaps the most difficult specimens to find are those with sharp clear crystals, like that shown in Figure 208.

Chalcanthite—$CuSO_4 \cdot 5H_2O$

Copper sulfate with water; kål kån'thīt. An attractive vivid blue mineral occurring in oxidized zones of copper deposits. Its ready solubility in water, and the need to preserve it from moist atmospheres, prevent greater acceptance in collections. Large fine masses, sometimes fibrous, and more rarely, in distinct sharp wedge-shaped crystals, occur in Arizona mines and also in mines near Imlay, Pershing County, Nevada; Chuquicamata, Chile, etc. Copper sulfate is a favorite substance for demonstrating the growth of crystals from water solution and, with care, crystals of considerable perfection to several inches can be grown. Some persons cannot resist "decorating" natural matrix specimens with synthetic chalcanthite, and cases are known where handsome specimens of this sort have been sold to the unwary as genuine.

Coquimbite—$Fe_2(SO_4)_3 \cdot 9H_2O$

A colorful hydrous iron sulfate forming fine pale violet crystals; kō kĭm'bīt. Splendid crystals in singles and scattered on matrix, have come in abundance from oxidized zones of ore deposits in Chile where it is associated with other sulfates as roemerite, copiapite, paracoquimbite, etc. More lately, as small tabular hexagonal violet crystals to ½″ on matrix, from San Rafael Swell, Emery County, Utah. In dry air, this species loses water and assumes a white coating. It is common practice to preserve the specimens in air-tight containers or to immerse the entire piece in dilute plastic solution to prevent loss of water or damage through contact with water.

ANHYDROUS SULFATES CONTAINING HYDROXYL OR HALOGEN

Brochantite—$Cu_4(OH)_6(SO_4)$

Antlerite—$Cu_3(OH)_4(SO_4)$

Brochantite is a copper hydroxyl sulfate, occurring abundantly in Arizonan and Chilean deposits as vivid emerald-green acicular crystals; brō shȧnt'ĭt. Antlerite, ant'lĕr ĭt, is similar in composition and appearance, but is less common. Its occurrences parallel those of brochantite. Prized mainly by micromounters and those who specialize in the collection of copper minerals.

Linarite—$PbCu(OH)_2(SO_4)$

An azure-blue lead-copper hydroxyl sulfate, found in small sharp crystals in a number of copper deposits in the United States and elsewhere; lĭn'ȧ rīt. The Mammoth Mine at Tiger, Arizona, provides handsome specimens displaying thin films filling crevices in brecciated rock, and in places, druses of small to large euhedral crystals, some to 1″ by ½″. In previous mining, the Mammoth Mine provided very large crystals, some to 4″ length, but only small crystals, seldom over ⅛″, are available today. A good micromount material. Confused with azurite which it closely resembles in color, but linarite does not effervesce in acids.

COMPOUND SULFATES

Hanksite—$Na_{22}KCl(CO_3)_2(SO_4)_9$

A complex sodium-potassium chloro-carbonate-sulfate; hȧnk'sĭt. Abundant only in the saline beds of Searles Lake, San Bernardino County, California. Occurs in hexagonal crystals, commonly resembling doubly-terminated quartz crystals; also in tabular to stubby hexagonal prisms terminated by large $c[0001]$ faces, and beveled by the dipyramid $o[10\bar{1}2]$ (Figure 210). Also interpenetrant. Dull to greasy luster. Faintly yellow to grayish or greenish, commonly with inclusions forming distinct cores of greenish hue. Largest crystals 3″ in length. Collectors obtain crystals by digging in surface mud or material pumped from wells. To prevent alteration, crystals are immersed in mineral oil or heavily coated with plastic spray.

Caledonite—$Cu_2Pb_5(OH)_6(CO_3)(SO_4)_3$

A dark green, copper-lead hydroxyl sulfate-carbonate; kăl'ĕ dō nĭt. Forms tufts and carpets of prismatic micro-crystals on matrix rock in oxidized zones of copper deposits. The largest crystals, to ½″ length, are said to have come from the Mammoth Mine, Tiger, Arizona. Relatively rare, prized for micromounts.

FIG. 210. Hanksite crystals from the bed of Searles Lake, San Bernardino County, California. Crystals are pale yellowish-gray in hue, or sometimes greenish because of inclusions. Forms include the hexagonal prism, dipyramid, and basal pinacoid. Largest crystal about ¾″ long.

CHROMATES

Of the very few mineral chromates, only one, *crocoite,* is commonly collected; the others are quite rare or occur only in very small inconspicuous crystals, occasionally collected by micromounters.

Crocoite—Pb(CrO₄)

Name: From the Greek, *krokos,* "saffron," in allusion to the color resembling the deep orange produced by saffron dye; krō′kō ĭt.

Crystals: Monoclinic. Most cabinet specimens display slender crystals bounded by prism *m*[110] and elongated along the *c* axis; cross-sections are approximated square. Rarely terminated because crystals usually pass across openings in cavities to meet opposite walls or other crystals. Usually striated on *m*[110] parallel to the *c* axis. Commonly hollow parallel to the *c* axis, particularly in larger individuals. Aggregates form jackstraw masses of slender crystals, showing no twinning relationships. Also granular massive.

Physical Properties: Very brittle; fracture conchoidal to uneven. Distinct cleavage parallel to prism faces *m*[110]. H. 2½-3. G. 6.0.

Optical Properties: Transparent to translucent. Intense orange or red-orange; rarely, yellow, red. Streak orange-yellow. Luster adamantine; distinctly silvery on slightly altered crystal surfaces. Biaxial positive. R.I., $\alpha = 2.29$, $\beta = 2.36$, $\gamma = 2.66$ (lithium light). Birefringence 0.37. Pleochroism weak, in several shades of basic color. Transparent specimens have been cut into small brilliant gems.

Chemistry: Lead chromate.

Distinctive Features and Tests: Color and crystal habit. When crushed and mixed with sodium carbonate, fuses at 1½, yielding a lead globule on charcoal.

Occurrences: In oxidized zones of ore deposits containing lead and chromium, associated with pyromorphite, cerussite, wulfenite and vanadinite. On the whole, a very rare mineral, practically all specimens being supplied from the Adelaide, West Comet, and Dundas Extension mines in the Dundas district, and at Heazlewood and Whyte River of Tasmania,

Australia. Sulfur-yellow massicot sometimes accompanies the Dundas crocoite. Crystals mostly on brown earthy limonite, either as scattered needles or as dense criss-cross aggregates, and sometimes, in masses of nearly solid crocoite with angular cavities between prisms. The largest single crystals, sometimes terminated, reached lengths of $3\frac{1}{2}''$ and diameters to $\frac{3}{8}''$. Some matrix specimens measured several feet across. Good Tasmanian specimens are scarce and very high priced. Excellent crocoite also occurred in mines of the Beresov District, near Sverdlovsk in the U.S.S.R., Urals; some are nearly indistinguishable from the Tasmanian. Good crystals were found at Goyabeira, near Congonhas do Campo, Minas Gerais, Brazil. The vivid color and slender prisms of crocoite have always made this species a collector's prize, and good specimens never go begging for buyers. The average specimen usually shows slender prisms, not over $\frac{1}{16}''$ diameter, and not over $1''$ to $1\frac{1}{2}''$ long, sprinkled thickly on limonite gossan. It is an extremely fragile mineral and care must be used in handling.

PHOSPHATES, ARSENATES, VANADATES

Numerically, the phosphates, arsenates and vanadates are very impressive, the total being nearly 230, but few species are economically important, while the vast majority occur as minute crystals, in inconspicuous masses, or in other forms which may be easily ignored by the prospector interested in economic minerals or the amateur interested in large and colorful specimens for his collection. On the other hand, the micromounter finds a fertile field among the numerous species because so many are found only in small crystals whose fullest appreciation requires the use of magnification. As before, only those species primarily of interest to the collector will be described fully, with brief notes on others of interest.

ANHYDROUS PHOSPHATES, ETC.

Triphylite Group

TRIPHYLITE SERIES

Triphylite—Li(Fe,Mn)(PO₄) (Iron-rich end member)

Lithiophilite—Li(Mn,Fe)(PO₄) (Manganese-rich end member)

Both species are lithium, iron-manganese phosphates, in which iron and manganese occupy the same sites in the crystal structure, forming a continuous series between the end members shown; trĭf′ĭ lĭt, lĭth′ĭ ō fĭ lĭt. Both occur as primary minerals in granitic pegmatites, but very rarely in distinct crystals, and then usually simple in form and with curved or rough faces (Figure 211). Commonly, alteration causes obliteration of the original color (blue-gray in triphylite; brown, yellow-brown, brown-red in lithiophilite) by creation of manganese oxides which thoroughly permeate crevices in the crystals. When alteration is carried to its extreme, crystals appear as sooty black masses within pegmatite, usually surrounded by haloes of black stains where the manganese oxides have migrated outward into the surrounding pegmatite. In some instances, notably at the Stewart Lithia Mine, Pala, San Diego County, California, rude crystals and formless masses are found to $12''$ diameter, within which cavities and porous areas occur containing a variety of alteration products as follows: unaltered material in the cores, surrounded by a layer of dark red-brown sicklerite, then hureaulite, also reddish brown;

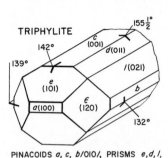

TRIPHYLITE

PINACOIDS *a, c, b*/010/, PRISMS *e,d, l,*
AND *Є* (epsilon).

FIG. 211.

next, buff to brownish-yellow salmonsite, followed by white to pinkish hureaulite ("palaite"). Blue, lilac, rose-red and purple strengite, purpurite, and heterosite masses are scattered through the salmonsite layers; also found are very small films and aggregates of bright yellow stewartite, and dark purplish-blue, almost black, vivianite. The alteration just described is aided by the infiltration of surface waters. When later hydrothermal activity is responsible for alteration, after triphylite-lithiophilite is formed within the pegmatite, the following species, as at Branchville, Connecticut and at Varuträsk, near Boliden, Sweden, may occur: eosphorite, reddingite, fairfieldite, dickinsonite, phosphoferrite, wolfeite, triploidite, and fillowite. The collector's opportunities to obtain good crystal specimens of primary triphylite-lithiophilite are rather slim, but from the standpoint of the micromounter, or the student interested in genesis of minerals, specimens accompanied by their alteration products are rich sources of study material, and at times, colorful too. Good specimens with alteration products have been obtained from a number of pegmatites in the Pala District of San Diego County, California, and from the famous complex pegmatite at Branchville, Fairfield County, Connecticut; also from other pegmatites in the states of Connecticut and Maine. Good opaque triphylite crystals of bluish-gray color to several inches diameter were found some years ago in a pegmatite near Chandler's Mill, Newport, New Hampshire. From the pegmatite at Varuträsk, Sweden, and from pegmatites near Huhnerkobel and Hagendorf in Bavaria, Germany.

HETEROSITE-PURPURITE SERIES

Heterosite—(Fe,Mn)(PO₄) (Iron-rich end member)

Purpurite—(Mn,Fe)(PO₄) (Manganese-rich end member)

Iron-manganese phosphates, commonly found in granitic pegmatites as alteration products of triphylite-lithiophilite; hĕt′ĕr ō′sīt; pŭr′pŭ rīt. Colors are distinctive: deep rose to reddish-purple, sometimes very intense. Good cleavage parallel to a[100] often evident on massive pieces. No euhedral crystals are known, and the usual occurrence is in the form of irregular masses, discolored by black alteration products upon the exteriors or as veinlets inside the masses. Large cleavages, commonly to 3″ to 4″ diameter, make colorful additions to the collections. Good to fine specimens come from Newry, Oxford County, Maine; from Palermo Mine, North Groton, New Hampshire; in the Custer pegmatite district, Pennington County, South Dakota; from Chanteloube, France in pieces to 3″ by 4″; also fine from Usakos, near Kleinspitzkopje, South-West Africa.

Beryllonite—NaBe(PO₄)

A very rare sodium-beryllium phosphate; bĕr ĭl′ŏ nīt. Occurs only in Stoneham and Newry, Oxford County, Maine. At Stoneham it was found in disintegrated pegmatite material in alluvial slopes of McKean Mountain and repeated efforts have not tracked it to pegmatite *in situ*. Fragments and rude crystals, to 2″, are always colorless. Perfect cleavage parallel to *b*[010]. Gems have been cut from clear pieces but are not notable for brilliancy because of the low refractive index of circa 1.55. Pieces showing distinct faces are very rare and highly prized, as are even formless fragments.

Xenotime—Y(PO₄)

Yttrium phosphate; zěn′ō tīm. Relatively rare, occurring as small zircon-like crystals in granitic pegmatites or in granitic rocks. Usually brown, but distinguished from zircon, with which it may easily be confused, by perfect cleavage parallel to *a*[100] and lower hardness. Crystals rarely exceed 1½″ length. Good crystals from pegmatites at Setesdal, Iveland, Norway; also at Tvedestrand, Kragerö, etc. Also well-shaped crystals from Madagascar. A mineral sought for by the single-crystal collector and one not easily found.

Monazite—(Ce,La,Y,Th)(PO₄)

Name: From the Greek *monazein*, "to be alone," in allusion to its isolated crystals and their rarity when first found; mŏn′ă zĭt.

Crystals: Monoclinic. Commonly in isolated small euthedral crystals, usually with faces somewhat curved. Larger crystals often crude or composed of subparallel crystals. Mostly somewhat tabular, compressed along the *a* axis, with wedge-shaped terminations on both ends of the *c* axis (Figure 212). Sometimes twinned on plane *a*[100].

Physical Properties: Brittle; conchoidal to uneven fracture. Distinct cleavage on *a*[100] in some specimens, but obscured in others where the crystal structure has been damaged by radioactivity; in such metamict specimens the fracture also tends to be good conchoidal rather than uneven. H. 5-5½. G. 4.6-5.4, increasing with thorium.

Optical Properties: Transparent to translucent. Red-brown, brown, yellowish-brown; commonly coated by thin surface layers of paler hue. Streak white. Luster waxy, or resinous to adamantine. Biaxial positive. R.I., variable according to composition, $\alpha = 1.785\text{-}1.800$, $\beta = 1.787\text{-}1.801$, $\gamma = 1.840\text{-}1.849$. Birefringence 0.049-0.055.

Chemistry: A rare earth phosphate. Thorium often replaces cerium or lanthanum. The radioactivity of thorium accounts for the metamict state observed in many specimens.

Distinctive Features and Tests: Crystal form; reddish-brown color. Infusible.

Occurrences: Crystals occur only in granitic pegmatites, usually as well-formed individuals enclosed in feldspar and associated with zircon, xenotime, fergusonite, samarskite, apatite, and columbite. Good crystals in pegmatites in North Carolina, sometimes in very large individuals as one measuring 11″ by 9½″ by 6½″ and weighing 58¾ pounds from Mars Hill, Madison County. A number of pegmatites in Colorado have yielded fine large crystals, notably from the Trout Creek Pass district in Chaffee County, sometimes in "museum" specimens 8″ across; from the Quartz Creek (Ohio City) district in Gunnison County where euhedral crystals to 2″ length have been found; on Centennial Cone in Jefferson County, also at the Burroughs Mine, as very

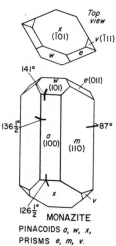

FIG. 212.

large crystals from Prairie Divide area, northern Larimer County, and in Eight Mile Park, Fremont County, as rude crystals to 1½″. Tin deposits of Llallagua, Bolivia, provide excellent crystals of monazite notable for absence of thorium; most are flesh-pink in color, twinned, and associated with other species in outstanding micromount specimens. Numerous localities in the southern part of Norway furnish excellent small sharp crystals to 1¼″, commonly doubly-terminated with fine sharp faces. Similar crystals from various localities in Madagascar, principally from Mt. Vohombohitra, north of Ankozobe, and at Ampangabe and Ambatofotsikely; crystals ordinarily reach 2½″ in length, but much larger ones are known; twins of excellent form are also obtained. Bonazite is chiefly of interest to the collector of single crystals, matrix specimens seldom being offered.

HYDRATED PHOSPHATES

Phosphophyllite—$Zn_2(Fe,Mn)(PO_4) \cdot 4H_2O$

Hydrous zinc iron-manganese phosphate; fŏs fŏf′ĭ līt. An exceptionally rare species, previously found only in a pegmatite at Hagendorf, Oberpfalz, Germany, as small transparent bluish-green crystals, seldom over ¼″ diameter. Lately, in splendid euhedral crystals to 2″ by 1″ by ¾″, and perfectly transparent, in vugs in massive sulfides from Poopo, Bolivia. This astonishing new occurrence has furnished a considerable number of beautiful single crystals, small crystal groups, and even matrix specimens. A few specimens have been cut into beautiful faceted gems. It is still an extremely rare species and the Bolivian material quickly disappeared from the market despite the high prices asked.

Vivianite Group

Vivianite—$Fe_3(PO_4)_2 \cdot 8H_2O$

Name: After English mineralogist J. G. Vivian; vĭv′ĭ ăn ĭt.

Crystals: Monoclinic. Predominantly prismatic, elongated along the c axis, and flattened on b[010] (Figure 213). Crystals seldom sharply terminated (Figure 214), but prism faces often flat and lustrous. In small to large prismatic crystals on matrix, in jumbled groups but also in stellate groups, and massive bladed or fibrous. Crystals rarely without conspicuous, partly-developed cleavages.

Physical Properties: Flexible in thin sections and sectile. Perfect cleavage parallel to b[010] and very easily developed. When thin sections are bent, yielding occurs along this cleavage, the thin laminations bending about 45°, then breaking with a fibrous fracture. H. 1½-2. G. 2.68.

Optical Properties: Transparent. Colorless when first removed from the ground, rapidly darkening to some shade of blue or green, and with further exposure to light, to dark purplish-blue, dark bluish-green, or bluish-black. Most specimens sold are already quite dark. Streak very pale purplish or greenish. Luster vitreous; sometimes pearly on cleavage plane. Biaxial positive. R.I., $\alpha = 1.579$-1.616, $\beta = 1.602$-1.656, $\gamma = 1.629$-1.675. Birefringence 0.050-0.059. Pleochroism strong, dark green material (Poopo, Bolivia) giving: dark purplish-blue, dark green, pale green.

Chemistry: Hydrated iron phosphate. Ferrous iron (Fe^{+2}) is normal in fresh material, but converts to ferric iron (Fe^{+3}), the latter being responsible for the typical dark green or blue hues after exposure to light.

Distinctive Features and Tests: Highly perfect, easily developed cleavage; flexibility of thin laminations; color; softness. Fuses at 2, yielding an iron-gray bead which is magnetic. Easily soluble in acids.

Occurrences: As a secondary mineral in some ore veins, and commonly, in clays and other sedimentary deposits where it forms concretions or is associated with organic matter

Fig. 214. *Right:* Vivianite. Bladed crystals of deep green color from Trepča, Yugoslavia. Size 1½″ tall. Crystal forms are those shown in Figure 213.

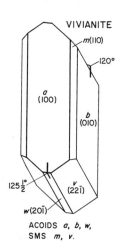

VIVIANITE

m(110)

120°

a
(100)

b
(010)

125½° *v*
(22Ī)

w(20Ī)

ACOIDS *a, b, w,*
SMS *m, v.*

FIG. 213.

as wood, fossil bones, peat and lignite. Interesting crusts and radiated masses of bladed crystals occur in many places in sedimentary formations of the Atlantic coastal plain of the United States, as at Mullica Hill, Gloucester County, Allentown and Shrewsbury in Monmouth County, New Jersey; also in Maryland, near Washington, D.C. New Jersey localities provide vivianite replacing or growing from fossil belemnites. Excellent crystals formerly from upper portions of the orebody in Ibex Mine, Leadville, Lake County, Colorado. Recently, from Bingham Canyon, Tooele County, Utah, as dark blue lathlike crystals on rock, the largest crystal to 5″ by ½″. In splendid blue-green crystals to 2″ by ½″ from within vugs in rock, Blackbird District, Lemhi County, Idaho, associated with ludlamite and siderite. Interesting nodules of radiating blades to 2″, in diatomite sediments in bed of Lake Britton, Shasta County, California. Magnificent dark green crystals occur in upper vein portions of tin deposits of Poopo and Llallagua, Bolivia, yielding single prismatic crystals of intense color to 6″ or more in length; a fine terminated crystal of 2″ by 2″ by 5″ has been recorded from the San Jose vein at Llallagua. Excellent crystals formerly from tin veins in St. Agnes and Truro districts, Cornwall, England. Also fine deep green crystals from Trepča, Yugoslavia (Figure 214). Large crystals from the Ashio mines of Shimotsuke Province, Japan. Fine crystals to 1½″ by ½″ from the Wannon River Falls, Victoria, Australia. Recently, in very large etched crystals, some to 48″ in length (!), discovered in a bog in the region of N'gaoundéré, Cameroons.

The goal of collectors in respect to vivianite specimens, is to obtain single crystals from the Bolivian localities. Most are sharp, smooth-faced, and uncommonly perfect in addition to being very large. Furthermore, most are transparent and many, flawless. Attempts have been made to cut vivianite into faceted gems, but as far as is known, very few have succeeded in view of the extreme softness and cleavability.

Erythrite—$(Co,Ni)_3(AsO_4)_2 \cdot 8H_2O$ (Cobalt-rich end member)

Annabergite—$(Ni,Co)_3(AsO_4)_2 \cdot 8H_2O$ (Nickel-rich end member)

Names: From the Greek, *erythros,* in allusion to the red color; ĕ rĭth′ rīt. From Annaberg, a famous mining town in Saxony, Germany; ăn′ă bĕrg′ĭt.

Varieties: On the basis of substitutions for cobalt and nickel: *calcian, zincian, ferroan, magnesian.*

Crystals: Monoclinic. Good crystals rare, especially of annabergite which is found almost entirely in earthy masses, crusts or powders. Erythrite commonly occurs in slender prismatic to acicular crystals, elongated along the *c* axis, forming globular tufts of micro crystals; also in bladed aggregates (Figure 215), with crystals flattened along the *b* axis, with the face *b*[010] prominent, and striated parallel to the *c* axis, and terminated by *y*[101] and *v*[$\bar{2}$21].

Fig. 215. Erythrite. Deep purplish-red bladed crystals on massive cobaltite from Bou Azzer, Morocco. Size 1½″ by 1″.

Physical Properties: Sectile. Perfect and easily developed cleavage parallel to *b*[010], yielding flexible laminations. H. 1½-2½. G. 3.06 (erythrite).

Optical Properties: Transparent to translucent. Erythrite usually distinctive deep purplish-red in larger crystals or dense tufts, but pale pink to nearly colorless as nickel content equals that of cobalt. Toward annabergite, when nickel predominates, the color becomes pale green or intense yellow-green. Streak whitish, tinged by basic color. Luster vitreous; pearly on *b*[010]. Erythrite biaxial positive. R.I., $\alpha = 1.626$, $\beta = 1.661$, $\gamma = 1.699$. Birefringence 0.073. Pleochroic in shades of red (erythrite).

Chemistry: Hydrous arsenates of cobalt-nickel. A complete series extends between end members.

Distinctive Features and Tests: Vivid, purple-red of erythrite and tendency to form thin crusts or tufts is enough to identify this mineral. The color of cuprite (*chalcotrichite*) is somewhat similar but is more brown or orange in tinge. Crystals with micaceous cleavage may be confused for purplish chlorites, but the latter do not give the deep blue borax bead furnished by erythrite. Fuses at 2 to a gray globule, giving off arsenic fumes. A small fragment in the borax bead stains deep blue. Because arsenic attacks platinum, an iron wire loop should be used for the bead test.

Occurrences: As alteration products of arsenides of cobalt and nickel, in outcrops or within short distances of the surface. Fine encrustations of micro crystals formerly from mines in the Blackbird District, Lemhi County, Utah. Common in the Timiskaming District, Ontario, on outcrops of nickel sulfide veins but seldom in good specimens. Lately, as fine acicular micro crystals to ⅛″ length, forming velvety carpets to several inches across in fissures in rock, from Mina Sara Alicia, near Alamos, Sonora, Mexico. Perhaps the best specimens came many years ago from within cavities in rock in the mines of Schneeberg, Saxony, Germany, particularly from the Grube Rappold, where acicular needles to 4″, forming stellate sprays, were recorded. Lately, in magnificent bladed crystal aggregates of singular size and perfection, from Bou Azzer, Morocco, associated with nickel-cobalt arsenides and sulfides; better specimens display blades to 1″, criss-crossed in fine groups in openings in solid sulfides. Annabergite specimens are frequently offered by mineral dealers, but as mentioned before, are seldom more than earthy crusts. Erythrite specimens from Saxony and Bou Azzer are highly prized, and usually deliver good prices. Large matrix specimens to 4″ or so, are very difficult to obtain. The tufted types are favorite micromount material, even smaller specimens providing a wealth of fine sharp needles in handsome sprays.

Variscite Group

Variscite—Al(PO₄)·2H₂O (Aluminum end member)

Strengite—Fe(PO₄)·2H₂O (Iron end member)

Names: From *Variscia,* an ancient name for the Vogtland district of Germany where the mineral was first identified; văr′ĭs ĭt. After German mineralogist J. A. Streng (1830-1897); strĕng′ĭt.

Varieties: Iron substituting for aluminum: *ferrian variscite;* also vice versa, an *aluminian strengite.*

Crystals: Orthorhombic. Variscite crystals rare; specimens providing distinct individuals much prized by micromounters. Usually pseudo-octahedral with large $o[111]$ faces, modified by $c[001]$. Cryptocrystalline variscite is abundant in nodules, veinlets, or fillings in breccia (Figure 216). Strengite crystals common but usually microscopic, and predominantly pseudo-octahedral, like variscite, but also tabular, compressed along the c axis, or stout prismatic with $a[100]$ and $b[010]$ prominent. Strengite usually forms crusts composed of radiating acicular crystals.

Physical Properties: Massive types fairly tough, breaking with porcelain-like conchoidal fractures of dull luster. Translucent variscite masses break with somewhat splintery to good conchoidal fracture. Good cleavage parallel to $b[010]$ but seen only in crystals. H. 3½-4½. G. 2.57 in crystals of variscite but near 2.53 in massive types; strengite, 2.87.

Optical Properties: Transparent in crystals; translucent to nearly opaque in massive forms. Variscite usually some shade of yellow-green, from deep green in translucent massive types of great compactness, to pale green, almost white, in earthy porous types; wide variations in shade noted within single specimens, usually dark in centers of nodules to light along fringes. Streak white. Luster dull to waxy; crystals vitreous. Strengite hues usually red, carmine, violet, but also colorless. Streak white, tinged with basic color. Luster vitreous on crystals; otherwise dull. Biaxial negative in variscite.

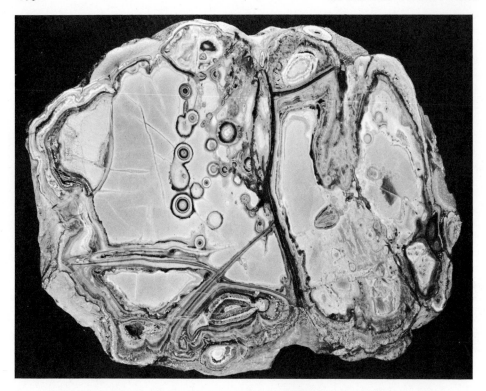

FIG. 216. Polished half of a variscite nodule from Fairfield, Utah. Size 6¼″ by 4¼″.

R.I., $\alpha = 1.563$, $\beta = 1.588$, $\gamma = 1.594$. Birefringence 0.031. Indistinct refractometer reading on massive variscite, near $n = 1.58$. Pleochroism not perceptible. Strengite biaxial positive. R.I., $\alpha = 1.707$, $\beta = 1.719$, $\gamma = 1.741$. Birefringence 0.034.

Chemistry: Hydrous aluminum-iron phosphates forming a complete series, but most examples are close to end members in composition.

Distinctive Features and Tests: Variscite nodules from Fairfield, Utah, unmistakable (Figure 216). From other localities, sometimes resembles turquois except in respect to color. Distinguished by reactions in the flame, variscite losing color at red heat, becoming pale violet when cool; turquois decrepitates and darkens. Strengite crystals change to dark brown opaque earthy masses on heating. Infusible.

Occurrences: Variscite forms in near-surface deposits from downward infiltration of phosphate-charged waters into aluminous rocks. Associates are wavellite, "eyes" of gray millisite with surrounding rings of wardite and colorless gordonite, yellow earthy crandallite, cryptocrystalline apatite, chalcedony, goethite and metavariscite. Strengite is also a near-surface secondary mineral forming from the alteration of phosphates containing iron, as triphylite in granitic pegmatites, or from phosphatic species in near-surface deposits of limonite iron ores; common associates are beraunite, cacoxenite, vivianite, metastrengite, frondelite, dufrenite, etc. The finest specimens of variscite are the ovoid to nearly spherical nodules to 12″ diameter, found in breccia openings in sedimentary rocks in Clay Canyon, near Fairfield, Utah County, Utah (Figure 216): the majority of specimens do not exceed about 3″ to 6″ diameter, and many with similar exteriors prove to be vuggy or contain very little variscite on being cut open. The best specimens are largely variscite, handsomely veined and spotted with "eyes," and with only thin surrounding bands of associated species; they are now difficult to obtain. Other localities in Utah provide less attractive material, as the breccia-fillings of deep green material from near Lucin in Box Elder County; also from 9 miles west of Stockton in the foothills of the Stansbury Mountains, Tooele County. Also fine material from Nevada at several localities in Esmeralda County and Mineral County. Micro crystals with metavariscite crystals occur as crusts lining small vugs in massive variscite at the Lucin, Utah, locality; also as incrustations on rock from Pencil, Montgomery County, Arkansas. An unusual pale green massive *ferrian variscite, redondite,* occurs in sulfide ores in the Cole Shaft, Bisbee, Cochise County, Arizona; masses reach $1\frac{1}{2}''$ diameter. Dark purple micro crystals of strengite occur in small cavities in pegmatite at Kreuzberg and Hagendorf, Bavaria, Germany; also in altered triphylite, in the Stewart Lithia Mine and other mines at Pala, San Diego County, California. Pale lavender crystals to $\frac{1}{16}''$, forming radiate aggregates on limonite, occur on Rock Run, Cherokee County, Alabama.

Variscite is prized for the handsome nodules which are usually sliced through and polished; some has been used for cabochon gems. Fine nodules are relatively inexpensive in smaller sizes, but exceptional specimens in the range of $8''$ to $10''$ are costly.

Scorodite—Fe(AsO$_4$)·2H$_2$O

A hydrous iron arsenate forming a series with mansfieldite, in which aluminum predominates instead of iron; skôr'ō dīt. Relatively rare in distinct micro crystals. Localities for micromount material are Lölling, Carinthia, Austria; St. Stephens, Cornwall, England; Laurion, Greece. Also near Ouro Preto, Minas Gerais, Brazil, where crystals of green color to $\frac{3}{8}''$ diameter were once found. Minute pale blue crystals on rock occur at Bossabut, Creuse, France.

ANHYDROUS PHOSPHATES
WITH HYDROXYL OR HALOGEN

Adelite Group

Conichalcite—CaCu(OH)(AsO$_4$)

A vivid green calcium-copper hydroxyl arsenate forming lustrous small crystals prized by micromounters; kŏn'ĭ kăl'sīt. Suitable material provided by vivid green crusts on rock from the American Eagle and other mines in the Tintic district, Juab County, Utah. Also fine in the Higgins Mine, Bisbee, Cochise County, Arizona, and more recently, from an unspecified source in Sonora, Mexico.

Austinite—CaZn(OH)(AsO$_4$)

Calcium zinc hydroxyl arsenate forming small sharp colorless to pale yellow crystals prized by micromounters; aws'tĭn ĭt. Fine specimens from Gold Hill Mine, Gold Hill, Tooele County, Utah, where small white prismatic crystals associated with adamite occur on limonite. Also as bright green drusy crystals encrusting rock in Mina Ojuela, Mapimi, Durango, Mexico, and Tsumeb, South-West Africa.

DESCLOIZITE SERIES

Descloizite—$(Zn,Cu)Pb(OH)(VO_4)$ (Zinc-rich end member)

Mottramite—$(Cu,Zn)Pb(OH)(VO_4)$ (Copper-rich end member)

Names: After French mineralogist Alfred Des Cloizeaux (1817-1897); dā kloi′ zīt. After Mottram St Andrews, an early locality from which specimens were first described, in Cheshire, England; mŏt′răm īt.

Varieties: Nine variety names have been given to species in this series, but it is sufficient to say that the series is divided in half, the appropriate species name being given according to predominance of zinc or copper.

Crystals: Orthorhombic; usually in micro crystals or as more or less parallel crystal aggregates (descloizite) forming plumose growths as in Figure 217. Very seldom in sharp

FIG. 217. Dark brown plumose aggregates of descloizite crystals on massive descloizite from Tsumeb, South West Africa. Size 3″ by 1¾″.

crystals with good faces, then pyramidal with faces of $o[111]$ prominent, yielding flattened pseudo-octahedral crystals.

Physical Properties: Brittle; conchoidal to uneven fracture. No cleavage. H. 3-3½. G. variable, usually 6.2(descloizite) to 5.9(mottramite).

Optical Properties: Transparent to opaque. Brown, reddish-brown to black in descloizite, becoming green in some mottramites. Streak orange to brownish-red to yellow; also greenish. Luster greasy. Biaxial negative. R.I., descloizite, $\alpha = 2.185$, $\beta = 2.265$, $\gamma = 2.35$; mottramite, $\alpha = 2.21$, $\beta = 2.31$, $\gamma = 2.33$. Birefringence, 0.165-0.12. Pleochroic.

Chemistry: Copper-zinc lead hydroxyl vanadates, forming a complete series.

Distinctive Features and Tests: Associations; plumose growths. Easily fused at 1½-2 on charcoal, eventually yielding a globule of lead. Powdered material dissolves in hydrochloric acid, staining solution yellow-green.

Occurrences: Secondary minerals in the oxidized zones of ore deposits containing vanadium. The brown crusts of micro crystals so commonly associated with vanadinite and wulfenite are usually a species in this series as are also the opaque dark-brown to gray-brown surface layers which sometimes form on vanadinite crystals. Micro crystals are abundant in ore deposits at many places in Arizona and New Mexico; also in Mexico but good specimens are rare. Some good specimens come from Pim Hill near Shrewsbury, Shropshire, England, in the form of thin crusts on rock. The finest material came from a number of localities in South-West Africa, as from Otavi, where large clusters of dark brown mottramite crystals, many individuals over an inch in size, were recovered in early mining; from Grootfontein (descloizite) in magnificent large brown-black crystallized groups; showy plumose groups from Friezenberg and similarly from Abenab and Tsumeb, the latter locality providing fine groups of green mottramite; also fine and large from Uitsab.

Herderite—CaBe(F,OH)(PO$_4$)

A calcium-beryllium fluorine-hydroxyl phosphate; hĕr′dĕr ĭt. Hydroxyl-herderite forms a series with herderite, representing the end member in which the hydroxyl ion predominates over fluorine. Species in this series are low-temperature hydrothermal minerals formed in granitic pegmatites, and are quite rare except in a few pegmatites in New England. Herderite occurs at Stoneham, while hydroxyl-herderite is found at Hebron, Paris, and Newry, Oxford County, Maine; also in choice ½″ crystals at Rumney, Grafton County, New Hampshire. The best crystals, some quite sharp and several inches long, have been found in Maine, notably in the Fisher Quarry, Topsham, Androscoggin County, Maine, but also fine from Stoneham.

AMBLYGONITE SERIES

Amblygonite—(Li,Na)Al(F,OH)(PO$_4$)

Montebrasite—(Li,Na)Al(OH,F)(PO$_4$)

Natromontebrasite—(Na,Li)Al(OH,F)(PO$_4$)

The three species named are lithium-sodium aluminum fluorine-hydroxyl phosphates; ăm blĭg′ō nĭt; mŏn′tĕ brā′zĭt; nā′trō—. Almost exclusively formed in granitic pegmatites as small to large grains, some of which may be many feet in diameter, and rarely, as single crystals with very rude faces. The finest crystals came from a pegmatite (Lower Pit) on Plumbago Mountain, Newry, Oxford County, Maine; crystals reached 4½″ diameter and displayed lustrous sharp faces and were largely transparent. More recently, an unusual lath-like crystal of yellow color, about 6″ length, came from Minas Gerais in Brazil. Clear material suitable for faceted gems has also come from Brazil, some being fine yellow; also from Sakangyi, Burma. Ordinary amblygonite is associated with feldspar, quartz, lepidolite and other species in pegmatites in which lithium is an abundant element.

Brazilianite—NaAl$_3$(OH)$_4$(PO$_4$)$_2$

Sodium aluminum hydroxyl phosphate; bră zĭl′yăn ĭt. A rare species found formerly only in Brazil in pegmatite at Conselheira Pena, Minas Gerais, and later at the Palermo Mine and Charles Davis Mine near North Groton, Grafton County, New Hampshire. In the last two years a large supply has come from Brazil, including crystals of astonishing size and perfection, some associated with diamond-shaped books of silvery muscovite mica in handsome matrix specimens. The finest crystals range from deep, slightly greenish-yellow equant euhedrons, to some which are nearly olive green; the largest are about 4½″ long and nearly 3″ broad. Even larger crystals were found but were not doubly-terminated. Spear-shaped crystals, similar to those of New Hampshire, came from Mantena, Minas Gerais, Brazil. New Hampshire crystals reach 1″ but most are considerably smaller. Because of expense, brazilianite specimens are not generally included in many collections and even when expense is a secondary consideration, most available specimens are little more than cleavage fragments.

Olivenite Group

Olivenite—$Cu_2(OH)(AsO_4)$

Name: From the German *Olivenerz*, literally "olive" ore, in allusion to typical color; ō lĭv′ĕn īt.

Crystals: Orthorhombic. Usually small simple prismatic individuals, commonly of diamond-shaped cross-section; sometimes elongated along the *a* axis or the *c* axis (Figure 218). Also acicular and in fibrous radiate masses.

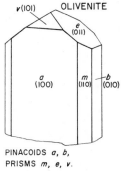

PINACOIDS *a*, *b*,
PRISMS *m*, *e*, *v*.

FIG. 218.

Physical Properties: Brittle; conchoidal to irregular fracture, H. 3. G. 3.9-4.5.

Optical Properties: Translucent to opaque. Usually some shade of olive-green; also greenish-brown, brown, yellowish. Streak colorless. Luster vitreous to somewhat adamantine. Biaxial positive. R.I., $\alpha = 1.772$, $\beta = 1.810$, $\gamma = 1.863$. Birefringence 0.091.

Chemistry: Copper hydroxyl arsenate. Phosphorus may substitute for arsenic, forming a partial series toward libethenite; iron sometimes substitutes for copper.

Distinctive Features and Tests: May be easily confused with libethenite and less easily with ludlamite which is brighter green and more transparent. Fuses readily on charcoal, emitting garlicky arsenic fumes. Dissolves easily in acids, staining solutions blue-green.

Occurrences: A secondary mineral in oxidized portions of ore deposits, commonly associated with adamite, malachite, azurite, limonite, and scorodite. As very small crystals forming druses on rock in the Magma Mine, Superior, the Mammoth Mine at Tiger, and the Copper Creek district of Pinal County, Arizona. At the Nickel Plate Mine, Blackbird District, Lemhi County, Idaho, as drusy linings in vuggy quartz. Sharp prismatic crystals on rhyolite from Majuba Hill, Pershing County, Nevada. In good to fine specimens in mines in the Tintic district, Juab County, Utah. Excellent specimens from Wheal Unity Mine, Gwinnear, Cornwall, England. Green crystals from Gap Garonne, Le Pradet, near Hyères, Var, France. Crystals to $\frac{1}{16}$″ diameter, in fine druses on chalcocite, from Tsumeb, South-West Africa. Specimens are usually drusy; many provide excellent micromount material.

Libethenite—$Cu_2(OH)(PO_4)$

Very similar to olivenite in size and forms of crystals, occurrence, associates, habit, color, etc. A copper hydroxyl phosphate; lĭ bĕth′ĕn īt. Good micro crystals occur in the open pit at Inspiration and Castle Dome, Gila County; also in the Coronado Mine, Morenci District, Greenlee County, Arizona. Dark green crystals to $\frac{1}{4}$″ on quartz from the original locality near Libethen, Neusohl, Romania. Also from Cornwall, England; Chuquicamata and Coquimbo, Chile; Mindouli, Congo, and elsewhere. As with olivenite, this species provides crystals mainly for micromounters and specimens for the collector specializing in copper species.

Adamite—$Zn_2(OH)(AsO_4)$

Name: After French mineralogist Gilbert-Joseph Adam (1795-1881); ăd'ăm ĭt.

Crystals: Orthorhombic. Usually small, forming druses of numerous tightly interlocked individuals with wedge-shaped $t[120]$ faces uppermost. Commonly elongated along the b axis (Figure 219) and forming radiate growths or spheroids implanted on matrix (Figure 220).

Physical Properties: Brittle; uneven to subconchoidal fracture. Good cleavage parallel to $d[101]$, poor parallel to $b[010]$. H. $3\frac{1}{2}$. G. 4.3-4.5.

Optical Properties: Transparent to translucent. Usually some shade of yellow-green, often intense; also green, brownish-yellow; rarely, colorless or white, or violet to rose. Streak colorless to faintly colored. Luster vitreous to

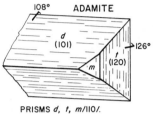

PRISMS *d, t, m//10l.*

FIG. 219.

somewhat oily. Some specimens strongly fluoresce pale yellow under ultraviolet. Biaxial negative or positive. Indices variable according to composition; R.I. (Mapimi), $\alpha = 1.722$, $\beta = 1.742$, $\gamma = 1.763$. Birefringence 0.020.

Chemistry: Zinc hydroxyl arsenate. Copper substitutes for zinc, forming a partial series toward olivenite; cobalt and iron also substitute for zinc.

Distinctive Features and Tests: Wedge-shaped drusy crystals; color; fluorescence. Colors more vivid than olivenite. Distinguished from drusy smithsonite by lack of bubbling in hydrochloric acid. Fuses with difficulty at 3.

Occurrences: A secondary mineral in oxidized portions of ore deposits, usually coating limonite gossan in association with smithsonite, calcite, olivenite, hemimorphite, malachite, and azurite. Good micro crystals occur in the Gold Hill Mine, Gold Hill, Tooele County, Utah; also in the Iron Blossom Mine. Tintic district, Juab County. In California as small colorless crystals at Chloride Cliff, Amargosa Range, Inyo County. Magnificent vivid green specimens, probably the world's best, and strongly fluorescent, occur in large drusy patches on limonite with scorodite, mimetite and wulfenite, or alone, at Mina Ojuela, Mapimi, Durango, Mexico (Figure 220). Matrix pieces reach 6″ by 6″ size and may be covered completely by sparkling drusy crusts to $\frac{1}{4}$″ thick; also in fine spherical aggregates.

FIG. 220. Adamite. Drusy coating on limonite from Mina Ojuela, Mapimi, Durango, Mexico. The sparking reflections are from pairs of $t[120]$ faces. Size $3\frac{3}{4}$″ by $2\frac{3}{4}$″.

The largest crystals reach $\frac{1}{2}$″, but those of $\frac{1}{8}$″ to $\frac{3}{16}$″ are more usual. Pink and green crystals occur at Cap Garonne, Le Pradet, near Hyères, Var, France. Fine fluorescent specimens from within vugs in smithsonite at Laurion, Greece. Also in excellent specimens from Tsumeb, South-West Africa.

Augelite—$Al_2(OH)_3(PO_4)$

A very rare aluminum hydroxyl phosphate; aw′jĕ līt. The finest crystals are transparent colorless, to 1″ diameter, from the Champion sillimanite mine, near Laws, Mono County, California. Many are flawless and some have been cut into gems. Also in small sharp crystals, forming druses in tin veins at Machacamarca, Potosi, and at Oruro, Bolivia.

Apatite Group

APATITE SERIES

The apatites are essentially phosphates of calcium with chlorine, hydroxyl, fluorine or carbonate acting also as anions, and therefore placed just before the phosphate radical in formulas. A wide range of substitutions is possible among these anions, resulting in a wide variety of species. The series is divided into fluorapatite, chlorapatite, hydroxylapatite, and carbonate-apatite, but the latter is relatively rare in specimens of interest to the collector, while of the previous species mentioned, fluorapatite furnishes the majority of specimens usually labeled as "apatite" in collections. It is not easy to distinguish between series members, except that some occur in characteristic aggregate habits. The discussion following is therefore a general one, omitting the rarer carbonate-apatite, but discussing the other three species together.

Apatite—$Ca_5(F,Cl,OH)(PO_4)_3$

Name: From the Greek, *apate*, "deceit," because often mistaken for other species; ȧp′ă tīt.
Varieties: A massive type commonly found in limestone caverns is called *francolite,* and is largely carbonate-apatite. Another cryptocrystalline type, sometimes forming chalcedony-like masses, is called *collophane.*

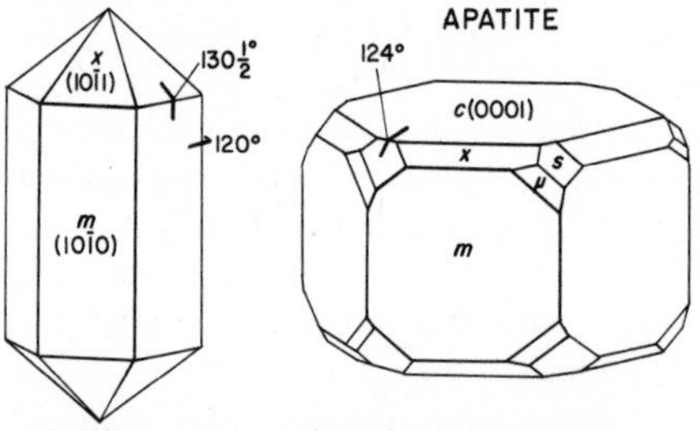

FIRST ORDER PRISM *m*, FIRST ORDER DIPYRAMID *x*, SECOND ORDER DIPYRAMID *s*/11$\bar{2}$1/, THIRD ORDER DIPYRAMID *μ*/21$\bar{3}$1/ (*mu*), PINACOID *c*.

FIG. 221.

Crystals: Hexagonal; common. Predominantly short to long hexagonal prisms (Figure 221. Prominent forms are first order prism $m[10\bar{1}0]$ with base $c[0001]$. The second order prism $a[11\bar{2}0]$ commonly truncates the edges of the first order prism and pyramid $x[10\bar{1}1]$ truncates edges between base $c[0001]$ and prism $m[10\bar{1}0]$, or may draw to a point. Crystals from pegmatites tend to be tabular but also prismatic, while crystals from the Alpine vugs are usually tabular. Faces vary considerably in smoothness, some being glassy and transparent, but others striated or etched (Figure 222).

Crystals enclosed in marble are commonly rounded or pitted, as if partly melted. Twinning is very rare. Transparent crystals are frequently cut into faceted gems.
Physical Properties: Brittle. Some crystals crumble readily when extracted from matrix because of numerous fractures. Fracture smooth conchoidal in clear crystals, but mostly

uneven to small conchoidal. Poor cleavages parallel to $c[0001]$ and $m[10\bar{1}0]$. H. 5. G. 3.1-3.2.

Optical Properties: Transparent to translucent, rarely opaque. Wide variety of hues: white, colorless and gray, many shades of green, blue, yellow; also reddish, pink, violet. Hues vary from very pale to very dark. Streak white. Luster vitreous to resinous, and on

FIG. 222. Apatite. *Top left:* transparent yellow crystal from Cerro Mercado, Durango, Mexico. *Top right:* white crystal, observed on the c(0001) face, on dark green chlorite and white adularia, from Schwarzenstein, Italian Alps. Size 3″ by 2″. *Center:* simple hexagonal terminated prism from Minas Gerais, Brazil. *Bottom left:* tabular crystal from Clark Mine, Rincon, San Diego County, California. *Bottom center:* corroded hexagonal prism of pink color from Himalaya Mine, San Diego County, California, and blue prism from gem gravels of the Mogok District, Burma. *Bottom right:* doubly terminated crystals from Bedford Township, Ontario, Canada.

some specimens, distinctly oily. Very slender acicular inclusions parallel to c axis sometimes cause strong chatoyancy; good examples have been cut into catseyes. Uniaxial negative. Considerable variation in refractive indices according to composition, but the following are typical of fluorapatites: O = 1.636, E = 1.633. Birefringence 0.002-0.004. Dichroism weak in yellow and pink apatite but often strong in blue and green examples, e.g., Durango (yellow)—greenish-yellow, yellow; Himalaya Mine (pink)—pink, slightly paler pink; Mogok (blue)—blue, colorless; Ontario (green)—blue-green, olive-yellow. Many specimens are fluorescent under ultraviolet.

Chemistry: Calcium fluorine-chlorine-hydroxyl phosphates. Manganese, cerium, and strontium substitute in small amounts for calcium.

Distinctive Features and Tests: Crystals; softness; luster. Distinguished from similar beryl and tourmaline crystals by hardness tests. Fuses with difficulty at 5-5½. Slowly dissolves in hydrochloric acid, the surface turning white and porous on removal from acid. *Occurrences:* Very widely distributed in various types of rocks and mineral deposits as in marbles, in pegmatites, in ore veins, and in the vugs characteristic of Alpine occurrences. Only a few important specimen localities can be mentioned. Large crystals enclosed in marble occur in many deposits in Renfrew, Lanark, Frontenac and Haliburton counties, Ontario; also in Ottawa County, Quebec. Individuals to 500 pounds are recorded, while single crystals, many doubly-terminated, from 2″ to 12″ long and ½″ to 3″ diameter, have been offered as mineral specimens. Fine matrix specimens are provided by green crystals in pink calcite, the latter chipped away carefully to expose the apatite crystals. All Canadian crystals are severely checked and cracked, and most crumble when removed from matrix, thus making undamaged crystals rare despite their abundance *in situ*. Occasionally, clear fragments can be cut into faceted gems. Canadian apatites are easily recognized by the curious rounding of prism and termination edges (Figure 222). Excellent crystals to 3″, with actinolite and asbestos at Pelham, Hampshire, Massachusetts. Splendid transparent, violet to purple, tabular to stubby, prismatic crystals occur in pegmatite cavities in New England, notably at Mount Apatite, Androscoggin County, Maine; the record crystal from this locality is 1½″ by 1¾″ by 1½″ (the "Roebling" apatite in the Smithsonian collections). Most crystals are far smaller, usually as stubby hexagonal prisms nearly as wide as long, and from ¼″ to ⅜″ diameter; the finest specimens display isolated crystals on masses of silvery cookeite. Fine pink prismatic crystals, closely resembling tourmaline, occur in the Himalaya Mine, Mesa Grande, San Diego County, California; individuals to 4″ length have been found, but most range between 1″ and 1½″ length. The pink hue is fugitive; when crystals are exposed to light, the pink disappears, to be replaced by colorless or pale greenish-gray to greenish-blue. Other pegmatite mines in San Diego County provide specimens, notably large dark blue crystals, commonly with thin exterior zones of opaque chatoyant material, in the White Queen Mine, Hiriart Hill, Pala, and in the San Pedro Mine (Figure 223). An enormous quantity of splendid euhedral transparent yellow crystals occurs in the open pit iron mines at Ciudad Durango, Durango, Mexico, where literally millions of crystals have been supplied to the specimen market over a period of at least four decades. Usual crystals are smooth lustrous transparent yellow individuals, about ¾″ in length but ranging upward to some as large as 4″ in length and 1½″ in diameter; matrix specimens are commonly available, displaying prisms implanted on reddish or blackish vuggy hematite. Fine large crystals, but usually opaque, occur in Scandinavia as at Gellivara and Nordmark in Sweden; occasionally fine blue and white from Snarum, Norway. Beautiful purplish tabular crystals to ¾″ diameter on matrix from tin mines in central Europe as near Ehrenfriedersdorf, Saxony, Germany, and from Schlaggenwald, Bohemia, Czechoslovakia. Splendid sharp white to colorless tabular crystals to 2½″ diameter from Switzerland as at Alpe Della Sella, at Gletsch, and in the St. Gotthard region generally; also fine white and colorless from Schwarzenstein, with chlorite and adularia, in the Tyrol of Austria, and with chlorite at Bolzano, Italy. The yellow crystals of Jumilla, Murcia, Spain, occur as terminated prisms with calcite on volcanic rock. Perhaps the finest apatite specimens of all come from tin veins in Bolivia, notably at Llallagua, where white tabular complex crystals, largely transparent with smooth sharp faces, occur on masses of ore; crystals commonly reach diameters to 1″ and over, and matrix specimens of 6″ diameter heavily sprinkled with tabular crystals were once found. Also from Potosi and Catavi, Bolivia, in pale purple tabular crystals, some to 2″ diameter. At times, fine sharp green crystals from Minas Gerais, Brazil. Fine blue crystals come from gem gravels of Mogok, Burma.

FIG. 223. Tabular apatite crystals from San Pedro Mine, Hiriart Hill, Pala, San Diego County, California. Exteriors are chatoyant white material but interiors are dark blue and transparent. Size of largest crystal 1½″ diameter and ¾″ thick. *William C. Woynar Collection, San Diego.*

Most eagerly sought specimens are purple crystals on matrix, preferably from Maine localities but also from the central European localities. Next in desirability are clear tabular crystals from Bolivia or Europe, followed by well-formed crystals from marble. The abundance of Mexican yellow apatites makes them a drug on the market, and only fine clear crystals several inches in length are sought for by experienced collectors. The pink apatite crystals from California are nearly unique, but so few are produced that it is difficult to obtain examples.

PYROMORPHITE SERIES

Pyromorphite—$PbCl(PO_4,AsO_4)_3$ (Phosphate-rich end member)

Mimetite—$PbCl(AsO_4,PO_4)_3$ (Arsenate-rich end member)

Names: From the Greek, *pyr,* "fire," and *morphe,* "form," in allusion to the fact that a melted globule assumes a crystalline shape on cooling; pī′rō̆ môr′fīt. From the Greek, *mimetes,* "an imitator," in allusion to its resemblance, to pyromorphite; mĭ′mĕ tīt.

Varieties: With calcium substituting for lead: *calcian;* also a *vanadian* variety. The distinctive curved crystals from Cumberland, England and elsewhere, are locally called *campylite.*

Crystals: Hexagonal; usually simple short hexagonal prisms, commonly with hollow centers descending some distance from the terminations. Also in crystals terminated by the pyramid $x[10\bar{1}1]$ with large or small base $c[0001]$ (Figure 224). Many crystals are barrel-shaped individuals forming curved aggregates resembling hemispheres (*campylite*); also tapered, like spindles, and in parallel groupings (Figure 225). Rarely, undistorted in fine, sharp, lustrous crystals.

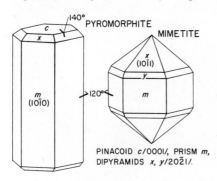

PINACOID *c/0001/,* PRISM *m,*
DIPYRAMIDS *x, y/20$\bar{2}$1/.*

FIG. 224.

Physical Properties: Very brittle; uneven to subconchoidal fracture. H. $3\frac{1}{2}$-4. G. 7.04 (pyromorphite), 7.24 (mimetite).

Optical Properties: Translucent to opaque. Pyromorphite usually in shades of brown, green, yellow or orange; mimetite usually yellow, yellow-brown, brown, orange. Streak white or faintly tinged with basic color. Luster resinous to nearly adamantine. Uniaxial negative. R.I., variable, higher in mimetite, $O = 2.058$-2.147, $E = 2.048$-2.128. Birefringence 0.010-0.019.

Chemistry: Lead chlorine phosphate-arsenates. A complete series extends between end members.

Distinctive Features and Tests: Cavernous hexagonal prismatic crystals; lack of transparency; luster. Species fuse easily at $1\frac{1}{2}$, pyromorphite rounding quickly and forming a white opaque bead on cooling, which when observed with a magnifying glass, is seen to be covered by numerous slightly curved faces joining at sharp angles to each other. Mimetite decrepitates, then quickly fuses into a globule, which after a minute or two of flaming, suddenly sputters, emits arsenic fumes, reduces its size drastically, then quiets; if the flame is taken away at this moment, a small bead of lead will be found beneath the crust of the fused globule. Soluble in nitric acid.

Occurrences: Secondary minerals formed in the oxidized ore veins containing lead, commonly associated with cerussite, smithsonite, hemimorphite, anglesite, pyrite, vanadinite, wulfenite and mottramite. Pyromorphite is more abundant both in respect to quantity and quality of crystallized specimens. Fine green and yellow pyromorphite crystals, with galena and cerussite, from Society Girl Claim, Moyie, Steele mining district, British Columbia. Magnificent druses studded by sharp green prisms from numerous mines in Coeur d'Alene district, Shoshone County, Idaho, especially in the Hercules Mine, as large crystals to $1\frac{1}{4}$" x $\frac{3}{8}$"; grayish-violet crystals to $\frac{3}{4}$" from Caledonia Mine; large cavernous crystals from Lookout Mountain Mine; also fine from Little Giant and Blackbear mines. Most Idaho specimens display simple hexagonal green prisms of $\frac{1}{2}$" length, many cavernous, on rusty massive quartz. Bright orange-yellow globular mimetite on matrix from 79 Mine, Banner district, Gila County, Arizona; similar masses, some to several inches in thickness, from Bilbao Mine, Ojo Caliente, Zacatecas, Mexico; pale greenish-gray pyromorphite crystals, with wulfenite, from Ojuela Mine, Mapimi, Durango.

Fine specimens from Wheal Alfred, Phillack, Cornwall, England; splendid globular aggregates of mimetite from Dry Gill, near Roughten Gill, Caldbeck Fells, Cumberland, the type locality for *campylite.* Specimens from the latter place show rounded aggregates of lustrous brown hue and sometimes vivid yellow, measuring to $\frac{1}{2}$" diameter or better,

FIG. 225. Pyromorphite and vanadinite. *Top:* pale brown curved pyromorphite crystals on galena and quartz matrix from Grube Rosenberg, Bad Ems, Rheinland, Germany. Size 4″ by 2″. *Bottom:* dark red-brown cavernous vanadinite crystals on limonite matrix from Villa Ahumada, Sierra Los Lamentos, Chihuahua, Mexico. Largest crystals about ½″ in length.

implanted on quartz. Yellow types sometimes contain small cavities in which sharp acicular crystals suitable for micromounts are present. Exceptionally fine pale brown barrel-shaped pyromorphite crystal aggregates, sprinkled on quartz breccia fragments, in Grube Rosenberg, Bad Ems, Rheinland, Germany; some specimens are many inches across and covered with crystals to $\frac{1}{2}''$ or more in length. Extremely large pyromorphite crystals, usually brown-stained and partly altered to galena, to $\frac{7}{8}''$ diameter, at Kantenbach near Bernkastel, Mosel, Germany; very similar in size and appearance are crystals from Beresovsk, Urals, U.S.S.R. (Figure 226). Friedrichssegen, near Ems, Hesse-Nassau,

FIG. 226. Pyromorphite crystals from Beresovsk, Ural Mountains, U.S.S.R. These unusually large crystals are about $\frac{3}{4}''$ in diameter. *Photo courtesy Schortmann's Minerals, Easthampton, Massachusetts.*

Germany, is also noted for splendid specimens, including clusters of brown $\frac{3}{4}''$ crystals, or singles sprinkled on rock fragments, or crystals growing in parallel position. Johanngeorgenstadt in Saxony, Germany, provides tabular crystals of mimetite to $\frac{1}{2}''$ length, displaying pyramidal faces. Vivid green isolated crystals of pyromorphite from Pribram, Czechoslovakia. Other localities furnishing excellent specimens are various mines in South-West Africa and Broken Hill, New South Wales.

Pyromorphite is an excellent cabinet mineral, and good specimens do not go begging for buyers. Most commonly represented in collections are the green specimens from Idaho, now quite scarce, the *campylite* of Cumberland, not so scarce, and to a much lesser extent, the large German crystals, some of which are no longer available except from old collections. The Bad Ems, Germany, locality has been recently productive of fine specimens.

Vanadinite—Pb$_5$Cl(VO$_4$)$_3$

Name: After its composition; vă nàd′ ĭ nīt.
Varieties: Arsenic substituting for vanadium: *arsenian,* commonly called *endlichite.*

Crystals: Hexagonal. Usually sharp-edged and smooth-faced, but also cavernous and curved and sometimes in parallel groupings (Figure 225). The majority of crystals are simple, stubby hexagonal prisms; some are considerably elongated. The base $c[0001]$ is sometimes truncated along prism intersections by the pyramid y $[20\bar{2}1]$. Seldom massive.

Physical Properties: Very brittle; uneven to conchoidal fracture. H. $2\frac{1}{2}$-3. G. 6.5-7.1.

Optical Properties: Translucent to nearly opaque; rarely transparent. Principally in shades of red or brown, or combinations of same; also brownish-yellow, and yellow. Streak white to yellowish. Luster resinous to nearly adamantine. Uniaxial negative. R.I., $O = 2.416$, $E = 2.350$. Birefringence 0.066.

Chemistry: Lead chlorine vanadate. Phosphorus and arsenic substitute for vanadium, and small amounts of calcium, copper or zinc for lead.

Distinctive Features and Tests: Crystals less distorted, more lustrous and perfect than pyromorphite, and often vivid red or tinged with red. Easily fuses at $1\frac{1}{2}$ on charcoal, decrepitating slightly, then rapidly melting; after some moments, the globule suddenly grows larger and froths, then sinks into the charcoal, leaving a number of very small lead beads. Easily dissolves in nitric acid, staining the solution yellow; dissolves in hydrochloric acid, staining the solution green.

Occurrences: A secondary mineral in oxidized ore deposits containing lead, associated with pyromorphite, mimetite, wulfenite, descloizite, cerussite, anglesite, quartz, and limonite. The finest United States specimens come from numerous sources in Arizona and New Mexico. In the latter state, formerly in museum specimens from Fairview Claim, Black Canyon district, Dona Ana County; exceptional from Georgetown district mines, Grant County; especially fine in yellow, red and brown crystals to $\frac{5}{8}''$, from Sierra Grande Mine, and others, in Hillsboro district, Sierra County; also from Dewey Mine and others near Palomas Gap, Caballos Mountains. In Arizona, the following are important localities: Globe district, Gila County, especially fine brilliant red hexagonal crystals, liberally encrusting breccia at Apache Mine; also at 79 Mine. Apache vanadinites are world-famous, and although the crystals seldom exceed $\frac{1}{2}''$, they are smooth-faced and brilliant, and pure red in hue; occasionaly druse-coated breccia blocks to $12''$ are obtained. The source is not exhausted. Similar crystals, sometimes partly altered in specimens near the outcrop, occur on breccia fragments in the Old Yuma Mine, Tucson Mountains, near Tucson, Pima County; this mine is more famous for its exceptional wulfenites. Fine crystals from Mammoth-St. Anthony Mine, Tiger, Pinal County; also very fine in Red Cloud Mine, Trigo Mountains, Yuma County. In California, good specimens have been obtained from El Dorado Mine, near Indio, Riverside County, and from Vanadium King Mine, Camp Signal, near Goffs, San Bernardino County. Superb red-brown hexagonal crystals to $\frac{1}{2}''$ on calcite matrix, and also tapered pale brown crystals of the *endlichite* variety, from Villa Ahumada, Sierra Los Lamentos, and the Santa Eulalia district, Chihuahua, Mexico. Small brown hexagonal vanadanite crystals occur in the Wanlockhead mines, Leadhills district, Scotland. Astonishingly large crystals, reportedly to $5''$ length, of vanadinite coated with thick brown alteration crusts of descloizite, occur at Djebel Mahseur, near Oudjha, Morocco; specimens from this locality usually show thick rude hexagonal prisms to $1''$ to $1\frac{1}{2}''$ length, forming druses; the bright orange interiors are invisible except on fractured surfaces. Good specimens of vanadinite are not easy to obtain, especially in large crystals; the average good specimen displays simple hexagonal crystals about $\frac{1}{4}''$ to $\frac{1}{2}''$ in length.

LAZULITE SERIES

Lazulite—(Mg,Fe)Al$_2$(OH)$_2$(PO$_4$)$_2$ (Magnesium-rich end member)

Scorzalite—(Fe,Mg)Al$_2$(OH)$_2$(PO$_4$)$_2$ (Iron-rich end member)

Names: From an older German name, *Lazurstein,* "blue stone," in allusion to color; lȧz'ū līt. After Brazilian mineralogist E. P. Scorza; skôr'zȧ līt.

Crystals: Monoclinic Usually in veinlets, granular aggregates, and blebby masses enclosed in other rocks, but good crystals have been found at a number of localities. The latter resemble simple dipyramids (Figure 227). Commonly twinned on composition plane *a*[100]. Crystal faces seldom smooth and sharp.

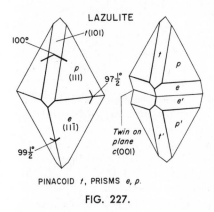

PINACOID *t*, PRISMS *e, p.*

FIG. 227.

Physical Properties: Brittle to very brittle, difficult to remove crystals intact from matrix. Fracture uneven. Poor to good cleavage parallel to *m*[110]. H. 5½-6. G. about 3.1 (lazulite) to 3.4 (scorzalite).

Optical Properties: Mostly translucent to opaque; very rarely, clear and transparent. Distinctive deep blue in masses, appearing almost black. Very small grains, veinlets, etc., appear much lighter blue. Streak white. Vitreous to dull luster. Biaxial negative. R.I., $\alpha = 1.604$-1.639, $\beta = 1.626$-1.670, $\gamma = 1.637$-1.680. Birefringence 0.033-0.051. Strong pleochroism: dark blue, green-blue, colorless (Laws).

Chemistry: Magnesium-iron aluminum hydroxyl phosphates. Probably a complete series extends between end members but lazulite members are more abundant.

Distinctive Features and Tests: Can be confused with lazurite. Small fragments of ⅛" to ¼" redden on edges, but do not fuse, then swell and split into small puffy sheets which blow away in the flame. Slowly soluble in hot acids.

Occurrences: In highly metamorphosed rocks, particularly quartzites; also in quartz veins and in some granitic pegmatites. Commonly associated with andalusite, kyanite, quartz, sillimanite, dumortierite, rutile, pyrophyllite, garnet, mica, and sometimes corundum. From Graves Mountain, Lincoln County, Georgia, as sharp pale blue, thoroughly shattered crystals to 1½", of dipyramidal habit, completely enclosed in sugary granular quartz with kyanite and rutile; crystals sometimes can be freed from matrix without crumbling, furnishing individuals to 1". Large deep blue masses to 6" across occur in Champion sillimanite mine, Laws, Mono County, California; some can be cut into cabochons. Fine crystals in massive quartz and calcite occur near Werfen, Salzburg, Austria. Crystals to 6" length from Westana, Sweden. Small transparent gem crystals occur in diamond-bearing gravels of Minas Gerais, and elsewhere in Brazil, some reportedly cut into small gems. Good crystals from any locality are rarities and much prized by single crystal collectors; common offerings are masses of rock enclosing crystals, but usually with the crystals sheared off flush with rock fracture surfaces.

Turquois Group

TURQUOIS-CHALCOSIDERITE SERIES

Turquois—$CuAl_6(OH)_8(PO_4)_4 \cdot 4H_2O$

Name: The name means *Turkish,* perhaps applied to Persian material brought via early trade routes through Turkey, and hence assumed to be from that country; tĕr′koiz, also tĕr′kwoiz.

Crystals: Triclinic; extremely rare. Predominantly in cryptocrystalline nodules or seam

FIG. 228. Blue turquois nodule with white kaolinite, Fox Turquois Mine, near Cortez, Lander County, Nevada. The botryoidal surface, typical of much nodular turquois, is well-developed in this unusually large specimen. Size 3¼″ by 2¼″.

fillings (Figure 228). Acute pyramidal micro crystals occur only at the Virginia locality.

Physical Properties: Brittle, especially in exceptionally compact massive material. Fracture crudely conchoidal, usually earthy to porcelainous in texture, but sometimes smooth and waxy in luster. H. variable, 4-6, earthy or porous types lower. G. variable, 2.6-2.9.

Optical Properties: Compact material translucent on thin edges. Pale green, blue-green, rarely, fine blues, commonly shading within same specimen because of variations in compactness. Hues deepen with impregnation by waxes, oils, and plastics, but frequently, impregnating oils or waxes decompose and cause dull unattractive greenish hues. Mean refractive index about 1.62, appearing as a vague line on the refractometer.

Chemistry: Hydrous copper aluminum hydroxyl phosphate. Divalent iron substitutes for copper. A partial series extends to chalcosiderite.

Distinctive Features and Tests: Not likely to be confused with other species except variscite which turns distinctive violet when heated in the flame, or chrysocolla, the powder of the latter staining a borax bead deep green but turquois leaving the bead colorless. If crushed and strongly heated first, turquois dissolves readily in hydrochloric acid.

Occurrences: A secondary mineral formed by infiltration of surface waters into phosphatic rocks containing copper and aluminum. Predominately in nodules enclosed in white or limonite-stained kaolinite, or thin seam fillings in igneous rocks as trachyte. Abundant in the southwestern United States, but uncommon elsewhere, and rare in Mexico. Splendid nodules from numerous localities in Nevada; one of 152 pounds is recorded from near Battle Mountain, Lander County, and later, one of 178 pounds from No. 8 Mine, Lynn district. New Mexico, Arizona, and Colorado provide fair to fine turquois in seam and nodular types, with production continuing from some deposits. The famous crystallized turquois comes from a small copper prospect near Lynch Station, Campbell County, Virginia, where it occurs as thin drusy coatings of pale blue micro crystals on quartz breccia. The largest specimens are about 3″ by 3″, and exhibit velvety carpets of microscopic crystals on one or more surfaces of the quartz blocks. This material is prized by micromounters and commands very high prices. In respect to color, the best turquois has been obtained for centuries from numerous excavations in brecciated trachyte and clay slate on the southwest slope of the Ali-Mirsa-Kuh Mountains, near Nishapur, Khorasan, Iran. Also fair to good from the Sinai Peninsula of Egypt, and from Chuquicamata, Chile. Aside from the crystallized Virginian material, fine large nodules as shown in Figure 228, are prized as colorful collection pieces; also prized are bright-hued seams in rock, but good specimens of this kind are difficult to obtain.

Ludlamite—$(Fe,Mg,Mn)_3(PO_4)_2 \cdot 4H_2O$

A rare hydrous iron-magnesium-manganese phosphate, forming small green wedge-shaped crystals as drusy coatings in oxidized portions of some ore deposits; lŭd′lăm ĭt. In the last decade, fine specimens with single crystals to ½″, came from the Blackbird district of Lemhi County, Idaho. Matrix specimens coated with drusy crystals reached sizes of 5″ by 5″. Sometimes micro crystals on matrix can be obtained from the original locality at Wheal Jane, Truro, Cornwall, England. Small crystals suitable for micromounting occur in Palermo Mine, North Groton, New Hampshire.

Wavellite—$Al_3(OH)_3(PO_4)_2 \cdot 5H_2O$

Name: After William Wavell (d. 1829) of England who discovered the mineral; wā′vĕl ĭt.

Crystals: Orthorhombic; always microscopic. Predominately in acicular radiating aggregates as crusts on rocks or as distinct spheres (Figure 229). Wedge-shaped tips of Arkansas wavellites are terminated by $p[101]$.

Physical Properties: Brittle; fracture fibrous in crusts. Perfect prismatic cleavage parallel to $m[110]$; good $p[101]$. H. 3½-4. G. 2.36.

Optical Properties: Translucent. White, greenish-white, very pale gray, brown, yellowish-brown, yellow and green. Streak white. Luster vitreous; pearly on fracture surfaces. Biaxial positive. R.I., $\alpha = 1.520$-1.535, $\beta = 1.526$-1.543, $\gamma = 1.545$-1.561. Birefringence 0.025-0.026. Strong pleochroism in green Arkansas material: blue-green, pale yellow. Polished masses give vague refractometer readings between 1.53-1.55.

Chemistry: Hydrous aluminum-hydroxyl phosphate. Iron sometimes substitutes for hydroxyl in small amounts.

Distinctive Features and Tests: Green Arkansas material is unmistakable, but thin crusts of much less vivid coloration from many other localities are not as distinctive. Dissolves readily in acids. In the flame, puffs and separates into layers; infusible.

Occurrences: A secondary mineral formed in near-surface deposits in a variety of rocks, and more rarely, in upper portions of ore veins. Commonly associated with turquois. Undoubtedly the finest specimens are from narrow openings in brecciated and partly al-

FIG. 229. Wavellite aggregates on altered novaculite from Montgomery County, Arkansas. Size 3″ by 2″.

tered novaculite which is widespread in the Arkansas counties of Garland, Montgomery, and Hot Springs; the best examples occur near Pencil, Montgomery County, as spheroidal aggregates to $1\frac{1}{4}''$ diameter. Because crevices are usually very narrow, full spherical development is impossible, but occasionally, as in Figure 229, good balls are found. Fine colorless micro crystals to $\frac{1}{32}''$ thickness occur at Moores Mill, Cumberland County, and at Hellertown, Northampton County, Pennsylvania. Thin crusts of very pale gray color in the King Turquois Mine, near Villa Grove, Saguache County, Colorado. Common in tin veins of Llallagua, Bolivia, in superb micro crystals, some reaching lengths of $\frac{1}{4}''$, as tufts and usually associated with vauxite, paravauxite, and childrenite; also in large pale green stalactitic masses. In crusts from Tracton and Kinsale, County Cork, Ireland; near Barnstable, Devonshire, England. In radiate disks in slate, Back Creek, Tasmania, Australia.

Torbernite and Metatorbernite Groups

Torbernite—$Cu(UO_2)_2(PO_4)_2 \cdot 8$-$12H_2O$

Name: After Swedish chemist Torbern Bergmann (1735-1784); tôr′bĕrn ĭt.

Crystals: Tetragonal. Usually tabular, flattened along the c axis, and square in outline. The broad faces seen on practically all crystals are $c[001]$. Many crystals extremely thin, commonly forming "books" of slightly divergent individuals. Also in scaly aggregates.

Physical Properties: Brittle; thin laminations somewhat flexible. Perfect micaceous cleavage on $c[001]$. H. 2-$2\frac{1}{2}$. G. 3.22.

Optical Properties: Translucent; sometimes transparent. Vivid yellowish-green, from pale to fairly dark. Streak greenish. Luster vitreous, but strong pearly on $c[001]$. Uniaxial negative. R.I., O = 1.59, E = 1.58. Birefringence 0.01. Dichroism distinct: green, blue.

Chemistry: Hydrous phosphate of copper and uranium. Substitutions of arsenic for phosphorus and lead for copper sometimes occur. Loss of water in dry atmospheres causes alteration to metatorbernite, and probably most cabinet specimens are this species.

Distinctive Features and Tests: Green square mica-like crystals seldom over $\frac{1}{8}''$ across.

Greener than autunite, but more certainly distinguished from the latter by lack of fluorescence under ultraviolet.

Occurrences: A secondary mineral of oxidized ore veins containing uranium, and probably derived from alteration of uraninite; sparingly in granitic pegmatites. As small micro tablets in granitic pegmatites of New England and North Carolina; also in pegmatites of Black Hills district, South Dakota. Excellent specimens, displaying numerous octagonal platey crystals to $\frac{1}{8}''$ on pale altered trachyte, from Mina Candelaria, Moctezuma, Sonora, Mexico; solid coverings to several square inches have been found. The finest specimens, often showing sprinklings of $1''$ square tablets on limonite gossan occurred in upper portions of various copper mines in Cornwall, England, notably near Gunnislake, near Callington, and from Redruth. Some English crystals are nearly bipyramidal in habit. Exceptionally fine from a number of places in Australia, as Mt. Painter in Flinders Range, South Australia; South Alligator Gorge, and along the Adelaide River, Northern Territory. Abundantly at Kasolo, Chinkolobwe, and other places in Katanga, Congo. Most cabinet specimens display only small crystals of torbernite, the exceptional specimens from England and Australia being difficult to obtain. All specimens are very fragile, the platey crystals often being loosely attached and falling off at the slightest jar. It is sometimes desirable to dip specimens in thin solutions of transparent lacquer to cement crystals to matrix.

Autunite—$Ca(UO_2)_2(PO_4)_2 \cdot 10\text{-}12H_2O$

Name: After the locality at Autun, Saône-et-Loire, France; aw'tŭn īt.

Crystals: Tetragonal. Thin to very thin tabular crystals, compressed along the c axis, closely resembling torbernite. Commonly in divergent fan-like growths (Figure 230), or in masses of jumbled crystals. Small crystals, in the micro range, are nearly perfect, but quality rapidly falls as crystals grow larger.

Physical Properties: Flexible in thin laminations. Perfect and easily developed cleavage on $c[001]$. H. $2\text{-}2\frac{1}{2}$. G. $3.1\text{-}3.2$.

Optical Properties: Transparent to translucent. Predominantly vivid lemon-yellow, greenish-yellow, or pale green; but less dark than torbernite. Streak yellowish. Luster vitreous; strong pearly on $c[001]$. Strongly fluoresces pale yellow-green under ultraviolet. Usually biaxial negative, although it should be uniaxial negative. R.I., $O = 1.577\text{-}1.578$, $E = 1.553\text{-}1.555$. Birefringence $0.023\text{-}0.024$.

Chemistry: Hydrous phosphate of calcium and uranium. Barium and magnesium substitute for calcium. No series exists between autunite and torbernite. Water is readily lost, causing alteration to meta-autunite; probably all cabinet specimens are the latter.

Distinctive Features and Tests: Vivid coloration; tabular crystals; fluorescence evident even in daylight. The very strong fluorescence under ultraviolet is matched only by the easily-distinguished willemite. Minute, brilliantly green fluorescent specks seen on specimens of granitic pegmatite, are usually autunite.

Occurrences: A secondary species resulting from the alteration of uranium-bearing minerals in oxidized zones of ore deposits and weathered portions of pegmatites. Sprinklings of minute tabular crystals are commonly observed in crevices in granitic pegmatites in New England and North Carolina, but only rarely in the pegmatites of California and Arizona. Within the last decade, magnificent specimens, completely eclipsing any found before, were discovered in seams in granitic rock associated with pale brown kaolinite at the Daybreak Mine, Mt. Spokane, Spokane County, Washington. The specimen illustrated in Figure 230, is a very modest one alongside the thick masses and crusts which

FIG. 230. Autunite. Platey crystals from the famous occurrence in the Daybreak Mine, Mount Spokane, near Spokane, Washington. As with most autunite, the loss of water has caused alteration to meta-autunite. Size 2″ by 1¼″.

reached dimensions of 12″ broad and several inches thick! Before it was recognized that prompt preservation steps were required, many early specimens lost water and disintegrated ruinously. Dealers now promptly clean and dry specimens and soak them in thin plastic lacquer solutions to preserve them, this being more satisfactory than spraying. On larger masses, individual crystals form divergent aggregates or "cockscombs" of at least 1½″ broad, although the usual run is from ½″ to 1″ broad. The coloration of Spokane crystals is distinctive, being vivid light yellow-green on edges, but deeper green to almost blackish-green near the centers. Other localities producing fine specimens are Mt. Painter, Flinders Range, South Australia, and Katherine, Northern Territory, Australia. Also fine from a number of localities in France as at Margac, Haute Vienne, and Autun, Saône-et-Loire. From Redruth and St. Austell, Cornwall, England, in fine specimens. The fragility of specimens, and the general rarity of this species, increases the cost, and high-quality examples may seem shockingly overpriced to the uninitiated.

TUNGSTATES AND MOLYBDATES

Only 16 species are included in these classes, with a very few occurring in desirable cabinet specimens. Molybdenum and tungsten are not abundant elements, but are extremely useful in technology. Both elements are widespread but seldom occur in concentrations conducive to economical mining; consequently, good specimens of their compounds are relatively scarce.

Wolframite Group

WOLFRAMITE SERIES

Huebnerite—MnWO₄ (Manganese end member)

Wolframite—(Fe,Mn)WO₄ (Iron-manganese intermediate member)

Ferberite—FeWO₄ (Iron end member)

Names: After German mineralogist Adolph Huebner; hūb'nĕr ĭt. After the German, *Wolfram,* tungsten; wool' frăm ĭt. After Rudolph Ferber of Germany; fĕr'bĕr ĭt.
Crystals: Monoclinic; commonly tabular to prismatic (Figure 231). Huebnerite usually in lath-like crystals, elongated along *c* axis and flattened along *a* axis; the resulting broad *a*[100] faces predominate and show striations parallel to the *c* axis. Wolframite mostly stubby prisms but also flattened along the *a* axis in thick tabular individuals and similarly striated. Ferberite frequently elongated along the *b* axis. Striations assist in orienting crystals. Quality of faces often varies in same crystal, faces of one form being very smooth, but those of another striated, etched, or roughened in some respect. Twinning common along *a*[100], producing contact twins evidenced by deep reentrants atop

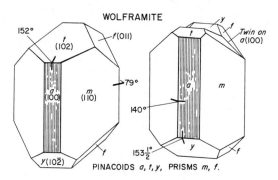

FIG. 231.

terminations (Figure 231). Rarely, one twin is turned about, producing crystals resembling orthoclase Carlsbad twins. Also granular massive.
Physical Properties: Brittle; fracture uneven to conchoidal. Perfect cleavage parallel to *b*[010]; parting often occurs parallel to external faces, resulting in "peeling" (Schlaggenwald). H. 4-4½. G. 7.1 (huebnerite) to 7.5 (ferberite).

Optical Properties: Huebnerite reddish-brown, translucent; ferberite black, opaque. Streak reddish-brown to black. Luster submetallic.

Chemistry: Manganese-iron tungstates, forming a complete series.

Distinctive Features and Tests: High gravity; crystal form; single cleavage; color. Cleavage and crystal form distinguish from cassiterite or columbite-tantalite, especially since the latter species is found primarily in granitic pegmatites. Wolframite and ferberite fuse easily at 2½-3 to magnetic globules with small facets developing upon cooling; huebnerite is not as easily fused.

Occurrences: In high temperature hydrothermal ore veins and quartz veins in or near granitic rocks; also in medium-temperature veins. Commonly associated with cassiterite, quartz, tourmaline, topaz, hematite, scheelite and zinnwaldite. The best United States specimens occur in numerous deposits in Colorado, particularly from Boulder, Gunnison, San Juan, and Park counties. Crystals range from ½″ to 2″, but specimens from Quartz Creek district, Gunnison County, have been recorded to 5″ lengths; fine crystals from Adams Lode, Bonita Mountain, near Silverton, San Juan County. Good small crystals on quartz from Erie and Enterprise mines near Austin, Nye County, Nevada. Recently in red-brown crystals in quartz, from Townesville, Vance County, North Carolina. Splendid single and twin crystals to 2″, sharp and lustrous, at Zinnwald and Schlaggenwald, Bohemia, Czechoslovakia, which for many years has furnished cabinet specimens. Large blocky prisms with characteristic wedge-shaped terminations as well as some resembling Bohemian crystals, occurred in considerable abundance in the tin veins of Bolivia, particularly at Llallagua; individuals reached lengths from 2″ to 3″ and exceptional specimens showed sharp lustrous black crystals perched on transparent slender prisms of quartz.

The finest specimens are Bolivian, followed by Zinnwald, Bohemia specimens, and then by Colorado specimens. Good crystals are not easy to obtain, although considerable matrix material from Colorado has recently come on the market, showing red-brown ¼″ crystals of huebnerite imbedded in calcareous rock, from which the calcite is removed by acid.

Scheelite—Ca(WO$_4$)

Name: After Swedish chemist Karl Wilhelm Scheele (1742-1786); sheel′ĭt.

Variety: Molybdenum substituting for tungsten: *molybdian.*

Crystals: Tetragonal. Sharp to crude dipyramidal crystals resembling octahedrons (Figure 232). Prominent faces usually e[112] or p[101]. Sometimes twinned on plane m[110] as penetration twins. Smooth faces rare; most crystals etched (Figure 233). Usually massive as scattered grains, veinlets, etc.

Physical Properties: Brittle; fracture conchoidal to uneven. Cleavage parallel to p[101] distinct but interrupted. H. 4½-5. G. 6.1, lowering to about 5.9 with molybdenum.

Optical Properties: Transparent to translucent. Colorless, white, yellowish-white, yellow, reddish-yellow, brownish; rarely, greenish or grayish. Streak white. Luster subadamantine. Bright blue fluorescence under ultraviolet, inclining to yellowish-white with increasing molybdenum. Uniaxial positive. R.I., O = 1.920, E = 1.937. Birefringence 0.017. Dichroism weak in orange specimens: orange-yellow, slightly greenish-yellow. Transparent crystals have been cut into attractive faceted gems displaying considerable dispersion.

Chemistry: Calcium tungstate. Molybdenum commonly substitutes for tungsten in small amounts.

Distinctive Features and Tests: Dipyramidal crystals; high gravity; softness; fluorescence.

Practically infusible. Powdered scheelite boiled in hydrochloric acid forms a yellow precipitate of hydrous tungstic oxide.

Occurrences: A high-temperature mineral in hydrothermal quartz veins, in pegmatites, and in contact metasomatic deposits, especially when the contact is with limestone; also

FIG. 233. *Right:* Orange transparent scheelite crystal in aplite with black tourmaline and white apatite. From Santa Cruz, Sonora, Mexico. Crystal oriented with c axis vertical. Crystal is 1¼″ diameter.

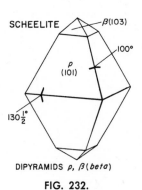

FIG. 232.

rarely, in low-temperature hydrothermal veins. Associates may be cassiterite, wolframite, topaz, fluorite, tourmaline, mica, in quartz veins, and garnet, epidote, diopside, tremolite, axinite, and idocrase in contact metasomatic deposits. Good "textbook" crystals formerly from the tungsten mine at Long Hill, Trumbull, Fairfield County, Connecticut. Fine orange-brown crystals to 1″ occur in the Mineral Mountains, near Milford, Beaver County, Utah, and faceted gems have been cut from clear pieces. Exceptionally fine, often partly flawless, in brownish-orange hues, from Boriana Mine, Hualpai Mountains, Mohave County, and in Cohen Mine, Cabezas Mountains, Cochise County, Arizona; single crystals to 4″ by 3″ have been recorded from Boriana, and to 2″ from Cohen. Mines near Darwin, Inyo County, California, produced sharp crystals to 1″; mines in the Greenhorn Mountains, Kern County, produced considerable quantities of pure white to colorless etched crystals and formless masses, some to 1½″ diameter, capable of being cut into flawless brilliant gems of many carats weight. The latter localities occasionally provided matrix specimens of quartz crystals with epidote and scheelite crystals attached. Fine brown sharp crystals in Caldbeck Fells, Cumberland, England. Numerous specimen localities in the mining districts of Saxony, as at Furstenburg, producing fine sharp yellow crystals in singles and groups; in crystals to 1½″ from Riesengrund, Riesengebirge; also fine white crystals to ⅝″ from Zinnwald, Bohemia, Czechoslovakia. Sharp small crystals of exceptional perfection from Traversella, Italian Piedmont; also in various localities in the Swiss Alps. Sometimes in good small crystals from Ponferrada, Leon, and Estepona, Andalucia, in Spain. Tan, brown, and gray crystals to 1″, from Mina Perdida, Huancaya, Peru. A number of localities in Japan provide handsome specimens, and from an unidentified source, an enormous crystal of 13″ from tip to tip is recorded by the Japanese mineralogist Wada. Very large crystals also come from Korea, as at Taehwa, and rudely dipyramidal crystals to about 6″ length, have been recorded as well as many smaller, better formed, examples. Good scheelite crystals are difficult to obtain, and those showing transparency with vivid orange-brown color are particularly desired; matrix specimens are rarer than single crystals. Large crystals are very expensive.

Wulfenite—Pb(MoO$_4$)

Name: After Austrian mineralogist Franz Xaver von Wulfen (1728-1805); wool′fĕn ĭt.

Varieties: According to substitutions for lead: *calcian,* and for molybdenum: *vanadian, tungstenian.*

Crystals: Tetragonal; predominately in thin to thick tabular crystals which may be simple rectangles or truncated on edges into octagons (Figure 234). Also prismatic. Tabular crystals compressed along *c* axis; largest faces are *c*[001].
Crystals occur as scattered singles on matrix, or in clusters, and sometimes so thickly as to completely cover the matrix material. A dusting of pale brown to brown descloizite is frequently observed. Wulfenite is less common in massive types.

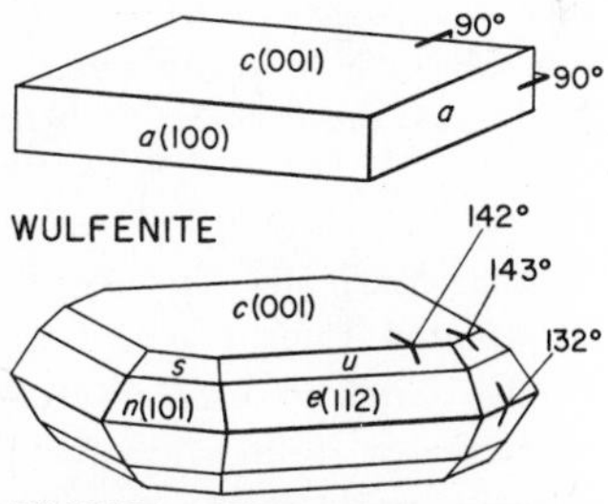

Physical Properties: Brittle; weak and fragile. Fracture subconchoidal to uneven. Distinct cleavage parallel to *n*[011]. H. 2½-3. G. 6.5-7.0.

Optical Properties: Transparent to translucent. Usually in some shade of orange-brown; also vivid yellow, yellowish-gray, greenish-brown. Sometimes color zoned parallel to *c*[001], such that a dark "filling" lies between paler layers in sandwich pattern. Streak white. Luster resinous to adamantine. Uniaxial negative. R.I., O = 2.405, E = 2.283. Birefringence 0.122.

WULFENITE

PINACOID *c*, PRISM *a*, FIRST ORDER
DIPYRAMIDS *e*, *u*/114/, SECOND ORDER
DIPYRAMIDS *n*, *s*/103/.

FIG. 234.

Chemistry: Lead molybdate. Calcium substitutes for lead in considerable amounts, as does tungsten for molybdenum; some substitution of vanadium for molybdenum has been noted.

Distinctive Features and Tests: Tabular, vividly colored crystals of bright luster. Promptly decrepitates in the flame. Fuses easily at 2; with continued flaming, the bead suddenly boils and emits sparks, then quiets and yields small globules of lead. Soluble in concentrated sulfuric acid; decomposed by nitric acid with production of molybdic oxide.

Occurrences: A secondary mineral in oxidized zones of ore deposits containing lead and molybdenum. Associated with descloizite, pyromorphite, vanadinite, mimetite, cerussite, limonite, calcite, and hemimorphite. Found in fine crystals in many deposits; those of the southwestern United States producing some of the best. Outstanding bright orange-red in Red Cloud and Hamburg mines, Trigo Mountains, Yuma County, Arizona, as tabular crystals to 2″ of exceptional luster and often transparent; also from Old Yuma Mine, near Tucson, Pima County, as single crystals sprinkled in narrow crevices in brecciated rock and closely resembling Red Cloud crystals but more orange; splendid orange-yellow thin tabular crystals, about ¼″ by ½″ across, from Rawley Mine, near

FIG. 235. Wulfenite. *Left:* brownish-orange tabular crystals on calcite with dark brown descloizite from Villa Ahumada, Sierra Los Lamentos, Chihuahua, Mexico. Size 3½" by 2". *Right:* extremely thin tabular pale brown crystals from the Glove Mine, Santa Cruz County, Arizona. Size 2" by 2". In both specimens, crystals are flattened along the c axis, making the broadest faces those of c[001].

Theba, Maricopa County; in vivid yellow crystals from Total Wreck Mine, Empire Mountains, Pima County; with willemite at Mammoth-St. Anthony Mine, Tiger, Pinal County; and in magnificent reticulated clusters, probably the world's largest crystals, from Glove (Sunrise) Mine, Tyndall district, Santa Cruz County (Figure 235). Some Glove crystals reach 4" on edge, and consist of exceedingly thin tablets of yellowish color, sometimes coated with descloizite. Those in Figure 236 are rounded on prism edges and pale brown in color. Fine specimens also from New Mexico, particularly in mines of the Central District, Grant County, and from Caballos Mountains, Sierra County. Exceedingly handsome specimens, displaying orange to brown thick tabular crystals with uneven edges, occur in abundance at Villa Ahumada, Sierra de Los Lamentos, Chihuahua, Mexico. Matrix specimens, showing tabular crystals of very simple habit to 1" to 1½" diameter, on massive calcite, reach sizes of 12" across or better (Figure 235). The Erupcion Mine at Villa Ahumada produces crystals which are nearly cubic in habit. Also excellent from Mapimi, Durango, and from Magdalena, Sonora, Mexico. Fine crystals on matrix from Mies, Yugoslavia, and from Pribram, Czechoslovakia. Recently, large tabular crystals of greenish-gray color covered by drusy quartz from M'Foati, Congo.

Wulfenites are prizes in every collection, and fortunately, are obtainable in good specimens from Mexican sources. The vivid orange-red crystals from Arizona are eagerly sought by single crystal collectors. Glove Mine specimens, obtained from only one pocket during the course of mining, are now scarce and expensive.

SILICATES

About one third of all species are silicates, containing silicon-oxygen tetrahedrons linked to each other or connected by other atoms in a wide variety of crystal structures.

FIG. 236. Wulfenite from Glove Mine, Santa Cruz County, Arizona. Size 6″ by 4″. Crystals tabular on c[001] with unusual rounded faces in the prism zone.

Throughout the world, silicates furnish splendid mineral specimens, with quartz being most prolific, followed by the feldspars, pyroxenes and amphiboles, and the zeolites. Despite the sheer quantity of silicate material in the earth's crust, so much is used up as small inconspicuous grains in rock, that really fine specimens of many species are surprisingly rare—in some instances, rarer than sulfides, elements, halides, etc. which are quantitatively much scarcer. In accordance with the scheme previously discussed in Chapter 3, silicates are classified by crystal structure patterns taken by the silicate tetrahedrons as follows:

Tektosilicates: Frameworks of tetrahedrons extending in three dimensions; *examples:* Silica Group, Feldspar Group, Zeolite Group.

Phyllosilicates: Tetrahedrons in sheets; *examples:* Mica Group, Chlorite Series.

Inosilicates: (īnō—) Single and double chains of tetrahedrons; *examples:* Amphibole Group, Pyroxene Group.

Cyclosilicates: Tetrahedrons forming rings; *examples:* Beryl, Tourmaline.

Sorosilicates: Tetrahedrons linked in pairs; *examples:* Epidote Group, Hemimorphite.

Nesosilicates: Single tetrahedrons; *examples:* Garnet Group, Aluminum Silicate Group.

TEKTOSILICATES

Silica Group

All minerals with the composition SiO_2, are included in this group, namely, quartz, tridymite, cristobalite, coesite, stishovite, opal, and lechatelierite. Quartz is by far the most abundant; opal is common; tridymite and cristobalite are not uncommon, but coesite, stishovite and lechatelierite are rare. Of these species, only lechatelierite is amorphous because it is a quickly-cooled silica melt, perhaps best known to collectors in the form of twig-like fusions of sand caused by the intense heat of lightning strikes (*fulgurites*). Opal was formerly considered amorphous, but recent work shows that it is composed largely of microcrystalline aggregates of cristobalite, between which are many pores containing variable amounts of water and small quantities of mineral matter. It is convenient to treat opal as a separate species however, and this will be done in a following section. Quartz will be treated in detail, but not tridymite or cristobalite, which seldom yield good specimens. The other minerals named are too rare to merit more than mention.

Crystalline silicas assume structures according to the pressure/temperature prevailing at the time of formation. Even within a "single" species, as quartz, the silicon-oxygen atoms change position slightly to accommodate changes in pressure/temperature conditions. The shift-over occurs when ordinary quartz, such as one may find anywhere, is raised over 573°C. At this point, the atoms rearrange themselves quickly into the polymorph known as *high-quartz*, the ordinary low temperature form being known as *low-quartz*. If the temperature is lowered, the atoms again readjust their positions to the low polymorph. Similar changes occur in tridymite and cristobalite, the several polymorphs being known as *low-tridymite, high-tridymite, low-cristobalite,* etc. All quartz, tridymite, and cristobalite cabinet specimens are low modifications although the crystals displayed on them may have the forms and angles of some higher temperature polymorph. It is interesting to note how crystal symmetry changes as silicon-oxygen tetrahedrons adjust to higher pressure/temperature conditions. Low-quartz is hexagonal, but not as symmetrical as high-quartz, which is also hexagonal; low-tridymite is orthorhombic, but two additional polymorphs become hexagonal; low-cristobalite is tetragonal, but becomes isometric in high-cristobalite. Coesite is monoclinic, and stishovite is tetragonal. Thus, as the *same* atoms linked in the familiar silicon-oxygen tetrahedrons are subjected to changes in external conditions, they may assume symmetries within all crystal systems except triclinic.

Quartz—SiO_2

Name: From the German *quarz,* of uncertain origin; kwôrts.
Varieties: Divided into two groups on the basis of crystal size and mode of aggregation: *crystalline,* includes all varieties occurring in large crystals; *cryptocrystalline* or *colloform,* includes all varieties in which submicroscopic crystals form granular to fibrous aggregates. The best known varieties in each group are as follows.

Crystalline Quartz

Rock Crystal—Colorless; when filled with numerous gas and liquid inclusions, the name *milky quartz* is used.

Smoky Quartz—Pale brown to nearly black. Color caused by natural irradiation; destroyed by heat treatment and restored by artificial irradiation.

Citrine—Pale to deep yellow. Color caused by iron. Much used for gems under misnomer "topaz."

Amethyst—Pale to dark violet or purple. Color caused by impurity ions influenced by natural irradiation; color fugitive, disappearing when heat-treated. Some heat-treated amethyst turns deep reddish-yellow ("Rio Grande topaz"), or pale green.

Rose Quartz—Pale pink, sometimes medium pink. Rarely in euhedral crystals. Color cause uncertain. Some types weakly chatoyant.

Blue Quartz—Pale gray-blue. Color believed caused by scattering of light by numerous acicular inclusions of rutile. Slightly chatoyant.

Quartz with Inclusions—*Rutilated,* with distinct acicular rutile crystals; *tourmalinated,* with acicular tourmaline; *aventurine,* spangled by small mica crystals; *tigereye,* replacement of asbestos and strongly chatoyant. Also inclusions of kaolin, chlorite, goethite, hematite, amphibole, etc.

Cryptocrystalline Quartz

Chalcedony (kăl sĕd′ ŏ nĭ)—Submicroscopic rod-like crystals arranged in parallel position, forming dense wax-like masses. Pale gray, blue-gray, to nearly colorless. The following subvarieties are chalcedony but have earned their names through long usage.

Carnelian and Sard—Chalcedonies naturally impregnated by iron compounds, imparting yellowish-brown (sard) to brownish-red (carnelian) hues. Seldom vivid, but usually improved by heat-treatment.

Chrysoprase—Bright yellow-green, caused by a colloidal nickel compound, possibly garnierite.

Chrysocolla Chalcedony—Pale blue-green, caused by colloidal chrysocolla.

Agate—Banded chalcedony, often displaying striking color and translucency contrasts, and commonly heat-treated to improve color.

Onyx—Straight-banded agate. Used for carving cameos and intaglios.

Moss Agate—Translucent chalcedony enclosing wispy, moss-like growths of chlorite, goethite, hematite, manganese oxides, etc. Numerous fanciful terms applied to subvarieties.

Jasper—Opaque to subtranslucent chalcedony containing much finely-divided mineral matter as clay, iron oxides, etc. Numerous terms applied to subvarieties.

Chert—Chalcedony deposited in pore spaces in sedimentary rocks, commonly in limestones, rendering much of the rock hard and tough. Aggregations of nearly pure material form nodules of *flint.*

Replacements—Chalcedony commonly replaces wood, bone, shell, and other organic matter, sometimes so faithfully that minute structural details are preserved. Also replaces fluorite, calcite, aragonite, and many other species.

Crystals: Hexagonal, enantiomorphic. Abundant in excellent crystals in many localities. Short to long prismatic; usually elongated parallel to *c* axis (Figure 237). Terminated by rhombohedrons $r[10\bar{1}1]$ and $z[01\bar{1}1]$, sometimes equally developed and then like hexagonal pyramids, or one rhombohedron may develop at the expense of the other, then causing terminations to resemble trigonal pyramids. Faces *r* and *z* frequently display differences in luster, markings, and coatings or inclusions. Color in amethyst often appears in triangular segments under one rhombohedron but not the other. Small faces of $s[11\bar{2}1]$ and $x[5161]$ often present in traces, or rarely, as large faces. The faces of

$m[10\bar{1}0]$ *usually striated* at right angles to the c axis, affording a most valuable clue to identification of quartz crystals.

The so-called "distorted" crystals of quartz, particularly abundant in granitic pegmatite cavities, are shards, splinters, and sheets of fractured quartz overgrown by additional material which may not be sufficient to alter the original highly irregular shape of fragments, thus leading to bizarre crystals. Crystals also occur "twisted" as notably from Swiss localities (Figure 238); in "scepters," likely to be found in almost any type of quartz deposit (Figure 239); with faces selectively developed, resulting in crystals of triangular cross-section; with chisel-shaped terminations (Dauphiné); or other equally strange results. Most common in druses of very small to large crystals (Figure 240); also isolated, and frequently, doubly-terminated. Druse amethyst prefers to form stubby crystals which display terminal faces prominently but prism faces only in traces. Rose quartz crystals poorly formed, with mottled, pitted or stepped faces.

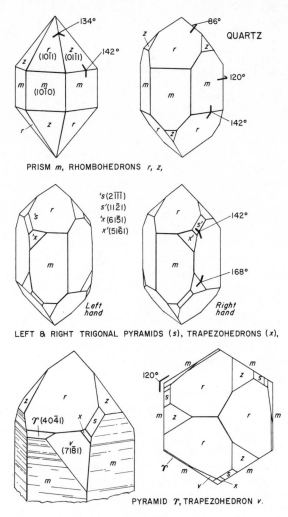

PRISM *m*, RHOMBOHEDRONS *r, z,*

LEFT & RIGHT TRIGONAL PYRAMIDS (*s*), TRAPEZOHEDRONS (*x*),

PYRAMID γ, TRAPEZOHEDRON *v.*

Fig. 237.

FIG. 238. Quartz crystals. *Left:* transparent smoky quartz crystal from the St. Gotthard region of Switzerland. Size 4" by 2". *Right:* "gedrehten," "gwindel," or "twisted" smoky quartz crystal from the Fellital-Val Giuf region of Switzerland. Size 3¼" by 2½". This is actually an aggregate of numerous crystals, each slightly offset from its neighbor thus creating curved m[10$\bar{1}$0] faces. The crystal illustrated is lying with its c axis in the horizontal direction.

Twinning very common, in fact, the untwinned crystal is a rarity. Three major twinning laws: Dauphiné, Brazilian, and Japanese. Dauphiné twinning evidences itself by irregular sutures running zig-zag across faces, approximately parallel to c axis, and crossing over terminal faces; such mark offsets between individuals twinned on the c axis (Figure 44, page 103). Brazilian twinning very seldom evident except on fracture surfaces where slight dips or differences in fracture surfaces, bounded by *straight* lines, mark junctions between twinned individuals, of which there may be many, as is so commonly displayed on amethyst fracture surfaces ("ripple" fracture). Brazilian twinning also evident in thin plates cut across c axis and placed between crossed Polaroids, when triangular segments, delineating individual crystals, assume markedly different colors. Japanese twinning unmistakable (Figure 46, page 105). A discussion of quartz twinning appears in the latter part of Chapter 4.

Physical Properties: Brittle, but tough in granular or cryptocrystalline varieties. Fracture excellent conchoidal in large flawless crystals (Figure 77, page 160), but uneven in crystals with numerous inclusions. Cleavage in quartz is seldom distinct, but most crystals tend to fracture along planes approximately parallel to rhombohedrons r and z. In some instances, this fracture becomes quite flat and then is a true cleavage. Cleavage of poorer quality and much harder to produce, has also been noted in planes parallel to the m faces. H. 7. G. 2.65; about 2.60 in cryptocrystalline varieties; variations up or down may

FIG. 239. Quartz crystals. *Left:* doubly-terminated rock crystal "candle," showing hexagonal cross-section at the bottom, gradually tapering and obliterating one set of $m[10\bar{1}0]$ faces until the cross-section at the top is triangular. From Minas Gerais, Brazil. Length 9¼". *Center:* magnificent scepter quartz displaying smoky quartz overgrown upon slightly smoky stem. From Minas Gerais, Brazil. Length 9". *Specimen from the Alice J. Walters Collection, Ramona, California. Right:* tapering amethyst crystals from Santa Margarita Mine, Guerrero, Mexico. Length 7".

FIG. 240. Rock crystal group from Madagascar. Largest crystal 5½″ long and displays bright reflection from a diamond-shaped $s[11\bar{2}1]$ face.

occur in jaspers and other impure forms of cryptocrystalline quartz but flawless crystalline quartz is nearly constant in value.

Optical Properties: Transparent to translucent; rarely opaque. Most commonly colorless or white, but also in numerous shades of brown, yellow, purple or violet; less commonly pink, and if inclusions are present (Figure 241), may occur in an extremely wide range of hues, especially in cryptocrystalline varieties. Color frequently zoned parallel to prism or rhombohedron faces, but also in sectors beneath r and z faces, as in amethyst, such

FIG. 241. Phantoms of clay minerals and chlorite enclosed in polished rock crystals from Minas Gerais, Brazil. Largest crystal 4″ in length.

that every other sector is colorless. Also colored in vague bands, streaks or veils, or in several hues at once. Coloration in agate and other cryptocrystalline varieties is influenced by impurities trapped during growth or by infiltration after formation, variations in hue from band to band being caused by variations in porosity. Streak white. Luster vitreous to slightly greasy on granular varieties and waxy on chalcedony (Figure 242). Strong chatoyancy in some varieties containing fibrous inclusions; also spangled aventurescence because of inclusions of mica, hematite, etc. Uniaxial positive. R.I., O = 1.554, E = 1.553. Birefringence 0.009. On the refractometer, chalcedony yields vague readings in the range $n = 1.530$-1.539. Uniaxial figure under crossed Polaroids displays "open" center where black cross fails to meet and is distinctive of this species. Some varieties of highly translucent chalcedony interfere with light and produce vivid spectral colors when thin polished slabs ($\frac{1}{16}$″ to $\frac{1}{8}$″), are held before a pinpoint source of light ("iris agate"). Chalcedony sometimes weakly fluorescent under ultraviolet.

Chemistry: Silicon dioxide. Flawless crystalline material is very pure, but much cryptocrystalline material rendered impure by inclusions.

Distinctive Features and Tests: Crystals; horizontal striations on prism faces of *m;* conchoidal fracture; hardness. Infusible; insoluble except in hydrofluoric acid and hot solutions of ammonium difluoride.

Occurrences: Abundant and widespread as an essential constituent of light-colored igneous rocks as granite, rhyolite, etc., but absent or only sparingly present in nepheline syenite or the dark-colored igneous rocks as basalt, gabbro, etc. However, silica is easily dissolved and transported in water under high pressure/temperature and thus finds its way into every kind of opening regardless of the nature of the enclosing rock. Practically all quartzite and sandstone is quartz. Also abundant in metamorphic rocks as schists and gneisses, particularly if derived from sedimentary rocks. Especially fine and large crystals occur in quartz veins, granitic pegmatite core units, and in many types of ore veins (Figure 240). Chalcedony and subvarieties are low temperature/pressure forms of quartz and most abundant in volcanic rocks and in limestones, but nearly absent in plutonic rocks as granite or in strongly metamorphosed rocks as schists and gneisses; also common in upper portions of veins or rocks as serpentines in which weathering has taken place. Only very few important or interesting occurrences can be mentioned.

Amethyst, smoky quartz, and several varieties of chalcedony occur in basalts of Nova Scotia, Canada, particularly along the northwest coast. Fine prismatic milky to clear crystals, sharp and elegant, and often to 10″ length, in singles and clusters, occur in a quartz mine at Black Rapids, Leeds County, Ontario. Colorful banded agates and other

FIG. 242. Translucent pale purplish-gray chalcedony forming botryoidal lining in cavity within volcanic rock. From near Julimes, Chihuahua, Mexico. Size 6½″ by 3½″.

chalcedonies occur in basalt on the north shores of the Great Lakes in Ontario. Excellent amethyst, some with bright red hematite inclusions, in crystals to $5\frac{1}{2}''$ diameter, forms druses in silver veins of the Thunder Bay region, Ontario; unbroken druses to $2\frac{1}{2}'$ length have been found. Extensive deposits of silicified wood occur in Red Deer Valley, Alberta, and a large variety of crystalline quartzes and chalcedonies at numerous places in southern British Columbia. Fine crystals of colorless, milky and smoky quartz, and massive rose quartz have been recovered from granitic pegmatites in New England, and North Carolina. Excellent amethysts from cavities in basalt sills of Massachusetts, Connecticut, and New Jersey, and from various localities in southeastern Pennsylvania, in Virginia, North Carolina, South Carolina and Georgia. The finest chalcedony pseudo-

morphs after coral and shells occur in the floor of Tampa Bay and nearby points north; coral geodes to nearly 24" diameter have been recorded, externally displaying the markings of coral colonies, but internally hollow and lined with botryoidal chalcedony in a variety of hues, sometimes overcoated by dusty quartz. Enormous quantities of limpid rock crystal have been mined for many years from quartz veins in sandstones in the Ouachita Mountains of west central Arkansas; crystals are generally prismatic to long prismatic, and occur mainly as druses lining cavities; crystals range in size from slender needles to individuals about two feet in length, but the average size is from 4" to 6"; a large druse, consisting of three fragments, and measuring about eight by six feet, occupies a special case in the Smithsonian's Natural History Museum, Washington, D.C. Rock crystal lines geodes from 2" to 24" diameter in sedimentary rocks in the Mississippi Valley area near the junction of Illinois, Iowa and Missouri (Figure 243). This area, and north into the Great Lakes region, also yields specimens of agates in alluvium. Westward of the Mississippi River, agates and petrified woods occur in gravels and *in situ*, in numerous places throughout the western United States (Figure 244). A notable deposit of chalcedony

FIG. 243. *Top:* chalcedony replacing tabular sixling twin of aragonite and coating quartz crystals. Size 3" diameter. From near Julimes, Chihuahua, Mexico. *Bottom:* half of a quartz-lined geode from within sedimentary rock, showing white calcite rhombohedrons. From the geode area at the junction of Iowa, Illinois and Missouri. Size $3\frac{1}{2}''$

replacing wood is preserved in the Petrified Forest National Park near Holbrook, Arizona. Large quantities of severely corroded amethyst crystals, associated with micro apatite crystals and scales of hematite, occur at Four Peaks in Mazatzal Mountains, Maricopa County, Arizona; another notable locality in this state is near Patagonia, Santa Cruz County, where slender tapering crystals to 12" length occur in a lead-zinc deposit, and occasionally form Japanese twins; one of the latter measured about 10" on

FIG. 244. Wood replaced by chalcedony. *Left:* oak from Eden Valley, Sweetwater County, Wyoming. Size 6¼″ tall, 3″ diameter. *Right:* silicified cypress from the state of Washington. Size 5″ by 1¼″ by 2½″. Both specimens display very faithfully replacement by quartz, the cellular structure being perfectly distinct.

each arm of the twin. Vivid blue-green chalcedony colored by chrysocolla occurs in copper mines of Arizona, sometimes replacing azurite crystals; good specimens are highly prized. Fine smoky quartz crystals, associated with microcline, and sometimes fine green amazonite, occur in small pegmatites in a large area in the Pikes Peak granite region to the west of Pikes Peak, Colorado; crystals to 24″ length have been found, but the finest are usually no more than 5″. Clear rock crystals to 2,000 pounds weight have been found in gravels near Mokelumne Hill, Calaveras County, California. Splendid agates, probably the most colorful in the world, occur over wide areas in Chihuahua, Mexico. Beautiful tapering phantom amethyst crystals occur in Guerrero (Figure 239), and drusy amethyst, usually pale in hue, is abundant in silver veins of Guanajuato, Mexico. Chalcedony is widespread in the volcanic rocks of Panama. An enormous quantity of quartz in all its varieties has come from Brazil and Uruguay, sometimes in crystals of astonishing size as a rock crystal, doubly terminated, of $5\frac{1}{2}$ tons weight, found in 1938 in the Diamantina district, Minas Gerais, Brazil. Serro Do Mar, in Rio Grande Do Sul, Brazil, once produced an enormous amethyst geode in basalt lined with crystals to about $1\frac{1}{2}''$ diameter; the geode was about 33 feet long, $5\frac{1}{2}$ feet wide, and an average of 3 feet high. Many museums display portions of this remarkable find, particularly the American Museum of Natural History in New York City. Exceptional rich pink quartz occurs near Galilea, Minas Gerais, Brazil, as drusy crystals on milky quartz, sometimes covering areas to $14''$ by $6''$. Rio Grande Do Sul and adjacent areas in Uruguay consistently produce many tons of chalcedony nodules as well as amethyst geode specimens; practically all commercial dyed agate is prepared from this material. The finest rose quartz crystals known, some of intense purplish-pink color and as much as $5/8''$ in length, form fracture linings in the quartz core of the Sapucaia pegmatite near Governador Valadares, Minas Gerais, Brazil.

Particularly fine European examples of quartz include the splendid groups and singles from numerous Swiss localities, ordinarily available in small druses of from several inches across to some about $12''$ across; in years past some exceptional finds have been made, as in a cavity on the Zinkenstock, near Grimsel, which produced over 50 tons of smoky crystals; other notable finds have been made near the Tiefen Glacier, in the canton of Uri, and elsewhere. All hues occur in Swiss material, from colorless to faintly smoky, to some splendid shining crystals of such deep color that they appear quite black (Figure 238); most are very free of flaws and inclusions. Swiss crystals are also prized for numerous rare forms displayed by their faces. Other European collector's items are the interesting corroded rock crystals of Porretta, Italy, and the brilliant small rock crystals lining cavities in Carrara marble. The mining districts of central Europe also provide excellent examples of quartz crystals as druses from within sulfide ore veins. Large and fine quartzes of every description are found at many localities on the Island of Madagascar (Figure 240). The classic Japanese twins occur at a number of localities in Japan, notably at Otomezaka, Yamanashi Prefecture, and at Narushima Island, Nagasaki Prefecture; individuals to $18''$ tall have been found, but the usual run is from $1\frac{1}{2}''$ to about $4''$ (Figure 46, page 105). Very large amethyst crystals, some to $12''$ length, occur in Korea.

Quartz is favored by collectors restricting their efforts to a single species, but even here, it is necessary to be selective because of the enormous variety. Even more discouraging in some respects, is the variety in chalcedonies and other cryptocrystalline quartzes which appears limitless. Nevertheless, quartz is and always will be a favorite among collectors because good specimens are always available but truly splendid pieces are not easy to get. The most expensive quartzes are Japanese twins, followed by exceptional Swiss specimens, rose quartz crystals, fine Uruguayan amethysts, in which the value is raised because of gem value, and quartzes with inclusions; among the latter may be mentioned the fine rock crystals which came from Brazil some years ago, containing large pyrite crystals to $5/8''$ diameter. Surprisingly rare, probably because most crystals are promptly cobbed into gem rough, are good citrine crystals.

Tridymite—SiO_2

Small, hexagonal, colorless to white, plate-like crystals, usually less than $\frac{1}{16}''$ diameter, found in crevices and cavities in volcanic rocks; trĭ′dĭ mīt. Collected primarily by micromounters. Originally found in andesite near Cerro San Cristobal, Pachuca, Mexico. Also in good specimens from the trachytes of the Siebengebirge region, Rhineland, Germany; Euganean Hills, north Italy; and in Puy de Dome, France.

Cristobalite—SiO$_2$

Small octahedral crystals, usually less than $\frac{1}{16}$" diameter; krĭs tō′bă lĭt. Found in cavities in volcanic rocks as at Cerro San Cristobal, Pachuca, Mexico, the type locality, and in the grayish spherulites, with yellowish fayalite crystals, enclosed in obsidian from numerous localities in the western United States, as in the noteworthy deposit near Coso Hot Springs, Inyo County, California. It is widespread but not easily detected except through magnified visual examination. A typical spherulite displaying euhedral cristobalite on interior walls is shown in Figure 77 (page 160).

Opal—SiO$_2$

Name: Possibly from the Sanskrit *upala,* meaning "precious stone" or "gem"; ō′păl.
Varieties: Common, without play of color, and of any hue or degree of translucency. When replacing wood, it is called *wood opal,* or if containing dendritic or mossy inclusions, it is called *moss opal. Fire opal* is the reddish to orange-red translucent material from Mexico which may or may not contain play of color. *Precious opal* is material of any basic hue which displays shifting flashes of intense color in red, green, orange, blue, purple, etc. It varies from barely translucent to nearly transparent. Sub-varietal names are *black opal,* when the body color is pale to deep gray or nearly black, *white opal,* when the body color is milky or nearly pure white. Much Mexican fire opal is also *precious opal. Hyalite* is colorless, transparent opal, usually occurring in botryoidal crusts (Figure 245). *Cachalong* and *hydrophane* are white, opaque, or barely translucent porous kinds which absorb considerable water; if the material becomes translucent after soaking, it is called *hydrophane;* the latter may or may not also show play of color. Over 148 terms descriptive of opal varieties have been assembled but many are trivial and not worth mentioning.

FIG. 245. Opal. *Left:* colorless opal *(hyalite)* on rhyolite from San Luis Potosi, Mexico. Size 2¼" by 1½". *Right:* dark red transparent opal cavity fillings in chalky trachyte from Queretaro, Mexico. Size 3" by 2".

Crystals: None are known and therefore opal is always massive, with considerable variation in properties. Sharp replacements of glauberite crystals by precious and common opal occur in Australian opal fields.

Physical Properties: Brittle to very brittle. No cleavage, but excellent and lustrous conchoidal fracture, often like glass in respect to perfection and providing an excellent recognition feature. H. variable, $5\frac{1}{2}$-$6\frac{1}{2}$. G. 1.99-2.25; lower in porous material; increases with absorption of water. Many specimens lose water on removal from the ground and in a matter of days to months, check or craze or even disintegrate. Water may be lost concentrically, as is common in Virgin Valley opal and Mexican opal, resulting in curved sections spalling from exteriors and leaving behind spherical or cylindrical nodules.

Optical Properties: Translucent to transparent; seldom opaque. Clear specimens usually filled with blue haze in incident light, but yellowish-brownish hazy in transmitted light. Usually pale in hue, but sometimes richly colored by inclusions which may cause assumption of almost every hue of the spectrum. Often mottled, streaked or banded in several colors and showing various degrees of translucency. Streak white. Luster bright vitreous to slightly waxy. The play of color in *precious opal* is believed caused by interference of light reflected from grids of cristobalite spaced at regular intervals, behaving somewhat like a ruled spectroscope grating. The hues are nearly monochromatic. Singly refractive; $n = 1.435$-1.455, varying with water content, and from place to place on the specimen. Extremely low value of $n = 1.23$ recorded for *hydrophane,* and extreme high of $n = 1.459$ for *hyalite.*

Chemistry: Silicon dioxide, but usually contains water in the range 4%-9% by weight; extremes to 20% recorded. Commonly contains small amounts of iron and aluminum in relatively pure material, and much foreign matter in impure types.

Distinctive Features and Tests: Perfect conchoidal fractures and brittleness, such that sharp edges can often be chipped by fingernail pressure. Infusible. Readily dissolves in hydrofluoric acid; attacked by hot concentrated hydrochloric acid.

Occurrences: A low temperature hydrothermal mineral, confined to surface or near-surface deposits in a wide variety of rocks, and in this respect, similar to *chalcedony quartz.* Occurs as nodules, seams, impregnations in porous rocks or sediments, as botryoidal crusts, particularly *hyalite,* and commonly replaces wood, shells and other organic matter. Deposited from hot springs (*siliceous sinter* or *geyserite*). Diatomite is a porous sedimentary rock composed of opalized skeletons of minute marine organisms (diatoms). Opal is abundant and widespread, especially *common opal; hyalite* is also widespread as very thin botryoidal crusts lining crevices in rock but seldom in thick crusts suitable for cabinet specimens; *precious opal* occurs in many places but seldom in large fine examples.

Thin films of fluorescent *hyalite* commonly occur in seams in pegmatites in New England and North Carolina; the greenish fluorescent hue is believed caused by colloidal particles of a uranium mineral. Magnificent large colorful *wood opal* logs occur widely in Washington, Idaho, Oregon, Utah, and Nevada; especially fine oak from Clover Creek, Lincoln County, Idaho, and from volcanic ash beds in south central Washington. *Hyalite* lines cavities in basalt, near Klamath Falls, Oregon. Small nodules of *precious opal* in volcanic rocks have been found in Idaho, Washington, and Oregon, but the prime locality is Virgin Valley, Humboldt County, Nevada, where opal replaces wood in volcanic ash and tuff beds, sometimes providing splendid examples to 7 pounds weight. Much shows a dark brown to black body color through which arise vivid flames of red, green, blue and purple; colorless and white types are also abundant. Excellent *precious opal* and *fire opal* occur in an extensive area of volcanic flows in Mexico, from San Luis Potosi to Guerrero; much fine material has been mined in Queretaro, and more recently, in Jalisco. Classic specimens of *hyalite* occur as thick encrustations lining cavities in volcanic rocks in the Cerritos area, Cerro del Tepozan, San Luis Potosi (Figure 245). *Precious opal* occurs in a number of localities in Honduras. *Common, fire* and *precious opal* have been mined to a limited extent in Brazil at several localities. Fine *precious opal* has been mined for centuries near Cernevica, Saros Comitat, Hungary. Fine specimens of *hyalite* from Waltsch, Bohemia, but now scarce and seldom offered. Many localities for *precious opal* are known in Australia, among them, Lightning Ridge and White Cliffs areas of New South Wales, the latter providing excellent glauberite pseudomorphs and perfect replacements of shells, bones, and other organic remains. Other opal-producing areas are in Queensland and South Australia. Exceptional *wood opal* from Bothwell, Tasmania.

The high value of *precious opal* makes acquisition of good specimens a costly and difficult matter because the tendency of miners is to dispose of cuttable specimens to buyers of rough, without much regard for mineralogical interest. A notable exception is Virgin Valley opal, most of which is sold for specimens but still commands high prices. Most desirable for collectors are glauberite and fossil pseudomorphs from White Cliffs; *hyalite* from Waltsch or San Luis Potosi, and *fire opal* in matrix from Mexico. *Precious opal* from Virgin Valley is readily available in astonishingly beautiful specimens, if one has the price. Attractive cabinet specimens are also provided by thin *precious opal* seam fillings in dark brown sandstone matrix (boulder opal) from Australia; if broken open along a seam, a spectacular display specimen may result.

Feldspar Group

Species in this group are closely related in composition and crystal structure, and taken together, are the most abundant of all minerals. The feldspars are placed in two sub-groups: *potassium feldspars,* and the sodium-calcium *plagioclase feldspars;* plǎj'ĭŏ klǎs. Potassium species are sanidine, orthoclase and microcline, each with the same formula: $K(AlSi_3O_8)$, but varying in crystal structure. Plagioclases form an unbroken series from sodium-rich albite to calcium-rich anorthite, with the general formula: $(Na,Ca)(Al,Si)$ $(AlSi_2O_8)$. The crystal structure of all feldspars is basically the same, leading to development of predominately squarish or blocky crystals, cleavages nearly at right angles, similar hardnesses, and properties which serve to distinguish feldspars from other minerals but cause difficulties in distinguishing one feldspar from another. Thus microcline and orthoclase look very much alike and produce nearly the same test results. Within the plagioclase series it is almost impossible in some instances to tell where a particular specimen belongs—to be absolutely sure it is necessary to employ the services of a professional mineralogist using professional laboratory equipment. For the amateur, a rough classification is sufficient, and there is no need for an exact classification. On the brighter side, it is well known that certain species prefer to form in certain environments as will be pointed out later, and finding a feldspar in a given environment is in itself a good clue to identity.

Among the potassium feldspars, sanidine is a high temperature polymorph of monoclinic orthoclase, but is much less abundant and will therefore not receive separate discussion. Microcline is triclinic, but its crystal structure is merely a slight modification of the monoclinic orthoclase structure, and it is difficult to distinguish from the latter species. The plagioclases are triclinic, and vary gradually in properties from albite to anorthite, the end members. Between them, a number of fences are arbitrarily put up to divide the series into five species, namely, albite (sodium rich), oligoclase, andesine, labradorite, bytownite, and anorthite (calcium rich). Divisions are marked by relative percentages of sodium and calcium, such percentages being placed at the lower right of the albite (Ab) and anorthite (An) symbols as shown below. If analysis of any specimen indicates the percentages shown, the specimen is labeled with the appropriate species name.

Albite	Ab_{100}-Ab_{90}	or:	An_0 -An_{10}
Oligoclase	Ab_{90} -Ab_{70}		An_{10}-An_{30}
Andesine	Ab_{70} -Ab_{50}		An_{30}-An_{50}
Labradorite	Ab_{50} -Ab_{30}		An_{50}-An_{70}
Bytownite	Ab_{30} -Ab_{10}		An_{70}-An_{90}
Anorthite	Ab_{10} -Ab_0		An_{90}-An_{100}

At higher temperatures, potassium and plagioclase feldspars form solid solutions and single crystals, but as temperatures lower to surface conditions, each species prefers to form its own crystals and does so in the process known as *exsolution*. The larger crystals still stand, but parts are distinctly one or the other, sometimes appearing plainly as streaks or criss-crosses of streaks resembling patterns in plaid fabrics. Such material is then known as *perthite*. Usually it is the potassium feldspar which retains the large crystal form, and the plagioclase which forms the streaks. Figure 248 shows a corroded microcline crystal from the Himalaya Mine, from which the plagioclase (albite) has been largely removed, leaving behind numerous cavities. The bladed albite at the base of the crystal may have come from the corrosion of the larger microcline crystal. *Perthite* is very abundant in granitic pegmatites, the potassium species usually being microcline.

Another intergrowth of feldspar characteristically occurring in granitic pegmatites is *graphic granite,* named from the patterns created in cross-sections by small rods of quartz enclosed within large feldspar crystals and then resembling cuneiform writing or Hebrew characters. Recent work has shown that some *graphic granite* quartz rods are interconnected, although through any given slice across the mass, they appear to be separate. Sometimes *graphic granite* protrudes into cavities and its faces then become dotted with small terminated quartz crystals growing in parallel position. *Graphic granite* is a sure field sign that a granitic pegmatite body is being examined.

Orthoclase—K(AlSi$_3$O$_8$)

Name: From the Greek *orthos,* at "right angles," and *klasis,* "fracture," in allusion to the right angle cleavages; ôr'thō klās; sanidine from the Greek *sanis,-ides,* "a board," in allusion to the tabular crystals commonly assumed by this species; săn'ĭ dĭn, săn'ĭ dĕn.

Varieties: Adularia (à dū-là'ĭ ă), a variety found in low temperature hydrothermal veins and distinctively crystallized in wedge-shaped or pseudo-rhombohedral individuals, usually of considerable transparency. *Moonstone* is any variety displaying a silvery to bluish light caused by reflection from numerous oriented inclusions; the effect is also called adularescence, after its being first noted in the *adularia* of the Swiss Alps.

Crystals: Monoclinic; predominantly as singles and groups of blocky crystals of rectangular or square cross-section, seemingly tetragonal in symmetry. Squarish prisms usually elongated along the *a* axis, showing prominently the pinacoids *b*[010] and *c*[001] (Figure 246). Tabular crystals commonly flattened along the *b* axis, showing large *b*[010] faces, and terminated at the ends of the *a* axis by pairs of narrow faces belonging to the prism *m*[110], and at the ends of the *c* axis by the pinacoids *c*[001] and *y*[$\overline{2}$01]; the latter faces form a blunt wedge. Twins are very common: *Carlsbad*—two tabular crystals as described above, seemingly pushed into each other part way, one being first turned around 180°. *Baveno*—twins usually quite square in cross-section, the twin crystals elongated along the *a* axis, and the twinning plane a diagonal one, running through opposite edges of the

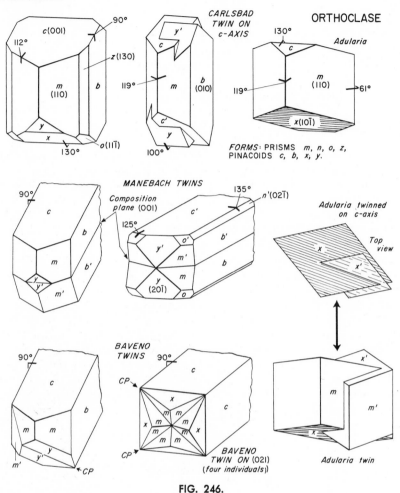

FIG. 246.

prismatic crystal. *Manebach*—the twinned crystal is commonly flattened along the *b* axis, making the *b*[010] faces the largest pair, and also commonly elongated along the *a* axis; the twinning plane is parallel to *c*[001]; on most crystals, a vee-shaped reentrant is placed on "top" of the Manebach twin, if the crystal is held so that the *a* axis is upright. Most orthoclase crystals are rough-faced and only rarely display sharp smooth faces. Sanidine crystals are mostly tabular rounded grains.

Physical Properties: Brittle; fracture uneven, often stepped, and rarely, conchoidal. Two prominent cleavages at right angles to each other, one perfect parallel to *c*[001] and easily developed, the other good, parallel to *b*[010] and not easily developed. H. 6. G. 2.56-2.59.

Optical Properties: Translucent to transparent. Mostly white to pink or brownish-pink; also colorless, yellow, pale brown. Streak white. Luster vitreous; somewhat pearly on cleavage planes. Biaxial negative. R.I., sanidine (Black Range), $\alpha = 1.521$, $\beta = 1.526$, $\gamma = 1.527$; birefringence 0.006; *moonstone* (Ceylon) $\alpha = 1.518$, $\gamma = 1.526$; birefringence 0.008; yellow transparent orthoclase (Madagascar), $\alpha = 1.522$, $\gamma = 1.527$; birefringence 0.005, and trichroic; pale greenish-yellow, pale yellow, nearly colorless.

Chemistry: Potassium aluminum silicate. Sodium sometimes substitutes for part of the potassium.

Distinctive Features and Tests: Distinguished from plagioclase feldspars by absence of twinning striations, particularly on fresh cleavage surfaces. Distinguished from microcline with difficulty, but occurrence helps because microcline is abundant in pegmatites while orthoclase most commonly occurs as phenocrysts in volcanic rocks. Right angle cleavages distinguish from other silicates. Fuses with difficulty at 5; soluble in hydrofluoric acid.

Occurrences: Sanidine common in rhyolites and trachytes as glassy formless grains; *adularia* typically in low-temperature hydrothermal deposits; orthoclase crystals usually occur as crude single and twin phenocrysts in igneous rocks, and sometimes in miarolitic cavities within granites. White *adularia* crystals, to 4″, occur with *uralite*, epidote and quartz at Sulzer, Prince of Wales Island, southeastern Alaska (Figure 247). Blocky rough-surfaced orthoclase singles and twins, to 2½″, occur in granitic rocks near Penticton, British Columbia. Similar crystals, to 3″, occur in altered quartz monzonite porphyries of the Breckenridge and Tenmile districts, Summit County, Colorado. Also in Colorado, sanidine crystals to several inches length, some being *moonstone,* in rhyolite on Ragged Mountain, Gunnison County. As with the large sanidine *moonstone* crystals to 12″ to 24″, found on the west slope of the Black Range, near Noonday Canyon, Grant County, New Mexico, outer parts of crystals are opaque white, but inner parts transparent; exceptional *moonstone* has been provided by New Mexican crystals. Volcanic rocks in New Mexico also provide good sharp phenocrysts of orthoclase to 3″, from north of Embudo Canyon on the west side of the Sandia Mountain, Bernalillo County, from Organ and South Canyon Districts of Dona Ana County, as 3″ crystals in quartz monzonite in Burro Mountains District of Grant County, in Luna and Otero counties and fine Carlsbad twins in monzonite porphyry in Red River District, Taos County. Small euhedrons, often Carlsbad twinned, occur at Crystal Pass, Ray, Pinal County, Arizona. Fine white *adularia* (*valencianite*) in simple crystals resembling Swiss crystals, and in size from very small to nearly an inch across, occur in mines of the Silver City District, Owhyee County, Idaho; many are of fine cabinet quality. Sanidine single and Carlsbad twins, occur on Chataya Peak, Kawich Range, Nevada. Crude orthoclase crystals to 4″, in the Santa Lucia granite, Carmelo Bay area, Monterey County, California, and crystals to 7″ length, commonly Carlsbad twinned, in monzonite porphyry near Twentynine Palms, San Bernardino County.

Adularia is especially fine and abundant at numerous localities in Switzerland, and often forms crystals to 6″ diameter although the average run is about 2″; many are coated with green chlorite; others occur as druses or interesting jumbles of numerous crystals or sometimes stacked in parallel position. *Adularia* also occurs in North Italy and West Austria. Fine twinned opaque orthoclase crystals to 3″, simple in habit, occur in druses in cavities in granite at Baveno quarries, Lago Maggiore, Italy; this is the type locality for Baveno twins. The Czechoslovakian locality at Karlovy Vary (Karlsbad), furnishes splendid small examples of orthoclase and gives its name to Carlsbad twins. A most unusual orthoclase, often in completely transparent flawless crystals, and of pale to deep greenish-yellow color, suitable for gems, occurs with diopside and zircon in a pegmatite near Itrongahy, Madagascar; the color is caused by a small amount of iron substituting for aluminum. Some crystals have been obtained to 3″ diameter, nearly flawless, and with excellent flat faces. Fine orthoclase crystals come also from Omi and Mino, Japan.

Euhedral orthoclase phenocrysts are easy to obtain, but having grown completely within igneous rocks, few are of good quality; nevertheless they are prized for single crystal collections. The *adularias* of Switzerland and adjacent Alpine areas are always in demand,

FIG. 247. Feldspar specimens. *Top left:* twinned crystals of *adularia* with dark green crystals of *uralite* from Sulzer, Prince of Wales Island, Alaska. Size 4″ by 3½″. *Top right:* green crystal of microcline with white albite capping and *cleavelandite* rosettes from the Crystal Peak area of Teller County, Colorado. Size 5″ by 3″. *Bottom left:* aggregate crystals of pale brown microcline from within a pegmatite cavity in the Crystal Peak area, Teller County, Colorado. Size 3½″ by 3″. *Bottom right:* sharp microcline crystal with smoky quartz and small white crystals of albite from pegmatite, Pilgrimshain, near Striegau, Silesia, Poland. Size 2″ by 1¾″.

good specimens seldom go begging for buyers; similarly, the unusually sharp opaque Baveno crystals are also desirable. Perhaps the rarest orthoclase crystals of all are the yellow Madagascar crystals which are usually obtainable in fractured pieces but very seldom in euhedrons of good to fine quality; most good crystals do not exceed 1".

Microcline—$K(AlSi_3O_8)$

Name: From the Greek *mikros,* meaning "small," and *klinein,* "to incline," in allusion to the small inclination of the third crystallographic axis, throwing this species into the triclinic rather than the monoclinic system; mī'krō klīn.

Varieties: Greenish or bluish-green color, *amazonite.* Much microcline, with a plagioclase feldspar, forms *perthite,* especially in granitic pegmatites.

Crystals: Triclinic; similar to orthoclase in respect to forms, habits, and twinning. Microcline commonly occurs in cavities and its crystals may therefore be of much better quality than orthoclase. Perthitic intergrowth often evident on crystals, especially in *amazonite,* whose green color may be mottled or interrupted by plaid-like exsolved strips or streaks of white albite; commonly capped or coated by white albite, as notably in the *amazonite* occurrences of Colorado (Figure 247). Often severely corroded in pegmatite pockets as shown in Figure 248; also commonly coated with perfectly transparent material which may be largely microcline as in pegmatites of Pala, California.

Physical Properties: Similar to orthoclase. *Amazonite* may sometimes reach H. $6\frac{1}{2}$.

Optical Properties: Similar to orthoclase, but colors include green to blue-green of *amazonite;* also flesh-red, pink, white, colorless. Indices in *amazonite* slightly higher than for orthoclase: $\alpha = 1.522$, $\gamma = 1.530$. Birefringence 0.008.

Chemistry: Potassium aluminum silicate. Sodium replaces potassium, and if exceeding potassium, the mineral is known as anorthoclase.

Distinctive Features and Tests: Green or blue-green crystals are always microcline rather than orthoclase. The plaid pattern of *perthite* is distinctive.

Occurrences: Primarily in granitic pegmatites, often forming large single-crystal grains many feet in diameter; also in hydrothermal veins. Fair to good crystals, but seldom fine, occur in numerous pegmatites in Maine, New Hampshire, Connecticut, and North Carolina. Small but sharp white to pale green crystals, with albite, smoky quartz, and fluorite, from miarolitic cavities in the Conway granite of New Hampshire. Fair to good crystals, usually corroded, occur in the pegmatite gem mines of San Diego County, California, and in similar bodies in adjacent Baja California, Mexico (Figure 248). The finest crystals of *amazonite* occur in numerous small pegmatites in the Pikes Peak red granite, covering a wide area just west of Pikes Peak, Colorado; especially fine examples came from larger pegmatites near Crystal Peak, north of Lake George, Teller County (Figure 247). The best crystals were deep blue-green, sometimes capped with white albite, and occasionally in druses with brilliant, near-black, smoky quartz crystals. Individuals to 16" length have been recorded from within quartz cores, but free-standing cavity crystals seldom exceed about 5" length, and the average is closer to 2". Splendid gem *amazonite* from Rutherford Mine No. 1 and Morefield Mine, near Amelia, Amelia County, Virginia, as interlocked anhedral grains to 18" diameter, closely associated with massive milky quartz and bladed albite in pegmatite cores. Fine *amazonite* also in Lyndoch Twp., Renfrew County, Ontario, Canada, but not in euhedrons; also in Brazil. Euhedrons of fair to good color formerly from pegmatites in the Urals and Ilmen Mountains, U.S.S.R. Opaque white crystals of good quality to $3\frac{1}{2}$", occur at Tawara, Honshu, Japan.

The prize specimens of microcline are the coarse druses of *amazonite* from Colorado, especially when associated with smoky quartz; unfortunately very few are available and

FIG. 248. Feldspar. *Left:* corroded perthite crystal partly coated by rosettes of *cleavelandite.* From the Himalaya Mine, Mesa Grande, San Diego County, California. Size 7″ by 3½″. *Right:* large white blades of twinned cleavelandite from Rutherford Mine No. 2, Amelia, Amelia County, Virginia. Size 5½″ by 4″.

prices demanded for fine examples are very high. The *amazonite* from Amelia has furnished the best lapidary material known, some being unusually translucent, deep in color, and free of disqualifying cracks over many inches of area.

PLAGIOCLASE SERIES

Albite—Na(Al,Si)(AlSi$_2$O$_8$) (Sodium end member)

Anorthite—Ca(Al,Si)(AlSi$_2$O$_8$) (Calcium end member)

Names: From the Latin *albus,* in allusion to the common color; ăl′bīt. From the Greek *an* plus *orthos,* "not upright," in allusion to the oblique crystals; ăn ôr′thīt. Oligoclase, ŏl′ĭ gō klās, from the Greek *oligos* and *klasein,* "little cleavage," because thought to have less perfect cleavage than albite; andesine, ăn′dĕ zĭn, from being found in the Andes Mountains; labradorite, lăb′ră dôr ĭt, from the locality in Labrador; bytownite, bī′town ĭt, from the "Bytown" locality, now Ottawa, Ontario, Canada.

Varieties: Thin platey crystals of albite are known as *cleavelandite* (Figure 248). *Sunstone* is aventurescent oligoclase containing numerous oriented hexagonal scales of hematite, showing strong bright orange-yellow spangled reflections close to cleavage plane *c*[001], and less commonly, close to cleavage *b*[010]. Silvery or pale bluish reflections of considerable intensity are often noted in albite, then providing *moonstone;* when such reflections display a variety of hues the term *peristerite* is sometimes used. Strong blue, green, and less commonly, red, gold, and brown reflections from labradorite and andesine, provide handsome specimens usually labeled *labradorite* although the position of the specimen in the plagioclase series may not fall within the labradorite range.

Crystals: Triclinic. Albite-oligoclase provide practically all specimen crystals in the series, the others usually occurring only as anhedral grains in rock. Crystals mostly flattened along the *b* axis (Figure 249), forming thin blades standing upright on matrix, with

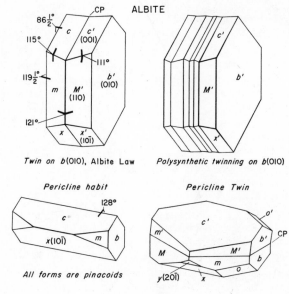

FIG. 249.

angular spaces between blades as in Figure 248; also in rosettes, rounded aggregates, and vuggy bladed aggregates. Albite-oligoclase predominately occur bladed (*cleavelandite*) and thus are easily recognized in granitic pegmatites in which they are sometimes abundant. Largest faces usually b[010]; edges of tabular crystals usually bounded by c[001] and x[$\bar{1}$01], plus blunt wedges formed by two m[110] faces. Polysynthetic twinning very common with b[010] the composition plane (*albite law*); less commonly along a plane nearly parallel to c[001] (*pericline law*). Twins form numerous thin crystals sandwiched together, resulting in very characteristic striations on cleavage planes c[001] and b[010]. Sometimes the twins are very thin and striations very narrow; at other times, twins may vary in thickness, but the perfectly straight junctions between them are still visible. Some crystals display striations because of being twinned according to both laws. Twinning also occurs according to Carlsbad, Baveno, or Manebach laws (see Orthoclase).

Physical Properties: Brittle; uneven to conchoidal fracture. Cleavages usually prominent on fractured specimens, perfect along c[001], and good along b[010]. H. 6. G. 2.62-2.76.

Optical Properties: Transparent to translucent. Transparency in cleavelandite reduced by development of numerous bright-reflective c[001] cleavages. Albite usually white or colorless, oligoclase same, but sometimes slightly greenish; andesine commonly colorless or white; labradorite, bytownite, anorthite commonly grayish, sometimes nearly black because of numerous inclusions; some labradorite is pale straw yellow. Streak white. Luster vitreous to somewhat greasy in labradorite; commonly pearly on cleavage planes. Biaxial positive or negative. Variations of refractive indices and specific gravities through the series is shown below.

Species	Refractive indices		Specific Gravity
Albite	$\alpha = 1.526$	$\gamma = 1.536$	2.62
Oligoclase	1.539	1.548	2.65
Andesine	1.550	1.558	2.68
Labradorite	1.560	1.568	2.70
Bytownite	1.568	1.578	2.73
Anorthite	1.576	1.588	2.76

Chemistry: Sodium-calcium aluminum silicates.

Distinctive Features and Tests: Bladed crystals (*cleavelandite*); striations on cleavage surfaces. In pegmatites or in miarolitic cavities in granite, the bladed character and usually much whiter hue of cleavelandite distinguish from potassium feldspars. Exact position in the series requires careful labratory determinations and is not feasible for the amateur. Fusible with difficulty at 5; sodium-rich members color flame strong yellow. Soluble in hydrofluoric acid.

Occurrences: Albite is found primarily in granitic pegmatites as bladed masses, sometimes of delicate bluish or greenish hues, or as bladed crystals coating walls of cavities. Bytownite and labradorite usually occur as small to large grains in gabbros and anorthosites. Andesine occurs in andesites and diorites. Oligoclase is common in monzonites and granodiorites, but is also found in granitic pegmatites. Anorthite commonly occurs in contact metamorphosed limestones.

Exceptional labradorite, noted for strong colors and suitability for gemstones and ornamental applications, occurs in anorthosites in eastern Labrador, sometimes in single crystal grains to 24″ across. Fine-grained labradorite widespread in New York State, especially in Essex County. *Peristerite* occurs in Ontario and has been used to a limited extent for gemstones. Excellent specimens of *cleavelandite* occur in core cavities of granitic pegmatites in New England, but the world's finest examples are from the Rutherford Mines, near Amelia, Amelia County, Virginia, where blades to 6″ by 4″ by 1″ line cavities in pegmatite, sometimes affording unbroken druses to four feet across; the average crystal size is, however, closer to 1″ to 2″ diameter (Figure 248). Some crystals are very transparent and a few nearly flawless. Fine *cleavelandite* also occurs in pockets of granitic pegmatites of San Diego County, California. *Cleavelandite* is commonly associated with the amazonite of Colorado. Occasionally, pegmatites of Minas Gerais, Brazil, produce good *cleavelandites,* approaching Amelia specimens in size and perfection. Fine smaller examples from several localities in the Alps and in the Tyrol. Transparent oligoclase, some of which has been faceted into small gems, occurs in the Hawk Mica Mine, Mitchell County, North Carolina. Excellent *sunstone,* the world's best, from Tvedestrand, Kragerö, and other localities in Norway. Magnificent gem labradorite, dark gray in body color and displaying vivid blue predominately, but also copper, yellow, red, etc., occurs in anhedrons to 6″ diameter in norite, at Ylamoa, near Lammenpaa, Finland. Andesine, bytownite and anorthite rarely supply suitable mineral specimens, but good anorthite crystals occur in the ejected material of Mount Vesuvius, Italy, and as black-coated crystals to 1″, from the island of Miyaki, Tokyo Bay, Japan.

Danburite—Ca(B₂Si₂O₈)

Name: From the extinct Danbury, Connecticut, locality; dȧn′bḗ ĭt.

Crystals: Orthorhombic. Prismatic, similar to topaz in habit, often with diamond-shaped cross section and wedge or pointed terminations. Prisms commonly bounded by four faces of form *l*[120] and terminated by steep wedge of *w*[041] (Charcas), but also by traces of

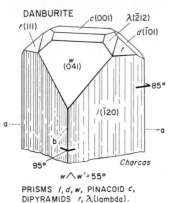

DANBURITE
r(111)   *c*(001)   λ(2̄12)
d(1̄01)
w(04̄1)
85°
l(1̄20)
a----   ----a
b
95°   Charcas
w∧*w*′=55°

PRISMS *l*, *d*, *w*, PINACOID *c*,
DIPYRAMIDS *r*, λ(lambda).

FIG. 250.

d[101] and other faces, as in Figure 250. Japanese crystals usually display *w* and *d* equally developed. Prism faces striated parallel to *c* axis. Also granular massive.

Physical Properties: Brittle; conchoidal fracture. No cleavage. H. 7. G. 3.00.

Optical Properties: Transparent to translucent. Ordinarily white to colorless, also faintly pink (Charcas) or straw yellow (Mogok); sometimes greenish because of inclusions (Japan). Streak colorless. Luster vitreous. Biaxial negative. R.I. $\alpha = 1.630$, $\beta = 1.633$, $\gamma = 1.636$. Birefringence 0.006.

Chemistry: Calcium boro-silicate.

Distinctive Features and Tests: Lacks cleavage and thus distinguished from very similar topaz crystals; specific gravity also distinctive. Fuses at 3½ to colorless glass, imparting greenish color to flame.

Occurrences: Primarily in metamorphosed limestones, but also in low-temperature hydrothermal veins with sulfides and calcite as at Charcas, San Luis Potosi, Mexico. In marbles near Russell, St. Lawrence County, New York, as pale reddish-brown prismatic crystals enclosed in calcite; single crystals to 8″ to 10″ reported. The locality at Danbury, Connecticut, is buried beneath the city. The finest danburites known come abundantly from mines at Charcas, San Luis Potosi, Mexico, in single crystals, and groups of crystals on massive danburite spanning 4″ to 10″. Crystals are sharp, clear at the tips, and afford faceting material, and range in size from about ½″ length to as much as 3″ long and 1″ thick (Figure 251). Some are faintly pink. Associated minerals are iron sulfides, calcite, and quartz, the latter forming druses of minute crystals completely coating danburite crystals.

FIG. 251. Danburite group. Colorless to white crystals from Mina San Sebastian, Charcas, San Luis Potosi, Mexico. Size 4″ by 3″. Crystals elongated parallel to c axis and displaying forms shown in Figure 250.

An unusual associate is micro apophyllite crystals. Exceptionally fine yellow etched crystals occur in gravels of Mogok, Burma, and from them handsome gems have been cut; faces are absent and fragments are rounded lustrous lumps reaching diameters of better than 1½″. Excellent, long prismatic crystals, many perfectly clear and colorless, occur associated with axinite in Toroku Mine, Higashi-Shodo, Miyazake Prefecture, and others in Kyushu, Japan; some crystals are phantomed with green and milky-white inclusions; crystals to 4″ occur, but the average is less than 2″. Danburite is a rare species, but fortunately, specimens are readily available from Charcas. Rarely, an old collection yields an example from New York, or perhaps single crystals from the Japanese localities.

Feldspathoid Group

Species in this group form in place of feldspars when a magma lacks enough silica to form the latter, hence the *feldspathoid* name. Species include leucite, nepheline, sodalite, cancrinite, and lazurite.

Leucite—$K(AlSi_2O_6)$

Name: From the Greek *leukos,* "white," in reference to the common hue of crystals; lū′sīt.

FIG. 252. Leucite trapezohedrons in pale gray volcanic tuff. From Ariccia, near Rome, Italy. Largest crystal 1" in diameter.

Crystals: Pseudo-isometric, probably tetragonal. The high-temperature form is isometric. Predominately in dull, rough-surfaced trapezohedrons (Figure 252).

Physical Properties: Brittle; conchoidal fracture. Very poor dodecahedral cleavage. H. 5½-6. G. 2.4-2.5.

Optical Properties: Translucent to transparent. Grayish-white, almost colorless in clear crystals, but with grayish tinge. Streak colorless. Luster vitreous to dull. Weak double refraction. R.I. $\alpha = 1.508$, $\gamma = 1.509$.

Chemistry: Potassium aluminum silicate.

Distinctive Features and Tests: Crystals closely resemble analcite, but are duller in luster and completely enclosed in rock instead of free in cavities as is usual with analcite. Softer than garnet. Infusible; garnet and analcite are *fusible*.

Occurrences: As phenocrysts in volcanic rock. Large, rude trapezohedrons to 2¼", from Wiesental, Bohemia, Czechoslovakia. The best crystals occur in porous tuffs and recent lavas in central Italy, and on Mount Vesuvius, near Naples; classic localities are Capo di Bove, Caserta, Roccamonfina, Ariccia, and the Alban Hills, near Rome. Crystals range from ¼" to 2" diameter; fresh crystals are smooth-faced and sometimes quite transparent, but larger crystals are usually dull and coated with a powdery alteration product.

Nepheline—(Na,K)(AlSiO$_4$)

Name: From the Greek *nephele,* "cloud," because it becomes clouded when put in strong acid; nĕf'ĕ lĭn.

Crystals: Hexagonal; rarely, in small stubby hexagonal prisms, terminated by $c[0001]$. Far more abundant as small to large grains and granular masses.

Physical Properties: Brittle; subconchoidal fracture. Good cleavage parallel to prism faces $m[10\bar{1}0]$. H. 6. G. 2.55-2.65.

Optical Properties: Transparent to translucent. Colorless, white, yellowish-white, or pinkish-white. Streak white. Luster vitreous to greasy. Uniaxial negative. R.I., O = 1.536-1.549, E = 1.532-1.544. Birefringence 0.004-0.005.

Chemistry: Sodium-potassium aluminum silicate.

Distinctive Features and Tests: Greasy luster; cleavage not as good as in feldspar nor at right angles as in scapolites. Easily decomposed by hydrochloric acid, yielding a clear jelly. Fuses at 4; a small splinter rounds into a clear glassy ball, coloring the flame yellow (scapolite melts to a white enamel).

Occurrences: As phenocrysts in igneous rock and in large grains and masses in pegmatites associated with nepheline syenites. In Italy, in metamorphosed limestone blocks at Monte Somma, Vesuvius, furnishing small glassy hexagonal crystals in cavities. Sharp hexagonal prisms from Capo di Bove, near Rome. In Hastings County, and elsewhere in Ontario, Canada, as large grains in pegmatites, and sometimes in large crude crystals; similarly in nepheline syenites at Litchfield, Kennebec County, Maine. Rarely represented in amateur collections due to scarcity of crystals.

Sodalite—Na$_4$Cl(AlSiO$_4$)$_3$

Name: In allusion to sodium content; sō'dă lĭt.

Crystals: Isometric; rare, then only in small dodecahedrons. Far more abundant as granular masses.

Physical Properties: Brittle; fracture uneven to subconchoidal. Good dodecahedral cleavage $d[110]$, easily developed and prominent on fracture surfaces of massive material. H. 6. G. 2.3; blue 2.28.

Optical Properties: Translucent to transparent. Commonly pale to deep blue; also white, gray, pink (*hackmannite*); blue material usually veined white or salmon. Streak colorless. Luster vitreous to greasy. R.I., $n = 1.48$. The variety *hackmannite* fluoresces brilliantly under ultraviolet; it also fades to white in daylight, but returns to pink when exposed to ultraviolet.

Chemistry: Sodium chlorine alumino-silicate.

Distinctive Features and Tests: Color; more translucent than lazurite and lazulite; lazurite also usually accompanied by pyrite. Decomposes and yields a jelly in hydrochloric acid. Fuses with difficulty in the flame, and then only on thin splinter edges; vivid yellow hue imparted to flame when splinter edges become white hot. After fusion, the enamel glows pale blue under ultraviolet (Brazil).

Occurrences: In nepheline syenite rock, often in large masses. Also as very small colorless crystals in metamorphosed limestone at Monte Somma, Vesuvius, Italy. Fine masses of sodalite suitable for ornamental purposes come from the Princess Quarry, Dungannon Township, Hastings County, Ontario, and also from other places in this and adjacent Renfrew County; fair to good massive from Kicking Horse Pass, and Ice River, British Columbia, Canada. Recently, large dark blue weathered masses from an unstated locality in Minas Gerais, Brazil. An old locality at Litchfield, Kennebec County, Maine, once produced good sodalite. The variety *hackmannite* occurs along the York River, near Bancroft, Hastings County, Ontario, Canada. The usual cabinet specimens are polished slabs of massive material.

Cancrinite—Na$_4$(HCO$_3$)(AlSiO$_4$)$_3$

Associated with sodalite and nepheline syenites, and sometimes used as an ornamental material when bright orange-yellow color. A sodium carbonate alumino-silicate; käng′krĭ nīt. Easily distinguished by color, associates, and bubbling in acids. Fine massive specimens occur in the sodalite deposits of Ontario and Maine.

Lazurite—(Na,Ca)$_4$(SO$_4$,S,Cl)(AlSiO$_4$)$_3$

Name: From the ancient Persian *lazhuward,* for the gemstone *lapis lazuli,* containing lazurite as the principal mineral; lȧz′ū rīt; lăp ĭs lăz′ ū lī.
Crystals: Isometric; very rare. Well-formed dodecahedrons only from the Afghanistan occurrence in rude singles to 1″, seldom displaying reasonably sharp faces. Predominately fine-granular.
Physical Properties: Brittle; fracture uneven. Poor dodecahedral cleavage on d[110] and not easily observed in ordinary granular material. H. 5-5½. G. 2.4-2.45. *Lapis lazuli,* a mixture of several minerals, usually including calcite and pyrite, may be as high as G. 2.7-2.9.
Optical Properties: Translucent at best. Intense rich blue, usually tinged with purple, but also greenish-blue, azure blue. Color modified by other minerals in *lapis lazuli.* Streak bright blue. Luster dull, somewhat greasy. Vague refractometer reading about $n = 1.50$.
Chemistry: A complex sodium-calcium alumino-silicate with sulfate, sulfur and chlorine appearing in various proportions.
Distinctive Features and Tests: Deep color; usually associated with pyrite and white veinings or mottlings of calcite; much finer-grained than sodalite or lazulite. In the flame, retains blue color after heating to redness, but fuses at 3½ to a yellowish bubbly glass; the latter crushes to a colorless powder. Soluble in hydrochloric acid, giving off rotten egg odor of hydrogen sulfide.
Occurrences: Uncommon except in a few deposits where it occurs in marble. Euhedral crystals only from the ancient but still productive Firgamu mines at the upper reaches of the Kokcha River, Badakshan, northeast Afghanistan, occurring as irregular masses and lenses in marble. Also as boulders in the bed of the Slyudyanka River, Sayan Mountains, near Lake Baikal, U.S.S.R., and in masses of considerable size in Chile in the Andes of Ovalle, the latter furnishing generally pale blue material, usually mixed with considerable quantities of white calcite. Dark blue material occurs in thin streaks and lenses on Italian Mountain, in the Sawatch Mountains of Colorado, and on Ontario Peak, San Gabriel Mountains, and Cascade Canyon, San Bernardino Mountains, San Bernardino County, California.

SCAPOLITE (WERNERITE) SERIES

Marialite—(Na,Ca)$_4$(Cl,CO$_3$,SO$_4$)[(Al,Si)$_4$O$_8$]$_3$ (Sodium end member)

Meionite—(Ca,Na)$_4$(Cl,CO$_3$,SO$_4$)[(Al,Si)$_4$O$_8$]$_3$ (Calcium end member).

Names: From the Greek, *skapos,* "shaft," in allusion to prismatic crystals; skȧp′ō līt. Origin of marialite uncertain; mȧ rē′ȧ līt. From the Greek, *meion,* "small," in allusion to the low pyramids on crystals; mī′ō nīt. Sometimes used in lieu of scapolite, the alternate name, *wernerite,* is after A. G. Werner (1750-1817), eminent German geologist; wĕr′nĕr ĭt.

Crystals: Tetragonal. Common as rude stubby prisms of square cross-section, terminated by low pyramidal faces and usually penetrated by many cracks (Figures 253, 254). Prominent forms are prisms of the first order, *m*[110] and the second order, *a*[100], and first order pyramid *r*[111]. Sometimes striated parallel to the *c* axis. Also corroded to formless masses or in coarse-granular aggregates.

Physical Properties: Brittle; fracture uneven to conchoidal. Distinct cleavages parallel to *m*[110] and *a*[100], but usually interrupted. H. 6. G. 2.6-2.8; increasing with calcium; gem varieties: 2.60-2.71.

Optical Properties: Transparent to translucent. Colorless, white, faintly yellow to distinct yellow, grayish, pale green, also pale blue, pink and pale violet. Streak colorless. Luster vitreous to somewhat resinous. Sometimes strongly chatoyant because of acicular inclusions parallel to *c* axis; rarely, aventurescent due to very small platy inclusions parallel

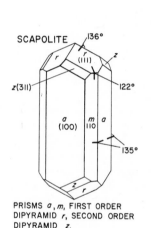

SCAPOLITE

PRISMS *a*, *m*, FIRST ORDER
DIPYRAMID *r*, SECOND ORDER
DIPYRAMID *z*.

FIG. 253.

FIG. 254. Scapolite crystals from Madagascar showing first order and second order prisms, and first order dipyramids. The c axis is upright. Note the crumbling of surface portions due to development of numerous cleavages and partings. Size 3″ by 1½″.

to prism faces. Uniaxial negative. R.I., O = 1.549-1.577, E = 1.540-1.550. Birefringence, variable, 0.009-0.027. Distinctly to strongly dichroic, yellow specimens giving colorless and yellow; violet giving blue, reddish-blue. Commonly fluorescent under ultraviolet (long wave best) orange, yellow, also pale violet, and white.

Chemistry: Sodium-calcium alumino-silicates with chlorine, carbonate and sulfate, forming an unbroken series between end members shown.

Distinctive Features and Tests: Stubby square prisms with poor quality faces; fluorescence; chatoyancy. Fuses at 3 to 4 to a white glass, coloring flame yellow; the glass is weakly fluorescent. Decomposed by hydrochloric acid but does not yield a jelly.

Occurrences: Typically in marble skarns as enclosed crystals or grains near contacts with igneous rock. Commonly associated with pale-colored pyroxenes, amphibole, apatite, zircon, sphene. Also in pegmatites, gneisses and schists. Abundant in the Grenville metamorphosed limestones at many points in Quebec, and in Ontario in Faraday Township, Hastings County, and near Eganville, Renfrew County, often in large crude pale yellowish-white prisms, some to 14″ by 9″ by 9″, and rarely, as blue fibrous masses. Similar occurrences in New York State produce fair to good crystals at Grasse Lake, near Rossie, near Gouverneur, St. Lawrence County, and at Diana, Lewis County. Many fair to good crystals, sometimes quite sharp, have been found in the Franklin marbles of Orange County, New York, and in adjacent areas in Sussex County, New Jersey, especially in the zinc deposit at Franklin. Good pink prisms to $\frac{1}{4}''$ by $2''$ occur in white marble at Campbell Quarry, Texas, Baltimore County, Maryland. Recently, in opaque white crystals, with dark green dipside, from metamorphic limestones in Oaxaca, Mexico; crystals to $10''$ tall have been found. Fine bright yellow gem material in severely corroded masses, yielding gems to 40 carats, occurs in Espirito Santo Brazil; also at Tsarasaotra near Ankozobe, Madagascar. Large sharp, translucent to transparent, striated prisms to $3''$ by $2''$ by $2''$, of white, gray, or colorless hues, occur near Betroka, Madagascar. Excellent facet and catseye material in rolled masses, in the gem tract of Mogok, Burma, yielding gems of white, pink, violet, and rarely, yellow hues, the catseyes being particularly fine. Splendid long sharp prisms of yellow color, and largely flawless, from Tremorgio, Switzerland, some reaching several inches length. Despite abundance in metamorphosed limestone, sharp crystals are difficult to obtain and few collections contain examples; probably the best crystals come from Betroka, Madagascar.

Zeolite Group

The word zeolite comes from the Greek, *zein,* "to boil," because of the bubbling of fragments in the torch flame as water is driven off. Zeolites are widespread and abundant as low-temperature species, usually among the last to form in any occurrence containing them. The finest examples occur in cavities in basalt and diabase, which rocks, as it so happens, are prized by the construction industry for the making of concrete. For this reason, specimens continue to be supplied from quarries near population centers where basalt and diabase bodies are exposed. Although most abundant in fine specimens in the rock mentioned, zeolites also occur in lesser quantities in upper portions of ore veins, in small amounts in granitic pegmatite cavities, and in near-surface portions of some sedimentary rocks. Common associates are apophyllite, datolite, prehnite, pectolite, calcite and quartz.

Approximately 30 zeolite species are recognized, but only a handful occur in good specimens. Close structural relationships do not exist among all members, but all possess a very open tektosilicate framework with relatively large openings in which various cations are held to balance the negative charges of the silicate framework, but which can be easily removed and others substituted. Similarly, the water can also be removed by gently heating the crystals without structural collapse. The channels within the zeolite crystal framework allow ion-exchange, as in water softeners, where "hard" water containing calcium ions is passed through a bed of synthetic zeolite and emerges with sodium in the water, making it "soft." The "hard" calcium ions are trapped in the zeolite. As a group, zeolites are most likely to be colorless or white, or only faintly tinged with color, and are also very brittle or very easily cleaved. Some, as laumontite, spontaneously give up water when removed from the ground, and very quickly crumble into powder unless preserved beforehand by dipping in plastic solution.

Heulandite—$(Ca,Na,K)(Al_2Si_7O_{18})\cdot 6H_2O$

Name: After English mineralogist H. Heuland; hū'lan dīt.
Crystals: Monoclinic; usually in warped coffin-shaped individuals with strong pearly luster

on the broad faces of $b[010]$ as shown in Figure 255. The sides of the "coffins" are bright but the irregular faces of $t[101]$ and $s[\bar{1}01]$ show no pearly luster. Each crystal is an aggregate of many, approximately parallel, tabular crystals, the middle of the group sagging and the narrow ends raised. Small crystals of about ⅛″ to ¼″ are like textbook drawings but larger crystals are often warped noticeably. In single crystals or clusters on matrix; also in druses; and rarely, granular massive.

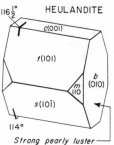

FIG. 255.

Physical Properties: Brittle; uneven stepped fracture. Cleaves readily parallel to $b[010]$ (the "top" of the coffin). H. 4, but about 3 on cleavage surfaces. G. 2.2.

Optical Properties: Transparent to translucent. Colorless, white; sometimes pale pink, yellow, or pale reddish-brown, and less usually, brownish-red. Streak colorless. Luster vitreous, but brilliant pearly on cleavage. Biaxial positive. R.I., $\alpha = 1.496$, $\beta = 1.497$, $\gamma = 1.501$. Birefringence 0.005. A thin cleavage flake in the polariscope displays a fine biaxial figure.

Chemistry: Hydrous calcium alumino-silicate. Sodium and potassium commonly substitute for calcium.

Distinctive Features and Tests: Coffin-shaped crystals; strong pearly luster; cleavage. A thick cleavage flake promptly turns white in the flame, exfoliating into numerous puffy laminations, eventually forming a wormlike mass, then fusing at 2½ to a white enamel. Gelatinizes in hydrochloric acid.

Occurrences: With other zeolites and zeolite associates, in basalts. Fine white crystals to 3″, abundant in basalt at Teigarhorn, Berufjord, Iceland. Also fine from Faeroe Islands. White to pale pink from a number of places in the northwest coastal basalts of Nova Scotia, as at Wasson's Bluff, Partridge Island, etc. Especially abundant and fine in the basaltic sills of the Watchung Mountains of northern New Jersey, where beautiful snow white crystals to 2″ length occur at Prospect Park and Garret Mountain, Paterson, Passaic County, and elsewhere (Figure 256). Pink to white crystals occur in basalt cavities near Cascadia, Linn County, also near Edwards, Tillamook County, and Ritter Hot Springs, Lane County, Oregon. Good brownish crystals to 1″ length, recently made their appearance with other zeolites from basalts in the region between Veronopolis and Bento Goncalves, Rio Grande Do Sul, Brazil. Sharp brick-red crystals at Bolzano, Italian Tyrol; also from various localities in the Austrian Tyrol and in Switzerland. Fine specimens from the mines of the Harz, Germany, notably from Andreasberg.

Harmotome—Ba(Al$_2$Si$_6$O$_{16}$)·6H$_2$O

Name: Of obscure derivation, possibly from the Greek *harmos*, for "joint" in allusion to twinned crystals; här′mŏ tōm.

Crystals: Monoclinic; always in interpenetrant twins (Figure 257), sometimes forming crosses. The crystals shown in Figure 258 display $b[010]$ faces, the blunt wedge-shaped faces being $m[110]$.

Physical Properties: Brittle; fracture uneven to subconchoidal. Cleavage easy on $b[010]$; distinct on $c[001]$. H. 4½. G. 2.44-2.50.

Optical Properties: Translucent. White, colorless, or slightly gray. Streak colorless. Luster vitreous. Biaxial positive. R.I., $\alpha = 1.503$, $\beta = 1.505$, $\gamma = 1.508$. Birefringence 0.005.

Chemistry: Hydrous barium alumino-silicate.

Distinctive Features and Tests: Cross-like twins whose ends resemble the tips of star drills. Whitens in the flame, puffing slightly, then crumbling into fluffy fragments which

FIG. 256. Pale pink heulandite crystals on matrix of altered basalt with datolite crystals. Small crystals show the "coffin" shape of Figure 255 but larger crystals are severely warped. Size 5¼" by 3". Prospect Park, Paterson, Passaic County, New Jersey.

blow away. Fuses at 3 to a white glass. Decomposes in hydrochloric acid but does not form a jelly.

Occurrences: With zeolites in basalt, trachyte, phonolite; also in gneiss and as a low-temperature mineral in some sulfide ore veins. This species is rare in United States zeolite localities but good specimens, sprinkled with small glassy crystals, have been found in diabase near Ossining, Westchester County, New York. In fine sharp grayish cruciform twins to ½″ diameter in a sulfide deposit with galena, pyrite, and pale yellow apophyllite in Korsnas Mine, Finland. At Kongsberg, Norway. Exceptional specimens with calcite and barite, from Bellesgrove Mine, Strontian, Argyllshire, Scotland and from Kilpatrick Hills, Dumbarton, and Campsie Hills, Stirling. In beautiful pale bluish-gray cruciform twins to ½″ from Andreasberg, Harz, Germany. Phillipsite is closely related to harmotome, but is confined in its occurrence to igneous rocks, usually forming crystals very similar to harmotome but smaller in size and more likely to be transparent.

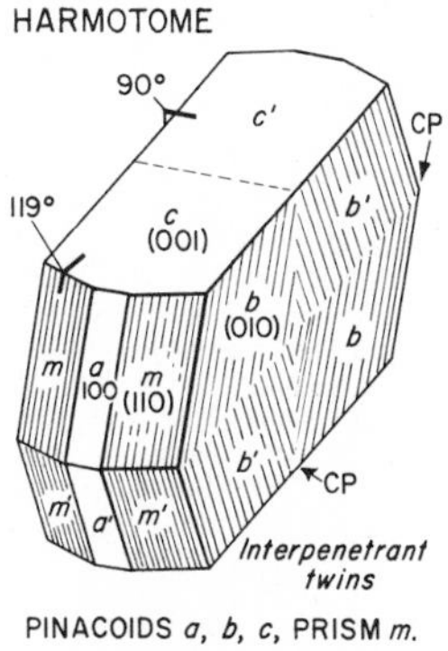

HARMOTOME

FIG. 257.

Stilbite—(Ca,Na)(Al$_2$Si$_7$O$_{18}$)·7H$_2$O

Name: From the Greek, *stilbein,* "to glitter," in allusion to luster; stĭl′bĭt.

Crystals: Monoclinic. Rarely, in distinct isolated crystals as in Figure 259, but more commonly, in bladed aggregates of numerous twinned individuals, pinched at the middle in hourglass fashion, swelling toward the extremities as in Figure 260. Doubly-terminated aggregates look like wheat sheaves and present an unmistakable appearance. The broad faces in such aggregates are *b*[010] and may be certainly identified by the presence of pearly

FIG. 258. Twinned harmotome crystals on matrix from Bellesgrove Mine, Strontian, Argyllshire, Scotland. Size 2½″ by 1¾″. The boat-shaped faces are *b*[010], the blunt wedges are formed by pairs of *m*[110] faces.

luster caused by perfect cleavage on this plan. Tips of aggregate crystals display a blunt wedge profile formed from faces of $m[110]$ (Figure 261). Twinning difficult to observe; cruciform twins on composition plane $c[001]$. Also in rosettes, ball-like aggregates, and massive bladed.

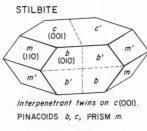

STILBITE

Interpenetrant twins on c(OOl).
PINACOIDS *b, c,* PRISM *m.*

FIG. 259.

FIG. 260. Sketch of stilbite "wheat sheave" aggregates. From Prospect Park, Paterson, Passaic County, New Jersey. Size about 2" by 2".

Physical Properties: Brittle; uneven fracture. Cleavage perfect and easily developed on $b[010]$. H. 4. G. 2.1-2.2.

Optical Properties: Translucent to transparent. White, pale to dark brown, also pink, reddish, and sometimes brownish-orange. Streak colorless. Luster pearly on cleaved surfaces, otherwise vitreous to dull. Biaxial negative. R.I., $\alpha = 1.482$-1.492, $\beta = 1.491$-1.500, $\gamma = 1.493$-1.502. Birefringence 0.011-0.010.

Chemistry: Hydrous calcium-sodium alumino-silicate.

Distinctive Features and Tests: Wheat sheaf aggregates in association with other zeolites. Promptly whitens in the flame, then puffs into curling, writhing irregular masses, greatly expanding in volume, but does not exfoliate into "worms" as does heulandite. Fuses at $3\frac{1}{2}$. Decomposed by hydrochloric acid without forming a jelly.

Occurrences: As with other zeolites, but also common in cavities of some granitic pegmatites, notably in the Pala area of San Diego County, California. Beautiful snow-white bladed crystals from Teigarhorn, Rodefjord, and Berufjord, and elsewhere in Iceland (Figure 261), often in crystals of 1" to 2". Fine specimens of white, orange or brown color in basalt sills of Nova Scotia, particularly at Cape d'Or, Horseshoe Cove, Partridge Island, and Two Islands in Cumberland County; also at Cape Blomidon and elsewhere in Kings County, and Annapolis County, etc. Exceptional specimens, usually pale brown to brownish-yellow, from many places in the Watchung basaltic sills of northern New Jersey, notably near Paterson, Upper Montclair, and Great Notch; sheaves to 3" radius have been recorded (Figure 260), also doubly terminated sheaves, sometimes completely loose and then making fine small specimens. Orange-brown aggregates on rock from Perkiomenville, Montgomery County, Pennsylvania. Excellent white crystals to 1", on matrix from near Mitchell, Wheeler County, from Edwards, Tillamook County, and Gable, Columbia County, and from many localities in Benton, Lane and Douglas counties, Oregon; fine specimens also from Washington, as near Washougal, Cowlitz County. Beautiful specimens from the silver veins of Guanajuato, Mexico. Recently, splendid rich brown crystal sheaves, some to 4", from Rio Grande Do Sul, Brazil, with green apophyllite, heulandite, etc. Beautiful sharp bladed pink crystals from the Faeroe Islands, and from Kilpatrick (to $1\frac{1}{2}$"), Dumbarton, and Kilmacolm, near Glasgow, Scotland. Striegau, Silesia, Poland, produces fine specimens associated with orthoclase crystals from cavities within

FIG. 261. Stilbite aggregate crystals on altered basalt with minor heulandite from Berufjord, Iceland. Size 2″ by 2″. Broad faces are c[001]; wedge terminations are pairs of m[110] faces.

granite. Also fine from Mitterberg, Salzburg, Austria, and from Aussig in Czechoslovakia. Bright orange sheaves from Mt. Painter, Southern Australia. Very large sheaves to 3″ to 4″, also spherical aggregates to 6″ diameter, of white to pale brown color, from basalt near Poona, India.

Stilbite is a good collector's mineral and excellent specimens are readily available from Oregon and New Jersey localities, often in handsome groups attractively associated with other minerals. The specimens from Iceland are difficult to obtain because the localities are not easily accessible, and no regular export is undertaken by Icelanders; the same is true of the celebrated localities near Poona, India.

Laumontite—(Ca,Na)(Al$_2$Si$_4$O$_{12}$)·4H$_2$O

Calcium-sodium alumino-silicate with water; law′mŏn tīt. Water is readily lost when specimens are removed from the ground, causing the crystals to crumble into powder unless first immersed in a solution of plastic or other sealant. Crystals usually very small white to pinkish square prisms with ends cut off by one pinacoidal face leaving sharp chisel-shaped terminations (Figure 262). Exceptional crystals, often to 1″, occur in Rio Grande Do Sul, Brazil (Figure 263). Also fine from Sunnyside, Cochise County, Arizona, in crystals to 1″ length. Small white crystals from New Jersey, Oregon, and Washington zeolite lo-

calities. Commonly coats other zeolites because it is last to form in cavities. The partly
dehydrated specimens usually found in collections are sometimes given the varietal name
leonhardite. Good cabinet specimens are rare, probably the best
being those from Brazil.

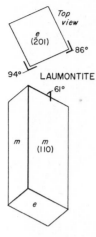

FIG. 262.

Chabazite—(Ca,Na,K)(Al$_2$Si$_4$O$_{12}$)·6H$_2$O

Name: From the Greek, *chabazios,* one of twenty stones mentioned
by the poet Orpheus in the poem *Peri lithon;* kăb'ă zīt, also shăb' ă
zīt.

Crystals: Hexagonal. Usually simple rhombohedral crystals resembling cubes, the angles between faces being close to 90°. Crystals
often display numerous offsets as in Figure 264. Also twinned on
the *c* axis, in penetration twins (*phacolite*), resembling low hexagonal dipyramids.

Physical Properties: Brittle; uneven fracture. Poor cleavage parallel
to rhombohedral faces *r*[10$\bar{1}$1]. H. 4. G. 2.05-2.15.

Optical Properties: Translucent, sometimes transparent. Commonly
pink, salmon, or red-brown; also white, colorless. Streak colorless.
Luster vitreous. Uniaxial negative. R.I., O = 1.48, E = 1.49. Birefringence 0.01.

Chemistry: Hydrous calcium-sodium-potassium alumino-silicate.

Distinctive Features and Tests: Rhombohedral crystals; reddish colors; poor cleavage. In
the flame, promptly loses color, but swells only slightly, then fuses at 3 to a white opaque
glass. Decomposed by hydrochloric acid.

Occurrences: With other zeolites in basaltic rocks. Small glassy crystals from Berufjord,
Iceland. Exceptionally large and fine flesh-red rhombohedrons to 1½″ on edge, from
Wasson's Bluff, Cumberland County, Nova Scotia (Figure 264), and elsewhere in this
province. Also fine from the New Jersey localities but not as abundant as other zeolites.
Fine translucent crystals to ⅜″ from Springfield, and Ritter Hot Springs, Lane County,
and near Vernonia and from Goble, Columbia County, Oregon. Numerous localities in
Scotland, especially on the Isle of Skye. From Aussig, Czechoslovakia. In ⅛″ to ¼″ twins
on basalt, near Melbourne, Victoria, and in showy sharp white crystals from western coastal
localities in Tasmania, Australia. Probably the best cabinet specimens are the handsome
large flesh-red crystals from Nova Scotia, or the similarly colored singles on matrix from
New Jersey localities. Large crystals are uncommon.

Gmelinite—Na$_2$(Al$_2$Si$_4$O$_{12}$)·6H$_2$O

A hydrous sodium alumino-silicate in which calcium sometimes substitutes for sodium;
mĕl'ĭ nīt. Usually in twinned crystals of pyramidal shape and commonly intergrown with
chabazite. Colorless, white, pink, reddish; distinct cleavage parallel to prism *m*[10$\bar{1}$0].
Less common than other zeolites, often in small inconspicuous crystals. In ½″ crystals at
Glenarm, and Magee Island, County Antrim, Ireland. At Cape Blomidon, Nova Scotia.
In the New Jersey occurrences as small complex brownish-pink crystals. Splendid pink
crystals to ⅞″ from Flinders Island, Victoria, Australia.

Analcime—Na(AlSi$_2$O$_6$)·H$_2$O

Name: From the Greek, *an-,* "not," and *alkimos,* "strong," in allusion to weak pyroelectricity; ăn ăl' sēm; formerly, *analcite,* ăn ăl'sīt.

FIG. 263. Laumontite crystals on basalt, from between Veronopolis and Bento Goncalves, Rio Grande Do Sul, Brazil. Size 4″ by 3″.

Crystals: Isometric; predominantly in crystals of simple trapezohedral form $d[110]$ as in Figure 265. Rarely, cubic crystals are found with trapezohedral truncations on corners. Also granular, and as phenocrysts in volcanic rock.

Physical Properties: Brittle; uneven to subconchoidal fracture. No cleavage. H. 5. G. 2.3.

Optical Properties: Translucent to transparent. Usually colorless or white, but also pale pink, and sometimes brilliant red because of hematite inclusions (Figure 266). Streak colorless. Luster vitreous. R.I., $n = 1.48$-1.49.

Chemistry: Hydrous sodium alumino-silicate.

Distinctive Features and Tests: Crystals resemble leucite and garnet, but are not likely to be confused for either species because of occurring principally in basalt. Fuses at $3\frac{1}{2}$ to a milk-white bleb, then to a clear glass, coloring flame yellow. It does not swell or exfoliate as other zeolites do. Partly fused material fluoresces distinctly in spots under ultraviolet, the color being pale greenish-white. Turns to jelly in hydrochloric acid.

Occurrences: Primarily with other zeolites in basalt cavities. Brilliant transparent crystals to 1″, at Five Islands, and to as much as $4\frac{1}{4}$″ diameter at Cape Blomidon, and from other zeolite localities in Nova Scotia. Excellent specimens at some northern New Jersey localities, notably in Paterson, Passaic County, as groups and druses displaying crystals from $\frac{1}{8}$″ to 1″ diameter, usually white, but sometimes flushed red by inclusions of hematite (Figure 266). Unusually fine clear crystals, some interpenetrantly twinned, occurred

FIG. 264. Rhombohedral crystals of brownish-red chabazite from Wasson's Bluff, Swan's Creek, Cumberland County, Nova Scotia. Size 2½″ by 1¾″.

in copper deposits of Keeweenaw Peninsula, Michigan, but are now rarely obtainable; striking specimens sometimes displayed bright copper inclusions in clear crystals. Excellent crystals from Table Mountain, near Golden, Jefferson County, Colorado. Good to fine at Ritter Hot Springs, Grant County, Oregon. In crystals to 1¼″ near Point Sal, California, and in ½″ crystals in andesite, at Mount Blanco, Death Valley, Inyo County; also in ⅜″ crystals in basalt near Malibu, Los Angeles County. Enormous, well formed, opaque white to pink trapezohedrons, to 3½″ diameter, occur in amygdaloid cavities at Seisser Alp, Italian Tirol; these are among the largest known. Exceptionally brilliant and transparent crystals, often of large size, sometimes in cubes modified by trapezohedrons, occur on the Cyclopean Islands, near Sicily. From Aussig, Czechoslovakia, and in the Harz, Germany. Opaque 1″ crystals from Isle of Skye, Scotland. Crystals to 2″ near Faskrudsfjord, Iceland. Fine sharp opaque white crystals to ⅜″, from Prospect, near Sydney, New South Wales, and clear glassy crystals from Flinders Island, Tasmania, Australia.

Natrolite—Na$_2$(Al$_2$Si$_3$O$_{10}$)·2H$_2$O

Name: From *natron,* "soda," in allusion to sodium content; nàt′rō līt.

Crystals: Orthorhombic; commonly slender to acicular and of square to octagonal cross-section; also short prismatic (Figure 267). Forms usually prism m[110], pinacoids a[100] and b[010], the latter sometimes flattening prisms along the b axis. Usually terminated by low pyramid o[111]. Also bladed massive and radiate fibrous.

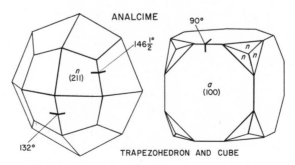

ANALCIME

146½°

n (211)

132°

90°

n n n

a (100)

TRAPEZOHEDRON AND CUBE

FIG. 265.

FIG. 266. Ball-like trapezohedrons of analcime, partly colored by bright red hematite inclusions, on altered basalt. From McBride Avenue Quarry, Paterson, Passaic County, New Jersey. Size 5″ by 4″.

Physical Properties: Brittle; uneven to conchoidal fracture. Perfect cleavage parallel to $m[110]$. H. 5½-6. G. 2.25.

Optical Properties: Transparent to translucent. White or colorless. Streak colorless. Luster vitreous to somewhat pearly on cleavage plane. Biaxial positive. R.I., $\alpha = 1.480$, $\beta = 1.483$, $\gamma = 1.490$. Birefringence 0.010.

Chemistry: Hydrous sodium alumino-silicate.

Distinctive Features and Tests: Crystals; prismatic cleavage. Natrolite crystals usually thicker than scolecite or mesolite; much that has been called "natrolite" in the New Jersey occurrences, is mesolite. Fuses at 3 to a colorless glass. Readily gelatinizes in hydrochloric acid.

Occurrences: Predominantly with zeolites, but also as veins in serpentine and other rocks where it forms in low-temperature environments. In the Nova Scotia zeolite localities. In fine short prisms to ⅟₁₆″ thick, at Snake Hill, Bergen County, New Jersey, and in clear prisms to ¼″ thick and 3″ long enclosed in calcite, in Houdaille Quarry, Bound Brook, Somerset County. In thick velvety layers lining crevices at Perkiomenville, Montgomery County, Pennsylvania. In excellent examples at Table Mountain, near Golden, Jefferson County, Colorado. At the zeolite localities near Springfield, Lane County, and reported in needles to 8″ long from near Medford, Jackson County, Oregon. In the Olympic Peninsula, Washington. Exceptionally thick prisms to ⅛″ thick and 3″ long, but few with terminations, occur in criss-cross

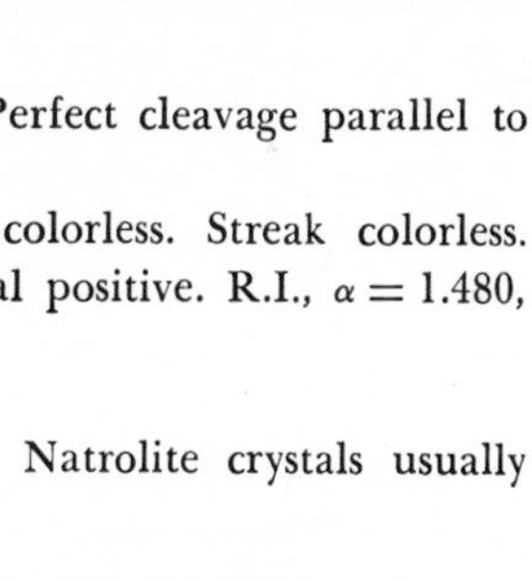

NATROLITE

117°   38°
o
(111)

89°   b(010)

m
(110)

DIPYRAMID o, PRISM m.

FIG. 267.

growths near Livingston, Montana. Perhaps the largest crystals known occur in contacts between schist and serpentine along the headwaters of Clear Creek, San Benito County, California, and at the upper end of the benitoite deposit near the headwaters of the San Benito River. At Clear Creek, octagonal colorless to white prisms measuring $5/8''$ thick and $1\frac{1}{2}''$ long, occur in narrow fissures penetrating the schist; some crystals are perfectly clear. At the benitoite mine, crystals to $\frac{1}{8}''$ by $\frac{1}{2}''$ by $1\frac{1}{4}''$, form clusters on massive granular or bladed natrolite. The prisms are usually flattened on $b[010]$ and corroded. Also at the benitoite mine occur short thick prisms, about $\frac{1}{2}''$ length, filled with crossite inclusions, and curiously warped. Good specimens occur in Europe at Aussig, Czechoslovakia; in the Alps; in the Auvergne of France; and as bright orange-white dense aggregates from Hohentwiel, Hegau, Germany; the latter specimens are very colorful and are sometimes polished to effectively display the attractive circular patterns.

Scolecite—Ca(Al$_2$Si$_3$O$_{10}$)·3H$_2$O

Similar to natrolite in respect to occurrence and crystals, and distinguished with difficulty from that species and mesolite. A hydrous calcium alumino-silicate; the name is from the Greek, *skolex,* a "worm," in allusion to the curling exfoliation in the flame; skŏl'ĕ sīt. H. 5. G. 2.27. Monoclinic. Biaxial negative. R.I., $\alpha = 1.513$, $\beta = 1.519$, $\gamma = 1.520$ (noticeably higher than natrolite). Birefringence 0.007. More likely to be found as slender hair-like crystals of snow-white color, or in finely fibrous chatoyant masses of radiate structure. The finest crystals occur with other zeolites in cavities in the Deccan basalts of Poona, India.

Mesolite—Na$_2$Ca$_2$(Al$_6$Si$_9$O$_{30}$)·8H$_2$O

Similar to the previous two species and found under the same geological circumstances. A hydrous sodium-calcium alumino-silicate. The name is derived from the Greek, *mesos-,* for "middle," because this species was supposed to fall between natrolite and scolecite in respect to composition and properties; mĕs'ō līt. H. 5. G. 2.26. Monoclinic. Biaxial positive. R.I., $\beta = 1.505$-1.507. Very low birefringence, behaving like a singly refractive mineral under the polariscope. In needles to $6''$ from Teigarhorn, Iceland. Fine mesolite, in exceedingly thin white needles, from many places in New Jersey. As thick chatoyant masses at Mount Pisgah, Lane County, New Era, Klackamas County, and Durkee, Baker County, Oregon. Also fine from Clifton Hill, near Melbourne, Victoria, Australia.

Thomsonite—NaCa$_2$(Al$_5$Si$_5$O$_{20}$)·6H$_2$O

A zeolite which seldom produces good crystals, occurring mostly in radiate fibrous masses displaying chatoyancy. Hydrous sodium-calcium alumino-silicate; tŏm'sŭn īt. H. 5. G. 2.3-2.4. Orthorhombic. Biaxial positive. R.I. variable, $\alpha = 1.511$-1.529, $\beta = 1.513$-1.531, $\gamma = 1.518$-1.545. Birefringence 0.007-0.016. Rosettes of whitish hue to $2''$ diameter, some resembling stilbite except for color, occur at localities in northern New Jersey. In long slender radiate masses from Peters Point, Nova Scotia. Beautifully colored fibrous masses occur in amygdaloidal basalt on the shore of Lake Superior near Grand Marais, Cook County, Minnesota, and have been used for cutting attractive cabochon gems; some is exceptionally compact, jade-like, and then forms gray-green translucent nodules known as *lintonite.* Fine specimens from Table Mountain, Jefferson County, Colorado, and at the zeolite localities in Oregon as at Lakeview, New Era, and Springfield. In Europe, at Weipert and Seeburg, Czechoslovakia and at Bernitzgrün, Vogtland, Germany.

PHYLLOSILICATES

As will be recalled, the prefix *phyllo-*, (fĭl′ō), refers to "leaf," in allusion to the thin cleavage flakes which can be produced by splitting the crystals of many species in this group. Some can be split into exceedingly thin flexible sheets, as in chlorite, or elastic sheets, as in muscovite. In others, notably the extremely fine-grained clays, the leafy structure is not evident, and for them, only the powerful electron microscope reveals the fact that the minute crystals are similar in habit to the larger ones grown by other species in this group. The single prominent cleavage displayed so characteristically by phyllosilicates is due to the crystal structure which consists of a sandwich of silicon-oxygen tetrahedrons, linked together in continuous sheets and containing between them, various ions as aluminum, magnesium, iron, and hydroxyl. The silicate tetrahedrons form nearly hexagonal rings within sheets, such submicroscopic geometry reflecting itself in the outward form of crystals, many of which look like reasonably perfect hexagons in profile, and suggest belonging to the hexagonal system when one first sees them. Apophyllite, included in the phyllosilicates, forms square crystals but its structure consists of tetrahedrons linked in fours instead of sixes, creating square rings and giving rise to tetragonal symmetry instead of the pseudohexagonal symmetry shown by the others. An important feature of all phyllosilicates is that the perfect cleavage is at right angles to the *c* axis, and it is along this axis that the crystals commonly develop into elongated prisms, or conversely, into tabular to wafer-like crystals.

Apophyllite—$KCa_4F(Si_4O_{10})_2 \cdot 8H_2O$

Name: From the Greek, *apos-*, "off," and *phyllon*, "leaf," meaning to flake off, in allusion to the behavior of crystals when heated in the flame; ă pŏf′ĭ lĭt.

Crystals: Tetragonal; usually cube-like or tabular, square in cross-section. Also prismatic or pyramidal in habit as in Figure 268. The square cross-section is produced by the four like faces of prism *a*[100]; these are usually very lustrous but striated parallel to the *c* axis. The basal plane *c*[001] is unmistakable because of the strong pearly reflections which arise from it; its faces may be rough or show numerous shallow offsets. Corners of cube-like crystals may be truncated by the dipyramid *p*[111], and in some crystals, its faces enlarge to form pointed terminations. Very commonly in fine crystals; less often in granular masses.

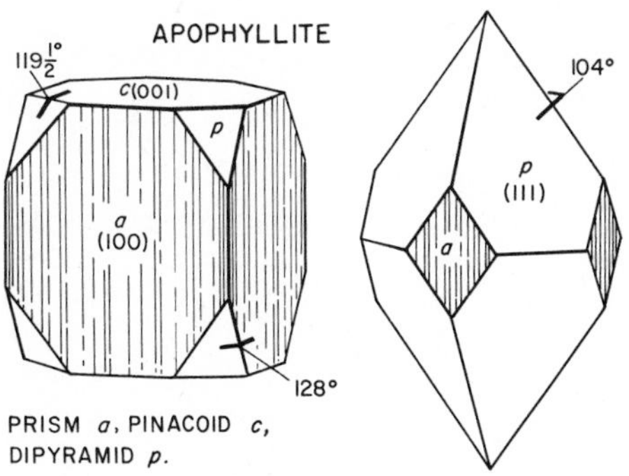

APOPHYLLITE

PRISM *a*, PINACOID *c*, DIPYRAMID *p*.

FIG. 268.

Physical Properties: Brittle; fracture uneven. Perfect cleavage parallel to *c*[001]; numerous partly-developed cleavages commonly impart the strong pearly luster mentioned. H. 4½ on *c*, elsewhere 5. G. 2.3-2.4.

Optical Properties: Transparent to translucent. Colorless, white; also faint pink or green.

Streak white. Luster vitreous, but strong pearly on c; some small crystals may not show the pearly luster. Uniaxial positive or negative. R.I., $O = 1.535$, $E = 1.537$. Birefringence 0.002. Good uniaxial figure appears on thin cleavage flake in the polariscope.

Chemistry: Hydrous potassium-calcium-fluorine silicate.

Distinctive Features and Tests: Cube-like crystals associated with zeolites; vertical striations on sides of crystals; strong pearly luster on one perfect cleavage plane. Fuses at 2, first swelling to about 5 times original volume. Melted portions fluoresce greenish-white under ultraviolet. Decomposed by hydrochloric acid.

Occurrences: Abundant with zeolites in cavities in basaltic rock; also in cavities in granite and gneiss, and sometimes abundant as a low-temperature hydrothermal mineral in sulfide ore veins (Guanajuato). Good crystals occur in the Bay of Fundy region of Nova Scotia, at Cape d'Or, Isle Haute, Swan Creek, in Cumberland County, and at Cape Blomidon, Kings County, etc. In fine pink specimens in mines of the Rossland area, Trail Creek Division, British Columbia. In excellent crystals, often handsomely set off by associates, at many places in the basaltic sills of northern New Jersey, notably Paterson, Passaic County, and Bergen Hill and Snake Hill in Bergen County. New Jersey crystals range in size from $\frac{1}{2}''$ clear glassy crystals to large white crystals $2''$ on edge and sometimes more. Fine pink, white, colorless, or yellow crystals to $2''$ on edge, often in large groups, formerly from French Creek Mines, Chester County, Pennsylvania. In recent years, extremely beautiful matrix specimens were obtained from diabase quarries in Virginia, west of the District of Columbia, notably near Centreville, Fairfax County, and also in Loudon County; crystals of tabular habit, some to $3''$ on edge, and implanted on vivid green prehnite, occur in fissures in mineralized joints in diabase. Formerly, in very fine glassy crystals from the copper deposits of the Keeweenaw Peninsula, Michigan. Good white crystals occur associated with other zeolites at Table Mountain, Jefferson County, Colorado. Large choice crystals in the Lambert Quarry, Kings Valley, Lane County, Oregon, and elsewhere in the basaltic rock of this state. Fine pearly white or pink crystals, often forming large druses, formerly abundant in the silver mines of Guanajuato, Mexico. Very large and fine, colorless, white or pale green crystals, recently came in quantity from Rio Grande Do Sul, Brazil; some matrix specimens display $1''$ to $3''$ pseudo-cubic crystals, rarely, to $6''$ on edge, associated with heulandite, stilbite, and other species (Figure 269). These specimens easily rival if not outclass, the specimens which came from basalts of the Deccan near Poona, India, many years ago. In Europe, fine specimens have been obtained from Långban, Sweden. At the Korsnas Mine, Finland, in small sulfur-yellow prismatic-pyramidal crystals. In splendid pink crystals from Andreasberg, Harz, Germany and from Aussig, Czechoslovakia. Also fine and large crystals from Teigarhorn, Berufjord, Iceland. From the Faeroes; on the Isle of Skye, Scotland; in Ireland, and elsewhere in the British Isles.

The top apophyllite specimens have long been conceded to be those from Poona, India, which reached $6''$ on edge, but the Rio Grande Do Sul crystals now appear to be better. Furthermore, those from India are not available except when old collections are broken up. The colorful association of prehnite noted in the Virginia specimens makes them vie for honors with the specimens from either Brazil or India. For sheer variety of associations, the specimens from the New Jersey localities are difficult to improve upon.

PHYLLOSILICATES IN CLAYS

Earthy or clayey minerals are often found in cavities of various descriptions, notably in the "pockets" of pegmatites, but also in rocks, as some granites, or gneisses which have been altered so drastically that the whole crumbles to a powder when touched (*saprolites*). Such clayey minerals commonly consist of several species, including one or more phyllo-silicates as kaolinite, halloysite, montmorillonite, etc., as well as very small particles of non-phyllosilicate species. It is not usually possible to identify them from external appearances, and indeed, to do so accurately requires expert work in a well-equipped

FIG. 269. Apophyllite crystals on basalt matrix with heulandite. Size 6″ by 5″. From between Veronopolis and Bento Goncalves, Rio Grande Do Sul, Brazil. Note the pearly luster on some c[001] faces.

mineralogical laboratory. Surface or near-surface alteration of granites and gneisses in which much feldspar is present, commonly results in the alteration of feldspar to kaolinite. In Cornwall, England, fine kaolinite pseudomorphs after twins of orthoclase are found embedded in the saprolitic granites excavated for clay as in the St. Austell district. An example is shown in Figure 34 (page 87). The clay-forming species endellite, halloysite, kaolinite and montmorillonite, have been identified in the pegmatites of the Pala area, San Diego County, California, where they form principally from the alteration of feldspars, spodumene, and tourmaline, and commonly occur as white, pink to red, or grayish masses in pockets. In some instances, well-formed "tourmalines" and "spodumenes" are found, complete with crystal forms and surface markings, but when scratched or scraped, prove to be clays. Submicroscopic muscovite crystals, and very small platey crystals of dull greenish chlorite, also commonly form clay-like masses in mineral deposits. In many instances, the distinctive hue and form of chlorite crystals make it possible to detect this species, particularly if a microscope can be used.

Serpentine—$Mg_6(OH)_8(Si_4O_{10})$

Name: Of obscure origin, possibly in reference to fancied snake-like markings or veinings in ornamental material; sĕr′pĕn teen.

Varieties: Two are recognized on the basis of crystal structure: *antigorite,* with platey structure, and *chrysotile,* with fibrous structure. There are other varietal names given to serpentine, some on the basis of color, others on pattern, but few on worthwhile mineralogical grounds. Long elastic fiber *chrysotile* is *asbestos* (Figure 270).

Crystals: Monoclinic. Distinct euhedral crystals unknown. Always cryptocrystalline massive, usually without distinctive features except in *chrysotile asbestos,* in which long fibers are prominent.

Physical Properties: Quite tough, except in *asbestos,* fibers of the latter being easily dislodged by scraping with the fingernail. Feels greasy when rubbed. H. variable, 2½-5. G. variable according to impurities, but selected specimens generally yield 2.5-2.6; translucent yellow-green material (*bowenite* or "Soochow jade"): G. 2.58-2.59; when translucent green (*williamsite*): G. 2.61.

Optical Properties: Translucent to nearly opaque. Ordinary massive varieties dark blackish-green, dark olive-green; also intense bluish-green, apple-green, yellow and yellowish-green. Frequently mottled or veined in a variety of hues, including brown, brown-red, etc. Much *chrysotile asbestos* is golden yellow. Streak colorless. Luster waxy or greasy; strongly silky in asbestos. Vague refractometer readings as follows: *bowenite,* 1.56, *williamsite,* 1.57.

Chemistry: Magnesium hydroxyl silicate. Iron may partially substitute for magnesium, and aluminum for magnesium and silicon.

Distinctive Features and Tests: Any rock mass of dark green color displaying softness in the range shown above and with a greasy feel, is probably serpentine. Harder than talc, softer than nephrite. *Chrysotile asbestos* is commonly associated with massive serpentine and marble. In the closed tube, gives off water when heated but *amphibole asbestos* does not. Infusible. Decomposed by hydrochloric acid but no jelly formed.

Occurrences: In small to large bodies derived from the hydrothermal alteration of magnesium silicate minerals, usually olivines, pyroxenes and amphiboles. Commonly in small quantities in impure marbles, and less abundantly as small grains in some calcareous schists; also in kimberlite. Associated with magnetite, chromite, magnesite, brucite, garnierite, zaratite, chondrodite, and chlorite. Possibly the finest specimens of long-fiber *asbestos* are mined in the Thetford Mines deposits, Megantic County, Quebec; solid pure masses taken from seams in impure serpentine rocks often run to 4″ or better in thickness, and many inches in the other dimensions (Figure 270). Bright golden yellow *chrysotile,* in narrow seams to ½″ thick, occurs in massive translucent yellow serpentine near Lake Valhalla, Montville, Morris County, New Jersey. Similar material occurs in the Salt River Valley *asbestos* deposits, Gila County, Arizona, sometimes replaced by cryptocrystalline quartz and then similar in color and chatoyancy to the tigereye of Africa. Pseudomorphs of chondrodite occurred in the Tilly Foster iron mine, near Brewster, Putnam County, New York, and also after pyroxene and amphibole in quarries near Easton, Northampton County, Pennsylvania. Similar pseudomorphs, probably after amphiboles, were once found in abundance near Oxbow, Jefferson County, near Gouverneur, St. Lawrence County, and in Orange County, New York. Fine green, highly translucent, gem quality *williamsite* occurs in chromite in serpentine almost on

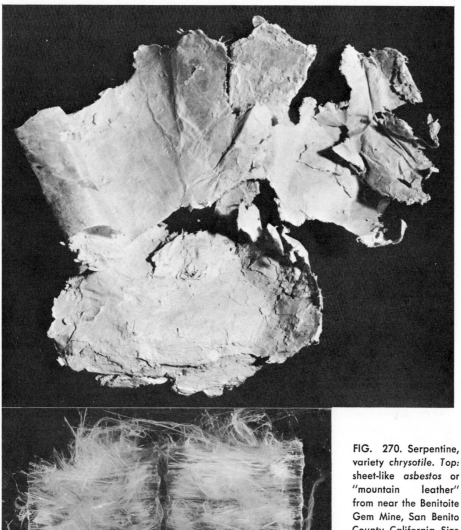

FIG. 270. Serpentine, variety *chrysotile*. Top: sheet-like *asbestos* or "mountain leather" from near the Benitoite Gem Mine, San Benito County, California. Size 5″ by 4″. Bottom: fibrous *asbestos* from Thetford Mines, Quebec. Size 4″ by 2½″. *Photo courtesy Schortmann's Minerals, Easthampton, Massachusetts.*

the border of Maryland and Pennsylvania near Rock Springs, in the latter state. Fine sheets of white *chrysotile* occur as coatings and fissure fillings in shattered serpentine near the benitoite mine, San Benito County, California (Figure 270). Pseudomorphs after olivine crystals to 3″ occur at Snarum, Buskerud, Norway. Distinctive and handsome mottled serpentine is extensively cut and polished into ornamental items from the deposits at the Lizard promontory, Cornwall, England. Fine ornamental material from localities in Greece and Italy, and elsewhere in Europe. Prized specimens are serpentine pseudos after iron-magnesium silicate minerals, the locality at Snarum, Norway, being especially noted. Good specimens of *chrysotile asbestos,* especially the vivid yellow types from Arizona, make attractive cabinet specimens.

Pyrophyllite—$Al_2(OH)_2(Si_4O_{10})$

A soft talc-like species, sometimes furnishing attractive specimens in which spherical aggregates of bladed silvery crystals are exposed by fractures. Aluminum hydroxyl silicate; pī rŏf'ĭ lĭt. Specimens usually show silvery, yellowish or greenish bladed crystals forming stellate groups to about 1″ diameter. Good examples occur at many places in North Carolina as at Staley, Randolph County, and in Guilford, Montgomery, and Orange Counties. Also good from Graves Mountain, Lincoln County, Georgia, and from Indian Gulch, Mariposa County, California.

Talc—$Mg_3(OH)_2(Si_4O_{10})$

The softest known mineral, but seldom in collections because of the scarcity of all but massive specimens, few of which are attractive. A magnesium hydroxyl silicate; talk. *Synonyms: steatite, soapstone,* particularly when massive. Interesting talc pseudomorphs after enstatite occur at Bamle and Snarum, Norway (Figure 34, page 87). Fine foliated green talc specimens are obtained from a number of localities, among which may be mentioned Greiner Alp, Zillertal, Austria, and Holly Springs, Cherokee County, Georgia.

Mica Group

The micas are very closely related in structure and properties. Although crystallizing in the monoclinic system, the *a* axis is tilted so little in respect to the *c* axis, that a goniometer measurement on the edge between prism faces and the perfect cleavage plane, *c*[001], seems practically 90°. Consequently, crystals develop into pseudo-hexagonal tablets, or elongated pseudo-hexagonal prisms. Sometimes a pair of prism faces is underdeveloped such that the cross-section of the cleavage plane is diamond-shaped. One perfect basal cleavage is characteristic. In some mica species, it is so well-developed that unbroken, exceedingly thin, elastic sheets can be easily split from crystals. There are large differences in hardness, a scratch test across a cleavage surface giving about 2½, while a test on the prism faces gives a higher value of about 4. Another effect of the sheet structure on properties is the easier transmission of light through the horizontal axial directions as compared to the direction through the cleavage plane; even in thin crystals, it is sometimes difficult to see through the cleavage plane yet possible easily to see light passing through the much thicker prism face direction. This effect is most pronounced in small undamaged crystals of no more than ½″ diameter. Strong to distinct pleochroism is also noted during such observations, in some brown muscovite, for example, the usual brown hue is obvious when looking down on the cleavage plane, but is replaced by a fine sea-green color when crystals are observed through the edges of tablets.

Identification of mica is ordinarily not much of a problem because of the perfect cleavage, shape of the crystals, and unmistakable elastic "snap" of cleavage flakes when they are bent and released. However, so-called "brittle" mica, margarite, breaks instead, but because it is almost always found with corundum, usually as glimmering crusts around corundum crystals, it need not be mistaken on this account. On the other hand, chlorite could be easily confused with mica except that chlorite tends to assume characteristic dark blue-green hues, and the cleavage flakes bend limply without snapping back.

Muscovite—(K,Na)Al$_2$(OH)$_2$(AlSi$_3$O$_{10}$)

Name: From the old name for Russia, *Muscovy*, where large sheets of mica were formerly obtained from pegmatites in the Urals, and popularly known as "Muscovy glass"; mŭs′kō vīt.

Varieties: Fine-grained massive kinds are known as *sericite* (sĕr′ĭ sīt) or *pinite* (pĭn′īt). The presence of trivalent chromium imparts a deep green color in the variety known as *fuchsite* (fūsh′īt) or *mariposite*. The name *roscoelite* has been applied to muscovite in which trivalent vanadium is present.

Crystals: Monoclinic. Compared to the large quantities present in rocks and mineral deposits, euhedral crystals are decidedly rare. They are usually pseudo-hexagonal prismatic (Figure 271), sometimes diamond-shaped in cross-section, then suggesting an orthorhombic mineral. The basal plane c[001] is always broad and always in evidence; it is also the cleavage plane. Prism faces consist of m[110] and pinacoid b[010], the latter often suppressed or missing, then giving the crystals the diamond shape spoken of. Usually tabular, flattened along the c axis; less commonly, in blocky crystals. Also in granular or scaly masses, sometimes very fine-granular (*sericite*). Smooth on c[001], but usually rough or striated elsewhere, the striations running parallel to the horizontal axes.

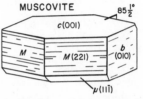

MUSCOVITE

PINACOIDS *c, b*, PRISMS *M, μ*(mu).

FIG. 271.

Physical Properties: Very easily cleaved into elastic sheets parallel to c[001]. When flexed, sheets eventually break with ragged edges. H. 2½ on cleavage surfaces, but as much as 4 across cleavage edges. G. 2.8-2.9.

Optical Properties: Transparent to translucent. Light passes more easily at right angles to the c axis than parallel to it. Usually some shade of rich brown, pale yellowish, greenish, yellow-green, reddish-brown and rarely, pink. Streak colorless. Strong pearly luster common on cleavage surfaces; otherwise vitreous; *sericite* often silky in luster. Strong silvery reflections from small scales as in gneisses and schists. Biaxial negative. R.I., $\alpha = 1.552$, $\beta = 1.582$, $\gamma = 1.588$. Birefringence 0.036. Pleochroism often distinct to strong; e.g., green muscovite (Amelia), colorless, yellow-green, gray-green. In the polariscope, a fine sharp biaxial figure is easily observed in small scales not more than $\frac{1}{32}''$ thickness and vivid colors appear in paper-thin scales.

Chemistry: Potassium-sodium hydroxyl alumino-silicate. Magnesium and iron substitute for aluminum; also chromium or vanadium. Sodium is commonly present substituting for potassium.

Distinctive Features and Tests: Brown varieties closely resemble phlogopite and are not easily distinguished except to note the type of occurrence in which found, phlogopite usually occurring in association with metamorphosed limestone, and muscovite in granitic rocks, in some strongly metamorphosed rocks as gneiss and schist, and in granitic pegmatites. Practically infusible, except in thin flakes and then with difficulty at 5. Not attacked by acids, except hydrofluoric.

Occurrences: Abundant in many kinds of granitic rocks and in siliceous metamorphic rocks but good specimens are obtained only from granitic pegmatites, from miarolitic

cavities in granites, or from cavities in some quartz veins. Only a few notable localities
will be mentioned. Curved aggregates, forming the so-called "ball" mica, occur in a
number of pegmatites in New England, notably in the Branchville pegmatite, near
Branchville, Fairfield County, Connecticut. Sharp euhedral crystals (Figure 272) occur

FIG. 272. Pale green muscovite "books" partly imbedded in *cleavelandite* cavity lining from
Rutherford No. 2 pegmatite, Amelia, Amelia County, Virginia. Size 3″ by 2″.

in cavities with *cleavelandite* in the Rutherford Mines, and others, in the vicinity of
Amelia, Amelia County, Virginia. At the Rutherford No. 2 mine, the crystals are often
exceptionally transparent and fine pale yellowish-green, also rose, and sometimes display both hues in the same crystal. Exceptional, thin-tabular, brown crystals have been
found in quartz veins in various places in North Carolina. Rarely, in good crystals,
from gem-bearing pegmatites in Minas Gerais, Brazil. Fine specimens with topaz and
quartz from Alabaschka near Mursinsk, in the Urals of the U.S.S.R. The world's record
for size is held by a single crystal from the Inikurti Mine, Nellore, India, which measured
15 feet in length and 10 feet in diameter, and delivered a total of 85 tons of muscovite.

Muscovite commonly contains inclusions, sometimes of considerable beauty, as bright red garnet crystals, deep green-blue spinels, magnetite, quartz, rutile and other species. *Paragonite* is compositionally similar to muscovite and also similar in appearance; it is uncommon, and among collectors, is best known as the matrix in which occur the blue kyanite and brown staurolite of the St. Gotthard region, Switzerland.

PHLOGOPITE-BIOTITE SERIES

Phlogopite—$KMg_3(OH)_2(AlSi_3O_{10})$

Biotite—$K(Mg,Fe)_3(OH)_2(AlSi_3O_{10})$

Names: From the Greek, *phlogopos,* "firelike," in allusion to the red-brown color; flŏg'ō pīt. After J. B. Biot, French naturalist; bī'ō tīt.

Crystals: Monoclinic. Biotite commonly in thin plates with indistinct forms, or in elongated plates resembling ribbons, as in many granitic pegmatites. Twinning as shown in Figure 273. Phlogopite often in well-formed prismatic crystals, sometimes developed along the c axis to several times the diameter of the prism; also in tapered crystals displaying pyramidal faces of $o[112]$ and $z[132]$. Well-developed phlogopite crystals sometimes resemble faceted cigar-like objects, tapering toward both ends and commonly assuming triangular cross-sections. Also in flaky masses and small rounded crystals.

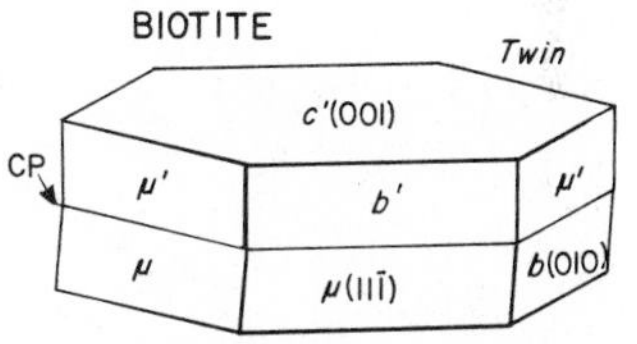

PINACOIDS b, c, PRISM μ.

FIG. 273.

Physical Properties: Tenacity and cleavage like muscovite, but the cleavage not as perfect. H. $2\frac{1}{2}$ on cleavage surface, 3 across cleavage. G. 2.8-3.4, rising with iron.

Optical Properties: Translucent; transparent in thin sheets. Biotite commonly dark brown to black; also dark green. Phlogopite usually rich red-brown to brown, but also yellow-brown, often displaying coppery reflections from the cleavage surfaces. Streak colorless. Luster splendent in biotite, almost submetallic in some specimens; also vitreous. Luster duller in phlogopite. Pearly luster on cleavage surfaces but not as strong as in muscovite. Biaxial negative. R.I., varying with composition, in the range: $\alpha = 1.54$, $\gamma = 1.68$. Birefringence 0.04 to 0.06.

Chemistry: Potassium, magnesium-iron hydroxyl alumino-silicates. Compositions vary considerably because of many substitutions.

Distinctive Features and Tests: The dark color and generally very small crystals of biotite usually distinguishes this species. Phlogopite is commonly found in metamorphosed limestone, while the coppery color of its $c[001]$ faces and the prismatic crystals are distinctive. Difficulty fusible at 5. Decomposed in boiling sulfuric acid.

Occurrences: Phlogopite occurs in sharp crystals in marbles, in small crystals or flakes in peridotites, and in large crystals in certain magnesium-rich pegmatites, as in the noted occurrences in Canada. Biotite is abundant in granitic rocks, in the outer zones of granitic pegmatites, and also in siliceous metamorphic rocks as gneisses and schists. Phlogopite abundant in Grenville metamorphic limestones and associated pegmatitic rocks of Ontario and Quebec, often in fine crystals, and not uncommon in extremely large ones. The world's record single mica crystal was uncovered at the Lacy Mine in Ontario many years ago, where a 90-ton phlogopite individual was taken out of a pegmatite; it measured 33 feet in length and about 14 feet in diameter! Ordinarily, the Canadian pegmatites furnish fine hexagonal prisms to about 3″ to 4″ diameter, showing coppery luster on cleavage planes, but seldom sharp except in smaller crystals. Similar specimens, though averaging smaller, occur in the marbles of St. Lawrence County, and in similar rocks of Orange County, New York, and in Sussex County, New Jersey. The marbles in and around Franklin, New Jersey, have provided sharp dark brown crystals in contacts along the famous zinc orebody. Good crystals, resembling those of Franklin, have recently appeared from localities in Madagascar, especially from Fort Dauphin. Biotite is common but not in good crystals, but exceptionally sharp thin tabular euhedrons, many of pale greenish or near-white hue, also darker, occur in cavities in the ejected limestone blocks of Monte Somma, Vesuvius, Italy; crystals average only $\frac{1}{2}''$ diameter, but form lustrous reticulated growths.

Lepidolite—$KLi_2(OH,F)_2(AlSi_4O_{10})$

Name: From the Greek, *lepidos,* "scale," in allusion to the scaly aggregates characteristic of much lepidolite; lĕ pĭd'ō lĭt.

Crystals: Monoclinic. Usually in scaly aggregates, sometimes with individual crystals no more than $\frac{1}{32}''$ diameter. Rarely, in large "books," or in sharp crystals. In granitic pegmatite cavities, much lepidolite forms small cylindrical books with rough sides as in Figure 274. Also noted as indented fringing overgrowths on muscovite. Commonly in

FIG. 274. Lepidolite crystals of cylindrical form partly coating a corroded perthite crystal and associated with blades of *cleavelandite.* Size 6″ by 3½″. From the Himalaya Mine, Mesa Grande, San Diego County, California.

sixling twins, composed of coffin-shaped individuals joined along the c axis.

Physical Properties: Easily cleaved parallel to c[001]. Sheets flexible and elastic. H. 2½ on cleavage surfaces, 3 across cleavage. G. 2.8-2.9.

Optical Properties: Transparent to translucent. Usually pale lilac to rose, but also yellow, gray, colorless. Color zoning parallel to prism and pinacoid faces. Streak colorless. Luster pearly on cleavage surfaces, elsewhere vitreous. Biaxial negative. R.I., $\alpha = 1.530$, $\beta = 1.553$, $\gamma = 1.556$. Birefringence 0.026. Good biaxial figure appears on cleavage flake in polariscope.

Chemistry: Potassium-lithium hydroxyl-fluorine alumino-silicate.

Distinctive Features and Tests: Much lepidolite is lilac in color, but not always so, and may be confused with similarly-hued muscovite. Crystals seldom smooth on edges of books, usually deeply indented or fringed in lace-like patterns. Fine-grained, massive material common in many granitic pegmatites, particularly near cores. Fuses at 2, actively bubbling and coloring flame crimson in streaks. When cooled, melt fluoresces white under ultraviolet. Flame test distinguishes from muscovite.

Occurrences: Confined to granitic pegmatites, usually as fine-granular masses near cores, as parallel overgrowths over muscovite, as cavity fillings and as stubby or tabular crystals in pockets. Commonly associated with colored tourmaline, morganite beryl, spodumene, amblygonite and other pegmatite species. In large quantities in some pegmatites of the New England region, but uncommon in the Amelia, Virginia, or North Carolina pegmatites. Abundant in the Black Hills, South Dakota pegmatite region, and especially so in the gem-bearing pegmatites of southern California. The fine-grained lepidolite masses of the Stewart pegmatite at Pala, San Diego County, California, sometimes reach dimensions of many feet and weigh many tons; much of this compact material is shot through with slender prisms of red tourmaline, furnishing attractive specimens. Also in San Diego County, superb sharp crystals formerly from the Little Three Mine near Ramona, in simple tabular individuals to ¾" diameter forming fine groups; abundant stubby crystals from the Himalaya Mine near Mesa Grande, implanted on feldspar and on tourmaline (Figure 274), or sometimes as overgrowths on muscovite crystals, forming specimens of wonderful luster, color and complexity; the latter reach 3" diameter. Large cleavage sheets from pegmatites near Penig, Saxony; also from various localities in Madagascar and Brazil. Fine ball-like and stalactitic growths, composed of numerous small individuals, have come from the Zambesi region of Mozambique, in masses to 5" diameter.

Chlorite Group

CHLORITE SERIES

Chlorite—(Mg,Fe,Al)$_6$(OH)$_8$[(Al,Si)$_4$O$_{10}$]

Name: From the Greek, *chloros,* "green," in allusion to the common color; klō′rīt.

Varieties: Because of extensive atomic substitution within the series, numerous varietal names have been assigned to members but only two are now commonly used, *penninite,* for thick prismatic crystals of dark green color, and *clinochlore,* for thin tabular crystals of hexagonal profile bearing a strong resemblance to the micas. The varietal name *kaemmererite* is sometimes used for chlorites colored rich violet because of the presence of chromium. The varietal name *diabantite* has been used for chlorites found with zeolites and other basalt cavity species in the New Jersey area, but many specimens labeled with this name have proven to be the pyroxene mineral babingtonite.

Crystals: Monoclinic. Commonly as thin tabular crystals of hexagonal outline, to thick tabular or prismatic with approximately triangular cross-sections. Crystals usually display the basal plane *c*[001] prominently, which is also the cleavage plane. Also abundant in minute scales as coatings or cavity fillings, or in globular clusters.

Physical Properties: Not brittle. Easily cleaved or crushed into numerous thin foliations which are flexible but *not* elastic, thus distinguishing chlorites from micas. Perfect, easily developed cleavage, parallel to *c*[001]. H. 2-2½. G. 2.6-3.3, rising with iron, but usually in the range of 2.7-2.9.

Optical Properties: Translucent to transparent in thin flakes. Color predominantly some shade of blackish-green, dark blue-green, olive-green, etc.; also bright green, yellow, violet, brown, and rarely, white. Streak colorless. Luster pearly on cleavage planes; elsewhere vitreous, or in scaly or earthy types, glimmering. Aggregates of microscopic scales commonly form greenish- or brownish-black, non-reflective, velvety coatings. R.I., usually in the range 1.57-1.60.

Chemistry: Essentially magnesium-iron hydroxyl aluminum silicates. Aluminum substitutes for magnesium or iron; some varieties contain appreciable quantities of chromium, nickel or manganese substituting for iron.

Distinctive Features and Tests: Blue-green, mica-like crystals. Confusion with mica eliminated by testing the chlorite scales for flexibility; the scales do not cleave from crystals as easily as in micas, nor do they snap back when bent. Harder than talc. Practically infusible except in thin scales, but usually whitens and swells slightly. Decomposed by boiling sulfuric acid.

Occurrences: Common in many kinds of rock and mineral deposits, but seldom in good crystals. Chlorites are usually of secondary origin, derived from alteration of pyroxenes, amphiboles, biotite, garnet, and other silicate minerals containing iron, aluminum and magnesium. Common in some metamorphosed siliceous rocks as in schists, also in partly-altered igneous rocks as basalt. Large and fine crystals sometimes occur in or near metamorphosed limestones. Large cleavage plates of clinochlore from Bagot, Renfrew County, Ontario; violet kaemmererite with chromite and serpentine, from various localities in Brome and Richmond Counties, Quebec. Fine clinochlore crystals to 2″ diameter, in thick clusters on serpentine, with magnetite, chondrodite, from the extinct locality at Tilly Foster iron mine, near Brewster, Putnam County, New York. Crystals to 2″ from Brintons Quarry, Chester County, and *kaemmererite* in fine dark violet crystal druses from Low's Mine and Wood's Chrome Mine, Lancaster County, Pennsylvania. Small sharp green crystals of tabular habit occur with melanite garnet near the benitoite locality, San Benito County, California. In sharp 2″ crystals from Zermatt, Switzerland, and from the Zillertal in Austria and the Pfitschtal, Trentino, Italy. Small sharp green crystals in curved aggregates accompany the grossularite garnet and diopside in the noted occurrences near Ala, Piedmont, Italy. Rosettes and velvety coatings of blackish-green chlorite commonly occur lining cavities in basalt and coat other species found in such cavities. Decidedly rare in good cabinet specimens.

INOSILICATES

Prominent in the crystal structures of inosilicates are the indefinitely long, single and double chains, composed of silicon-oxygen tetrahedrons. Single chains are linked by sharing of oxygen atoms between tetrahedrons, but double chains are additionally bonded with cross links. The direction of chains is parallel to crystal c axes with bonds between chains formed by metal ion linkages. This general pattern accounts for the tendency of many crystals to grow as acicular individuals, greatly elongated along the c axes. Prismatic, long-prismatic, bladed and even fibrous habits are commonly observed, but also stubby equant crystals.

Another prominent feature, of great use in identification, is the development of characteristic cleavages in the two great groups of minerals in the inosilicate subclass, namely, the *amphiboles* (ăm'fĭ bōl) and the *pyroxenes* (pĭr'ŏk seen). Such cleavages reflect the internal chain geometry and bond strengths, causing development of prismatic cleavages parallel to the c axes because the strengths of bonds between chains are not as great as within chains. Furthermore, the amphibole cleavages intersect each other at internal angles close to 87° and 93°, or practically at right angles, yielding cleavage splinters which appear to be almost rectangular in cross-section. On the other hand, the pyroxene cleavages are close to 124° and 56°, making cleavage splinters diamond-shaped in cross-section. Small broken fragments of minerals suspected to belong to one or the other group should be examined under magnification to see the angles produced between cleavage intersections, and thus place the unknown species into one or the other group. The cleavages described are diagrammed in Figure 80 (page 167), and are the result of *single-chain* structures in the pyroxenes, and *double-chain* structures in the amphiboles.

In addition to cleavage angles, further useful points to remember when faced with the problems of identification of members of the above groups is that the pyroxenes commonly appear in stubby or blocky crystals while amphiboles usually prefer long-prismatic, acicular, or even fibrous habits. Pyroxenes also predominate in the blackish basaltic or diabasic rocks, while amphiboles are more abundant in metamorphic rocks, especially in metamorphosed limestones near contacts with igneous rocks. However, the amphibole species hornblende is usually the dark, cleavable mineral seen in granodiorite and other granitic igneous rocks, aside from dark biotite, which can be recognized by its perfect cleavage.

The *pyroxenoids* are species whose chemical compositions resemble the pyroxenes, but crystal structures differ slightly but importantly in the way the chains are linked and arranged. Some species in this subgroup appear consistently in fibrous habits.

Amphibole Group

TREMOLITE-ACTINOLITE SERIES

Tremolite—$Ca_2Mg_5(OH)_2(Si_8O_{22})$

Actinolite—$Ca_2(Mg,Fe)_5(OH)_2(Si_8O_{22})$

Names: After the occurrence in Val Tremola in the Alps; trĕm'ō līt. From the Greek, *aktinos,* "ray," in allusion to the common radiate habit of prismatic crystals; ăc tĭn'ō līt.

Varieties: Straight-fiber, compact varieties, *asbestos;* felted fibers, *mountain leather;* very compact, felted, fibrous and tough, *nephrite jade.* Also *byssolite,* exceedingly long, acicular, uncompacted crystals (*see also byssolite* under *pyroxene*).

Crystals: Monoclinic, usually in fibrous, radiate aggregates (Figure 275); also compact fibrous as in *nephrite.* Crystals usually elongated, but may be stubby or blocky, especially when formed in marble.

Physical Properties: Brittle; fracture uneven to subconchoidal. Perfect cleavage, often interrupted, parallel to prism $m[110]$, forming fragments of diamond-shaped cross-section. H. 5-6. G. 2.98-3.35, rising with iron.

Optical Properties: Transparent to translucent. Tremolite white or faintly tinged with

FIG. 275. Yellow-green actinolite crystals forming radiate sprays with adularia, muscovite and quartz. From near Wrightwood, San Bernardino County, California. Size 4″ by 3″. Specimen in Elbert H. McMacken Collection, Ramona, California.

color, becoming green in various shades toward actinolite; also pink, brown, gray. Streak colorless. Luster vitreous; also weakly chatoyant in some crystals containing numerous inclusions oriented parallel to the c axis, and in some specimens of *nephrite*, notably in those from Kobuk, Alaska. Biaxial negative. R.I., usually within the ranges, $\alpha = 1.60$-1.66, $\gamma = 1.62$-1.68, rising with iron. Birefringence 0.02. *Nephrite* yields vague refractometer readings about $n = 1.62$. Distinct to strong pleochroism in some varieties, e.g., green crystals from Madagascar give dark olive-green, green, bright yellow-green.

Chemistry: Calcium magnesium-iron hydroxyl silicates. A complete series exists between end members shown. The pink variety of tremolite known as *hexagonite,* contains manganese substituting for iron.

Distinctive Features and Tests: Diamond-shaped prismatic cleavage sections; lighter hues than pyroxenes. Thin splinters fuse to glass near 3; the iron-rich varieties more easily, sometimes yielding magnetic globules. Insoluble except in hydrofluoric acid.

Occurrences: Tremolite occurs mainly in metamorphosed limestones; also in schists. Good crystals, often over 3″, occur abundantly in the Grenville marbles of Ontario as at Haliburton and Wilberforce, Haliburton County, and elsewhere within the area containing these formations. Similarly from marble at many places in St. Lawrence County, New York, notably from De Kalb, Russell, Gouverneur, and from near Edwards, in granular masses of transparent rounded crystals of attractive pink color (*hexagonite*), individuals sometimes reaching dimensions of $\frac{1}{2}$″ length; small flawless gems have been cut from larger grains. Also in the marbles of Orange County, New York, although these crystals may properly belong under hornblende as does the variety *edenite,* formerly considered to be actinolite. Dark green, bladed masses of actinolite to 3″ long, occur with *asbestos* at Pelham, Hampshire County, Massachusetts. Fine green blades of actinolite to 5″, in talc, near Chester, Vermont. The attractive specimen shown in Figure 275 is a mass of actinolite of yellow-green color from near Wrightwood, San Bernardino County, California. Fine sharp crystals of tremolite, of dark gray to white hue, occur in dolomite at Passo di Campolungo, Tessin, Switzerland. The variety *byssolite,* forming separate long, acicular crystals of greenish or bluish hue, is abundant in cavities in the Swiss Alps, occurring commonly as inclusions in quartz crystals and with epidote. Excellent *nephrite* occurs at numerous points near Jade Mountain along the Kobuk River, Alaska; in gravels of the Frazer River, British Columbia; over a wide area in Wyoming; in Khotan, Turkestan, and also in fine quality from New Zealand; numerous other occurrences are known. In general, the minerals in this series rarely furnish good cabinet specimens, even the best of crystals usually being far less perfect than those supplied by so many other species.

Hornblende—Na,Ca$_2$(Mg,Fe,Al)$_5$(OH)$_2$(Si,Al)$_8$O$_{22}$

Name: From the German miner's term, *Horn,* possibly referring to the color of horn, and *blenden,* "to deceive," because this material did not produce any useful metal; hôrn'blĕnd.
Varieties: When produced by alteration of pyroxene: *uralite.* Free-standing or lightly felted acicular crystals of bluish or greenish hue: *byssolite. Edenite* is a name formerly applied to light-colored hornblende containing little iron, and *pargasite,* a name given to varieties of light to dark green color containing some iron. *Common hornblende* is applied to nearly black varieties rich in iron.
Crystals: Monoclinic. Abundant in rocks as formless grains; also in stubby to blocky crystals, many of which appear to be nearly hexagonal in cross-section because of the combination of prism m[110] and pinacoid b[010] as in Figure 276. Crystals with m[110] predominant, have diamond-shaped cross-sections. Commonly twinned on plane a[100].

Physical Properties: Brittle; uneven to subconchoidal fracture. Perfect cleavage parallel to m[110] but usually interrupted and not easily developed in all specimens; cleavage fragments diamond-shaped in cross-section. H. 6. G. 3.0-3.4, rising with iron.

Optical Properties: Translucent to opaque. Black, dark green, brown. Thin splinters usually transmit light. Rarely transparent. Streak colorless. Luster vitreous. Biaxial negative. R.I., *edenite* (Franklin), $\alpha = 1.622$, $\beta = 1.630$, $\gamma = 1.645$; *pargasite* (Pargas), $\beta = 1.64$; *hornblende,* $\beta = 1.67$; indices rise with iron. Birefringence 0.023.

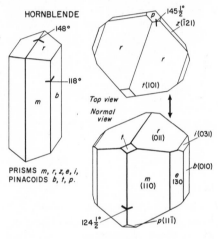

FIG. 276.

Chemistry: Complex compositionally because of substitutions; the presence of aluminum is the chief difference chemically between hornblende and tremolite-actinolite.
Distinctive Features and Tests: Diamond-shaped cleavages. Usually darker than tremolite-actinolite. Black tourmaline lacks cleavage. Fuses at 4 to a dark glass. Insoluble except in hydrofluoric acid.
Occurrences: Abundant in igneous and metamorphic rocks, and the principal species in the igneous rock known as amphibolite. Fine crystals occur in Ontario and Quebec associated with the Grenville marbles, a few localities are as follows: black to dark green in Bathurst, Lanark County, dark green prisms at High Falls and Ragged Chute on Madawaska River, Lanark County, Ontario; fine *edenite* from Grenville, Argentueil County, Quebec. Similar specimens from the marbles of St. Lawrence and Orange counties, New York, also in 2″, sharp crystals (*edenite*) from marble at Franklin, New Jersey. Handsome groups of lustrous black crystals to 1½″ from near Bilin and Schima in Bohemia, Czechoslovakia. Beautiful small sharp lustrous crystals from the lava of Vesuvius and elongated prisms to 1½″ from Roda near Predazzo, Italy. Good sharp hornblende crystals are not commonly offered in the specimen market because of the scarcity of productive localities. Probably the most colorful matrix specimens are those

from Canada where the crystals occur enclosed in calcite associated with other species as pyroxene, orthoclase, scapolite, apatite and sphene. The crystal groups of Norway are also very good while individuals from Schima, Czechoslovakia are prized by single-crystal collectors.

Other Amphiboles of Interest

Glaucophane (glaw'kō fān) and riebeckite (ree'bĕk ĭt) form a series compositionally similar to hornblende but containing principally sodium instead of sodium *and* calcium as shown in the formula for hornblende. Riebeckite commonly occurs in an *asbestos* variety comprised of dark blue fibers (*crocidolite*) which sometimes are as long as 10″; such fibers are usually coarser than those of either tremolite *asbestos* or serpentine *asbestos. Crocidolite,* replaced by quartz, forms the *tigereye* of the gem trade. Glaucophane is seldom available except in massive specimens. Crossite is close to riebeckite in many respects, and imparts the bluish color to the rocks enclosing the natrolite-benitoite veins at the headwaters of the San Benito River, California; it is also enclosed in some benitoite crystals, rendering them gray in hue and opaque.

Pyroxene Group

ENSTATITE-HYPERSTHENE SERIES

Enstatite—$Mg_2(Si_2O_6)$

Hypersthene—$(Fe,Mg)_2(Si_2O_6)$

Magnesium or magnesium-iron silicates; ĕn'stă tīt, hī'pĕr sthēn. Rarely in good crystals; mostly massive granular, sometimes in very large single-crystal grains. Large enstatite crystals of greenish color have been found at Bamle, Norway. A strong bronzy luster caused by numerous extremely small platey inclusions appears in various members of this series and lends them the varietal name *bronzite;* suitable specimens are sometimes cut into gems. The hypersthene associated with labradorite in Labrador also displays strong bronzy luster and was once cut into gems under the name "paulite."

DIOPSIDE-AUGITE SERIES

Diopside—$CaMg(Si_2O_6)$

Hedenbergite—$CaFe(Si_2O_6)$

Augite—$Ca(Mg,Fe,Al)[(Al,Si)_2O_6]$

Names: From the Greek, *di-*, and *opsis*, a "double sighting," because crystals commonly occur with two sets of prism faces which appear confusingly similar; dī ŏp'sĭd. After Ludwig Hedenberg, Swedish chemist; hĕd'en bĕrg ĭt. From the Greek, *augites*, "brightness," in reference to luster of crystals; aw'jĭt.
Varieties: Numerous varietal names have been assigned on the basis of chemical composition, appearance, etc., but few are worthy of repeating here. The following are still much in use: *chrome diopside,* bright green, containing chromium; *schefferite,* brown to black

pyroxene, containing manganese and iron; the name *jeffersonite* is commonly used for dark colored manganese-zinc pyroxene found in the zinc deposits of Franklin, New Jersey.

Crystals: Monoclinic; commonly short prismatic, as in Figure 277. The equal development of prism *m*[110] and pinacoids *a*[100] and *b*[010] often lends an octagonal cross-section to

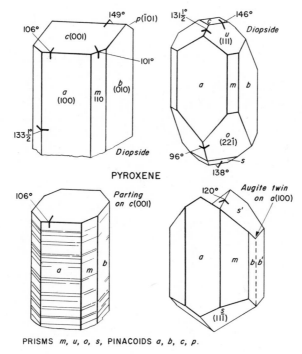

PRISMS *m*, *u*, *o*, *s*, PINACOIDS *a*, *b*, *c*, *p*.

FIG. 277.

crystals when viewed parallel to the *c* axis; the dark green crystals of Figure 278 display this well. Terminations are usually blunt wedges. Diopside crystals often rounded on edges, as if partly fused in the flame (Canada; New York), but faces often brilliantly smooth. Also long prismatic crystals of rectangular cross-section (diopside, Ala Valley), or rounded granular, massive. Twinning common, twin plane *a*[100]. Fine tubular inclusions parallel to *c* axis occur in some crystals (Ala Valley), causing strong chatoyancy, and then making them suitable for catseye gems.

Physical Properties: Brittle; uneven to conchoidal fracture. Crystals commonly disintegrate into numerous cube-like fragments when removed from matrix because of division along good cleavage planes parallel to prism *m*[110], intersected by an excellent parting parallel to *c*[001]. Parting is well-developed in the diopside of the Canadian-New York Grenville marbles, and often exceeds in flatness and broadness of surfaces, the quality of prismatic cleavages (Figure 277). H. 6. G. 3.25-3.55, rising with iron.

Optical Properties: Transparent to translucent. Iron-rich members usually black and opaque. Colorless, white, numerous shades of green, dark blue-green, yellowish-green, olive-green, etc.; also dark brown. Some shade of pale to medium green is most common in diopside, becoming darker, and finally black, in augite and hedenbergite. Streak colorless. Luster vitreous; silky chatoyancy in some diopside. Biaxial positive. R.I., transparent gem diopside generally yields, $\alpha = 1.670$, $\gamma = 1.701$; hedenbergite, $\alpha = 1.739$, $\beta = 1.745$, $\gamma =$

1.757; augite, $\beta = 1.69\text{-}1.71$; jeffersonite (Franklin), $\alpha = 1.713$, $\beta = 1.722$, $\gamma = 1.745$. Birefringence 0.018-0.032. Pleochroism strong in some dark-colored diopside, as green (Itrongahy), yielding green, paler green, rich olive-yellow; light colored varieties show indistinct pleochroism.

FIG. 278. Dark green diopside crystals in calcite with chalcopyrite, apatite and fluorite. Size $3\frac{1}{4}''$ by $1\frac{1}{2}''$. From Huddersfield, Quebec. Crystals are rounded on edges, as if partly fused. Elongated along the c axes. Prism zone bounded by prism $m[110]$ and pinacoids $a[100]$ and $b[010]$.

Chemistry: A series of calcium-magnesium or calcium-iron silicates in which augite additionally contains aluminum in substitution. Manganese, chromium and titanium may be present in small amounts in some varieties.

Distinctive Features and Tests: The inclination of $m[110]$ faces is nearly at right angles, thus leading to square, rectangular, or octagonal prisms which are also usually stubby. Green hues predominate in diopside, black in augite. As with other pyroxenes, the $m[110]$ cleavages are also nearly at right angles, distinguishing these species from amphiboles; an additional clue to identity is furnished by an often prominent parting parallel to $c[001]$. Fuses at 4 to a green glass; augite nearly infusible. Iron-rich varieties yield magnetic globules. Insoluble except in hydrofluoric acid.

Occurrences: Diopside occurs in marbles, often abundantly near contacts with igneous rocks, also in other types of metamorphic rocks rich in calcium; hedenbergite similarly occurs in calcareous metamorphic rocks and sometimes in high-temperature ore deposits. Augite is most abundant in the dark-colored igneous rocks as gabbros and basalts, but also in andesites. Good crystals of diopside are abundant at many places in the Grenville marbles of Ontario, Quebec and New York; also in marbles in Orange County, New York and the adjacent area in Sussex County, New Jersey. Fine crystals occur at Cardiff, Haliburton

County, Ontario, also in Hastings and Renfrew counties; crystals to 6″ length occur at Littlefield, Pontiac County, and fine sharp crystals in druses at Orford, Sherbrooke County, Quebec. Characteristically, marble-enclosed diopside is rounded on edges, many specimens being lustrous though formless masses of pale grayish-green color. Excellent transparent pale green diopside crystals were taken from marble near Richville, St. Lawrence County, New York, furnishing material for fine gems; crystals reached 3″ length and were over an inch thick. Fairly sharp crystals of dark brown *jeffersonite* were formerly obtained in quantity from the contacts of the zinc ore vein at Ogdensburg, Sussex County, New Jersey, single crystals to 12″ length having been recorded; average specimens displayed stubby individuals of about 1″ to 2″ diameter, forming druses later enclosed in white calcite; also from here and Franklin, crystals of *schefferite* to 2½″ diameter, and less abundantly, crude, nearly colorless diopside crystals. Pale green diopside crystals to 2″ occur in druses in massive diopside in the Bergen Park-Cresswell Mine, Jefferson County, Colorado; also from this state, sharp ½″ augite crystals in basalt in the Trail Creek-Gold Run Creek area of Grand County. Greenish-black *augite* crystals to ¾″, occur in great number in a crumbling basalt at Cedar Butte, near Tillamook, Tillamook County, Oregon. Diopside in rough crystals in marbles of the Crestmore area, Riverside County, California; large crystals are found in the New City quarry, south of Riverside. Fine green *chrome-diopside* in crystals to several inches, and small crystals sometimes clear enough to cut into gems, in the sulfide deposit at Outokumpu, Finland. Fine sharp diopside crystals formerly from Achmatovsk, Urals, U.S.S.R. Groups of sharp black crystals to 1½″ on edge occur as druses at Nord-marken, Wermland, Sweden. Transparent and sometimes gem quality rectangular pris-matic crystals of pale green color, some tipped by white chatoyant material, occur with grossularite crystals and chlorite at the celebrated localities in the Ala Valley, Piedmont, Italy; similarly from the Zillertal, Austrian Tyrol. Fine sharp black augite crystals in basalt at Bufaure, Italian Tyrol. The marbles near Campione, Switzerland, furnish diopside crystals to 8″ length. Sharp lustrous black augite crystals, usually doubly-terminated, and commonly to 1″ in size, occur on Mount Vesuvius, near Naples, Italy; also very fine on Mount Etna, Sicily. Gem diopsides occur in gravels in Ceylon and Mogok, Burma, the latter locality sometimes producing fine green catseye material; also in pegmatite at Itrongahy, Madagascar.

Aegerine—NaFe(Si₂O₆)

A sodium-iron silicate; named after the Teutonic god of the sea, Aegir; ē′jĭr ĭn. Acmite, an alternate name, is after the Greek, *akme,* a "point," in allusion to the pointed crystals (Figure 279). Forms prismatic, dark green, brown, or nearly black crystals, com-monly terminated by steep pyramids coming to a point, and thus distinguished from augite, which it otherwise resembles. Found in sodium-rich igneous rock, especially nepheline syenites. Notable localities are Magnet Cove, Garland County, Arkansas, yielding slender crystals to 3″ length and rarely, crystals of 8″ length. In Greenland at Narsarssuk Fjord, in black, stout prisms to 1½″ length. Sharp slender crystals in pegmatite at Rundemayr, Eker, near Kongsberk, Norway, some crystals to 12″ long have been found; also common in nepheline syenites along the Langesundfjord, as at Aroy, in dark green crystals to 1½″. Large crystals are commonly shattered because of being solidly enclosed in rock and good cabinet specimens seldom display crystals longer than 2″.

AEGERINE

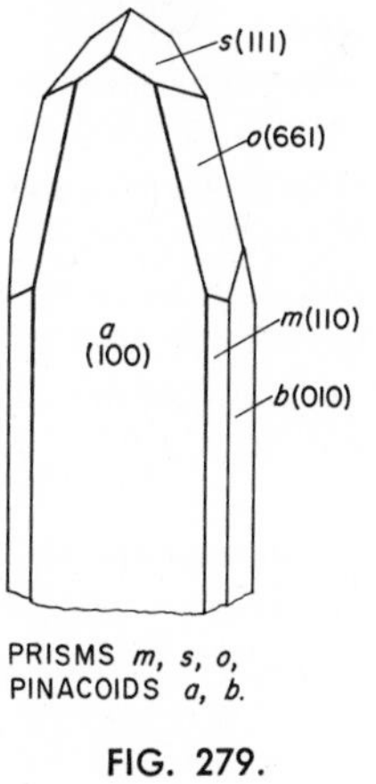

s(111)
o(661)
m(110)
a(100)
b(010)

PRISMS *m, s, o,*
PINACOIDS *a, b.*

FIG. 279.

Jadeite—NaAl(Si₂O₆)

Name: From the Spanish, *piedra de ijada,* "stone of the side," because supposed to cure kidney ailments if applied to the side of the body; jād′īt.

Varieties: Calcium and magnesium substitute for sodium and aluminum in *diopside-jadeite.* Partial replacement of aluminum by iron occurs in a dark green variety known as *chloromelanite.* There are numerous varietal names based on color and pattern in compact material used for gems and ornamental carvings, perhaps the most valuable kind being that colored intense emerald-green by chromium and then known in the trade as *imperial jade.*

Crystals: Monoclinic; very rare, and then only in very small slender prismatic individuals in vugs in massive jadeite. Predominantly fine-granular to coarse-granular massive, sometimes displaying individual crystal grains to $\frac{1}{2}''$ diameter. Also in waterworn alluvial boulders, sometimes tons in weight.

Physical Properties: Extremely tough in compact types. Fracture surfaces sugary in texture, and sparkling from exposed perfect cleavages parallel to $m[110]$. H. $6\frac{1}{2}$. G. 3.25-3.36, rising with iron; very pure material (Cloverdale) gives 3.25.

Optical Properties: Transparent to translucent. Colorless, white, slightly grayish, very pale grayish-blue, pale mauve, pale to intense blue-green, also gray-green; often stained by infiltration of iron oxides causing development of brownish, reddish, orange, or yellow hues in granular types. Streak colorless. Luster somewhat greasy on fracture surfaces of compact material; vitreous on polished surfaces. Biaxial positive. R.I., on single crystals (Cloverdale), $\alpha = 1.640$, $\beta = 1.645$, $\gamma = 1.652$; jadeite verging on *diopside-jadeite* (Guatemala), $\alpha = 1.654$, $\gamma = 1.669$. Birefringence 0.012. Refractometer readings on massive material from Burma generally fall in the range $n = 1.65\text{-}1.68$, with most specimens providing $n = 1.66$.

Chemistry: Sodium aluminum silicate.

Distinctive Features and Tests: Toughness; hardness. Higher specific gravity and refractive index distinguishes from nephrite and serpentine. Serpentine and nephrite display a far finer texture, looking much like congealed wax or lard; serpentine is difficultly scratched by the point of a knife, whereas jadeite is not scratched by a knife and its typical granular texture is usually apparent to the naked eye or is detectable under magnification. Fuses at $2\frac{1}{2}$ to a white glass, coloring the flame yellow.

Occurrences: Forms rock masses of considerable size as lenses, stringers and nodules in strongly metamorphosed sodium-rich rocks enclosed in serpentinous rocks. In the United States, jadeite of poor lapidary quality occurs in lenses and narrow vein-like bodies along contacts between schists and serpentines at a number of points along Clear Creek, San Benito County, California. Small euhedral crystals have been found in a boulder of jadeite near Cloverdale, California. Jadeite *in situ* has been found near Manzanal, Guatemala; it is not good lapidary quality. The premier source for carving quality jadeite is a series of *in situ* and alluvial deposits in Upper Burma, centered around the villages of Tawmaw and Hpakan. Interesting cabinet specimens consist of small unbroken alluvial boulders bearing upon them Burmese tax stamps and polished slots for examining the quality of the unaltered jadeite beneath the outer altered rind.

Spodumene—LiAl(Si₂O₆)

Name: From the Greek *spodoumenos,* "burnt to ash," alluding to the ashy color of early specimens; spŏd′ū mēn.

Varieties: Transparent gemmy crystals are commonly given the varietal names *kunzite,* when pink, violet, or purple, and *triphane* when colorless or yellow. The name *hiddenite* is usually restricted to an emerald-green variety colored by small quantities of chromium.

Crystals: Monoclinic; mostly lath-like, elongated along the *c* axis and flattened along the *a* axis such that two faces of *a*[100] are broadest, as in Figure 280. The latter faces are usually deeply grooved parallel to *c* axis by repeated combination of prism faces *m*[110]. Transparent gem crystals ordinarily consist of corroded cleavage blocks, splinters or laths, of approximately rectangular cross-section, showing only portions of striated *a*[100] faces; terminations usually irregularly rounded while etched cleavage surfaces display numerous shallow shield-shaped pits (Figure 281). Commonly twinned on composition plane *a*[100], the junction appearing in traces on opposite ends of the *c* axis. Long tapering straight or curved solution cavities of rectangular cross-section occur in gem crystals.

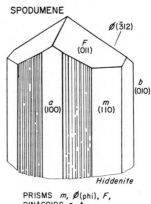

SPODUMENE

PRISMS *m*, ∅(phi), *F*,
PINACOIDS *a, b.*

FIG. 280.

Physical Properties: Gem varieties not very brittle, but ordinary, translucent spodumene often fractures and cleaves readily into numerous splinters. Fracture surfaces uneven to conchoidal. Cleavage parallel to *m*[110] nearly mica-like in respect to perfection and uninterrupted broadness of surfaces but developed with some difficulty in transparent gem material. It is noted in many instances that a partly-developed cleavage spontaneously continues splitting until the crystal falls in two. This may happen in a matter of minutes. H. $6\frac{1}{2}$-$7\frac{1}{2}$, the directions parallel to the *c* axis, softest, the directions at right angles on cleavage planes, intermediate, and the directions across the ends of the *c* axis, hardest. G. 3.17-3.23.

Optical Properties: Transparent to translucent. Gem varieties commonly flawless in large areas. White, grayish, or pinkish-white in ordinary varieties, but colorless, pink, violet, purple, bluish-green, yellowish-green, and yellow in gem specimens. Gem crystals often bicolored. Streak colorless. Luster vitreous brilliant in corroded gem crystals, but sometimes dull chatoyant in white translucent varieties because of numerous inclusions parallel to the *c* axis. Biaxial positive. R.I., indexes variable, colorless material giving $\alpha = 1.654$, $\gamma = 1.669$; *kunzite,* $\alpha = 1.660$, $\gamma = 1.675$; *hiddenite,* $\alpha = 1.662$, $\gamma = 1.677$. Birefringence 0.015. Pleochroism distinct to strong, depending on intensity of color; *kunzite,* colorless, violet pink, purplish-pink; *hiddenite,* emerald-green, bluish-green, yellowish-green. Yellow and yellowish-green examples only faintly pleochroic. Some examples fluoresce pale orange under ultraviolet.

Chemistry: Lithium aluminum silicate. Commonly, a small amount of sodium substitutes for lithium.

Distinctive Features and Tests: Flattened and striated "Roman sword" euhedral crystals; extensively corroded gem fragments. Hardness; cleavages. In the flame, fuses readily at $3\frac{1}{2}$, puffing slightly and throwing off small white flakes; edges of melt glow brilliantly, at which time the flame becomes streaked with crimson (lithium). Insoluble.

Occurrences: Confined to complex granitic pegmatites, where it forms euhedral, sharp, opaque or translucent white or pink crystals, usually altered in parts. Also as much larger, original crystals altered so extensively that only gemmy remnants are found, commonly enclosed in clay minerals. Large, slab-like, greenish-white crystals to 24″ length, have been found in pegmatites on Plumbago Mountain, Oxford County, Maine. At Huntington, Worcester County, Massachusetts, in fine, pink, opaque, altered, euhedral crystals to about 4″ length. *Hiddenite* occurs only in pegmatite veins in gneiss, near Hiddenite, Alexander

FIG. 281. Spodumene. Transparent gem fragments which are remnants of originally larger crystals. Top left crystal is from Minas Gerais, Brazil, but the others are from the San Pedro Mine, Hiriart Hill, Pala, San Diego County, California. Largest fragment is 6″ long.

County, North Carolina; crystals are commonly corroded, many being little more than striated splinters with rounded and flattened tips, however, some sharp crystals have been found (Figure 280), while the largest crystal on record is one of $2\frac{1}{2}″$ length, by $\frac{1}{2}″$ by $\frac{1}{3}″$. These crystals are now extremely scarce in any size and much prized by collectors. Even rarer are matrix specimens showing upright hiddenite crystals on gneiss matrix. Associated minerals are quartz, often in fine crystals, bright red reticulated rutile, pyrite, emerald and feldspar. The largest spodumene crystal on record was mined at the Etta Mine, Black Hills, South Dakota, and measured 42 feet in length, and 5′4″ in width;

it was estimated to weigh 90 tons. The Pala district of San Diego County, California, has been and still is a producer of gem spodumene, particularly *kunzite* and *triphane*, sometimes in crystals to 10″ or more in length. The most prolific localities are on Hiriart Hill, and include the Katerina, Vanderberg, and San Pedro mines; the finest material, of wonderful clarity, size and coloration, has come from the Vanderberg. The best-formed crystals came from the Pala Chief Mine nearby, usually as blunt "Roman sword" crystals, flattened on the *a*[100] faces, to 10″ length. Large quantities of spodumene in gem grade have come from numerous localities in Minas Gerais, Brazil; a recent find produced many pounds of material, possibly a half ton, among which was a dark purple-violet twinned crystal, weighing over 16 pounds, and largely gem quality. Pale green spodumene is common in Brazilian pegmatites but is uncommon in the California occurrences. Fine *kunzite* and other gem varieties of spodumene also occur at several localities in Madagascar. The best cabinet specimens of spodumene insofar as crystal form is concerned, are generally acceded to be the vivid *kunzite* from the Pala Chief Mine in California, followed by the Vanderberg specimens which are nearly as well-formed, but usually of better color. The *hiddenite* from North Carolina is always greatly in demand but is extremely rare. Crystals from Huntington, Massachusetts, although opaque, are also desirable.

Pyroxenoid Group

The *pyroxenoids* (pĭr ŏks′ĭ noid) include species compositionally similar to pyroxenes but differing in the way that the silicon-oxygen tetrahedron chains are arranged in the crystals. In the pyroxenes, the chains are parallel to the *c* axis, but in the pyroxenoids, the chains are parallel to the *a* axis. Only a few species are included in this subclass.

Wollastonite—Ca(SiO₃)

A calcium silicate, occurring in fibrous masses of elongated crystals; wŏl′lăs tŭn īt. Usually white, sometimes pale gray. H. 5. G. 2.9. Dissolves in hydrochloric acid and thus distinguished from tremolite and sillimanite which form similar fibrous masses. Pectolite is also confused with wollastonite but is usually associated with zeolites in basalt cavities while wollastonite is typically a product of strong metamorphosis of siliceous limestone. Fluorescent massive specimens from Franklin, Sussex County, New Jersey. In large white crystals at Diana, near Natural Bridge, Lewis County, New York. Common in the metamorphosed limestone of the Crestmore quarries near Riverside, Riverside County, California. Seldom provides interesting specimens.

Pectolite—Ca₂NaH(SiO₃)₃

Name: From the Greek, *pektos*, "compacted," in reference to the compact fibrous masses in which it usually occurs; pĕk′tō līt.
Crystals: Triclinic. Predominantly in compact radiate masses of very thin acicular crystals, elongated along the *b* axis. Rarely in distinct crystals visible to the naked eye. The radiate masses typically assume globular shapes (Figure 282).
Physical Properties: Very compact masses approach *nephrite* in toughness; ordinary compact material readily splits along the fibers. Individual fibers brittle. Cleavage perfect parallel to *c*[001] and *a*[100] but seldom visible in usual aggregates without magnification. H. not easily determined with accuracy because of variations in compactness; soft fibrous material strips away like asbestos, while hard compact material yields hardnesses

FIG. 282. Pectolite on basalt from Prospect Park, Paterson, Passaic County, New Jersey. The radiating structure is well shown here, while in the center of the specimen appear the spherical surfaces which face inward from cavity walls. Size 5″ by 3″.

of from 4 to 5. Some very soft material is altered to the montmorillonite clay variety *stevensite*. G. 2.86.

Optical Properties: Translucent. White, slightly grayish-white, pale yellowish-white, or very pale pink. Streak colorless. Luster strongly silky. Biaxial positive. R.I. (New Jersey), $\alpha = 1.600$, $\beta = 1.605$, $\gamma = 1.636$. Birefringence 0.036. Massive material yields a vague reading near $n = 1.62$.

Chemistry: Calcium sodium hydrogen silicate.

Distinctive Features and Tests: Ball-like aggregates; fibrous radiate fracture surfaces; zeolite associations. Fuses quietly at 2½-3, to a glass, coloring the flame yellow. Decomposed by hydrochloric acid without formation of jelly.

Occurrences: A secondary mineral formed predominantly in basalt cavities with zeolites, datolite, apophyllite and prehnite; occasionally in metamorphic rocks. Magnificent specimens, undoubtedly the world's finest, occur in abundance in the pillow basalt cavities of northern New Jersey; especially fine examples have been obtained from the Prospect Park quarry, and in previous years, from quarries on New Street, Paterson, Passaic County; also fine from Bergen Hill, Hudson County, and elsewhere in this state. Specimens show balls of pectolite in clusters, or sometimes forming solid pavements to 10″ by 8″ in cavities, and rarely, single balls with fibers to as long as 4″. Colorful specimens consist of matrix showing scattered balls of pectolite associated with prehnite and calcite, plus zeolites. Rarely, distinct crystals are observed. Coarse white to gray crystals occur at Magnet Cove, Garland County, Arkansas. Massive pectolite filling seams in rock has been observed near Middletown, Lake County, and at Elder Creek, Tehama County, California. Pectolite is uncommon in other zeolite localities as Nova Scotia, Iceland, etc.

Rhodonite—(Mn,Ca)(SiO₃)

Name: From the Greek, *rhodon,* "rose," in allusion to usual color; rō'dō nīt.

Varieties: The *calcian* variety is known as *bustamite;* bŭs'tă mīt; containing zinc substituting for manganese, *fowlerite;* fowl'ēr īt.

Crystals: Triclinic; rare. Predominantly in fine-grained masses. Blocky, tabular or square prismatic crystals, displaying prisms *m*[110] and *M*[110], and commonly compressed along the *c* axis, making *c*[001] faces broadest, as in Figure 283.

Physical Properties: Crystals brittle; fine-grained masses very tough, like jadeite. Fracture uneven to conchoidal in crystals, interrupted by cleavages; uneven in masses, with granular surfaces. Perfect cleavages parallel to prisms *m* and *M,* intersecting nearly at right angles. *Bustamite* sometimes cleaves perfectly, parallel to pinacoid *b*[010]. H. 6. G. 3.5-3.7; massive usually in range G. 3.60-3.70; transparent gem material (Broken Hill), G. 3.707.

Optical Properties: Translucent to transparent. Predominantly pale to deep pink; also red, brownish red, red-brown. Massive material very commonly streaked and veined by manganese oxide, often in handsome patterns. Streak colorless. Luster vitreous. Biaxial

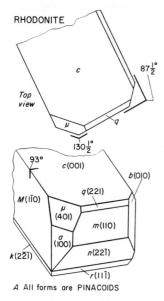

A All forms are PINACOIDS

FIG. 283.

FIG. 284. Rhodonite crystals in calcite with franklinite. From Franklin, Sussex County, New Jersey. Size 3″ by 2″.

negative or positive. R.I., variable, $\alpha = 1.716$-1.73, $\beta = 1.720$-1.74, $\gamma = 1.73$-1.744; *bustamite,* $\alpha = 1.66$, $\beta = 1.67$, $\gamma = 1.676$; transparent gem material (Broken Hill), $\alpha = 1.733$, $\gamma = 1.747$; massive material yields vague readings near $n = 1.73$ on refractometer. Birefringence 0.014. Pleochroism distinct, orange-red, brownish-red.

Chemistry: Manganese calcium silicate. Iron commonly substitutes for manganese, as does zinc in New Jersey varieties. Alters readily to black manganese oxides.

Distinctive Features and Tests: Pink blocky crystals; black veinings in massive material;

harder than rhodochrosite and not affected by hydrochloric acid as is that species. Fuses at 3 to black glass. Colors sodium carbonate bead blue-green.

Occurrences: Sometimes formed in large masses during metamorphism of sedimentary manganese ores; in contact metasomatic deposits; also formed by hydrothermal processes in ore veins. Massive rhodonite occurs in substantial deposits in many places in the world, but good gem material of cabochon or carving grade is not common. Crystals are decidedly rare. The premier locality for crystals of the variety *fowlerite* is the Franklin-Ogdensburg area, Sussex County, New Jersey, where two deposits of zinc ore in contact with marble, produced numerous fine specimens of thick, pink, tabular crystals painstakingly "worked out" by collectors from enclosing calcite. Rude crystals of *fowlerite* to 8″ by 2″ by 2″ have been found, but average cabinet specimens seldom show crystals more than 2″ long and perhaps ¾″ thick, arranged in groups on massive material (Figure 284). In some cavities, small, pink, transparent crystals, sharp, smooth, and of wedge-shaped habit, were found associated with brownish-yellow andradite and orange-yellow axinite. Ordinary *fowlerite* is associated with granular franklinite, brown andradite, brown willemite, and calcite. *Bustamite* is uncommon at Franklin. Small, bright, well-formed crystals of wedge-shaped habit occur in cavities in massive rhodonite near Pajsberg, Vermland, Sweden. Square prisms, sometimes perfectly transparent and flawless, to 1″ in length, occur in coarse-granular galena with feldspar at Broken Hill, New South Wales, Australia; exceptional crystals occasionally furnish small faceted gems.

The Franklin specimens are mostly collector's items of a past era, and such as happen to come upon the market are eagerly sought for by those who appreciate the color and form of these unique rhodonite crystals. Also much desired are the crystals from Broken Hill, especially sharp clear singles. The small crystals from Sweden are also rarities and in demand. Excellent polished specimens of massive rhodonite occur in many places in the United States, and more lately, splendid rich pink material has come from Australia.

Babingtonite—(Ca,Fe)(SiO$_3$)

A rare calcium iron silicate; bă′bǐng tŭn ĭt. Occurs in small dark green to black platy or prismatic crystals implanted on zeolites and associated minerals in basalt cavities. Commonly, it forms small rosettes which require magnification to see well. H. 6. G. 3.36. Small crystals, seldom over ⅛″, occur at many places in New Jersey. Fine small crystals at Blueberry Mountain, Woburn, Middlesex County, and in good crystals in Lane Quarry, near Westfield, Hampden County, Massachusetts. Good crystals also occur in pegmatites in the Baveno quarries, Lago Maggiore, Italy, and in the iron mines of Haytor, Devonshire, England.

Neptunite—(Na,K)(Fe,Mn,Ti)(Si$_2$O$_6$)

A rare complex silicate occurring as black prismatic crystals in very few places in the world; nĕp′tŭn ĭt. Occurs in simple monoclinic prisms of square cross-section, usually sharp and smooth, and showing dark red-brown reflections from thin splinters or chipped spots (Figure 285). Common forms are prism *m*[110], sometimes truncated by prism *a*[100]. Terminations usually consist of opposed pairs of *s*[111] and *o*[Ī11] faces, joining to give each termination a bluntly-pointed aspect. H. 5-6. G. 3.19-3.23. Conchoidal fracture; perfect prismatic cleavage. Streak reddish-brown. Biaxial positive. R.I., $\alpha = 1.69$, $\beta = 1.70$, $\gamma = 1.74$. Birefringence 0.050. Easily identified by the long square prismatic black crystals associated with benitoite in white natrolite on gray crossite-natrolite-serpentine matrix in the California occurrence or as long black prisms with raspberry-red eudialyte and other rare species in nepheline syenite from near Julianehaab, southwestern Greenland. The finest neptunites are those from the benitoite gem mine near the headwaters of the San Benito River in San Benito County, California; here they occur as sharp glossy black prisms implanted on matrix, or sometimes loose, with benitoite and minute honey-colored joaquinite crystals. Crystals to 2½″ have been found, but most matrix specimens seldom show individuals over 1″ long and about ¼″ square. Some large crystals contain inclusions of *crossite* and then become grayish-brown rather than black. Rare and desirable.

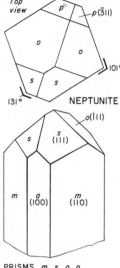

PRISMS *m, s, o, p,*
PINACOID *a.*

FIG. 285.

CYCLOSILICATES

The prefix *cyclo-* is well-chosen to describe the structure of crystals in this silicate subclass: they consists of silicon-oxygen tetrahedrons curving around to form closed rings, with atoms of metallic elements occupying places between them. The simplest pattern is the three-group triangular ring of benitoite, followed by the four-group ring of axinite. Six-group rings characterize such species as beryl and tourmaline. In effect, the SiO_4 groups are linked in the same way as in the chain structures of the inosilicates, except for looping around to join in closed rings. Axinite structure is complicated by the presence of (BO_3) groups forming flat triangles; additionally, (OH) groups are present, and the structure does not have the neat pattern of beryl or tourmaline. This is reflected in its triclinic symmetry, and crystals characterized by flatness, seemingly oddly-placed faces and sharp beveled edges. Another cyclosilicate, cordierite, is also less symmetrical, and crystallizes in an orthorhombic structure.

Although rings in beryl and tourmaline are arranged in flat sheets at right angles to the *c* axis, the excellent cleavage of the phyllosilicates, e.g., in mica, is missing because unlike mica, there are very strong bonds cementing the ring-sheets together. Nevertheless, some beryl and tourmaline crystals commonly crack, if not cleave, at right angles to the *c* axis, indicating greater strength within the planes containing rings and somewhat less between planes. The space in the center of rings is relatively large and accommodates a wide variety of ions, thus accounting for the wealth of elements in small quantities often noted in careful chemical analyses of these species, particularly tourmaline. The hemimorphism of tourmaline is caused by its SiO_4 tetrahedrons arranged in rings such that the bases of all tetrahedrons rest on a common plane with points upward (Figure 60, page 133). Thus if we could remove a tourmaline ring and examine it, we would find one side flat but the other studded with the tips of the tetrahedrons. Naturally, this difference in structure also means a difference in bonding forces, depending on which side of the ring is considered, and the result is that one end of the crystal tends to be flat and the other pointed. Usually this is evident in doubly terminated crystals where the pyramidal faces on one end are low, or even missing, while those on the other are steep. Pyroelectricity and piezoelectricity in tourmaline also result from hemimorphic structure.

Benitoite—BaTi(Si₃O₉)

Name: After the occurrence in San Benito County, California; bĕ nē′tō ĭt.
Crystals: Hexagonal. Nearly always in euhedral to subhedral crystals, approximately tri-

angular in shape, commonly flattened along the *c* axis, making many individuals tabular with broad *c*[0001] faces, as in Figure 286. Smaller crystals predominantly dipyramidal, the largest faces belonging to π(pi)[01$\bar{1}$1]. Also common are very narrow faces of prism μ(mu)[01$\bar{1}$0], with corners of triangular crystals truncated by faces of prism *m*[10$\bar{1}$0].

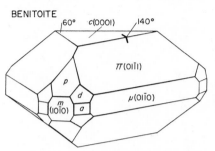

BENITOITE

PINACOID *c*, FIRST ORDER PRISM *m*, SECOND ORDER PRISM σ/11$\bar{2}$0/, FIRST ORDER DIPYRAMID *p*/10$\bar{1}$1/, SECOND ORDER DIPYRAMID *d*/22$\bar{4}$1/, THIRD ORDER PRISM μ, THIRD ORDER DIPYRAMID π.

FIG. 286.

Physical Properties: Brittle: conchoidal to uneven fracture. Cleavage pyramidal, poor or absent. H. 6-6½. G. 3.64-3.68.

Optical Properties: Transparent, also translucent. Deep blue to pale blue; rarely, colorless or white. Marked color zoning in many crystals, exterior portions, particularly along prism edges, being clear and deep blue, interior portions being pale blue, colorless or white. Streak colorless. Luster vitreous. Uniaxial positive. R.I., O = 1.757, E = 1.804. Birefringence 0.047, large. Strongly dichroic; deep blue, colorless. Strong pale blue fluorescence under ultraviolet.

Chemistry: Barium titanium silicate.

Distinctive Features and Tests: Association with white natrolite and black neptunite on blue-gray matrix; triangular crystals. Fuses at 4 to a dark blue, opaque glass. Fragments heated to red heat do not lose color.

Occurrence: Only one locality is known where it occurs as euhedral crystals implanted on walls of fissures in a much-altered serpentine, consisting principally of the amphibole *crossite,* impregnated with cryptocrystalline natrolite. Associates are black neptunite and minute brownish-orange crystals of joaquinite, all species usually being enclosed by fine-granular white natrolite (Figure 287). The natrolite is removed by soaking matrix specimens in dilute hydrochloric acid, during which treatment the natrolite gelatinizes. The gelatinous coating is periodically washed off in plain water and the specimen returned to the acid bath until the desired degree of crystal exposure is obtained. The finest crystals, some to 2″ diameter, are in natrolite-filled veins, but many small to large isolated crystals have been found in openings within *crossite-serpentine* immediately adjacent to the natrolite veins. Some crystals are so filled with *crossite* inclusions that they assume the bluish-gray color of this amphibole; singles to 2″ diameter have been recovered. Matrix specimens to 12″ square have been mined from the deposit, but average cabinet specimens range from several inches to about 6″ across. Crystals over ¾″ diameter are uncommon. Much benitoite has been cut into blue to colorless gems of exceptional beauty and dispersion. Specimens are now difficult to obtain because accessible portions of the deposit are buried under many tons of debris. A very rare mineral.

Dioptase—Cu$_6$(Si$_6$O$_{18}$)·6H$_2$O

Name: From the Greek *dia-* and *optazein,* "to see through," because its cleavage planes could be seen in the crystals; dī ŏp'tās.

Crystals: Hexagonal; usually in stubby hexagonal prisms, displaying *m*[10$\bar{1}$0] and terminated by rhombohedron *s*[1$\bar{1}$21] as in Figure 288. Sometimes in long prismatic micro crystals. Offsets develop in larger crystals and although brilliant, the faces are seldom flat. Also in granular aggregates and veinlets.

FIG. 287. Benitoite crystals on *crossite* matrix, with white natrolite and black neptunite crystals. Inset shows several loose benitoite crystals, looking down on the c axes, showing typical forms and the gray coloration imparted to those with numerous fibrous *crossite* inclusions. Size of large specimen 3¼" by 2". From the Benitoite Gem Mine, San Benito County, California.

Physical Properties: Brittle; fracture uneven to conchoidal. Perfect cleavages parallel to rhombohedron $r[10\bar{2}1]$; if a tip of a crystal is cleaved, the planes will truncate the nearly right-angle edge between adjacent $s[11\bar{2}1]$ faces. H. 5. G. 3.28-3.35.

Optical Properties: Commonly transparent; also translucent. Deep blue-green color, similar to the darkest emeralds in hue. Streak pale greenish-blue. Vitreous luster, somewhat greasy on fracture and cleavage surfaces. Uniaxial positive. R.I., O = 1.654, E = 1.708 (Mammoth-St. Anthony Mine). Birefringence 0.054, large. No dichroism.

Chemistry: Hydrous copper silicate.

Distinctive Features and Tests: Few minerals in oxidized portions of copper deposits possess the distinctive color, hardness, or rhombohedral terminations and cleavages of dioptase. Brochantite fuses on charcoal but dioptase blackens, crumbling into small pieces. Soluble in hydrochloric acid, staining the solution blue.

Occurrences: In druses, lining cavities in oxidized portions of copper deposits, or in openings in rocks immediately associated with such deposits. Fine micro crystals occur in cavities in pale blue chrysocolla with dark brown descloizite and orange wulfenite in the

Mammoth-St. Anthony Mine, Tiger, Pinal County, Arizona; similarly, from the Bon Ton Mine in the Clifton-Morenci district, Greenlee County, and in the Ox Bow and Summit Mines, Payson district, Gila County. Superb crystals, some to 2″ length and ¾″ diameter, from Mindouli and other places in the Niari River basin in Congo, Union of Central African Republics. Also splendid from Tsumeb, South-West Africa, in crystals to 1″ length, and from Guchab, Otavi, etc. From the Katanga of the Republic of the Congo. Small crystals to ⅛″ implanted on pale brownish limestone occur near Alytyn-Tube, Khirgiz Steppes, U.S.S.R.

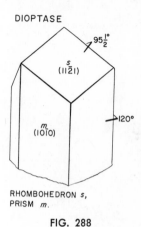

DIOPTASE

RHOMBOHEDRON s,
PRISM m.

FIG. 288

The beautiful color and brilliance of dioptase crystals, not to mention their general rarity, has placed great demands on productive sources, making prices of specimens very high as compared to other copper minerals. The best specimens are geodal druses, or small clusters of larger crystals. Some of the African material has been found enclosed in white granular calcite but poses problems in cleaning because of the solubility of dioptase in acid along with calcite; such specimens must be cleaned by painstaking mechanical removal of the calcite. Similarly, specimens of dioptase coated with limonite, often deeply implanted in crystal surfaces, cannot be cleaned by any known means without destruction of the dioptase.

Axinite—$(Ca,Mn,Fe)_3Al_2(OH)(BO_3)(Si_4O_{12})$

Name: From the Greek, *axine*, "an axe," referring to the sharp-edged tabular crystals resembling some cutting instruments; ăk′sĭ nīt.

Crystals: Triclinic; very commonly in tabular crystals bounded by knife-like edges, or in bladed aggregates. Broadest faces are usually pinacoids $b[010]$, $r[011]$, and $M[1\bar{1}0]$, as in Figure 289.

Physical Properties: Brittle; uneven to conchoidal fracture. Indistinct cleavage parallel to $b[010]$. H. 6½-7. G. 3.2-3.3.

Optical Properties: Transparent to translucent. Predominantly some shade of brown with overtones of reddish or purplish arising from the strong pleochroism; also brownish-gray, bluish-gray, grayish-violet; rarely, orange-yellow (Franklin) or pink. Streak colorless. Luster vitreous to somewhat greasy. Biaxial negative. R.I., $\alpha = 1.675\text{-}1.678$, $\gamma = 1.685\text{-}1.688$; yellow material (Franklin), $\alpha = 1.684$, $\beta = 1.692$, $\gamma = 1.696$. Birefringence 0.010. All specimens strongly pleochroic, brown material (Gavilanes) showing: rich violet, deep reddish-brown, and faint yellow; other hues have been reported.

Chemistry: Essentially calcium aluminum hydroxyl boro-silicate. Manganese or iron substitutes for part of the calcium.

Distinctive Features and Tests: Sharp-edged, wafer-like brown crystals; strong pleochroism. Fuses easily at 2½-3, and even more easily if fused with borax on charcoal; after fusion with borax, the resulting clear glassy bead is colored bottle green.

Occurrences: A high-temperature mineral formed in hydrothermal veins, in metamorphosed limestones at contacts with igneous bodies, and in fissures in granitic rocks or in certain granitic pegmatites associated with quartz, epidote, sphene and feldspars. In small bright yellow crystals, forming druses, in the contact metasomatic portions of the zinc ore deposits of Franklin, Sussex County, New Jersey. In large bladed masses from near Luning, Mineral County, Nevada. Exceptional specimens, rivaling those of

Bourg d'Oisans in respect to size, perfection and transparency of crystals, have come from near Coarse Gold, Madera County, California (Figure 290). Also in California, fine crystals occur in gold-bearing gravels from a number of places in the Mother Lode country; fine crystals in small pegmatites with quartz and epidote from various localities

FIG. 290. *Right:* Axinite crystal group from near Coarse Gold, Madera County, California. The flattened, wedge-shaped form of the crystals is shown well in this photograph. The largest crystal is about 2¾″ long and nearly flawless. *Photo courtesy Smithsonian Institution.*

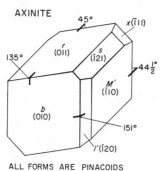

AXINITE

ALL FORMS ARE PINACOIDS

FIG. 289.

in Inyo, Kern and Mono counties; large pale brown crystals, some to 2½″ diameter, occur near Fallbrook, and in the Freeman Mine, Moosa Canyon, near Bonsall, San Diego County. Large crystals to 2″ across and ⅜″ thick from scheelite mines at Los Gavilanes, and elsewhere in northern Baja California, Mexico; material from the latter places has furnished fine faceted gems to several carats weight. Classic specimens, displaying mats of criss-crossed upright blades of sharp transparent brown axinite crystals, have long come from St. Cristophe, near Bourg d'Oisans, Isère, France; the associated species are epidote, quartz and prehnite. Fine crystals also occur lining fissures in greenstone and hornblende schists near Botallack, Wheal Cock, Trewellard, and Lostwithiel, Cornwall, England. Large dark brown and shining crystals in parallel position, sometimes forming groups to 4″ or more across, come from the danburite locality at Toroku Mine, Miyazaki Prefecture, Kyushu Island, Japan.

Consistently fine specimens have been furnished by the French locality, usually in druses to as much as 6″ across, but also in fine thumbnail and single crystal specimens. The yellow axinite from New Jersey is much sought after by collectors specializing in the minerals of this unique deposit. Lately, good single crystals have come from Baja California, and fine groups from the Japanese locality. The latter place has also furnished singles to 2″ across.

Beryl—Be$_3$Al$_2$(Si$_6$O$_{18}$)

Name: From the Ancient Greek, *beryllos,* signifying a green stone, but through later usage, applied only to beryl; bĕr′ĭl.

Varieties: On the basis of color: *emerald,* pale to dark blue-green; *aquamarine,* pale blue to medium blue, also bluish-green, and greenish-blue; *morganite,* pink, pale violet-pink, and pink-orange; *goshenite,* colorless; *golden beryl,* yellow. The term *cesium beryl* is applied to varieties containing cesium and other rare earth elements and include *goshenite* and *morganite.*

Crystals: Hexagonal; crystals common. Predominantly simple prismatic, combining the hexagonal prism $m[10\bar{1}0]$ and the pinacoid $c[0001]$; less commonly, small faces belonging to the hexagonal pyramid $p[10\bar{1}1]$ and dihexagonal pyramid $s[1121]$, modify terminations as in Figure 291. Habit is short to long prismatic, but cesium beryl appears mostly in

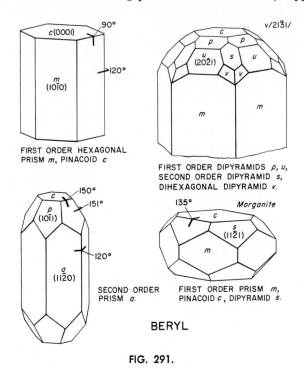

BERYL

FIG. 291.

tabular crystals compressed along the c axis, and usually corroded more or less severely. Some corroded remnants are highly irregular and may show no original faces, but are commonly bounded by curved surfaces of great brilliancy. Aquamarine is also found in corroded crystals, many with deep trumpet-shaped pits of hexagonal cross-section, descending from the pinacoidal faces $c[0001]$, and with rectangular shallow pits on prism faces, aligned parallel to the c axis (Figure 29, page 79). Tubular and planar inclusions often noted parallel to the c axis, sometimes sufficiently dense to create moderate chatoyancy.

Physical Properties: Strong but brittle. Fracture conchoidal to uneven. Poor cleavage parallel to $c[0001]$ but rarely observed. H. 8. G., in the broad range, 2.67-2.90; *emerald* usually 2.67-2.75; *aquamarine* usually 2.68-2.73; cesium beryls usually 2.80-2.90.

Optical Properties: Transparent to translucent. Wide color range but hues seldom intense. Much translucent to sub-translucent beryl enclosed in feldspar and quartz near margins of pegmatites, is commonly some dingy hue of greenish-yellow, green, or more rarely, blue; as crystals develop toward pegmatite cores, greater transparency is often noted as well as purity of hue. Aside from the hues noted under *Varieties,* a unique red beryl occurs in Utah, at the topaz locality in the Thomas Mountains. Most crystals consist of one color, but zoned crystals are not uncommon; some Uralian crystals show very sharp color zoning in narrow bands parallel to $c[0001]$. The orange color of cesium beryl usually disappears after heating or exposure to sunlight, a permanent pink hue remaining. A peculiar greenish hue, tinged with olive, is often converted by strong heating into a fine blue but much greenish-blue aquamarine, despite claims to the contrary, cannot be heat treated to drive off the green in order to achieve pure blue. The color of emerald is caused by small quantities of chromium. Streak colorless. Luster vitreous, sometimes silky when inclusions present. Uniaxial negative. R.I., in the general range, $O = 1.568$-1.602, $E = 1.564$-1.595; *emerald* range is $O = 1.571$-1.593, $E = 1.566$-1.586; *aquamarine* $O = 1.575$-1.586, $E = 1.570$-1.580; *morganite* $O = 1.590$-1.600, $E = 1.580$-1.590. Birefringence 0.004-0.007. Dichroism distinct except in yellow beryl; *emerald* gives yellowish-green, blue-green; *aquamarine* usually yields blue and colorless; *morganite,* pink and nearly colorless.

Chemistry: Beryllium aluminum silicate. Silicate tetrahedron rings of beryl are stacked over each other parallel to the c axis, the center spaces in the rings providing places for the ions of a number of elements to enter. In cesium beryl, cesium, sodium and potassium ions appear in these channels, with simultaneous substitution of lithium for aluminum, and aluminum for beryllium in order to balance electrical charges. The substitution of aluminum for beryllium results in beryls less rich in the latter element, and hence less desirable as ores of beryllium.

Distinctive Features and Tests: Hexagonal prismatic crystals; color. Much harder than apatite with which it may be confused; greater hardness also distinguishes pale translucent varieties from feldspar in which it is commonly enclosed. Fuses with difficulty at 5-$5\frac{1}{2}$. Insoluble in acids.

Occurrences: Mainly in granitic pegmatites as rude translucent crystals near margins, or in core cavities as transparent euhedrons. Emerald occurs in pegmatite-like veinlets traversing carbonaceous limestone in Columbia, in biotite schists in Russia, Brazil and Austria. Blue translucent crystals occur in pegmatites in Ontario, a notable locality being Lyndoch, Renfrew County. Abundant in many pegmatites of the New England region, mainly as ore-grade crystals, sometimes many tons in weight, but also in fine aquamarine, golden beryl, goshenite and morganite. Also common in the North Carolina pegmatites. Emerald also occurs with hiddenite in North Carolina at the locality at Hiddenite, Alexander County, also near Shelby, Cleveland County, where it occurs in small pegmatite veins, and in gneiss near Crabtree, Mitchell County; the largest crystals, several to $8''$ length, were found in the first two localities.

Fine aquamarine crystals occur in miarolitic pegmatite cavities atop Mount Antero, Chaffee County, Colorado, mostly as simple hexagonal prisms of several inches length associated with quartz, feldspar, phenakite and other species; the largest crystal of record measured $1\frac{1}{2}''$ thick and nearly $8''$ long. Gem pegmatites of San Diego County, California, are noted for morganite, and a few mines, for pale aquamarine; the finest morganite crystals occur in the Pala District, and lately, came in abundance from the White Queen Mine, Hiriart Hill, where tabular crystals of pale orange-pink hue to $4''$ diameter and $2''$ thickness were found. Many are well-formed, show little corrosion, and are implanted on pale bluish-white clevelandite and associated with quartz crystals (Figure 292). Deeper-hued crystals occur at the San Pedro Mine, Hiriart Hill, and some intense brownish-orange corroded crystals in the Himalaya Mine, Mesa Grande, San Diego County, sometimes to $4''$ across.

The premier locality for emerald crystals is Columbia where at least six productive deposits are known, and from them, much of the world's gem rough has been supplied. Matrix specimens are sometimes available and are extremely desirable although very expensive. Occasionally, fine matrix specimens are obtained from Gachala, Columbia,

FIG. 292. Beryl. *Top:* prismatic *aquamarine* crystals, the large crystal at the upper right being from Alabaschka, near Mursinsk, Ural Mountains, U.S.S.R., and the others from Minas Gerais, Brazil. The Uralian crystal is sharply color-zoned in bands of darker color parallel to c[0001]. Size 4¼″ long. *Bottom:* stubby *morganite* crystal of reddish-orange color in matrix of *cleave-landite* and quartz from the San Pedro Mine, Pala, San Diego County, California. *Specimen in William C. Woynar Collection, San Diego.* Size 5½″ wide. The *morganite* displays prism m[10$\bar{1}$0], pinacoid c[0001] and second order dipyramid s[11$\bar{2}$1], the latter truncating adjacent prism faces and creating triangular faces.

FIG. 293. Sketch of emerald crystals in biotite schist from the Takowaya River, Ural Mountains, U.S.S.R. Crystals are fully enclosed in schist and must be exposed by scraping. Size 4″ by 3″.

showing emerald prisms on showy pyrite druses. More commonly represented in collections are prismatic crystals, usually rude, enclosed in biotite schist from Conquista, Minas Gerais, Brazil, from Habachtal, Austria, and from Takowaya River, Urals, U.S.S.R. (Figure 293). Usually the schist is scraped away to reveal the crystals.

An enormous variety of beryl in good to outstanding specimens regularly appears on the market from Brazilian pegmatites, the most common examples being pale green or pale bluish-green crystals (Figure 292). Very rarely, matrix specimens appear and then are offered at extremely high prices. Productive pegmatites are principally centered at Governador Valadares and Teofilo Ottoni, Minas Gerais, and material finds its way into these communities. Very fine beryl also occurs in abundance in Madagascar, exceptional deep blue aquamarine and very good pink morganite being especially sought. Among the finest of all beryl specimens are aquamarine, golden beryl, and green beryl from various localities in the U.S.S.R. Some outstanding matrix specimens display slender transparent golden crystals to 2″ on corroded feldspar, notably from Mursinsk, near Sverdlovsk, Urals, and beautiful slender hexagonal prisms of green and blue from Adun-Tschilon Mountains, Transbaikalia. The latter crystals are sharply zoned in various colors parallel to the basal plane $c[0001]$ and are easily recognized on this account (Figure 292). Also very fine are pale blue prisms to 2″ on pocket matrix, from pegmatites in the Mourne Mountains, Ireland.

The finest beryl specimens are those displaying euhedral crystals on pocket matrix, but such are extremely rare and command very high prices. The majority of high quality specimens are detached crystals, colorful and largely transparent. Some cracked beryl crystals enclosed in quartz or feldspar also afford attractive specimens but are far less valued. The prices of specimen beryl increases sharply as crystals become clearer, sharper, free from disfiguring cracks and more deeply colored.

Cordierite—$(Mg,Fe)_2Al_4(Si_5O_{18})$

Names: After French geologist P. L. A. Cordier (1777-1861): kôr′dĭ ĕr ĭt. The alternate name, iolite, from the Greek, *ion,* meaning "violet," in allusion to the common color; ī′ō līt.

Crystals: Orthorhombic; rare. Short prismatic individuals of rectangular cross-section but seldom sharp. Repeated twinning on $m[110]$ forms crystals of nearly hexagonal cross-section. Some crystals contain minute oriented scales of hematite causing distinct reddish aventurescence. Extremely slender acicular inclusions result in chatoyancy and, rarely, asterism. Predominantly granular massive.

Physical Properties: Brittle; uneven to subconchoidal fracture. Poor cleavage parallel to $b[010]$; also a parting parallel to $c[001]$ in partly altered crystals. H. 7. G. 2.55-2.75, rising with iron. Transparent gem material usually in the range G. 2.57-2.61.

Optical Properties: Transparent to translucent. Commonly pale to dark violet; also colorless, gray, yellow, brown. Streak colorless. Luster characteristically oily but also vitreous. Sometimes chatoyant, asteriated or aventurescent. Biaxial negative. R.I., $\alpha = 1.53$-1.54, $\gamma = 1.54$-1.55. Birefringence 0.01. Pleochroism strong, in violet material, usually very dark blue-violet, dark violet, pale straw yellow.

Chemistry: Magnesium-iron aluminum silicate. Channels within the crystal structure accommodate small quantities of water and calcium.

Distinctive Features and Tests: Strong pleochroism usually evident without aid of dichroscope; look for dark violet and straw yellow hues. Fuses at 5-$5\frac{1}{2}$; does not lose color even after white-hot heating. Insoluble except in hydrofluoric acid.

Occurrences: Formed through medium to intense metamorphism in aluminum-rich rock, commonly in schists and gneisses. Sharp distinct crystals very rare, most specimens being altered in exterior zones. Partly-altered crystals to $1\frac{1}{2}''$, occur at Bodenmais, Bavaria, Germany. In crystals to about $2''$ diameter, from Orijarvi, Finland. Good glassy masses to several inches diameter, some with red hematite inclusions, from Kargerö, Norway. Rough-surfaced grains eroded from metamorphic rock from Madagascar, sometimes providing excellent gems. Similar material from gem gravels of Ceylon and Burma, and from Trichinopoly and Coimbatore districts, Madras, India. Massive material capable of being cut into small gems also occurs in Connecticut, Wyoming, and in the Yellowknife district of Northwest Territories, Canada.

Tourmaline—Na(Mg,Fe)₃Al₆(OH)₄(BO₃)₃(Si₆O₁₈)

Name: From the Singhalese, *turamali,* applied to water-rolled gem pebbles from Ceylon gravels; toor′mă lĭn, also -lĭn.

Varieties: On the basis of composition: *schorl,* black tourmaline rich in iron; *dravite* (drä′vĭt), magnesium-rich tourmaline, usually brown in color, but white if calcium substitutes for sodium; *lithia tourmaline,* containing lithium and aluminum substituting for iron and magnesium and commonly pink, red, green or blue in color. On the basis of color, the following varietal names are also used: red, *rubellite;* colorless, *achroite* (ā′krō ĭt); blue, *indicolite* (ĭn dĭk′ō lĭt).

Crystals: Hexagonal, very common. Usually short to long prismatic, either hexagonal in cross-section, or approximately triangular, terminated by low to steep pyramidal faces, with or without the pedion $c[0001]$, as in Figure 294. Six-sided prisms usually consist of the ditrigonal prism $a[11\bar{2}0]$. Nearly triangular crystals usually consist of three broad faces of prism $m[10\bar{1}0]$ with corners beveled by two faces of prism $a[11\bar{2}0]$, making a total of nine prism faces. When m and a are equally developed, nine equal-width faces appear. Crystals "frozen" within other minerals commonly lack striations, but those developed within openings, show numerous "phonograph record" striations parallel to the c axis and so consistently developed in this species, that it serves as a very useful recognition clue. Striated crystals are triangular in cross-section but prism faces bow outward because of oscillatory development of alternate faces of prisms *m* and *a,* some crystals then becoming nearly circular in cross-section (Figure 294). Doubly-terminated crystals usually show hemimorphic development of terminal faces such that one end will be terminated by lower pyramids than the other, or one end will display a broad pedion and the other a small one or none at all. Tubular inclusions parallel to the *c* axis occur in many crystals, especially lithia varieties from within pockets in granitic pegmatites; strong chatoyancy sometimes results. Also granular or fibrous massive.

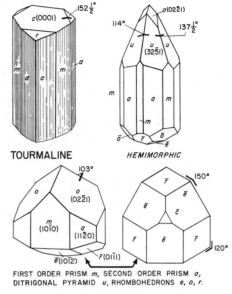

TOURMALINE   HEMIMORPHIC

FIRST ORDER PRISM *m,* SECOND ORDER PRISM *a,*
DITRIGONAL PYRAMID *u,* RHOMBOHEDRONS *e, o, r.*

FIG. 294.

Physical Properties: Brittle; fracture conchoidal to uneven. Crystals lacking flaws and inclusions break with conchoidal surfaces approaching the perfection of glass fractures. External portions of crystals are often checked with numerous small fractures more or less parallel to *c*[0001], sometimes extending through the prisms and causing them to break easily into tabular segments. In gem-quality crystals, fractures may extend part way and cause spalling of outer layers, leaving behind more or less spherical nodules of flawless material. H. 7½. G., generally in the range 3.0-3.3. Gem material gives the following ranges, pink-red, 3.03-3.05; green, 3.05-3.08; brown, 3.06; yellow, 3.10; blue, 3.10; black, 3.15-3.2. Strongly pyroelectric, specimens in sunlight quickly attracting dust particles; also piezoelectric.

Optical Properties: Transparent to opaque, but black varieties which seem opaque usually show dark red-brown or dark blue through thin splinters. White, colorless, brown, black; also pink, red, yellow, green, blue and numerous shades of each, sometimes blending imperceptibly into each other, or arranged in zones in the same crystal. Black tourmaline is by far most common. *Lithia tourmaline* often shows color zones concentric to prisms, with pink or red in cores, and green in exterior zones, but many variations in arrangement occur. Color zoning is a valuable recognition clue. Streak colorless. Luster vitreous, sometimes silky. Uniaxial negative, R.I., varies rather consistently with specific gravity, specimens of low gravity show low indices and vice versa. The range of indices is $O = 1.616\text{-}1.634$, $E = 1.630\text{-}1.652$. Birefringence variable, from 0.014 to 0.021. Dichroism distinct to strong depending on intensity of color and often apparent without the aid of the dichroscope in transparent crystals. Green crystals usually show olive-green, green; red or pink crystals, deep pink, pale pink; blue crystals, dark blue, pale gray-blue, etc.

Chemistry: Complex silicate of boron and aluminum. Composition varies widely because of substitutions, but the crystal structure pattern remains the same. Calcium substitutes for sodium, lithium and aluminum substitute for magnesium and iron, and some fluorine substitutes for hydroxyl.

Distinctive Features and Tests: Striated crystals of hexagonal to rounded-triangular cross-section; concentric color zoning; no cleavage. Much harder than apatite; further distinguished from beryl and apatite by very sharp striations, hemimorphic terminations and color. Brown varieties (*dravite*) fuse at 3, puffing on edges and forming a frothy white glass; *schorl* fuses at 4, small crystal prisms exfoliating along the *c* axes and form-

FIG. 295. Tourmaline. *Left:* at the top left is a black schorl specimen from the Hercules Mine, Ramona, San Diego County, California. Size 3½″. At top right is a dark green prism from Minas Gerais, Brazil, terminated by the pyramid o[022̄1]. The remaining crystals, most of which are doubly-terminated and show clearly the hemimorphic nature of this species, are from the Himalaya Mine in San Diego County. A few crystals show attached cylindrical crystals of lepidolite. *Right:* plumose schorl developing from originally simple crystals, with quartz and apatite. From a scheelite-bearing pegmatite near Santa Cruz, Sonora, Mexico. Size 4″ by 4″.

ing worm-like gray froths; *lithia tourmaline* becomes white and expands, forming a porcelain-like mass without melting. Insoluble in acids.

Occurrences: Schorl common as an accessory mineral in metamorphic rocks as in schists and gneiss; in some high temperature veins; often abundant in long prisms in granitic pegmatites. *Dravite* occurs in metamorphosed limestone. *Lithia tourmaline* occurs in complex pegmatites near the cores or within pockets, associated with quartz, beryl, apatite, microcline and cleavelandite, columbite, lepidolite, and other species typical of complex pegmatites containing lithium (Figure 295). There are very many localities for tourmaline and only a few providing the best specimens can be mentioned. *Schorl* abundant in granitic pegmatite districts, as in New England, South Dakota, Colorado, and California. Especially fine small *schorl* crystals, seldom over 1″, but in splendid lustrous groups, occur near Pierrepont, St. Lawrence County, New York (Figure 296); also in this county at Gouverneur and elsewhere, occur fine reddish-brown complex *dravite* crystals, some to 6″ length and 3″ diameter. Splendid gem-

FIG. 296. Sketch of schorl crystal group from Pierrepont, St. Lawrence County, New York. Size 2½″ by 2½″.

quality *lithia tourmaline* from pegmatites in Maine, notably at Mount Mica, Oxford County; greens from this locality are unmatched in respect to freedom from dingy olive-green coloration which so often proves a serious defect in green gem tourmaline from other localities. Excellent green and some pink crystals, often extremely thin in proportion to length, from Gillette Quarry, near Haddam Neck, Connecticut. Among the finest gem crystals and specimen groups known from any place in the world, are those from mines in San Diego and Riverside counties in California. From the Himalaya Mine at Mesa Grande, San Diego County, still come fine green and pink crystals, ranging from minute acicular crystals to many which are 5″ to 6″ in length, about 1″ diameter, and very commonly doubly-terminated as in Figure 295; also matrix specimens of singular beauty displaying gem crystals associated with quartz, microcline, cleavelandite and lavender lepidolite. The Pala District of San Diego County produced very large pink, red and blue, color-zoned prisms, many reaching 7″ and diameters of 3″ to 4″. Matrix specimens from such mines as the Tourmaline King, Tourmaline Queen, and Pala Chief were produced in small numbers but are the finest in existence. *Schorl* is common in the pegmatites of the Ramona district, San Diego County, as sharp prisms associated in pockets with spessartite garnet, smoky quartz and feldspars; the Little Three Mine also produced very dark green stubby crystals of nearly circular cross-section, some to 4″ diameter and only slightly longer. Recently, large, color-zoned *lithia tourmalines* to 4″ diameter, have been found in several pegmatites near Alamos, Baja California Norte, Mexico.

In Europe, premier localities are the extinct pegmatite quarries at San Piero, island of Elba, Italy, and the pegmatites near Mursinka in the Urals. Elba produced small transparent pink crystals primarily, seldom over 1″ length, perched on feldspar matrix with quartz and beryl; other colors were also obtained from the pockets. The Uralian pegmatites produced rich purplish-pink to red stubby prisms seldom over 2″ length. Fine doubly-terminated *dravite* crystals to 1½″ length, came from Dobrava, Carinthia, Austria.

Good doubly-terminated *schorl* crystals to 2″ length, occur near Kragero, Norway. Many very large and fine tourmalines, some of gem quality, have been found in Mozambique and Madagascar. Specimens of *rubellite,* in prisms to 20″, and composed of subparallel growths of smaller crystals, were found in a very large pegmatite near Lorençao Marques, Mozambique. The Madagascar localities are noted for producing hexagonal prisms to 4″ diameter and 8″ length, with remarkable triangular color zoning, such that thin slices, cut across the *c* axis, display numerous sharp triangles, and sometimes hexagons, of various colors, in concentric arrangement. Karibib, South-West Africa, has furnished fine green gem tourmalines. Green gem tourmalines occur in pegmatites near Spargoville, Western Australia, and excellent transparent crystals to 4″ by 1″ were found on Kangaroo Island, South Australia.

Practically all commercial gem tourmaline now comes from Minas Gerais, and several other states in Brazil, where numerous decayed pegmatites yield every known variety except *dravite,* but even the latter has been found on occasion in marbles. Matrix specimens of Brazilian tourmaline are uncommon, the greatest number being single crystals or groups of crystals. This is understandable in view of the fact that alteration of pegmatite feldspar has resulted in kaolinization, and only the resistant species as quartz, beryl, tourmaline, spodumene, etc., survive to the present time, but not on a matrix of feldspar as is commonly the case in pegmatites which have been less severely altered.

The tremendous variety in tourmaline makes this species an excellent choice for the specialist-collector, but the best specimens are far from cheap. Good specimens from the Himalaya Mine in California are commonly offered, but the very large crystals from the Pala mines are specimens of the past. Elba and Mursinka tourmaline is also rarely encountered, as well as the unique cross-sections of large Madagascar crystals. Brazilian tourmaline is usually available, but is quite expensive in sizes over 2″.

SOROSILICATES

The characteristic grouping of silicon-oxygen units in the sorosilicates consists of two tetrahedrons linked tip-to-tip by sharing the same oxygen atom. The resulting double-tetrahedron looks like an hourglass. These groups are not connected to each other either in sheets or in chains as was noted in species previously described, and consequently there is not the tendency for all species to adopt some characteristic habit or to display similar properties. The paired tetrahedrons are inter-connected by metal and hydroxyl ions and by water molecules, and considerable variety exists in chemical compositions and crystal structures, further contributing to variations in habit and properties among the species included in this subclass.

Hemimorphite—$Zn_4(OH)_2(Si_2O_7) \cdot H_2O$

Name: From the Greek prefix *hemi-,* meaning "half," and *morphe,* "form," literally "half-form," in allusion to the differences in crystal forms displayed on opposite ends of the *c* axis; hĕmĭ′môr fīt.

Crystals: Orthorhombic, hemimorphic. Usually thin tabular, compressed along the *b* axis such that the faces of *b*[010] are the broadest; also slightly elongated along the *c* axis (Figure 297). Commonly in fan-shaped aggregates of platey crystals, as in Figure 297, sometimes consisting of very small crystals and then forming compact mammillary masses or stalactites. Large crystals rare. Because crystals usually grow attached at one end of the hemimorphic *c* axis, the difference in faces is not ordinarily seen. Crystals striated parallel to the *c* axis.

Physical Properties: Brittle; fracture uneven to subconchoidal. Perfect cleavage, but interrupted, parallel to *m*[110]. H. $4\frac{1}{2}$. G. 3.4-3.5.

Optical Properties: Transparent to translucent. Predominantly white or colorless; also pale blue, gray or greenish. Streak colorless. Luster vitreous to subadamantine; commonly slightly silky because of inclusions parallel to *c* axis. Biaxial positive. R.I., $\alpha = 1.614$, $\beta = 1.617$, $\gamma = 1.636$. Birefringence 0.022.

Chemistry: Zinc hydroxyl silicate with water.

Distinctive Features and Tests: White bladed aggregates. Soluble in hydrochloric acid but does not bubble as does smithsonite; forms silica jelly. Practically infusible; strongly heated fragments become white, and after cooling, are very pale lemon-yellow color and weakly fluoresce white under ultraviolet (Mapimi).

Occurrences: In oxidized portions of zinc deposits, usually very close to the surface and associated with galena, smithsonite, sphalerite, cerussite and anglesite. In the early mining in the open pit at the Sterling Hill zinc orebody, Ogdensburg, Sussex County, New

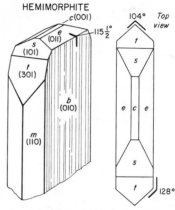

PINACOIDS *b*, *c*, DOMES *e*, *t*, *s*.

FIG. 297.

Jersey, very large mammillary masses of white material to 6″ to 12″ thick were found resting on marble, and produced numerous but not very attractive specimens. Fine crystals have been found at Elkhorn, Jefferson County, Montana. Especially fine white to colorless crystals to $\frac{1}{2}$″ length from various mines in the Leadville district, Lake County, Colorado. The finest North American specimens are from gossan in the Mina Ojuela, Mapimi, Durango, Mexico, where single blades, sharp and transparent, to 1″ length, were found planted thickly on limonite (Figure 298). Much of this material is still available. Abroad, fine specimens were found at Nerchinsk, Transbaikalia, Siberia; at Altenberg, Saxony, Germany; also from Iglesias in Sardinia, Italy, and in magnificent crystals from Djebel Guergour, Constantine, Algeria.

Ilvaite—CaFe$_2$O(OH)(Si$_2$O$_7$)

A calcium-iron hydroxyl silicate. Named after the island of Elba, Italy, formerly known as *Ilva*; ĭl′vă ĭt. A relatively rare species, but prized for cabinet specimens when the black glossy crystals are large and well-formed. Crystals usually stubby individuals of diamond-shaped cross-section, with prism faces striated parallel to the *c* axis as in Figure 299. Splendid crystals to 3″ length occur in the Laxey Mine, South Mountain, Owhyee County, Idaho. Also fine from the original Elba localities at Rio Marino and Capo Calamita. Excellent crystals from Seriphos Island, Cyclades, Greece, and from various mines in Japan.

Idocrase—Ca$_{10}$Mg$_2$Al$_4$(OH)$_4$(Si$_2$O$_7$)$_2$(SiO$_4$)$_5$

Name: From the Greek, *eidos*, "form," and *krasis*, "mixture," because the forms appeared to be those of several minerals to the namer of this species; ī′dō krās. The alternate name, vesuvianite, is from the locality at Vesuvius; vĕ sū′vĭ ăn ĭt.

Varieties: Very compact, jade-like, *californite;* massive blue, *cyprine.*
Crystals: Tetragonal, uncommon. Predominantly simple stubby prismatic crystals of square cross-section; also dipyramidal and in combinations as in Figure 300. Many crystals are box-like, with equally developed prism faces of $a[100]$, slightly truncated by prism

FIG. 298. White to colorless tabular crystals of hemimorphite, arranged in divergent sprays on limonite gossan, from Mina Ojuela, Mapimi, Durango, Mexico. Size 3″ by 3″. The hemimorphic character of crystals is not apparent because of attachment by one end of prisms, all of which are elongated parallel to the c axes. Crystals flattened parallel to the b axis, making broadest faces those of b[010].

$m[110]$, and terminated by pinacoid $c[001]$. Sometimes striated on $a[100]$ parallel to the c axis. Faces of $a[100]$ are smooth but covered by very low growth hillocks. Many crystals from within marble are rudely formed and rounded on edges as if partly fused. Also massive, and commonly, in cryptocrystalline aggregates.
Physical Properties: Brittle, but cryptocrystalline aggregates very tough. Fracture conchoidal to uneven, and in massive varieties, sugary in texture. Cleavage poor parallel

to prism faces of $m[110]$. H. $6\frac{1}{2}$. G. 3.3-3.5; *californite* range is 3.3-3.6, the white varieties being lower than the green.

Optical Properties: Translucent to transparent. Crystals usually some shade of characteristic dark brown-green hue, somewhat similar to epidote. Also yellow-green, yellow, brown, and rarely, blue. *Californites* white to dark green, and less commonly, fine yellow-green; also mottled and veined in shades of colors mentioned. Streak colorless. Luster vitreous to somewhat oily. Uniaxial negative or positive. R.I., usually in the ranges O = 1.706-1.726, E = 1.702-1.732. Birefringence 0.004-0.006. Dichroism distinct, yellow-brown material (Laurel) gives: reddish-brown, yellow-green.

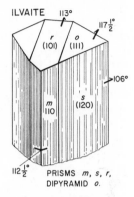

PRISMS *m, s, r,*
DIPYRAMID *o.*

FIG. 299.

Chemistry: Essentially calcium-magnesium-aluminum hydroxyl silicate. Considerable variation in composition occurs because iron and manganese substitute for magnesium, some calcium is replaced by sodium or potassium, and aluminum by iron or titanium.

Distinctive Features and Tests: Square cross-section of crystals; distinctive colors. *Californite* difficult to distinguish from cryptocrystalline grossularite, nephrite, or jadeite, but jadeite fuses more easily and colors flame bright yellow; nephrite is much harder to fuse while grossularite yields a bead of different hue. Epidote is magnetic after melting and also displays perfect cleavages. Idocrase fuses easily at 3, bubbling, then forming a brownish or greenish glass.

Occurrences: Primarily in metamorphosed impure limestone near contacts with igneous bodies, associated with andradite, grossularite, wollastonite, calcite and diopside. Less commonly, in serpentines associated with chromite. Gem quality yellow-brown material in rude prismatic crystals in pegmatite, in Grenville marble, from near Sixteen Island Lake, Laurel, Argenteuil County, Quebec. Beautiful micromount crystals of unusual rich purplish-pink color in massive material in Montreal Chrome Pit, Black Lake, Megantic County, Quebec. Also in good specimens from various places in Ontario. Fine crystals occur in marble at Marble Bay, Texada Island, British Columbia. Dark brownish-green square prismatic crystals to 1″, in groups in calcite, near Sanford, York County, Maine. Exceptional crystals from asbestos-bearing rock at Eden Mills, Lamoille County, Vermont. Large rude crystals from marble near Olmstedville, Essex County, New York. The variety *cyprine* occurs at Franklin, Sussex County, New Jersey. Large crude dipyramidal crystals to 5″ diameter occur at Magnet Cove, Garland County, Arkansas. Crystals to 3″ occur near Ludwig, Nevada. Fine crystals to $1\frac{1}{2}$″ in pale blue calcite at Scratch Gravel, northwest of Helena, Lewis and Clark County, Montana; similar crystals in blue calcite at the limestone quarries near Riverside, Riverside County, California.

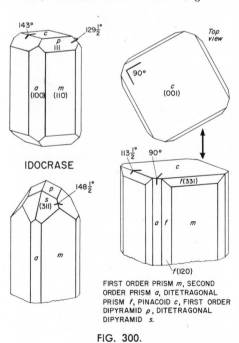

IDOCRASE

FIRST ORDER PRISM *m*, SECOND ORDER PRISM *a*, DITETRAGONAL PRISM *f*, PINACOID *c*, FIRST ORDER DIPYRAMID *p*, DITETRAGONAL DIPYRAMID *s.*

FIG. 300.

Californite is abundant at several places in California, notably in Siskiyou County; fine yellow-green material of exceptional quality and of excellent translucency occurs near Pulga, Butte County. Rude prismatic yellow crystals to 3″ diameter, occur with grossularite at Xalostoc, Morelos, and Lake Jaco, Chihuahua, Mexico. Lustrous dark olive-green square prismatic crystals to 1½″ by 2″ from various places near Kristiansand, and from Eiker, Oslo district, Norway. Similar crystals at Canzocoli, near Predazzo, and in the Ala Valley, Italy; the crystals from the latter place are beautiful rich green and transparent. Crystals to ¾″ diameter, usually terminated by pyramids, occur in altered limestone blocks ejected from Monte Somma crater at Vesuvius, Italy. Very fine green to brownish-green, stubby, prismatic crystals occur in a soft grayish rock at Achmatovsk, near Slatoust, and along the Wilui River, in Siberia. The latter place is one of the few furnishing crystals in matrix.

Idocrase crystals are prized by single-crystal collectors because they can be obtained in nearly textbook-perfect, doubly-terminated crystals of 1″ or better in length. Matrix specimens are rare, as are faceted gems cut from transparent material. Massive material from California is used extensively in amateur lapidary projects but much suffers from soft spots which prevent uniform polish.

Epidote Group

Zoisite—$Ca_2Al_3O(OH)(SiO_4)(Si_2O_7)$

A comparatively rare member of the Epidote Group, seldom yielding cabinet specimens of interest; zō′ĭ sīt. The pink massive variety, known as *thulite,* is used in lapidary work. A calcium aluminum hydroxyl silicate. Closely related to other members in the group but orthorhombic in crystallization while the others are monoclinic.

CLINOZOISITE-EPIDOTE SERIES

Clinozoisite—$Ca_2Al_3O(OH)(SiO_4)(Si_2O_7)$

Epidote—$Ca_2(Al,Fe)Al_2O(OH)(SiO_4)(Si_2O_7)$

Names: From the Greek, *klinein,* "to incline," and *zoisite,* after Austrian Baron Zois. Because of the inclination of one axis, clinozoisite is monoclinic instead of orthorhombic as in zoisite; klĭ′nō zō ĭ sīt. The derivation of epidote is from the Greek, but unfortunately is inexplicable; ĕp′ĭ dōt.

Crystals: Monoclinic, common. Usually short to long prismatic, elongated parallel to the *b* axis; also nearly equant or tabular, and in bladed or intermeshed aggregates of prismatic crystals. Much epidote is granular massive. In general, epidote crystals display long narrow faces of pinacoids *a*[100], *c*[001], and narrower faces of pinacoid *r*[$\bar{1}$01]. The terminations at either end of the *b* axis commonly consist of a pair of faces of the prisms *n*[$\bar{1}$11] and *o*[011], forming a blunt wedge, as in Figure 301. The pinacoidal faces mentioned are usually striated or grooved parallel to the *b* axis. Cross-sections of prisms are approximately diamond-shaped or nearly rectangular, but many crystals are highly irregular owing to oscillatory development of faces. Twinning on plane *a*[100].

Physical Properties: Brittle; uneven to conchoidal fracture. Perfect but interrupted cleavage parallel to *c*[001]. H. 7. G. 3.3-3.6, increasing with iron.

Optical Properties: Translucent to transparent. Clinozoisite commonly transparent in sections to ½″ thick, but epidote seldom transparent except in thin chips or in exceptional crystals, as those from Austria. In these, a brown color is seen through the faces of $a[100]$ but as the crystal is turned to place the faces of $c[001]$ foremost, the light is cut off almost completely, and the color becomes very dark green. Clinozoisite usually grayish-brown, pale to medium brown, sometimes with a greenish tinge; epidote is usually

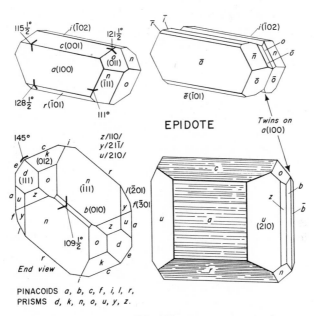

PINACOIDS *a, b, c, f, i, l, r,*
PRISMS *d, k, n, o, u, y, z.*

FIG. 301.

a distinctive, very dark yellow-green (pistachio). In almost flawless epidote crystals, light is absorbed so strongly that the pistachio color is not apparent and crystals appear black; however, in very slender crystals or in fibrous or granular masses, the pistachio color is vivid. Finely divided epidote is decidedly yellow in hue, tinged with green. Luster vitreous to slightly greasy. Streak colorless. Biaxial negative. R.I., variable, lower in clinozoisite, rising with iron to a maximum in epidote; the birefringences also rise. Brown clinozoisite (Baja California), $\alpha = 1.712$, $\gamma = 1.721$; birefringence 0.009. Dark green epidote (Knappenwand), $\alpha = 1.738$, $\gamma = 1.772$; birefringence 0.035. Strong pleochroism, usually dark brown, reddish-brown, and green in epidote; dark brown, greenish-brown, and yellow in brown clinozoisite. Gems are cut from epidote but are usually unsatisfactory because appearing so dark.

Chemistry: Calcium-aluminum hydroxyl silicates. In epidote, iron substitutes for aluminum.

Distinctive Features and Tests: Pistachio color of epidote; clinozoisite resembles epidote in crystals and associations, but is usually brownish and paler in hue. Epidote fuses at 3-4, more readily in iron-rich varieties, bubbling, then turning to a black cinder, the latter usually magnetic.

Occurrences: In low and medium grade metamorphic rocks, often in fine crystals in cavities, associated with actinolite (*byssolite*), feldspar, quartz, chlorite, axinite, apatite, sphene and black tourmaline. Also in fine crystals in contact zones of metamorphosed limestone, and in some granitic rocks where it forms large crystals in pegmatites (Baja California). There are many localities for crystals and only a few can be mentioned.

Fine lustrous stubby crystals to 1½″, in the Calumet Iron Mine, Chaffee County; crystals to 3″ from Epidote Hill, at an altitude of over 10,000 feet in the Elk Creek area, Park County, Colorado. Extremely large epidote crystals, some to 12″ length in the Peacock, the Decorah, and other mines of the Seven Devils district, Adams County, Idaho. Crystals of epidote are common in the Sierra Nevada of California; fine examples were found in Inyo and Kern counties, especially in the Greenhorn Mountains, Kern County, where crystals to 2″ by 2″ by 6″ have been uncovered. Some good crystals also occur in metamorphosed limestone near Riverside, Riverside County, California. Also in the McFall Mine, near Ramona, San Diego County, California, associated with quartz and grossularite. Recently, crystals of clinozoisite to 5″, were found in scheelite deposits in Los Gavilanes-Castillo del Real district of Baja California Norte, and epidote in the Pino Solo-Alamos district in small pegmatite bodies consisting of quartz, black tourmaline, chlorite and large crystals of sphene.

Magnificent groups and single crystals (Figure 302), associated with uralite, adularia, quartz, some of the latter Japanese-twinned, occur in abundance in contact zones in marble near Sulzer, Prince of Wales Island, southwestern Alaska. Tabular single crystals reach 3″ square. Groups to 12″ across were plentiful; this locality is not exhausted. In Europe, excellent crystals occur in Arendal, Norway, and in the Ala Valley of northern Italy. Especially fine crystals occur at the world's best locality, the Knappenwand, Untersulzbachtal, near Salzburg, Austria. Slender shining prismatic crystals, beautifully formed, and with perfect terminations, were once found in abundance (Figure 302). Many were

FIG. 302. Epidote specimens. *Top:* very dark green doubly-terminated crystal from the famous Knappenwand locality in the Untersalzbachtal near Salzburg, Austria. Size 3½″ long. Smaller crystal from Greenhorn Mountains, Kern County, California. The long axis of the large crystal is the *b* axis; ends are terminated by pairs of $n(\bar{1}11)$ faces. *Bottom:* group of tabular epidote crystals similar in forms to the striated crystal sketched in Figure 301. From Sulzer, Prince of Wales Island, Alaska. Size about 5″ by 4″. The specimen has been "smoked" with ammonium chloride to reduce reflections and emphasize forms. *Photo courtesy Smithsonian Institution.*

perched on matrices of grayish-green *byssolite* with small colorless apatite crystals. Singles to 12″ length and slightly over 1″ diameter were recovered. Good specimens from this locality are now virtually unobtainable. The finest epidotes are from the Austrian locality, followed by those from the Prince of Wales Island locality in Alaska, but neither has been productive in recent years and good examples are scarce and expensive.

Allanite—(Ca,Ce,La,Na)$_2$(Al,Fe,Mn,Be,Mg)$_3$O(OH)(SiO$_4$)(Si$_2$O$_7$)

An extremely complex silicate, varying greatly in composition; ăl′ăn ĭt. Because it usually contains some radioactive thorium, crystals are damaged internally (metamict), and are very brittle, breaking or crumbling into small pitch-like particles whenever re-moval from enclosing rock is attempted. Good crystals, of lathlike shape, and with sharp edges, occur almost exclusively in granitic pegmatites and in a few magnetite deposits, but as pointed out before, it is difficult to extract them. Because of radioactivity, en-closing rock is commonly streaked or stained black immediately around allanite crystals; this, plus the thin lathlike shape, serves to identify them. Good crystals have been found at Madawaska, Ontario. Also in good crystals in asbestos and actinolite, at Pelham, Hampshire County, Massachusetts. Found as very slender blades to 16″ in albite in the Rutherford Mines, Amelia, Amelia County, Virginia. Fine specimens from Arendal, Norway.

NESOSILICATES

Unlike other silicates, the nesosilicates contain only isolated tetrahedrons of silicon-oxygen, without direct linkages through corner atoms of oxygen. The tetrahedrons are spaced evenly with metal ions between, generally resulting in very simple, densely-packed structures. Characteristically, they are hard minerals, of higher specific gravities than other classes of silicates, and commonly form stubby, blocky or equidimensional crystals. Some species are important rock-forming minerals, while many provide handsome cabinet specimens.

Olivine Group

OLIVINE SERIES

Forsterite—Mg$_2$(SiO$_4$)

Fayalite—Fe$_2$(SiO$_4$)

Names: From the olive-green color; ŏl′ĭ vĕn. After the island of Fayal, Azores, where specimens were found; fā′ăl ĭt. After Johann R. Forster, German naturalist; fôr′stĕr ĭt. The name *peridot,* applied to green gem varieties, is of uncertain origin; pĕr′ĭ dŏt or pĕr′ĭ dō.
Crystals: Orthorhombic, uncommon. Usually in rounded grains enclosed in volcanic rocks. A few localities provide good crystals, notably St. Johns Island (Zebirget) in the Red Sea, while small sharp crystals are common in scoriaceous volcanic rocks at other places. The St. Johns crystals are tabular, compressed along the *a* axis, making the broadest faces those of the pinacoid *a*[100], which faces are usually striated parallel to the *c* axis because of oscillatory combination with *m*[110], as in Figure 303. The wedge-shaped terminations usually display shining faces of prism *d*[101] and matte-surfaced faces of prism *k*[021] and basal pinacoid *c*[001]. Rude tabular gem crystals, partly corroded and bearing impressions suggesting acicular or bladed mineral formerly in contact, pos-

sibly wollastonite, come from Burma and reach 3″ in length. Inclusions of black spinel are observed in Arizona specimens.

Physical Properties: Brittle; conchoidal fracture, often as perfect as glass. Cleavage sometimes distinct parallel to b[010]. H. 6. G. (forsterite) 3.22-4.39 (fayalite). *Peridot* typically G. 3.34.

Optical Properties: Transparent to translucent. Usually some shade of green, from an intense pure yellow-green to olive-green to greenish-brown; also white (forsterite), brown and black. Streak colorless. Luster vitreous to oily. Weak silky to aventurescent sheen in some crystals. Biaxial positive (forsterite) to negative (fayalite). R.I., rising with substitution of iron for magnesium, forsterite: $\alpha = 1.635$, $\beta = 1.651$, $\gamma = 1.670$; gem *peridot* typically: $\alpha = 1.654$, $\beta = 1.671$, $\gamma = 1.689$; fayalite: $\alpha = 1.835$, $\beta = 1.877$, $\gamma = 1.886$. Birefringences also rise with iron, from 0.035 to 0.051. Pleochroism weak, usually in shades of basic color; more distinct in brown varieties.

Distinctive Features and Tests: Tabular striated crystals; lack of strong pleochroism. Infusible; cracks at white heat but does not lose original color. Easily soluble in hot hydrochloric acid with formation of silica jelly.

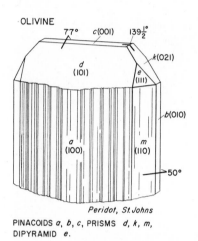

PINACOIDS a, b, c, PRISMS d, k, m, DIPYRAMID e.

FIG. 303.

Occurrences: Predominantly in dark, silica-poor, igneous rocks as gabbro, peridotite, basalt; the rock dunite is largely olivine. Also in some metamorphosed limestone, and in the famous locality on St. Johns Island, in vein-like deposits in brecciated serpentine. Rude crystals of forsterite (*boltonite*) to 1″ occur in marble at Bolton, Worcester County, Massachusetts, and in well-formed crystals in the Crestmore quarries, Riverside, Riverside County, California. Small grains of gem quality peridot occur on Timothy Mountain, near Lac LaHache, British Columbia; similarly in Kilbourne Hole, near Afton, Dona Ana County, New Mexico, and at several localities in the San Carlos Indian Reservation, Gila County, and at Buell Park, Apache County, Arizona. Occasionally, attractive matrix specimens are obtainable, showing segregations of peridot to 6″ or more in diameter, enclosed in gray basalt (Figure 304). Sharp micro crystals (fayalite) occur with cristobalite in lithophysae in obsidian, near Coso Hot Springs, Inyo County, California. As gemmy grains in some cinder cones of the Hawaiian Islands. Large crystals, completely altered to serpentine, but preserving considerable sharpness, occur at Snarum, Norway. Forsterite occurs in the ejected material of Monte Somma, Vesuvius, Italy; also in the Alban Hills near Rome. The world's finest crystals, and highly prized by single-crystal collectors, are the sharp, transparent, tabular crystals from St. Johns Island in the Red Sea, Egypt, opposite the Egyptian port of Bernice. Beautiful yellow-green crystals to 4″ diameter have been recorded but the amateur is fortunate if he can obtain good specimens of 1″ to 1½″, purse permitting. Very large crystals, some to 3″, occur in Bernardino Valley, Mogok district, Burma, and yield fine faceted gems. Many contain very small elongated platey inclusions, sometimes causing chatoyancy or aventurescence; the color is excellent yellow-green. This occurrence is believed to be in marble. Perhaps the sharpest crystals of species in this series occur in volcanic ash of recent eruptions as perfect but minute glassy individuals suitable only for micromounting; a good locality is debris littering the slopes of extinct Koko Volcano near Honolulu, Oahu, Hawaii.

FIG. 304. Segregation nodule of olivine (*peridot*) in pale gray vesicular basalt from the San Carlos Indian Reservation, Gila County, Arizona. Size 3″ by 3″. Each nodule of olivine contains numerous small grains which are released to the soil when the rock weathers and are collected for the gem material which they contain.

Miscellaneous Species of the Olivine Group

Tephroite—(Mn,Mg,Zn)$_2$(SiO$_4$)

A rare silicate, found in abundance only at Franklin-Ogdensburg, Sussex County, New Jersey, where it sometimes forms ½″ blocky crystals in marble; tĕ′frō ĭt. Crystals are bluish-green in daylight but appear pink in artificial light. Massive tephroite is gray, brown, reddish-brown, and flesh-red; it is easily mistaken for reddish or brownish willemite (*troostite*) but can be distinguished by its right-angle cleavages.

Larsenite—PbZn(SiO$_4$)

Also rare, but has been found in white acicular micro crystals in vugs at Franklin, New Jersey; lăr′sĕn ĭt. Calcium larsenite is chemically similar except that calcium substitutes for lead; it occurs only massive with larsenite at Franklin, and is noted for its very strong lemon-yellow fluorescence under ultraviolet, rivaling if not exceeding the similar fluorescence of willemite.

Monticellite—CaMg(SiO$_4$)

A rare mineral formed by contact metamorphism in impure limestones; mŏn′tĭ sĕl′ĭt. In poor prismatic crystals to 3″ with garnet and diopside at Crestmore, Riverside County,

California; more commonly massive or in grains in blue marble at this and other localities nearby.

Phenakite Group

Phenakite—Be₂(SiO₄)

Name: From the Greek, *phenakos*, "deceiver," because once mistaken for quartz; fēn'ă kīt.
Crystals: Hexagonal. Predominantly in well-developed crystals ranging from disk-like individuals compressed along the *c* axis, and terminated by low rhombohedral faces (Brazil), to stubby, prismatic crystals (Colorado), as in Figure 305. Also in rounded, poorly-formed crystals (Urals). Colorado crystals sometimes grow in twins elongated parallel to the *c* axis with prisms striated in the same direction, and with reentrants on the terminations.

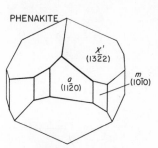

PHENAKITE

x'
(132̄2)

a
(112̄0)

m
(101̄0)

FIRST ORDER PRISM *m*, SECOND
ORDER PRISM *a*, THIRD ORDER
RHOMBOHEDRON *x'*.

FIG. 305.

Physical Properties: Brittle, but rather tough. Conchoidal fracture. Poor cleavage parallel to prism *a*[112̄0]. H. 7½-8. G. 2.97-3.00.
Optical Properties: Transparent; also translucent. Colorless or white, but also tinged bluish by inclusions (Colorado). Streak colorless. Luster vitreous. Uniaxial positive. R.I., O = 1.651-1.667, E = 1.653-1.668. Birefringence 0.015.
Chemistry: Beryllium silicate.
Distinctive Features and Tests: Flattened disk-like crystals from other localities commonly show vertical striations instead of horizontal striations as on quartz. Harder than quartz; lacks good cleavage and thus distinguished from topaz. Infusible. Insoluble.
Occurrences: In granitic pegmatites and miarolitic pegmatites in granite; also in mica schist with emerald and chrysoberyl (Urals). Small crystals have been found in pegmatite at Lords Hill, Oxford County, Maine; and with topaz on Baldface Mountain, Carroll County, New Hampshire. Crystals to 2″ from Morefield Mine, Winterham, Amelia County, Virginia. From miarolitic cavities in granite atop Mount Antero, Chaffee County, Colorado, in crystals to 1⅜″ length, as singles or twins; most are white or colorless, but some tinged blue because of inclusions, or sometimes faintly yellow. Small colorless crystals accompany quartz and amazonite in some pegmatite pockets of the Pikes Peak red granite area, west of Colorado Springs, Colorado. The finest specimens are large, very brilliant, tabular, clear and colorless crystals from San Miguel di Piracicaba, Minas Gerais, Brazil; crystals reach 3″ diameter and form clusters to 7″ across; this locality appears to be exhausted. The largest known crystals, usually rude translucent white prisms to 1″ wide and 6″ to 8″ long, occur in pegmatite at Kragerö, Norway. Crystals to 4″ diameter, with one recorded as weighing more than a pound, occur in biotite schist, associated with emerald and chrysoberyl (*alexandrite*), in emerald mines along the Takowaya River, near Sverdlovsk, Urals, U.S.S.R. Many crystals are rounded but some display good faces. Occasionally, clear crystals were cut into gems. Chlorite-coated crystals to 1″ length, were found in vugs along the Rientallucke, Fellital, Switzerland.

Willemite—Zn₂(SiO₄)

Name: After Willem I, king of the Netherlands; wĭl'ĕm ĭt.
Varieties: Very large, red-brown crystals from Franklin, New Jersey, are called *troostite* although there are no important differences between this and other willemite.
Crystals: Hexagonal. Stout to slender hexagonal prisms of *a*[112̄0] terminated by pina-

coid $c[0001]$ and several rhombohedrons as $r[10\bar{1}1]$ and $e[01\bar{1}2]$, as in Figure 306. Larger crystals usually rude, with curved prismatic faces, rounded terminations, and pitted surfaces. Micro crystals sharp and rich in forms. Also compact fibrous, granular. Blood-red or silvery platey inclusions present in some massive types, causing feeble chatoyancy. ***Physical Properties:*** Brittle; uneven to conchoidal fracture. Poor cleavages parallel to prism $a[11\bar{2}0]$ and pinacoid $c[0001]$. H. $5\frac{1}{2}$. G. 3.89-4.19.

Optical Properties: Translucent to transparent. Colorless, pale green, yellow, orange; ordinary massive material commonly apple-green, olive-green, brownish-red, brown. *Troostite* red-brown, chocolate-brown. Streak colorless. Luster resinous to vitreous; also slightly aventurescent or chatoyant. Uniaxial positive. R.I., O = 1.691-1.694, E = 1.719-

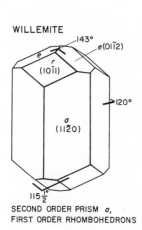

WILLEMITE

SECOND ORDER PRISM *a*,
FIRST ORDER RHOMBOHEDRONS

FIG. 306.

FIG. 307. Hexagonal prisms of brown willemite, variety *troostite*, in calcite, garnet and franklinite matrix, from Franklin, Sussex County, New Jersey. Size 6″ by 5″.

1.725. Birefringence 0.028-0.031. Dichroism weak. Strong yellow-green fluorescence under ultraviolet; phosphorescence also strong, sometimes lasting 15 minutes; absent in reddish or brownish specimens.

Chemistry: Zinc silicate. Considerable manganese and some iron replace zinc. Manganese in very small amounts activates fluorescence.

Distinctive Features and Tests: Vivid fluorescence; resinous luster; associations. Franklin granular zinc ore often also displays vivid red fluorescence of associated calcite. Nearly pure material infusible; reddish types from Franklin fusible with difficulty at $4\frac{1}{2}$-5. Soluble in hydrochloric acid.

Occurrences: Abundant as a primary and secondary mineral in the unique deposits of Franklin and Ogdensburg, Sussex County, New Jersey; occurs elsewhere as micro crystals in oxidized zones of ore deposits containing zinc. The only significant cabinet specimen localities are those of New Jersey, where magnificent masses, large crystals of *troostite* (Figure 307), some to 2″ by 6″, and also smaller crystals of other varieties, occur in marble, associated with black franklinite, red zincite, garnet, and calcite. The finest crystals, some 3″ by ¼″ and perfectly transparent, occurred in crevices, and from them, very fine gems have been cut. Much vivid light green material is also gemmy but is noted more for its exceptional fluorescence, noticeable even in broad daylight. Despite abundance, most material is massive and even poorly-formed crystals are now rare and highly prized. Micro crystals forming thin, pale blue crusts and associated with wulfenite, vanadinite, quartz, and mimetite, occur in the Mammoth-St. Anthony Mine, Tiger, Pinal County, and in the Apache Mine, near Globe, Gila County, Arizona. In micro crystals at Altenberg, near Moresnet, Belgium; at Mindouli, Congo, and in other zinc deposits. Willemite forms such small inconspicuous crystals that probably much is overlooked and it may be a more common mineral than is supposed.

Aluminum Silicate Group

Andalusite—Al$_2$O(SiO$_4$)

Name: After the Spanish province of Andalusia; ăn′dă lū′sīt.

Varieties: Highly impure prismatic crystals of rude form with cross-like inclusions, *chiastolite;* kī ăs′tō līt. The name is from the Greek, *chiastos,* for "marked with an X," in allusion to the cross-like figures made by the inclusions when crystals are cut and polished (Figure 309).

Crystals: Orthorhombic. Predominantly in rude crystals seldom showing good forms except in the transparent, gem-quality specimens from Brazil. The latter are prismatic, square or rectangular in cross-section, and always etched. *Chiastolite* crystals are very rough and approximate the shape of cigars. Common forms are prism *m*[110], faces of which are nearly at right angles to each other, and the pinacoid *a*[100] which appears on some Brazilian crystals as a narrow face between a pair of *m*[110] faces, as in Figure 308. Crystals elongated parallel to the *c* axis. Hair-like inclusions in some Brazilian gem crystals.

Physical Properties: Brittle but tough. Fracture conchoidal in gem crystals to uneven in other types. Cleavage perfect parallel to *m*[110] (Brazil), but far less so in ordinary crystals. *Chiastolite* shows no cleavage but is inclined to fracture readily across the *c* axis. H. 7½; *chiastolite* nearer 3½-4½. G. 3.1-3.2; usually lower in *chiastolite.*

Optical Properties: Usually translucent; rarely transparent (Brazil). Brown, red-brown, pink, white, and white or pale gray with black inclusions (*chiastolite*); gem material pale olive or gray green, also pink. Streak colorless. Luster vitreous. Biaxial negative. R.I., indices generally in the range $\alpha = 1.634\text{-}1.641$, $\gamma = 1.644\text{-}1.648$; gem material (Brazil, Ceylon) $\alpha = 1.633$, $\gamma = 1.644$. Birefringence 0.007-0.011. Remarkable pleochroism in gem crystals, easily visible without a dichroscope, such that crystals appear greenish through prism faces but deep brownish-red near the ends of the *c* axes. In the dichroscope, the following are observed (Brazil): reddish-brown, olive-green, pale straw yellow.

Chemistry: Aluminum silicate. Manganese is sometimes present.

Distinctive Features and Tests: Internal figures of *chiastolite.* Easily-visible pleochroism of gem crystals and squarish shapes distinguish from tourmaline. In some pegmatites, pink columnar aggregates of andalusite may be mistaken for translucent *lithia tourmaline,* but broken pieces display nearly right-angle cleavages which tourmaline does not have. Infusible. Insoluble.

Occurrences: Mainly in contact-metamorphosed shales, commonly as *chiastolite* (Figure 309). Also in other metamorphic rocks, especially muscovite schists, and in some pegmatites. There are many localities for good specimens of *chiastolite,* perhaps the best being

Lancaster, Worcester County, and Westford, Middlesex County, Massachusetts; also from Kern, Mariposa, and Madera counties, California; and abroad, in exceptional specimens from near Bimbowrie, South Australia. Prismatic crystals occur at Standish, Cumberland County, Maine; in large crystals near Leiperville and Upper Providence, Delaware

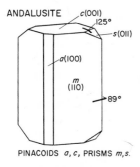

ANDALUSITE

PINACOIDS *a,c*, PRISMS *m,s.*

FIG. 308.

FIG. 309. Andalusite, variety *chiastolite*. Prismatic crystals enclosed in chlorite schist, from Chowchilla, Madera County, California. The large crystal at the upper left has been ground and polished to show the typical cross markings of this variety. Untouched fractured prisms appear at top right and lower right. Size 3½″ by 3″. *Specimen in Elbert H. McMacken Collection, Ramona, California.*

County, Pennsylvania, and lately, in rude brown crystals with some smooth faces from the Cargo Muchacho Mountains, Imperial County, California. Gem crystals, commonly of good form but with etched surfaces, and also rolled pebbles, occur near Santa Tereza, Espirito Santo, and Arassuahy, Minas Gerais, Brazil; crystals to 2″ length were found but most are less than 1″. Gem pebbles occur in the gravels of Ceylon. When properly cut, faceted gems show distinct spots of red on opposite sides with greenish or yellowish between.

Sillimanite—$Al_2O(SiO_4)$

Aluminum silicate; sĭl′ĭ măn ĭt. Usually in compact fibrous aggregates of nearly jade-like toughness, and only rarely in isolated crystals. The alternate name, fibrolite (fī′brō līt), is commonly used in England. Interesting dark brown crystals to several inches in length and capable of being cut into sharp catseye gems have been found recently in Oconee County, South Carolina. Perfectly transparent rolled pebbles of pale blue hue, and very rarely, crystals with distinct faces, occur in the gem gravels of Mogok district, Upper Burma. Ordinary fibrous sillimanite occurs in white, brownish or greenish silky aggregates in strongly metamorphosed schists and gneisses; occasionally it is chatoyant.

Kyanite—Al$_2$O(SiO$_4$)

Name: From the Greek, *kyanos*, "dark blue," referring to the common color; kī'ă nīt.
Crystals: Triclinic. Predominantly in bladed crystal aggregates, and less commonly, in distinct euhedral crystals. Bladed aggregates diverge in sprays, one end usually being thicker. Crystals elongated along the *c* axis; the largest faces nearly always the pinacoid

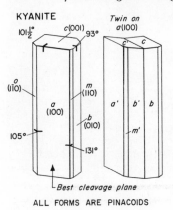

KYANITE

Twin on a(100)

Best cleavage plane

ALL FORMS ARE PINACOIDS

FIG. 310.

a[100], on which the perfect easy cleavage lies, and which usually displays a pearly luster because of development of numerous cleavages (Figure 310). The next broadest faces are ordinarily the pinacoid *b*[010], on which the next best cleavage lies. The acute edge made between *a* and *b* faces is truncated by the pinacoid *m*[1Ī0]. Crystals often filled with fibrous white inclusions parallel to the edge between faces of *a*[100] and the pinacoid *c*[001]. Good faces of the latter form seldom appear. Repeated twinning common parallel to *a*[100], evidenced by numerous reentrant striations on the *b*[010] face parallel to the *c* axis.

Physical Properties: Flexible rather than brittle, crystals usually bending somewhat, then breaking across the *c* axis with development of a fibrous to stepped surface. Cleavage perfect parallel to *a*[100] and easily developed; less perfect and much more difficult to develop parallel to *b*[010]; a stepped parting parallel to *c*[001]. The hardness is remarkable to say the least, the direction parallel to the *c* axis on the broadest faces, *a*[100], tearing easily before the point of a steel needle, and giving a hardness of about 4½, while the direction at right angles is so hard that the needle slides without engaging; this direction is about 6½. Slightly harder on *b*[010] and *c*[001], perhaps as much as 7. G. 3.65-3.69.

Optical Properties: Transparent to translucent. Usually some shade of pale to deep blue with overtones of violet; also green, colorless, white, grayish. Commonly color-zoned with blue cores and greenish or colorless exterior zones. Streak colorless. Luster vitreous to pearly. Biaxial negative. R.I., $\alpha = 1.715$, $\beta = 1.726$, $\gamma = 1.732$. Birefringence 0.017. Distinct pleochroism; blue crystals yield dark violet blue, blue, and colorless; green crystals yield pale green, colorless. Suitable material has been cut into gems.

Chemistry: Aluminum silicate.

Distinctive Features and Tests: Bladed habit; cleavages; differences in hardness. Infusible. Insoluble.

Occurrences: Confined to schists and gneisses, and to quartz veins or pegmatites cutting them. Abundant in New England and in the Appalachian Mountains. Fine pale blue blades to 3″ length, without terminations, associated with white quartz, apatite, garnet, biotite and plagioclase, from Baker Mountain near Farmville, Prince Edward County, Virginia. Exceptional terminated stubby crystals of green color to 3½″ long, occasionally clear enough to yield faceted gems, from Yancey County, North Carolina. Blue blades at the rutile locality on Graves Mountain, Lincoln County, Georgia. Splendid blue blades to 6″ long, sometimes of rich sapphire blue and clear enough for gems, from

Capelinha, Minas Gerais, Brazil, and green crystals elsewhere in this state (Figure 311). Classic specimens, displaying slender pale blue transparent prisms with rich brown transparent staurolite crystals in white paragonite mica schist, have been obtained for many years from the Pizzo Forno locality in Switzerland (Figure 311). Large bladed masses,

FIG. 311. Kyanite. *Left:* dark blue bladed crystal from Minas Gerais, Brazil. *Right:* blue crystals with slender brown staurolite crystals in paragonite mica schist from Pizzo Forno, Switzerland. Size 3″ by 2″.

often quite clear, occur near Sultan Hamud, Tanganyika. Clear fragments occur in the gem gravels of Ceylon. Ordinary bladed kyanite is easily available, but sharp crystals are decidedly rare. Specimens from Switzerland are very attractive, and are standard cabinet items.

Topaz—$Al_2(OH,F)(SiO_4)$

Name: From the Greek, *topazion,* a name formerly applied to some gemstone whose identity has now been lost; tō′paz.

Varieties: The name *pycnite* is sometimes applied to columnar massive material.

Crystals: Orthorhombic; common. Mostly in stubby to medium-long prismatic crystals of diamond-shaped cross-section. Less commonly long prismatic, or in coarse to fine granular masses, and columnar aggregates. Crystals range in size from very small to some of several hundred pounds weight (Brazil). Usually the faces of prisms $m[110]$ and $l[120]$ are present, with terminations consisting of combinations of the dipyramid $o[111]$ and prisms $f[011]$ or $y[021]$ as in Figure 312. The basal pinacoid $c[001]$ is also common, and in some crystals, is large. Very commonly, faces of certain forms are etched or corroded differently, sometimes perfectly smooth faces being found adjacent to a set of another form which are rough and rounded. Such corrosion is not consistent but varies in crystals from different deposits. Topaz is susceptible to deep corrosion and jagged masses are found with no trace of original faces (Figure 313). Gas and liquid inclusions abundant in many crystals; healed cleavages are often marked by sharp planes of inclusions.

Physical Properties: Brittle; fracture conchoidal. Practically all broken crystals display the perfect cleavage parallel to the basal pinacoid $c[001]$; this cleavage is easily developed and appears in traces even on waterworn pebbles. H. 8. G. 3.5-3.6, rising with fluorine.

Optical Properties: Transparent to translucent. Usually colorless to faintly blue; also medium blue, like fine aquamarine, and various shades of yellow to rich yellow- or brownish-orange; rarely, pink or violet-pink. Many brown crystals fade when exposed

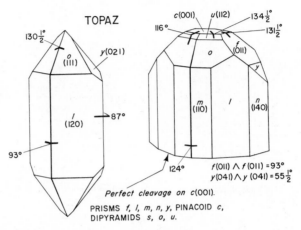

FIG. 312.

to daylight, turning colorless or pale blue. Color zoning common, usually as zones parallel to prism faces m and l. A rich brown color can be induced by x-ray irradiation but is not permanent; heat-treatment of orange crystals from Ouro Preto, Brazil, results in pink. Streak colorless. Luster vitreous. Biaxial positive. R.I., $\alpha = 1.607\text{-}1.630$, $\beta = 1.610\text{-}1.631$, $\gamma = 1.618\text{-}1.638$. Birefringence 0.008-0.011. Pleochroism distinct; in heat-treated pink gems, pink and colorless; in natural pink: pink and shades of orange-yellow; ordinary yellow crystals: faint yellow and deeper yellows; blue gives blue and colorless.

Chemistry: Aluminum hydroxyl-fluorine silicate.

Distinctive Features and Tests: Stubby crystals with wedge-shaped terminations; prominent single cleavage; hardness. Infusible. Insoluble.

Occurrences: The finest specimens are obtained from granitic pegmatite cavities, or from miarolitic cavities in granites. Smaller crystals occur in high-temperature quartz veins or in cavities in rhyolite. In Maine pegmatites, as at Lords Hill, Oxford County, also Fisher Quarry, Topsham, Sagadahoc County. In New Hampshire, the granitic rock in the vicinity of Conway, Carroll County, and extending toward Baldface Mountain has provided fine crystals, some pale blue and sherry, from miarolitic cavities; the summit of South Baldface, the Lovejoy gravel pits (crystals to $3\frac{1}{2}''$), and Redstone Quarry near Conway have all been productive of good examples associated with smoky quartz, albite, and microcline. A few fine crystals were found in the old tungsten mine at Trumbull, Fairchild County, Connecticut. In pale blue poorly-formed crystals from Morefield pegmatite, Amelia County, Virginia. A considerable region of granitic rock containing small to large miarolitic cavities centering on the Llano Uplift in Mason County, has produced many hundreds of pounds of fine topaz crystals, usually pale to medium blue, some to 4″ long. Much material was taken from stream bed gravels but also from outcropping pegmatite bodies. The larger proportion of crystals are clear and can be cut into gems, but sharp crystals are uncommon. Pegmatites in the Pikes Peak red granite west of Colorado Springs, covering a large area in Teller, Douglas, Park, and El Paso counties, have furnished sherry specimens, and also some blue crystals. Devil's Head in Douglas County, Colorado, has furnished excellent etched gem-quality blocky crystals to nearly 5″ length. Small but textbook-perfect crystals, colorless or sherry, occur with quartz, hematite, pseudobrookite, fluorite, calcite, the rare mineral bixbyite, garnet, and very rarely, small tabular hexagonal crystals of a unique red beryl, in cavities in pale gray rhyolite on Thomas Mountain, Juab County, Utah. Magnificent pale blue crystals to 3″ length occur in the Little Three pegmatite near Ramona (Figure 313), and at the

FIG. 313. Topaz crystals. *Left:* a highly corroded colorless specimen from Tres Barras Mine, near Teofilo Ottoni, Minas Gerais, Brazil. Size 3¼″ by 2¾″. *Center:* pale blue crystal from the Little Three Mine, Ramona, San Diego County, California. The large triangular face is y(041), and is matched by a like face on the reverse side of the crystal. *Right:* pale blue crystal from Sakangyi, Burma. The cleavage plane parallel to the basal pinacoid c[001] is readily apparent in the left and right specimens.

Ware Mine, Aguanga Mountain, San Diego County, California, associated with cleavelandite, tourmaline, lepidolite, etc.

Practically all gem topaz of brownish-orange or yellow-orange is now obtained from unique deposits near Ouro Preto, Minas Gerais, Brazil, imbedded in goethite-stained kaolin with quartz crystals; fine slender terminated crystals from this deposit are found in almost every collection but matrix specimens are very rare. Numerous other localities in Minas Gerais and other Brazilian states have provided colorless, pale to medium blue, and yellowish or sherry topaz crystals in great abundance, including some which are completely etched or corroded as shown in Figure 313. Specimens over 400 pounds weight have been recorded. Magnificent topaz matrix specimens, easily the best in existence, came from pegmatites near Mursinsk in the Ural Mountains, north of Sverdlovsk, U.S.S.R., as blocky flawless blue crystals perched on matrix of cleavelandite, microcline, smoky quartz and large fan-shaped books of mica. Many specimens were coated with cookeite mica, which was painstakingly removed by scraping. Crystals to 3″ by 2″ were common. In splendid yellow crystals from the Adun-Tschilon Mountains, near Nerchinsk, Transbaikalia, U.S.S.R., many to 5″ to 6″ length, sharp and transparent. Also in gold-bearing gravels of Sanarka district, Chkalov, U.S.S.R. A classic locality at Schneckenstein, near Muldenberg, Vogtland, Germany, furnishes small pale yellow crystals to ¾″ diameter from cavities in topaz-rock with quartz and black tourmaline. Beautiful clear crystals in miarolitic cavities in granite in the Mourne Mountains, County Down, Ireland; also in large crystals, some to a pound in weight, from pegmatites in the Cairn Gorm Mountain and the granitic rocks of Banffshire and Aberdeenshire, Scotland. Topaz is relatively uncommon in the Madagascar pegmatites, but some good crystals to 3″ diameter were obtained from near Ampangabe. Sharp crystals from ¾″ to as much as 5″, also water-worn specimens, occur in abundance in the cassiterite gravels of Jos, Nigeria. Very fine sharp colorless or pale blue crystals occur in pegmatites near Kleine Spitzkopje, South-West Africa. Large colorless to pale blue crystals, and some sherry specimens, in pegmatite at Sakangyi, Burma. Very fine crystals to 3″, from Japan in Omi and Mino provinces. Small gemmy crystals from Pine Creek, Northern Territories, and from Mt. Cameron, Bell Mount, Flinders Island, Victoria, Australia.

The largest topaz crystals are certainly those from Minas Gerais, Brazil, and recently, a considerable number of very pale yellowish crystals, some sharp but others severely corroded, have come from Tres Barras Mine, near Teofilo Ottoni, sometimes in individuals weighing nearly 40 pounds. However, the best cabinet specimens are those on matrix, but are very difficult to obtain because topaz cleaves so easily that crystals often part company with matrices. Taking first honors for matrix specimens are those from the Urals, especially from near Mursinsk. Rarely, good matrix specimens are available from the Little Three Mine near Ramona, California. Small but fine matrix specimens are sometimes obtainable from the Schneckenstein locality in Germany.

Staurolite—$Fe_2Al_9O_7(OH)(SiO_4)_4$

Name: From the Greek, *stauros*, "cross," because commonly twinned in cross-like forms; stôr′ō lĭt.

Crystals: Orthorhombic. In single or twinned crystals, or in clusters of several crystals, embedded in schist or gneiss. Single prismatic crystals show pseudo-hexagonal cross-sections formed by broad faces of the prism $m[110]$ and slightly narrower faces of the pinacoid $b[010]$. Terminations are usually broad faces of pinacoid $c[001]$, modified by small triangular faces of the prism $r[201]$. Twins of two types are common; the most abundant examples are furnished by a pair of individuals twinned on plane [231], crossing at nearly 60° angles; the second type, nearly right angle twins, is formed by a pair of crystals twinned on plane [031], as shown in Figure 314. The latter furnish the "cross stones" or "fairy crosses," often sold as good luck charms.

Physical Properties: Brittle; uneven to conchoidal fracture; partly-altered specimens display an earthy fracture surface. Cleavage poor along $b[010]$. H. 7 in unaltered specimens, to as low as 3-4 in altered crystals. G. 3.7-3.8; lower in altered crystals.

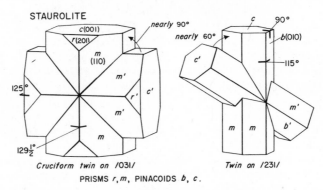

STAUROLITE

Cruciform twin on /031/

PRISMS *r, m*, PINACOIDS *b, c*.

Twin on /231/

FIG. 314.

Optical Properties: Opaque to translucent; rarely transparent. Rich brown with a tinge of red or orange in very small crystals. Streak colorless. Luster vitreous to dull. Biaxial positive. R.I., variable, generally in the range: $\alpha = 1.739$-1.747, $\beta = 1.744$-1.754, $\gamma = 1.750$-1.762. Birefringence 0.011-0.015.

Chemistry: Iron-aluminum hydroxyl silicate. Some iron is replaced by magnesium or manganese, and some aluminum by ferric iron.

Distinctive Features and Tests: Unmistakable twins. Any brown prismatic crystal enclosed in schist, phyllite, or gneiss may be tourmaline, andalusite or staurolite, but crystal form distinguishes staurolite from tourmaline, while andalusite commonly forms much cruder crystals and the typical enclosures of the *chiastolite* variety. Infusible. Insoluble.

Occurrences: Exclusively in metamorphic, aluminum-rich rocks as schists, gneisses, phyllites, associated with muscovite or paragonite mica, kyanite, sillimanite, garnet and tourmaline. Commonly encloses small almandite garnets. Abundant in mica schist at many places in New England as at Windham, Cumberland County, Maine, where it occurs in crystals to 2″; at Sugar Hill, Franconia, and Lisbon, Grafton County, New Hampshire, etc. In twins, in Patrick County, Virginia, and as very fine singles and twins to 1¼″ from Fannin County, Georgia. Fine twins from near Pilar, Rio Arriba County, New Mexico (Figure 315). Sharp single crystals, often slender prismatic and sometimes transparent,

FIG. 315. Staurolite crystals from Pilar, New Mexico, showing two modes of twinning. Crystals are obtained from within mica schist, some of which still adheres to specimens. Largest crystal about 1″ diameter. The inset photo shows a small single crystal from Fannin County, Georgia. *Photo courtesy Scott Williams Mineral Company, Scottsdale, Arizona.*

and occasionally cut into very small gems, occur with blue kyanite in paragonite schist near Pizzo Forno, Switzerland. The premier localities for cabinet specimens are Pizzo Forno, Switzerland, and for singles and twins, Fannin County, Georgia, and Pilar, New Mexico.

Garnet Group

Among the most popular collection species are the several isometric silicates known generally as garnets but given specific names according to chemical composition. The group is arbitrarily divided into six species, but extensive substitutions take place. These substitutions affect compositions and properties and make it difficult, in many instances, to be sure from the outward appearance alone where any given specimen falls within the group. However, certain species characteristically occur in certain environments, and if this is known, it is often possible to label specimens accordingly without being far wrong. Thus the bright red garnet granules found in certain volcanic rocks are usually pyrope; orange-red or brown-orange crystals found in metamorphosed limestone are grossularite, etc. The theoretically pure garnets marking divisions within the group, along with important properties, are listed below.

Species	Formula	Specific gravity	R.I.
Almandite	$Fe_3Al_2(SiO_4)_3$	4.32	1.705
Pyrope	$Mg_3Al_2(SiO_4)_3$	3.58	1.83
Spessartite	$Mn_3Al_2(SiO_4)_3$	4.19	1.80
Grossularite	$Ca_3Al_2(SiO_4)_3$	3.59	1.735
Andradite	$Ca_3Fe_2(SiO_4)_3$	3.86	1.895
Uvarovite	$Ca_3Cr_2(SiO_4)_3$	3.78	1.86

Rarely does any example of garnet furnish an ideal chemical composition to the analyst, and the vast majority of specimens show more or less substitution, placing them between a pair of the species shown above. Series exist between pairs of theoretically pure species, as for example, the complete series which extends between almandite and pyrope, between almandite and spessartite, and between grossularite and andradite. However, only limited substitution takes place between the calcium garnets (grossularite, andradite), on the one hand, and the iron, manganese garnets (pyrope, almandite, spessartite) on the other.

All garnets are fairly hard but brittle and inclined to shatter into small cubic fragments. Specific gravities are moderately high. They are resistant to alteration and commonly appear as raised knobs on enclosing rocks where the rocks are exposed to weathering. Garnet is a very common constituent of sand. The similar crystal structure in all species results in the formation of ball-like individuals or spherical masses when completely enclosed in rock. Crystal forms are primarily the dodecahedron d[110] or trapezohedron n[211]. Many crystals are one form or the other, but combinations occur, resulting in beautifully faceted crystals, as shown in Figure 316. The hexoctahedron s[321] commonly bevels edges of the dodecahedron, while the latter may be truncated by the trapezohedron. In some crystals, all three forms are present.

Almandite—Fe₃Al₂(SiO₄)₃

Name: After Alabanda, a town of ancient Caria (Asia Minor); ăl′măn dīt. The alternate name, *almandine,* is still much used; ăl′măn dēn.

Crystals: Isometric; abundant. Often finely-formed but also in crude spherical single crystals extensively shattered internally which fall apart when removed from matrix. Oriented inclusions occur in some crystals, affording four-rayed star stones. Also in disseminated grains, granular material.

Physical Properties: Brittle; conchoidal fracture. No cleavage but shattered crystals usually divide into nearly cubic fragments. H. $7\frac{1}{2}$. G. 3.95-4.32.

Optical Properties: Transparent to translucent. Very dark red, brownish-red, or purplish-red. Streak colorless. Luster vitreous. R.I. $n = 1.78$-1.83.

Distinctive Features and Tests: Color; crystals. Fuses at 3 to black magnetic globule. Insoluble.

Occurrences: Typically in metamorphic siliceous rocks as schist and gneiss; common in border zones of granitic pegmatites, sometimes in large crystals. A great many localities are known furnishing fine specimens and only a few can be mentioned. Sharp lustrous crystals to 1″ diameter, from gray-brown biotite schist along the Skeena and Stikine rivers, British Columbia, and near Wrangell, southeastern Alaska (Figure 317). Fine dodecahedral crystals to $1\frac{1}{2}$″, at Roxbury, Connecticut. Dodecahedral crystals to 3″, partly altered to chlotite, from Michigamme, Michigan. Star garnets from Emerald Creek and surrounding areas, Benewah County, Idaho; crystals to 4″ diameter reported. Large green-coated dodecahedrons to 6″ diameter and 14 pounds weight from chlorite schist in Sedalia Mine, near Salida, Chaffee County, Colorado. Gem quality from deposits in Jaipur, Kishinggarh, Rajasthan, and Hyderabad, India, and from Lindi Province, Tanganyika, from the Mazibika River, Rhodesia, and from various places in Madagascar.

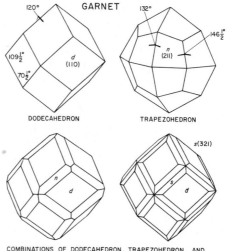

COMBINATIONS OF DODECAHEDRON, TRAPEZOHEDRON, AND HEXOCTAHEDRON.

FIG. 316.

Pyrope—Mg₃Al₂(SiO₄)₃

Name: From the Greek, *pyropos,* "fiery-eyed," in allusion to the red hue; pī'rŏp.

Varieties: Beautiful rose-red to pale purplish-red *rhodolite* (rō'dō līt) compositionally intermediate between almandite and pyrope but closer to the latter.

Crystals: Isometric. Rarely in well-formed crystals; usually in small rounded grains. Black augite or spinel inclusions occur in some crystals.

Physical Properties: Brittle; conchoidal fracture. No cleavage. H. $7\frac{1}{4}$. G. 3.65-3.80 (gem material). *Rhodolite* 3.84.

Optical Properties: Transparent. Deep red; also red tinged with orange or purple. *Rhodolite* is relatively pale violet-red. Streak colorless. Luster vitreous. R.I., $n = 1.730$-1.75 (gem material); *rhodolite* 1.76.

Distinctive Features and Tests: Rounded grains or masses; color. Fuses at 4 to a black, non-magnetic globule. Insoluble.

Occurrences: Pyrope is far less common than almandite, occurring in silica-poor igneous rocks as peridotite, or in serpentine derived from alteration of peridotite; also in some strongly metamorphosed schists and gneisses. Fine *rhodolite* occurs in the gravels of

FIG. 317. Garnet specimens. *Left:* dodecahedral crystals of brownish andradite from Stanley Butte area, Graham County, Arizona. Size 2″ by 2″. *Right:* dark red crystal in mica schist from near Wrangell, Alaska. Forms are the dodecahedron modified by the trapezohedron. Size 3½″ by 2½″.

Cowee Creek and on Mason Mountain, Macon County, North Carolina, sometimes affording attractive matrix specimens. Rounded grains of pyrope, seldom over ½″ diameter, occur in peridotites in Kentucky, in Arkansas at the diamond-bearing pipes near Murfreesboro, Pike County; and in various localities in Apache County, Arizona, San Juan County, Utah, and McKinley County, New Mexico. Similar deposits occur near Trebnitz, Bohemia, Czechoslovakia, and for many years have furnished most of the small gems used in garnet jewelry. Also in the De Beers and Kimberley mines, Union of South Africa; in the New England District of New South Wales, and from Anakie and Ruby Vale, Queensland, Australia.

Spessartite—$Mn_3Al_2(SiO_4)_3$

Name: After an occurrence in the Spessart District, Bavaria, Germany; spĕs′ĕr tīt.
Crystals: Isometric. Predominantly in corroded irregular masses covered with striations forming diamond-like patterns corresponding in angles to dodecahedral faces; less commonly, in good dodecahedral or trapezohedral crystals, but rarely lacking the striations mentioned.
Physical Properties: Brittle; conchoidal fracture. H. 7¼. G. 4.12-4.20.
Optical Properties: Transparent. Usually some shade of orange-red, orange, or yellow-orange. Streak colorless. Luster vitreous. R.I., $n = 1.79$-1.81.
Distinctive Features and Tests: Color; corroded crystals. Fuses with bubbling at 3 to a pale reddish-brown glass which is slightly magnetic.
Occurrences: The finest crystals come from cavities within granitic pegmatites; also enclosed in feldspar and quartz in such pegmatites, usually darkening in hue and grading into almandite toward the borders of any pegmatite body. Granular massive spessartite occurs also with rhodonite in some manganese-rich, metamorphic rocks and in gas cavities in some rhyolites. Large corroded crystals of spessartite to 4″ diameter, occur in albite at Rutherford Mine No. 2, Amelia, Amelia County, Virginia, furnishing beautiful specimens and some excellent gem material. Crystals to 1″, some well-formed, occur in several pegmatites of the Ramona District, San Diego County, California; the Little Three and Hercules mines are famous for specimens displaying euhedral spessartite crystals with black tourmaline and smoky quartz on albite-microcline matrix (Figure 318). Sharp, very dark,

FIG. 318. Spessartite crystal at base of black tourmaline and associated with rosettes of *cleave-landite* albite and quartz, from the Hercules Mine, Ramona, San Diego County, California. Size about 2½″ across. Forms are the dodecahedron almost equally but irregularly developed with the trapezohedron. *Photo courtesy Smithsonian Institution.*

red-brown, trapezohedrons to ¾″, occur in small gas cavities in rhyolite near Ely, White Pine County, Nevada; smaller but better crystals occur with topaz in rhyolite at Ruby Mountain, near Nathrop, Chaffee County, Colorado. Some fine spessartite has come from Pocos de Cavalos, Ceara, Brazil, and from Tsilaizina, Madagascar. Compared to other garnet species, spessartite is rare; the finest matrix specimens are those from the Ramona District, California. The corroded masses from Amelia are also very desirable.

Grossularite—$Ca_3Al_2(SiO_4)_3$

Name: From the Latin, *grossular,* "gooseberry," because some grossularite crystals are pale green like that fruit; grŏs′ū lĕr ĭt.

Varieties: A number of varietal names are applied, usually on a color basis, as *essonite* or *hessonite, cinnamon stone, hyacinth, rosolite,* etc.

Crystals: Isometric. Predominantly in unmodified dodecahedrons. Also granular to cryptocrystalline massive. Translucent crystals commonly show a peculiar granulated texture as if composed of numerous small rounded grains pressed into crystal form; transparent specimens show a swirled effect instead, much like that seen in water when a thick syrup is added and briefly stirred. These appearances are very characteristic of grossularite but may require magnification to see well.

Physical Properties: Brittle; uneven to subconchoidal fracture. Very tough in fine-grained masses. H. 7. G. 3.604-3.65; massive material, 3.36-3.55.

Optical Properties: Translucent to transparent, but the degree of transparency is seldom as high as in spessartite with which it may be confused because of color. A wide range of hues: white, rarely, colorless; yellowish, grayish or greenish white; pale to medium green, but seldom vivid; orange-red, reddish-orange, and yellowish-orange, reddish-brown; rarely, pink. Some rich green massive material has been found. Streak colorless. Luster somewhat greasy. R.I., $n = 1.734$-1.748; massive material shows an indistinct reading on the refractometer near $n = 1.72$.

Distinctive Features and Tests: Pale-colored dodecahedral crystals. Massive material not easily distinguished from massive idocrase. Fuses at 3 with slight bubbling, the melt forming a gray, non-magnetic, glassy bead.

Occurrences: Predominantly in metamorphosed impure limestone, associated with wollastonite, diopside, chlorite, idocrase and calcite. Also in some carbonate-rich schists and serpentinous rocks where crystals formed on fissure walls are commonly covered by later calcite. Only a few localities can be mentioned because of the abundance of this garnet. Fine colorless crystals to $1/4''$ diameter, lining cavities in massive grossularite, occur at Wakefield and Hull townships, Gatineau County, and in the Southwark Asbestos Pit, near Black Lake, Megantic County, Quebec. Lustrous pale brown crystals to $3''$ diameter occur in calcite-filled seams in schist near Minot, Androscoggin County, Maine, and form fine matrix groups when the calcite is removed. Fine crystals in marble, near the Calumet Iron Mine, Chaffee County, Colorado. From numerous localities in California, and

occasionally in excellent crystals from marbles and in contact zones of scheelite deposits. In the mines of Los Gavilanes, Baja California Norte, in calcite and quartz with scheelite, diopside, axinite and clinozoisite. A classic locality for white and pink dodecahedrons, some to $4''$ diameter, is near Xalostoc, Morelos, Mexico (Figure 319): many crystals are remarkably smooth-faced with sharp edges. Similar crystals occur with idocrase near Lake Jaco, Chihuahua, Mexico. Beautiful druses of transparent orange-red crystals associated with chlorite and pale green transparent diopside crystals occur on the Mussa Alp, Ala Valley, Piedmont, Italy; this is a classic locality and its specimens are seen in all important collections. Sharp dodecahedrons and trapezohedrons to $1''$, of oil-green color, at the junction of the Vilui and Achtaragda rivers, Jakutsk, Siberia, U.S.S.R. Rolled pebbles, but also some sharp dodecahedral crystals of gem quality, occur in gem gravels of Ceylon. Massive grossularite, in-

FIG. 319. Unusually large pink dodecahedron of grossularite from Rancho San Juan, Xalostoc, Morelos, Mexico. Diameter $2\frac{1}{2}''$.

tensely green in color, presumably because of the presence of chromium, and of fine lapidary quality, occurs at Buffelsfontein and Turffontein, north of Pretoria, Transvaal, Union of South Africa.

Andradite—$Ca_3Fe_2(SiO_4)_3$

Name: After Portuguese mineralogist J. B. de Andrada e Silva (1765-1838); ản'drả dīt.
Varieties: Transparent yellow-green to intense green, close in hue to emerald, *demantoid; melanite* is the black variety; *schorlomite,* a black titanium-rich andradite.
Crystals: Isometric; often in fine sharp dodecahedral crystals. Also massive. *Demantoid*

rarely occurs in well-formed crystals and characteristically contains microscopic inclusions of fibrous actinolite, spreading from a center within the crystal to form distinctive "horsetails." Practically every demantoid gem shows them. Platey inclusions oriented parallel to dodecahedral faces, imparting a strong metallic sheen, occur in Stanley Butte crystals.

Physical Properties: Brittle; fracture conchoidal to uneven. H. $6\frac{1}{2}$. G. 3.83, variable. *Demantoid* usually 3.82-3.85; *melanite* rises to 3.90.

Optical Properties: Translucent to transparent; sometimes opaque. Wide range of hues, from black, brownish-black, brown, and olive to brownish-green; also vivid emerald green, yellow, or brownish-red. Streak colorless. Luster vitreous to slightly greasy. R.I., *melanite*, $n = 1.89$; *demantoid*, $n = 1.888$-1.889.

Distinctive Features and Tests: Much andradite is olive-green or brown-green, the hues being quite different from those of grossularite. Fuses at $3\frac{1}{2}$ to black magnetic globule. Insoluble.

Occurrences: Commonly a product of metasomatic alteration of impure limestone by introduction of iron-bearing mineral matter; *melanite* and *demantoid* occur in serpentinous rocks. Large rude dodecahedrons of distinctive yellow-brown color, to 3″ diameter, of the variety formerly called *polyadelphite,* occur in zinc deposits of Franklin, Sussex County, New Jersey; nearby, on Balls Hill, very brilliant black dodecahedrons were found, one of which is recorded as nearly 7″ diameter; at Sterling Hill, Ogdensburg, as rough brown dodecahedrons as well as sharp red crystals to as much as 5″ diameter. The French Creek Mines of Chester County, Pennsylvania, are noted for dark olive-green striated trapezohedrons to $2\frac{1}{4}$″ diameter, and the iron mine at Cornwall, Lebanon County, for striated trapezohedrons to 1″ diameter, of dark brown-olive color, many quite sharp. Some fine *melanite* crystals occur in Magnet Cove, Garland County, Arkansas. Splendid groups of sharp brown to bronzy-green dodecahedrons have been collected for years from cavities in marble skarn near Stanley Butte, Graham County, Arizona; crystals reach 2″ diameter but most average about $\frac{5}{8}$″ as in Figure 317; basically all are brown but those displaying greenish hues do so because of platey inclusions in layers just beneath surfaces of crystals. Fine brown andradite crystals to 2″ diameter, with quartz, calcite and epidote, occur on Garnet Hill, on Moore Creek, near Walker, Calaveras County, California. Splendid *melanite* occurs near the benitoite locality, San Benito County, California, as brilliant crystals to $\frac{1}{4}$″ diameter forming druses in shrinkage cracks in altered serpentine, associated with calcite, chlorite, and diopside; areas to more than a foot square are covered by crystals. Sharp *melanite* dodecahedrons to $\frac{5}{8}$″ diameter occur at Monte Somma, Vesuvius, and at Frascati, Alban Hills, near Rome, Italy. *Demantoid* in rounded grains or very small sharp crystals, enclosed in asbestos, occurs in serpentine in Ala Valley, Piedmont, Italy. The celebrated locality for gem *demantoid* is in the Urals, where it occurs as rolled masses in gold washings of the Sissersk District, Nizhni-Tagil, and in serpentine on the Bobrovka River.

Kimzeyite—$Ca_3(Zr,Ti,Mg,Fe,Nb)_2[(Al,Fe,Si)O_4]_3$

A new garnet recently identified and described as very small brown dodecahedrons, modified by the trapezohedron, from Magnet Cove, Garland County, Arkansas; kĭm′zē ĭt. The occurrence is in carbonate rock in the Kimsey calcite quarry. G. 4.0 R.I., $n = 1.94$.

Uvarovite—$Ca_3Cr_2(SiO_4)_3$

A relatively rare chromium garnet noted for vivid green color and smallness of crystals. Named after Count S. S. Uvarov (1765-1855), Russian statesman and ardent amateur mineral collector; oo vȧr'ō vīt. Consistently associated with chromite as film-like druses in crevices or fissures in rock close to chromite deposits. Crystals seldom exceed ⅛″ diameter and most specimens show individuals far less in size. Fine specimens are rare and highly prized. Recently, in excellent specimens from Magog, Stanstead County, Quebec, as bright green micro crystals on bladed grayish diopside; also in outstanding specimens, with crystals to ⅛″ diameter, in serpentine, from Thetford Mines, Megantic County. Fine druses coating chromite ore, from numerous chromite mines near Riddle, Grant County, Oregon, and south into northern California counties, especially at Red Lodge Mine, Yuba County. Excellent uvarovite crystals, but also very small, formerly came from near Bissersk, Ural Mountains, U.S.S.R. The largest crystals, some to an inch in diameter, but crude and only translucent, have come from Outokumpo, Finland.

Zircon—Zr(SiO₄)

Name: Adapted from the French but of obscure origin; zĭr'kŏn.

Varieties: On the basis of color in gem varieties, *jargoon,* colorless, and *hyacinth,* bright reddish-orange or brownish-red. *Cyrtolite* is a radioactive zircon-like mineral described under *Chemistry* below.

Crystals: Tetragonal; common. Predominantly in small simple short prismatic individuals, of square cross-section, displaying four faces of prism a[100] or m[110], terminated by the pyramid p[101] as in Figure 320. Crystals usually occur as singles completely en-

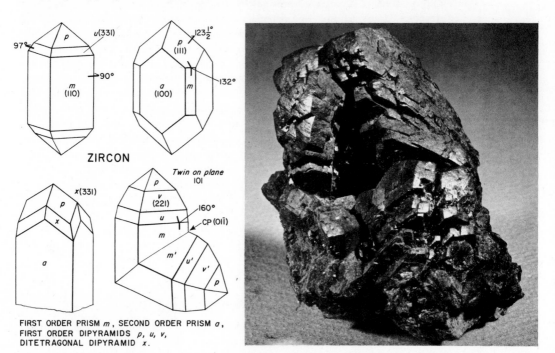

FIG. 320.

FIG. 321. Zircon, variety *cyrtolite,* from Dungannon Township, Hastings County, Ontario. Removed from within feldspar from a pegmatite. *Cyrtolite* typically forms numerous sub-parallel crystals as shown, covered by the forms indicated in the upper right sketch of Figure 320. The color is dark reddish-brown. Size 2½″ by 2″.

closed in rock; many are fairly sharp and lustrous, but others are rounded slightly, as though partly melted in a flame. The variety cyrtolite; sĭr'tō lĭt (Figure 321), habitually forms subparallel aggregates as shown. Twinning sometimes occurs on plane [112], then yielding twins diverging at an angle close to 120°.

Physical Properties: Brittle; uneven to conchoidal fracture; perfect conchoidal in *metamict* zircon (see *Chemistry* below). Indistinct cleavage parallel to prism faces of $m[110]$; cleavage absent in metamict zircon. H. $6\frac{1}{2}$-7; lower in metamict zircon, about 6. G. variable, 3.94-4.71.

Optical Properties: Transparent to nearly opaque. Most commonly some shade of brown or reddish-brown; also grayish, greenish, or violet-gray. Blue and white are produced by heat treatment of brown material. Streak colorless. Luster adamantine to resinous in metamict types. Uniaxial positive. R.I., variable, but generally in the range: $O = 1.84$-1.924, $E = 1.85$-1.992. Birefringence, variable, 0.01 to 0.059. Pleochroism weak, except in heat-treated blue, where it is strong colorless and blue. See below for further optical properties. Commonly fluorescent pale yellow-green or orange, sometimes strongly.

Chemistry: Zirconium silicate. Many crystals contain radioactive thorium, and possibly uranium, replacing part of the zirconium, resulting in destruction of normal crystal structure through bombardment by alpha particles, explaining the wide variations in physical and optical properties. For convenience, the degree of metamictization is indicated by the terms *high, intermediate* and *low* zircon. *High zircon* has not been affected by radioactivity and therefore displays normal properties, including the ability to form sharp x-ray patterns. *Low zircon* is extremely affected by radioactivity, such that its internal structure is amorphous. The *intermediate zircon* shows properties lying between the two. When alpha particle damage is extreme, crystals become black or blackish-brown, almost opaque, and fracture with perfect conchoidal surfaces of pitchy appearance; commonly, the exteriors are altered to a pale-brown, earthy crust. Presumably, *crytolite* is very altered in this respect because only interior portions of crystals show dark pitchy appearance. *Low zircon* does not give x-ray patterns unless strongly heated, but this expedient often fails; they also are much lower in specific gravity and refractive indices. *Intermediate zircon* is abundant in the gem gravels of Ceylon as transparent but always slightly clouded, rolled pebbles, principally in unattractive shades of green. The properties are compared below.

Type	Specific Gravity	Refractive Index	Birefringence
High zircon	4.65-4.71	$O = 1.924$-1.933 $E = 1.983$-1.992	0.058-0.059
Intermediate zircon	4.10-4.65 (about)	$O = 1.84$ (about) $E = 1.85$ (about)	0.010 (about)
Low zircon	3.94-4.10	$n = 1.78$-1.84	almost nil

Distinctive Features and Tests: Crystals; high specific gravities; color. In clear specimens, the large birefringence is easily detected with a magnifying glass. Can be confused with idocrase but the latter fuses. Zircon is infusible but brownish kinds lose color when subjected to red or white heat. The powder of some zircon introduced in a borax bead sometimes causes the bead to fluoresce pale green under ultraviolet although the original crystal may not. Insoluble.

Occurrences: Abundant as very small crystals in some granite, syenite and nepheline-syenite; rarely, in metamorphosed limestone and in some schists or gneisses; also in granitic pegmatite as *cyrtolite*. Brown glossy crystals of extraordinary size were once found abundantly in Ontario, notably in Dungannon Township, Hastings County, and in Sebastopol Township, Renfrew County. Crystals to 15 pounds weight have been recorded. Smaller crystals, to ½″, occur with wollastonite, pyroxene and graphite in Grenville, Argenteuil County, Quebec, and in fine crystals in apatite veins of Buckingham Township, Ottawa County. Fine 2″ crystals of lustrous black color were found in the Hill Mine, Balls Hill, near Franklin, Sussex County, New Jersey. Small but sharp brown crystals to ½″ occur in pegmatite on St. Peter's Dome, southwest of Colorado Springs, El Paso County, Colorado. Good crystals, often perfectly formed, occur at several localities near Tuxedo, Henderson County, North Carolina. Sharp, doubly-terminated crystals to ⅝″, in the Wichita Mountains, Oklahoma. Excellent crystals from many places in Norway as in the Langesundfjord district; also at Arendal, Kragerö, and elsewhere. In exceptional sharp brown crystals of high luster from near Miask, Ilmen Mountains, U.S.S.R., where stubby crystals to 1″ are common. Brown crystals to 4″ occur at Mount Ampanobe, Madagascar. In the Blantyre District of Nyassaland, Rhodesia, as brown euhedral crystals to ¾″. Abundant in the gem gravels of Ceylon, often in sharp crystals and to several inches length; similarly, in gem gravels of Mogok, Burma, and abundantly in gravels in various places in Thailand and Indochina. Remarkably fine crystals to 1¼″ are found in the Harts Range, Northern Territories, Australia, and of good quality at Uralla, Sapphire, and Inverell, New England District, New South Wales. A sharp lustrous prism measuring 4″ by 3″, weighing 4 pounds, has been reported from an unspecified locality in Australia. Crystals to 2¾″ occur at Jido, Kokai-gun, Korea. Lustrous sharp single crystals over ¾″ are rare and much prized for single-crystal collections. The Canadian matrix specimens, also those from Miask, are often attractive but now very difficult to obtain. The Canadian locality which once produced extremely large crystals is now lost.

Humite Group

A closely-related group of magnesium fluorine-hydroxyl silicates; hŭm′ĭt. Included are norbergite and chondrodite, which species occasionally afford good mineral specimens although well-formed crystals are rare; nôr′bĕrg ĭt, kŏn′drŏ dīt. Norbergite is abundant as rounded blebs of pale brownish-yellow color in the metamorphic limestone of the Sparta-Franklin district, Sussex County, New Jersey, extending into Orange County, New York. Astonishingly fine, deep brownish-red, sharp crystals of chondrodite, some to 2″ diameter, occurred at the Tilly Foster Iron Mine, near Brewster, Putnam County, New York; some crystals were perfectly transparent and were cut into garnet-like gems. Small sharp crystals of all members of this group occur in metamorphosed limestone blocks ejected from Monte Somma, Vesuvius, Italy.

Sphene—CaTiO(SiO$_4$)

Name: From the Greek, *sphen,* "wedge," alluding to the shape of typical crystals; sfēn. The alternate name, titanite, tĭ′tă nīt, is passing out of use.

Crystals: Monoclinic. Predominantly as sharp-edged tabular crystals on which the broadest faces are usually prisms $n[111]$ and $m[110]$, or prism n combined with pinacoids $c[001]$ and $a[100]$, as in Figure 322. The very large crystals from Baja California are twinned on $a[100]$, most examples being rather thin tabular, with broadest faces being an opposed pair of $a[100]$ faces. Crystals from Switzerland are remarkable for their complexity and richness of faces, and for "cross" interpenetration twins. Sphene occurs mostly as scattered crystals but also as granular masses.

Physical Properties: Brittle; conchoidal fracture. Cleavage parallel to *m*[110] sometimes distinct. H. 6. G. 3.4-3.55; gem material usually 3.52-3.54.

Optical Properties: Transparent to translucent. Some dark brown sphene is nearly opaque. Brown, yellow, green, gray; also in color-zoned crystals, commonly with paler material surrounding darker interior zones. Brown material heat-treats to brownish-

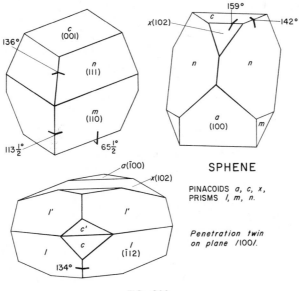

FIG. 322.

orange. Streak colorless. Luster adamantine. Biaxial positive. R.I., variable according to composition, $\alpha = 1.843\text{-}1.950$, $\beta = 1.870\text{-}1.970$, $\gamma = 1.943\text{-}2.092$. Birefringence 0.10-0.142, very large and easily observed in clear crystals with a magnifying glass. Distinct to strong pleochroism in shades of basic color.

Chemistry: Calcium-titanium silicate. Usually contains iron and sometimes chromium, niobium and sodium.

Distinctive Features and Tests: Wedge-shaped crystals; adamantine luster; strong birefringence in clear crystals; twins. Fuses at 5 only on thinnest edges of dark material; pale gem material practically infusible. Partly soluble in acids.

Occurrences: Usually as very small grains in granitic rocks, but also in cavities in granitic and metamorphic rocks, sometimes in fine crystals; in metamorphosed limestones. Euhedral crystals in cavities commonly associated with chlorite, epidote, amphibole, feldspar and quartz; associates in marbles are usually pyroxene, scapolite, calcite and apatite. Very large and very dark brown lustrous crystals with "melted" surfaces occur with apatite, calcite, scapolite, etc., in the Grenville marbles of Ontario; a notable locality is Eganville, Renfrew County, where tabular crystals to 80 pounds have been found, but the best are seldom more than 3″ to 4″ diameter. Similar sphene is widespread in the Grenville formation elsewhere in Ontario and Quebec, and into St. Lawrence County, New York, as at Rossie, Oxbow, Gouverneur, etc., sometimes in crystals to 8″ diameter. Smaller crystals, also dark brown, of tabular wedge-shaped habit, occur sparingly in the Franklin marble of Orange County, New York and Sussex County, New Jersey. Beautiful, sharp, transparent, gem-quality, twinned crystals to 2″, of bright yellow color, occurred in the Tilly Foster Iron Mine, near Brewster, Putnam County, New York. Excellent gem-quality crystals were found many years ago in Mullens Quarries, near Bridgewater, Chester County, Pennsylvania. Fine brown translucent crystals to 1″ diameter, with adularia and black tourmaline, occur near Whitehall, Jefferson County, Montana.

At present the newly discovered sources in Baja California Norte, Mexico, continue to produce gem-quality crystals of very large size from many small pegmatitic quartz-epidote veins over a wide area centered about El Alamo; a new locality 30 miles east of San Quintin is producing transparent crystals of a chromian variety to 1¼″ diameter, of vivid green color approaching that of emerald. The largest El Alamo crystals, of brownish hue, are rude, about 4″ to 5″ diameter, wedge-shaped, and about ¾″ thick; however, the sharpest crystals are considerably smaller and also better from the standpoint of gem material. Colors range from dark brown, greenish-brown, olive-green, and rarely, vivid pale green or yellowish-green. Crystals usually twinned such that the sharp edges show narrow reentrants. Associated species are clear quartz, commonly penetrated by *schorl,* also epidote, chlorite, and adularia.

Fine sharp crystals occur at numerous localities in Switzerland, but others are commonly imbedded in chlorite and therefore rough and pitted; the interpenetration twins from this country are especially prized by collectors. Sharp crystals of sphene are by no means common, perhaps the best being those carefully extracted from marble in the Canadian occurrences. The Swiss twins and matrix specimens are seldom beautiful but nevertheless are costly.

SILICATES OF UNDETERMINED STRUCTURE

Prehnite—$Ca_2Al_2(OH)_2(Si_3O_{10})$

Name: After Dutch Colonel van Prehn, who brought it to Europe in 1774; prăn′ĭt or prĕn′ĭt.

Crystals: Orthorhombic (Figure 323). Distinct crystals rare, by far the greatest quantity of this mineral appearing in botryoidal crusts composed of minute fibrous radiate crystals.

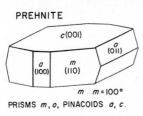

PREHNITE

m m=100°

PRISMS *m, o*, PINACOIDS *a, c*.

FIG. 323.

Sometimes crystal terminations appear to be edges of cubes, but more commonly, the crust surfaces are covered with small curved aggregate crystals shaped like the ends of blunt chisels.

Physical Properties: Aggregates brittle but tough, breaking with the fiber and exposing the cleavage surfaces parallel to *c*[001]. H. 6½. G. 2.80-2.95; gem-quality translucent material usually in range 2.88-2.94.

Optical Properties: Translucent. Usually medium to pale green or yellow-green, sometimes nearly white; rarely, yellow or dark gray-green, the latter hue owing to inclusions of byssolite. Streak colorless. Luster vitreous to pearly on fracture surfaces. Biaxial positive. R.I., $\alpha = 1.616$, $\beta = 1.626$, $\gamma = 1.649$; a refractometer reading on ordinary fibrous material gives about $n = 1.63$. Birefringence 0.033.

Chemistry: Calcium aluminum hydroxyl silicate.

Distinctive Features and Tests: Green botryoidal crusts, commonly pierced by rectangular openings which formerly were occupied by anhydrite. Harder than botryoidal hemimorphite and smithsonite, and lower in specific gravity. Fuses at 2½, bubbling and swelling to several times original volume, then forming a brownish glass. Insoluble.

Occurrences: Principally in cavities in basalt and diabase, associated with calcite, apophyllite, pectolite, datolite and zeolites; also in cavities in some gneisses, syenites and granites, then usually associated with epidote. Rather abundant in numerous, widespread localities. Very fine specimens from the many quarries in the Watching Mountains, northern New Jersey, perhaps best of all, in coatings to 14″ across, from lower New Street Quarry, West Paterson, and from Prospect Park Quarry, Paterson, Passaic County (Figure 324). Also fine from Farmington, Connecticut, and from Westfield, Massachu-

FIG. 324. Prehnite crusts originally coating anhydrite crystals. The latter dissolved after formation of prehnite, leaving behind cavities corresponding to their shapes. The large flat area at the upper left of the specimen was once a spray of slightly divergent anhydrite crystals. Size 7″ by 4½″. Prospect Park, Paterson, Passaic County, New Jersey.

setts. Superb specimens, many with single and groups of white apophyllite crystals sprinkled on them, from quarries in fairfax and Loudon counties, Virginia, in slabs to 12″ across and also in spherical growths to 1″ diameter. Good specimens from copper deposits in the Keeweenaw Peninsula, Michigan. Exceptional, nearly transparent, white to pale green crystals to ½″ across, at Motta Naira, Graubunden, Switzerland. From Bourg d'Oisans, Dauphine, France, in rounded aggregates, some of which are nearly spherical and make excellent specimens. Pale green and nearly yellow masses, sometimes highly translucent, occur at various places in basalt at Renfrewshire and Dumbartonshire, Scotland. Fine yellowish masses from Prospect, New South Wales. From various localities in the Rhine Palatinate of Germany. The best specimens are the rich green masses from Paterson, New Jersey, and those from Virginia, especially if apophyllite is associated.

Datolite—Ca(OH)(BSiO₄)

Name: From the Greek, *dateisthai,* "to divide," because granular aggregates crumble readily; dăt'ō līt.

Crystals: Monoclinic. Predominantly in blunt wedge-shaped crystals, the faces of which present a confusing array of planes in various shapes and sizes as shown in Figure 325. Druses abundant, also in scattered crystals on matrix, in coarse-granular aggregates, and in the unusual occurrences in the Keeweenaw Peninsula, Michigan, as very compact, fine-grained masses of porcelainic texture.

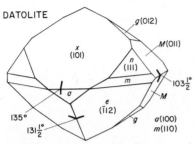

DATOLITE

g(012)
x (101)
M(011)
n (111)
m
103½°
a
ε (1̄12)
135°
131½°
g
a(100)
m(110)
M

PRISMS *g, m, M, n, ε*(epsilon), PINACOID *a,*

FIG. 325.

Physical Properties: Brittle, weak; coarse-grained aggregates crumble readily. No cleavage. H. 5½. G. 2.8-3.0.

Optical Properties: Crystals usually transparent; also translucent. Predominantly very pale yellowish-green, small crystals appearing nearly colorless. Massive material from Keeweenaw is white when pure but commonly tinged red by copper inclusions; more rarely, it is orange or yellow. Streak colorless. Luster vitreous. Biaxial negative. R.I., $\alpha = 1.626$, $\beta = 1.654$, $\gamma = 1.670$. The large birefringence of 0.044 is easily evident when examining crystals with a magnifying glass.

FIG. 326. Datolite crystals coating altered basalt. From Prospect Park, Paterson, Passaic County, New Jersey. Size 6″ by 4″. The datolite crystals are pale yellow-green in hue and highly modified. Some small spheres of prehnite are associated.

Chemistry: Calcium hydroxyl borosilicate.

Distinctive Features and Tests: Glassy, faint greenish, wedge-shaped crystals with odd-shaped faces seemingly placed at random. Fuses at $2\frac{1}{2}$ with bubbling and swelling, first to a white fluffy mass, then to a white or clear bead; flame color strong yellow-green. Insoluble.

Occurrences: As a secondary mineral in cavities in basalt and diabase, associated with calcite, prehnite, apophyllite, drusy quartz and zeolites; also in copper deposits of Keeweenaw, Michigan, and in low-temperature portions of some ore veins; rarely, in crevices in granites, serpentines, and other rocks. In fine crystals in the Smith & Lacey Mine, Loughborough, Frontenac County, Ontario. The world's largest crystals occur in fissures in basaltic rock at the Lane Quarry, near Westfield, Hampden County, Massachusetts, single crystals often reaching $2''$ diameter and individuals to $3''$ have been recorded; large slabs coated with crystals, some quite transparent and capable of being cut into gems, commonly occur. Fine crystals from Rocky Hill near Hartford, and in the Roncari Quarry near East Granby, and from other Connecticut localities south to New Haven. Very fine in basalt quarries of northern New Jersey, as around Paterson, Passaic County, but crystals seldom over $3/4''$ diameter (Figure 326). Common in cavities in copper deposits of Keeweenaw, Michigan, where beautiful glassy crystals were found; crystals to $2\frac{1}{2}''$ diameter have been recorded from the Osceola Mine. The unique nodular datolite from Keeweenaw

FIG. 327. Massive porcelainous datolite in nodular form from copper mines in the Keeweenaw Peninsula of Michigan. Basically the fine-grained datolite is white but is locally discolored by small specks of native copper. Nodules cut and polished. Size of largest specimen is $3''$ by $2''$.

reaches sizes to $12''$ diameter, but the average runs far lower, to $3''$ to $5''$. Most nodules are pure white or reddish near borders, colored by minute inclusions of native copper as shown in Figure 327. Productive mines have been Quincy, Franklin, Iroquois, Clark, and Delaware. Nodules are usually sliced and polished. Good crystals occur in silver veins of Guanajuato, Mexico. In Europe, in good crystals at Andreasberg, Harz, Germany; on the Seiser Alp, Trentino, Italy, and from Roseberry, Montagu County, Tasmania.

Appendix A

Identification Tables

The following tables present information on species discussed in Descriptive Mineralogy, and should be useful for quickly narrowing the list of possibilities when the reader is faced with the problem of identifying an unknown species. In general, the procedures outlined in Chapter Ten should be applied to the problem, and as will be noted, the following tables are arranged accordingly. Properties such as color and luster, habits, fracture or cleavage, etc., which are determinable by visual examination, appear first, followed by tables for properties which require tools or equipment for their determination. It is again emphasized that a careful, unhurried visual examination of any specimen does more to extract valuable clues to identity than any other process, and in the majority of cases, is sufficient to narrow the list of suspects to a very few.

Table 1

Color-Luster

METALLIC LUSTER

Silver or tin white: Silver, antimony, arsenic, allemontite, mercury, arsenopyrite, loellingite, bismuth, cobaltite (pinkish cast), skutterudite, smaltite, chloanthite.

Grayish silver white: Some galena and stibnite.

Steel gray: Platinum, bournonite, tarnished arsenopyrite.

Blue gray: Galena, stibnite, molybdenite, tetrahedrite.

Dull to dark gray: Chalcocite, platinum, graphite, argentite, some tarnished galena, tennantite, enargite.

Purplish gray: Covellite, bornite.

Copper red: Copper, niccolite (much paler).

Brownish red: Bornite (on fresh fracture surface).

Bronze and brass yellow: Pyrrhotite, chalcopyrite, millerite (sometimes greenish), pyrite, marcasite (pale in hue or sometimes greenish).

Gold yellow: Gold, chalcopyrite (distinguished by brittleness).

SUBMETALLIC LUSTER

Black: Hematite, ilmenite, rutile, pyrolusite, uraninite, manganite, goethite, wolframite, magnetite, columbite, tantalite, argentite, chalcocite, graphite, enargite.

Reddish or brownish black: Rutile, goethite, tantalite, huebnerite.

Blackish red: Pyrargyrite, rutile.

ADAMANTINE LUSTER

Black: Cassiterite, rutile, zircon, sphalerite, diamond, cerussite.

White or colorless: Diamond, anglesite, cerussite, zircon, scheelite, cassiterite.

Brown: Sphene, monazite, sphalerite, zircon, some corundum (due to inclusions), wulfenite, scheelite, vanadinite, pyromorphite, descloizite, cassiterite, rutile, diamond, cerussite.

Red-brown: Rutile, monazite, sphalerite, zircon, vanadinite, descloizite, cuprite.

Red: Proustite, realgar, cinnabar, cuprite, rutile, vanadinite.

Orange: Crocoite, zircon, wulfenite, scheelite, mimetite, orpiment (yellowish).

Yellow: Sphene, wulfenite, mimetite, sphalerite, orpiment, rutile (in very fine acicular crystals).

Green: Zircon, andradite, vanadinite, sphene, sphalerite.

RESINOUS, OILY OR GREASY LUSTER

White or colorless: Apatite, nepheline, witherite, halite, gypsum.

Brownish-black: Wurtzite.

Brown: Axinite, willemite, pyromorphite, serpentine.

Orange or Yellow: Sulfur, orpiment, sphalerite, willemite, serpentine.

Green: Mottramite, willemite, apatite, pyromorphite, serpentine.

Blue: Sodalite.

VITREOUS LUSTER

Black: Spinel, quartz, opal, hornblende, tourmaline, olivine, fayalite, andradite.

Gray: Corundum, smithsonite, strontianite, kernite, wavellite, quartz, grossularite.

White or colorless: Corundum, diaspore, fluorite, calcite, smithsonite, dolomite, hemimorphite, aragonite, strontianite, kernite, ulexite, colemanite, barite, celestite, gypsum, wavellite, leucite, danburite, albite, opal, quartz, orthoclase, microcline, scapolite, heulandite, harmotome, stilbite, chabazite, analcime, natrolite, apophyllite, tremolite, jadeite, spodumene, beryl, idocrase, forsterite, phenakite, topaz, grossularite.

Brown: Corundum, diaspore, chrysoberyl, calcite, siderite, smithsonite, andradite, dolomite (pale), barite, quartz, opal, stilbite, axinite, tourmaline, idocrase, epidote, clinozoisite, fayalite, andalusite, topaz, staurolite, almandite, grossularite.

Red: Corundum (various shades from pink to red), diaspore (pink), spinel, fluorite, tourmaline, rhodochrosite, erythrite, strengite, opal, orthoclase (pinkish), rhodonite, andalusite, almandite, pyrope, spessartite, grossularite, microcline (flesh-pink), scapolite (pink), chabazite, spodumene, beryl.

Orange: Corundum, calcite, stilbite (orange brown), topaz, spessartite, grossularite.

Yellow: Corundum, chrysoberyl, fluorite, calcite, smithsonite, aragonite, colemanite, barite, apatite, quartz, opal, orthoclase, scapolite, spodumene, beryl, topaz, andradite, datolite.

Green: Corundum, spinel, chrysoberyl, fluorite, smithsonite, vivianite (dark), beryl, olivenite, adamite, apatite, wavellite, opal, microcline, tremolite, actinolite, fayalite, olivine, epidote, idocrase, tourmaline, dioptase, hornblende, diopside, jadeite, spodumene, grossularite, andalusite, datolite, andradite.

Blue: Corundum (many shades), spinel, fluorite, smithsonite, azurite, barite (pale), celestite (pale), vivianite (dark), apatite, lazulite (dark), scorzalite, sodalite, benitoite, beryl, tourmaline, kyanite, topaz.

Purple and Violet: Corundum, diaspore, spinel, fluorite, vivianite (dark), strengite, apatite, quartz, spodumene, cordierite, kyanite, almandite (reddish-violet).

PEARLY LUSTER
(Usually due to partly-developed cleavages)

White or colorless: Brucite, gypsum, apophyllite, stilbite, heulandite, muscovite, albite.
Brown: Stilbite, muscovite, phlogopite, biotite.
Red: Erythrite.
Yellow green: Torbernite, autunite.
Green: Torbernite, kyanite, chlorite, vivianite, microcline (*amazonite*).
Blue: Kyanite.
Purple: Vivianite.

SILKY LUSTER

Black: Goethite, hematite.
White: Ulexite, gypsum, natrolite, scapolite, tremolite, pectolite, aragonite, calcite, cerussite, kernite, apatite.
Brown: Barite, calcite.
Pink: Scapolite, gypsum.
Yellow: Quartz (*tigereye*), serpentine (*chrysotile*), gypsum.
Green: Malachite, actinolite, prehnite, wavellite, serpentine, tourmaline, diopside.
Blue: Quartz (replacement of *crocidolite*).

WAXY LUSTER

Blue: Turquois.
Green: Turquois, variscite, some opal.

EARTHY LUSTER
(Usually finely divided or porous aggregates)

Black: Graphite, pyrolusite, uraninite, manganite, goethite.
White: Clay minerals, anglesite (associated with galena).
Brown: Goethite (limonite—very abundant and widespread), descloizite, staurolite, andalusite (*chiastolite*).
Red: Hematite.
Green: Variscite, turquois, annabergite, chlorite.
Blue: Lazurite.

TABLE 2

SOME HABITS OF CRYSTALS AND CRYSTAL AGGREGATES

ISOMETRIC SYSTEM

Cubes: Galena, pyrite, argentite, cuprite, uraninite, halite, fluorite, diamond.
Octahedrons: Diamond, gold (flattened plates), cuprite, spinel, magnetite, franklinite, fluorite.
Dodecahedrons: Copper, magnetite, lazurite, garnets, diamond.
Pyritohedrons: Pyrite, cobaltite.
Trapezohedrons: Analcime, garnets.
Formless masses, nuggets, wires, etc.: Gold, silver, copper, platinum, argentite, bornite, skutterudite, smaltite, chloanthite, uraninite, sodalite, lazurite.

TETRAGONAL SYSTEM

Square dipyramids: Cassiterite, scheelite (some closely resemble octahedrons), wulfenite (rare), apophyllite (rare), zircon.

Square prisms: Apophyllite, scapolite, idocrase, zircon.

Tetrahedral (sphenoids): Chalcopyrite.

Tabular prisms: Torbernite and autunite (very thin), wulfenite.

Pseudotrapezohedral: Leucite.

Pseudocubic: Apophyllite.

Massive: Pyrolusite, cassiterite (*"stream"* or *"wood tin"*).

HEXAGONAL SYSTEM

Hexagonal prisms: Pyrargyrite, proustite, corundum, apatite, pyromorphite, mimetite, vanadinite, quartz, beryl, calcite, millerite (acicular).

Hexagonal dipyramids: Corundum.

Hexagonal tabular: Graphite (very thin), molybdenite (thin), pyrrhotite, covellite (very thin), corundum (ruby), hematite, ilmenite, apatite, beryl.

Hexagonal pyramids: Wurtzite.

Rhombohedrons: Cinnabar, hematite, calcite, siderite, dolomite, chabazite.

Scalenohedrons: Calcite, rhodochrosite, smithsonite.

Trigonal prisms: Dioptase, tourmaline, willemite.

Trigonal tabular: Benitoite, phenakite.

Massive: Arsenic, allemontite, bismuth, antimony, niccolite, brucite, nepheline.

ORTHORHOMBIC SYSTEM

Prismatic: Stibnite, loellingite, enargite, bournonite, chrysoberyl, aragonite, cerussite, barite, celestite, anglesite, olivenite, adamite, wavellite, danburite, natrolite (acicular).

Dipyramidal-prismatic: Sulfur, olivine (*peridot*), andalusite, topaz, staurolite.

Tabular: Sulfur, chalcocite, marcasite (commonly in "cockscombs"), enargite, hemimorphite, columbite, tantalite, strontianite, barite, celestite, anglesite.

Bladed: Diaspore, goethite, strontianite.

Pseudohexagonal tabular: Chalcocite (sixling twins), chrysoberyl (sixling twins), cerussite (sixling twins), witherite, aragonite.

Cogwheel twins: Bournonite.

Spear-shaped: Descloizite, mottramite.

Massive: Goethite, variscite, wavellite, cordierite, olivine, prehnite.

MONOCLINIC SYSTEM

Prismatic: Realgar, manganite, azurite, kernite, gypsum, vivianite, tremolite, actinolite, spodumene, epidote, clinozoisite.

Blocky: Wolframite, orthoclase, hornblende, diopside.

Bladed: Vivianite, erythrite, huebnerite, tremolite, sphene.

Wedge-shaped: Orpiment, azurite, gypsum, monazite, wolframite, sphene.

Acicular: Tremolite, actinolite, crocoite.

Pseudorhombohedral: Arsenopyrite, colemanite, datolite.

Pseudodipyramidal: Lazulite.

Pseudohexagonal tabular: Muscovite, phlogopite, biotite, lepidolite, chlorite.

Massive: Malachite, annabergite (earthy), scorzalite, serpentine, actinolite, jadeite.

TRICLINIC SYSTEM

Blocky: Microcline, rhodonite.
Bladed: Albite, kyanite.
Wedge-shaped: Axinite.
Acicular: Ulexite.
Massive: Ulexite, pectolite, turquois, rhodonite.

TABLE 3

CONSPICUOUS FRACTURES, CLEAVAGES AND PARTINGS

FRACTURES

Excellent conchoidal fracture: Cinnabar, realgar, pyrite, proustite, cuprite, hematite, uraninite (*pitchblende*), diaspore, spinel, chrysoberyl, cerussite, azurite, apatite, quartz, opal, danburite, leucite, benitoite, beryl, olivine, phenakite, garnets, datolite.

CLEAVAGES AND PARTINGS

Cubic: Galena, halite, cobaltite.
Octahedral: Diamond, fluorite.
Dodecahedral: Sphalerite, sodalite.
Rhombohedral: Millerite, corundum (parting), calcite, siderite, rhodochrosite, smithsonite, dolomite, dioptase.
Basal (Hexagonal, $c[0001]$): arsenic, allemontite, antimony, bismuth, graphite, molybdenite, pyrrhotite (parting), covellite, corundum (parting), brucite.
Basal (Orthorhombic, $c[001]$): Sulfur (imperfect), barite, celestite, topaz.
Basal (Tetragonal, $c[001]$): Torbernite, autunite, apophyllite.
Basal (Monoclinic, $c[001]$): Muscovite, phlogopite, biotite, chlorite, lepidolite, clinozoisite, epidote.
Prismatic: Wurtzite, cinnabar, arsenopyrite, aragonite, strontianite, kernite, ulexite, barite, celestite, lazulite, wavellite, andalusite, scapolite, natrolite, tremolite, actinolite, hornblende, diopside, spodumene.
Pinacoidal: Realgar, orpiment, stibnite, loellingite, enargite, manganite, diaspore, goethite, kernite, ulexite, colemanite, gypsum, vivianite, erythrite, strengite, adamite, wolframite, huebnerite, ferberite, orthoclase, microcline, albite, heulandite, harmotome, stilbite, rhodonite, kyanite.

TABLE 4

HARDNESS AND RELATED PROPERTIES

Hardness	Species	Remarks
1	Talc	Greasy feel
1-2	Graphite	Greasy feel; soils fingers
1-1½	Molybdenite	Greasy feel; marks paper
1½-2	Covellite	Somewhat sectile
1½-2	Realgar	Sectile
1½-2	Orpiment	Sectile; flexible cleavage laminae
1½-2	Vivianite	Sectile; flexible cleavage laminae
1½-2½	Sulfur	Very brittle; slightly sectile
1½-2½	Erythrite	Sectile; flexible laminae

TABLE 4 *Continued*

Hardness	Species	Remarks
2	Stibnite	Somewhat sectile; very easily cleaved
2	Gypsum	Sectile; easily flexible laminae
2-2½	Proustite	Slightly sectile; very brittle
2-2½	Cinnabar	Very brittle
2-2½	Argentite	Perfectly sectile
2-2½	Bismuth	Sectile; brittle
2-2½	Kernite	Very brittle
2-2½	Autunite	Flexible laminae
2-2½	Torbernite	Brittle laminae
2-2½	Chlorite	Flexible laminae
2½	Galena	Brittle; easy perfect cubic cleavage
2½	Pyrargyrite	Very brittle
2½	Halite	Very brittle; easy perfect cubic cleavage
2½	Brucite	Sectile; flexible laminae
2½	Ulexite	Brittle; splinters readily
2½-3	Gold	Malleable; sectile
2½-3	Silver	Malleable; sectile
2½-3	Copper	Malleable; sectile
2½-3	Bournonite	Brittle
2½-3	Chalcocite	Brittle
2½-3	Anglesite	Very brittle
2½-3	Crocoite	Very brittle
2½-3	Vanadinite	Very brittle
2½-3	Wulfenite	Brittle
2½-3	Phlogopite	Elastic laminae; softer on cleavage face
2½-3	Biotite	Elastic laminae; softer on cleavage face
2½-3	Lepidolite	Elastic laminae; softer on cleavage face
2½-4	Muscovite	Elastic laminae; softest on cleavage face
2½-5	Serpentine	Tough; some varieties fibrous or splintery; hardness varies according to varieties
3	Enargite	Brittle; easily cleaved
3	Bornite	Somewhat sectile; brittle
3	Cerussite	Very brittle
3	Olivenite	Very brittle
3-3¼	Calcite	Brittle; very easily cleaved
3-3½	Antimony	Brittle
3-3½	Millerite	Brittle
3-3½	Witherite	Brittle
3-3½	Barite	Quite brittle; easily cleaved
3-3½	Celestite	Brittle; easily cleaved
3-3½	Mottramite	Brittle
3-3½	Descloizite	Brittle
3-4	Allemontite	Brittle
3-4	Heulandite	Brittle; easily cleaved
3-4½	Tetrahedrite	Brittle
3½	Arsenic	Brittle
3½	Strontianite	Easily cleaved

TABLE 4 *Continued*

Hardness	Species	Remarks
$3\frac{1}{2}$	Adamite	Brittle
$3\frac{1}{2}$-4	Sphalerite	Brittle; easily cleaved
$3\frac{1}{2}$-4	Chalcopyrite	Brittle; more so than pyrite
$3\frac{1}{2}$-4	Wurtzite	Brittle
$3\frac{1}{2}$-4	Cuprite	Very brittle
$3\frac{1}{2}$-4	Siderite	Brittle; less easily cleaved than calcite
$3\frac{1}{2}$-4	Rhodochrosite	Brittle; less easily cleaved than calcite
$3\frac{1}{2}$-4	Dolomite	Brittle; not easily cleaved
$3\frac{1}{2}$-4	Malachite	Rather tough in fibrous masses
$3\frac{1}{2}$-4	Azurite	Brittle
$3\frac{1}{2}$-4	Wavellite	Brittle
$3\frac{1}{2}$-4	Mimetite	Very brittle
$3\frac{1}{2}$-4	Pyromorphite	Very brittle
$3\frac{1}{2}$-$4\frac{1}{2}$	Pyrrhotite	Brittle
$3\frac{1}{2}$-$4\frac{1}{2}$	Strengite	Brittle
$3\frac{1}{2}$-$4\frac{1}{2}$	Variscite	Rather tough in massive form
4	Manganite	Brittle
4	Fluorite	Brittle; easily cleaved
4	Aragonite	Brittle; easily cleaved
4	Chabazite	Brittle
4	Stilbite	Brittle
4-$4\frac{1}{2}$	Platinum	Tough; malleable; sectile
4-$4\frac{1}{2}$	Smithsonite	Brittle; rather tough in massive form
4-$4\frac{1}{2}$	Wolframite	Brittle
4-5	Pectolite	Brittle; splintery; very tough in certain massive types
4-6	Turquois	Brittle to somewhat tough
$4\frac{1}{2}$	Colemanite	Brittle; easily cleaved
$4\frac{1}{2}$	Hemimorphite	Very brittle; compact types tougher
$4\frac{1}{2}$	Harmotome	Brittle
$4\frac{1}{2}$-5	Scheelite	Brittle
$4\frac{1}{2}$-5	Apophyllite	Brittle; very easily cleaved
$4\frac{1}{2}$-$6\frac{1}{2}$	Kyanite	Brittle; slightly flexible in thin laminae
5	Apatite	Brittle
5	Dioptase	Brittle; easily cleaved
5	Analcime	Brittle
5-$5\frac{1}{2}$	Niccolite	Brittle
5-$5\frac{1}{2}$	Goethite	Brittle; quite tough in compact types
5-$5\frac{1}{2}$	Monazite	Brittle
5-$5\frac{1}{2}$	Lazurite	Brittle
5-6	Ilmenite	Brittle
5-6	Uraninite	Brittle
5-6	Actinolite	Brittle; splintery
5-6	Tremolite	Brittle; splintery; compact masses very tough
$5\frac{1}{2}$	Cobaltite	Brittle
$5\frac{1}{2}$	Loellingite	Brittle
$5\frac{1}{2}$	Datolite	Brittle; tough in massive types
$5\frac{1}{2}$	Willemite	Moderately brittle

TABLE 4 *Continued*

Hardness	Species	Remarks
5½-6	Arsenopyrite	Brittle
5½-6	Skutterudite	Brittle
5½-6	Lazulite	Moderately brittle
5½-6	Scorzalite	Moderately brittle
5½-6	Natrolite	Brittle; tougher in compact masses
5½-6	Leucite	Brittle
5½-6½	Opal	Very brittle
6	Magnetite	Quite brittle
6	Orthoclase	Brittle; rather easily cleaved
6	Albite	Brittle; crumbles easily
6	Nepheline	Brittle
6	Sodalite	Quite brittle
6	Scapolite	Brittle; often crumbles
6	Hornblende	Brittle; splintery cleavages
6	Diopside	Brittle; splintery cleavages
6	Rhodonite	Brittle; compact masses very tough
6	Olivine	Rather brittle
6	Sphene	Brittle
6-6½	Marcasite	Quite brittle
6-6½	Pyrite	Brittle
6-6½	Columbite	Very brittle
6-6½	Tantalite	Brittle
6-6½	Rutile	Brittle
6-6½	Pyrolusite	Brittle; compact masses rather tough
6-6½	Benitoite	Brittle
6-7	Cassiterite	Brittle but somewhat tough
6-7	Zircon	Brittle to quite brittle
6½	Hematite	Brittle; tough in compact masses
6½	Microcline	Brittle; easily cleaved
6½	Jadeite	Extremely tough
6½	Idocrase	Brittle; very tough in compact masses
6½	Andradite	Brittle
6½	Prehnite	Moderately tough in compact masses
6½-7	Diaspore	Very brittle
6½-7	Axinite	Quite brittle
6½-7½	Spodumene	Brittle
7	Quartz	Brittle; compact types tough
7	Cordierite	Brittle
7	Danburite	Brittle
7	Epidote	Brittle; compact masses tough
7	Clinozoisite	More brittle than epidote
7	Staurolite	Brittle
7	Grossularite	Brittle; compact types tough
7¼	Pyrope	Brittle
7¼	Spessartite	Brittle
7½	Andalusite	Rather tough; *chiastolite* soft
7½	Tourmaline	Brittle; often crumbles

TABLE 4 *Continued*

Hardness	Species	Remarks
7½	Almandite	Brittle; often crumbles
7½-8	Phenakite	Quite tough
8	Spinel	Brittle
8	Topaz	Brittle; easily cleaved
8	Beryl	Brittle
8½	Chrysoberyl	Rather tough
9	Corundum	Rather tough in gem varieties; other types part readily
10	Diamond	Rather tough

TABLE 5
STREAKS

METALLIC (SHINING)

Gold-yellow: Gold.

Silver-white: Silver, arsenic, bismuth, allemontite.

Copper-red: Copper.

Grayish-white: Platinum.

NON-METALLIC

Black: Pyrolusite, argentite, graphite (shining), argentite (shining), covellite (shining), skutterudite, smaltite, chloanthite, tetrahedrite, bournonite, ilmenite, magnetite, columbite, ferberite.

Greenish-black: Chalcopyrite, millerite, pyrite.

Brownish-black: Niccolite (pale), pyrite, marcasite, geikielite, uraninite, wolframite.

Gray-black: Chalcocite, bornite (pale), galena, pyrrhotite, covellite (shining), stibnite, cobaltite, loellingite, marcasite, arsenopyrite (dark), enargite, bournonite.

Gray: Antimony, graphite (shining), stibnite, molybdenite (bluish to greenish).

Brown: Sphalerite (pale to colorless), wurtzite, tetrahedrite (dark), rutile (pale).

Brownish-red: Cuprite (shining), hematite, pyrophanite, manganite, descloizite, huebnerite.

Brownish-yellow: Goethite.

Red: Cinnabar, pyrargyrite (dark), proustite (bright), hematite, tennantite, tantalite (dark).

Orange-red: Realgar.

Orange-yellow: Crocoite.

Orange: Descloizite.

Yellow: Orpiment (pale), vanadinite (very pale), autunite (very pale).

Green: Malachite (pale), vivianite (very pale), mottramite (very pale), torbernite (pale).

Blue: Azurite (pale), lazurite, dioptase (pale greenish).

Purple: Vivianite (very pale).

TABLE 6

SPECIFIC GRAVITIES

G.	Species	G.	Species
1.71	Borax	2.8-3.4	Biotite
1.91	Kernite	2.8-3.4	Phlogopite
1.96	Ulexite	2.86	Pectolite
1.99-2.25	Opal	2.87	Strengite
2.05-2.15	Chabazite	2.95-3.0	Aragonite
2.07	Sulfur	2.97-3.00	Phenakite
2.1-2.2	Stilbite	2.98-3.16	Tremolite
2.1-2.2	Graphite	3.00	Danburite
2.17	Halite	3.06	Erythrite
2.2	Heulandite	3.0-3.3	Tourmaline
2.25	Natrolite	3.0-3.4	Hornblende
2.28-2.30	Sodalite	3.1-3.2	Lazulite
2.3	Analcime	3.1-3.2	Apatite
2.32	Gypsum	3.1-3.2	Autunite
2.3-2.4	Apophyllite	3.1-3.2	Andalusite
2.36	Wavellite	3.16-3.35	Actinolite
2.39	Brucite	3.17-3.23	Spodumene
2.4-2.5	Lazurite	3.18	Fluorite
2.45-2.50	Leucite	3.2-3.3	Axinite
2.42	Colemanite	3.22	Torbernite
2.44-2.50	Harmotome	3.22-3.80	Forsterite
2.5-2.6	Serpentine	3.22-4.39	Olivine
2.53-2.57	Variscite	3.25-3.36	Jadeite
2.55-2.65	Nepheline	3.25-3.30	Diopside
2.55-2.75	Cordierite	3.28-3.35	Dioptase
2.56-2.59	Microcline	3.30-3.45	Hedenbergite
2.56-2.59	Orthoclase	3.3-3.5	Diaspore
2.6-2.8	Scapolite	3.3-3.4	Scorzalite
2.6-2.9	Turquois	3.3-3.4	Clinozoisite
2.62	Albite	3.3-3.6	Idocrase
2.62-2.76	Plagioclase	3.3-4.3	Goethite
2.65	Quartz	3.4-3.5	Hemimorphite
2.65	Oligoclase	3.4-3.6	Rhodochrosite
2.67-2.90	Beryl	3.4-3.6	Epidote
2.68	Vivianite	3.4-3.55	Sphene
2.68	Andesine	3.45-3.55	Augite
2.7-2.9	Chlorite	3.49	Orpiment
2.71	Calcite	3.50-3.53	Diamond
2.71	Labradorite	3.5-3.6	Topaz
2.73	Bytownite	3.5-3.7	Rhodonite
2.76	Anorthite	3.56	Realgar
2.8-2.9	Muscovite	3.58-3.98	Spinel
2.8-2.9	Lepidolite	3.60-3.65	Grossularite
2.80-2.95	Prehnite	3.6-4.05	Malachite
2.8-3.0	Datolite	3.64-3.68	Benitoite
2.8-3.0	Dolomite	3.65-3.69	Kyanite

TABLE 6 *Continued*

G.	Species	G.	Species
3.65-3.80	Pyrope	5.3	Hematite
3.7-3.8	Staurolite	5.5	Millerite
3.71-3.75	Chrysoberyl	5.5-5.8	Chalcocite
3.72-3.76	Strontianite	5.55	Proustite
3.77	Azurite	5.7	Arsenic
3.8-3.9	Siderite	5.83	Bournonite
3.80-4.39	Fayalite	5.85	Pyrargyrite
3.82-3.90	Andradite	5.9	Mottramite
3.89-4.19	Willemite	5.9-6.1	Scheelite
3.9-4.1	Sphalerite	6.0	Crocoite
3.9-4.5	Olivenite	6.07	Arsenopyrite
3.94-4.71	Zircon	6.1-6.9	Skutterudite
3.95-4.32	Almandite	6.14	Cuprite
3.96-3.98	Celestite	6.2	Descloizite
3.98	Wurtzite	6.33	Cobaltite
4.0-4.1	Corundum	6.36-6.39	Anglesite
4.0-4.5	Smithsonite	6.5-7.0	Wulfenite
4.1-4.3	Chalcopyrite	6.5-7.1	Vanadinite
4.12-4.20	Spessartite	6.5-9.7	Uraninite
4.20-4.25	Rutile	6.55	Cerussite
4.29	Witherite	6.57-8.00	Tantalite
4.3	Manganite	6.61-6.72	Allemontite
4.3-4.5	Adamite	6.7	Antimony
4.3-4.6	Barite	6.8-7.1	Cassiterite
4.4-4.5	Enargite	7.04	Pyromorphite
4.4-5.0	Pyrolusite	7.1	Huebnerite
4.6-4.65	Pyrrhotite	7.2-7.4	Argentite
4.6-4.7	Molybdenite	7.24	Mimetite
4.6-4.76	Covellite	7.3	Wolframite
4.6-4.9	Tennantite	7.4	Loellingite
4.6-5.4	Monazite	7.5	Ferberite
4.63	Stibnite	7.58	Galena
4.7	Ilmenite	7.78	Niccolite
4.89	Marcasite	8.1	Cinnabar
4.9-5.1	Tetrahedrite	8.95	Copper
5.02	Pyrite	9.7-9.8	Bismuth
5.07	Bornite	10-11	Silver
5.15-6.57	Columbite	14-19	Platinum
5.18	Magnetite	15-19	Gold

TABLE 7

REFRACTIVE INDICES AND RELATED PROPERTIES

NOTES: Indices rounded to two decimal places. High and low indices, as O,E, in uniaxial species, and alpha, gamma, in biaxial species are connected by a dash "-". Ranges are indicated by the word "to". Singly refractive species are indicated by the capital letter "I" for isotropic; uniaxial species by the letter "U", and biaxial species by the letter "B". Plus and minus are used for optic signs.

Indices	*I,U,B*	*Sign*	*Birefringences*	*Species*
1.43	I			Fluorite
1.44 to 1.46	I			Opal
1.45-1.47	B	−	0.025	Borax
1.45-1.49	B	−	0.034	Kernite
1.48	I			Sodalite
1.48-1.49	U	−	0.01	Chabazite
1.48 to 1.49	I			Analcime
1.48-1.49	B	+	0.010	Natrolite
1.48-1.49 to 1.49-1.50	B	−	0.011 to 0.010	Stilbite
1.49-1.66	U	−	0.172	Calcite
1.49-1.52	B	+	0.03	Ulexite (single crystal)
1.49-1.50	B	+	0.005	Heulandite
1.50				Lazurite (massive)
1.50 to 1.51				Ulexite (massive)
1.50-1.68 to 1.57-1.76	U	−	0.179	Dolomite, ankerite
1.50-1.51	B	+	0.005	Harmotome
1.51	U	+	0.001	Leucite
1.52 to 1.53	B	−	0.005-0.008	Orthoclase or microcline
1.52-1.67	B	−	0.15	Strontianite
1.52-1.53	B	+	0.009	Gypsum
1.53-1.54 to 1.58-1.59	B	±	0.050 to 0.052	Plagioclases
1.53 to 1.54	I			Chalcedony
1.53-1.55	B	+	0.027	Wavellite (single crystal)
1.53-1.54 to 1.54-1.55	B	−	0.01	Cordierite
1.53-1.56	B	−	0.026	Lepidolite
1.53-1.69	B	−	0.155	Aragonite
1.53-1.54 to 1.54-1.55	U	−	0.004 to 0.005	Nepheline
1.54				Wavellite (massive)
1.54	U	±	0.002	Apophyllite
1.54-1.55 to 1.55-1.58	U	−	0.009 to 0.027	Scapolite
1.54 to 1.68	B	−	0.04 to 0.06	Phlogopite, biotite
1.54	I			Halite
1.55-1.59	B	−	0.036	Muscovite
1.55	U	+	0.009	Quartz
1.56 to 1.57				Serpentine (massive)
1.56-1.58 to 1.59-1.60	U	+	0.02	Brucite
1.56-1.57 to 1.59-1.60	U	−	0.004 to 0.007	Beryl
1.57 to 1.60	B	+	0.02	Chlorite
1.58-1.63 to 1.62-1.65	B	+	0.050 to 0.059	Vivianite
1.58				Variscite (massive)

TABLE 7 *Continued*

Indices	I,U,B	Sign	Birefringences	Species
1.59-1.61	B	+	0.028	Colemanite
1.60-1.82	U	−	0.219	Rhodochrosite
1.60				Witherite (massive)
1.60-1.62 to 1.66-1.68	B	−	0.02	Tremolite, actinolite
1.60 to 1.65				Nephrite
1.60-1.64	B	−	0.033	Lazulite
1.61-1.62 to 1.63-1.64	B	+	0.008 to 0.011	Topaz
1.61-1.64	B	+	0.022	Hemimorphite
1.62-1.63 to 1.63-1.65	U	−	0.014 to 0.021	Tourmaline
1.62-1.85	U	−	0.227	Smithsonite
1.62				Pectolite (massive)
1.62				Turquois (massive)
1.62				Nephrite (usual reading)
1.62-1.63	B	+	0.093	Celestite
1.62-1.65	B	−	0.023	Hornblende
1.63-1.7	B	+	0.073	Erythrite
1.63-1.67	B	−	0.044	Datolite
1.63				Prehnite (massive)
1.63-1.64	B	−	0.006	Danburite
1.63-1.64	U	−	0.002 to 0.004	Apatite
1.63-1.87	U	−	0.240	Siderite
1.63-1.64 to 1.64-1.65	B	−	0.007 to 0.011	Andalusite
1.64-1.67	B	+	0.035	Forsterite
1.64-1.65	B	+	0.012	Barite
1.64-1.68	B	−	0.051	Scorzalite
1.65				Datolite (massive)
1.65-1.67	U	+	0.015	Phenakite
1.65 to 1.68				Jadeite (massive)
1.65-1.66 to 1.67-1.68	B	+	0.015	Spodumene
1.65-1.69	B	+	0.035	Peridot
1.65-1.71	U	+	0.054	Dioptase
1.66-1.68 to 1.73-1.75	B	+	0.014	Rhodonite
1.67-1.70	B	+	0.018	Diopside
1.68-1.69 to 1.68-1.69	B	−	0.010	Axinite
1.69-1.72 to 1.69-1.73	U	+	0.028 to 0.031	Willemite
1.70-1.71 to 1.73-1.73	U	+	0.004 to 0.006	Idocrase
1.71-1.74	B	+	0.034	Strengite
1.71-1.72	B	+	0.009	Clinozoisite
1.71 to 1.80	I			Spinel
1.71-1.75	B	+	0.032	Augite
1.72-1.73	B	−	0.017	Kyanite
1.72-1.75	I			Gahnite
1.72-1.76	B	−	0.020	Adamite
1.73				Rhodonite (massive)
1.73 to 1.75	I			Pyrope
1.73-1.84	B	+	0.108	Azurite
1.73-1.75	I			Grossularite

TABLE 7 *Continued*

Indices	I,U,B	Sign	Birefringences	Species
1.74-1.77	B	+	0.035	Epidote
1.74-1.76	B	+	0.018	Hedenbergite
1.74-1.75 to 1.75-1.76	B	+	0.011 to 0.015	Staurolite
1.75-1.76	B	+	0.008 to 0.010	Chrysoberyl
1.76-1.80	U	+	0.047	Benitoite
1.76	I			Rhodolite
1.76-1.77 to 1.77-1.78	U	−	0.008	Corundum
1.77				Malachite (massive)
1.77 to 1.80	I			Hercynite
1.77-1.86	B	+	0.091	Olivenite
1.78-1.83	I			Almandite
1.79-1.84 to 1.80-1.85	B	+	0.049 to 0.055	Monazite
1.79-1.81	I			Spessartite
1.80-2.08	B	−	0.275	Cerussite
1.84-1.89	B	+	0.051	Fayalite
1.84-1.85 to 1.92-1.99	U	+	0.01 to 0.059	Zircon
1.84-1.94 to 1.95-2.09	B	+	0.10 to 0.142	Sphene
1.88-1.89	B	+	0.017	Anglesite
1.89	I			Andradite
1.92-1.94	U	+	0.017	Scheelite
1.92	I			Galaxite
1.96-2.25	B	+	0.29	Sulfur
2.00-2.10	U	+	0.098	Cassiterite
2.05-2.06	U	−	0.010	Pyromorphite
2.13-2.15	U	−	0.019	Mimetite
2.19-2.35	B	−	0.165	Descloizite
2.21-2.33	B	−	0.12	Mottramite
2.28-2.40	U	−	0.122	Wulfenite
2.29-2.66	B	+	0.37	Crocoite
2.37	I			Sphalerite
2.40-3.02	B	−	0.62	Orpiment
2.35-2.42	U	−	0.066	Vanadinite
2.42	I			Diamond
2.54-2.70	B	−	0.166	Realgar
2.61-2.9	U	+	0.287	Rutile
2.79-3.09	U	−	0.295	Proustite
2.85	I			Cuprite
2.88-3.08	U	−	0.203	Pyrargyrite

Table 8

Commonly Fluorescent Species

NOTE: The abbreviations SW and LW refer to short wave and long wave ultraviolet lamps respectively.

White or near-white: Scheelite (SW) (strong), witherite, aragonite, calcite, brucite, anglesite (SW), colemanite (LW), ulexite.

Red or pink: Corundum, particularly *ruby* (strong), spinel (LW), nepheline, calcite (sometimes strong), chrysoberyl.

Orange: Zircon (often strong), spodumene, sodalite, pectolite (SW), nepheline, scapolite (often strong), anglesite, aragonite.

Yellow: Zircon, opal, scapolite, sodalite (sometimes strong), apatite, hemimorphite, cerussite (LW), anglesite, gypsum, chalcedony.

Green: Chalcedony (SW) opal, willemite (often intense), chabazite, gypsum, autunite (intense).

Blue: Diamond, scheelite (SW) (strong), fluorite (strong), witherite, brucite, benitoite (strong), danburite (LW).

Appendix B

Selected References
and Reading Material

PART 1. MINERALOGY AND PETROLOGY

BATEMAN, A. M.—*Economic Mineral Deposits*. 2nd Ed. New York: John Wiley and Sons, Inc., 1950. Thorough treatment of mineral deposits and how they came into being.

———— —*The Formation of Mineral Deposits*. New York: John Wiley and Sons, Inc., 1951. Essentially a condensation and simplification of the previous work and much better suited to the amateur.

BERRY, L. G., and B. MASON—*Mineralogy: Concepts, Descriptions, Determinations*. San Francisco: W. H. Freeman and Co., 1959. A popular college-level textbook, laying stress upon atomic structures of minerals. Handsomely illustrated with numerous fine drawings. An excellent book in all respects.

BRAGG, W. L.—*The Crystalline State*. London: G. Bell and Sons, Ltd., 1933.

———— —*Atomic Structure of Minerals*. Ithaca: Cornell University Press, 1937. Classical treatments of crystal structures as revealed by x-ray investigations.

BUCKLEY, H. E.—*Crystal Growth*. New York: John Wiley and Sons, Inc., 1951. Highly technical but portions are of interest to the amateur.

CAMERON, E. N., *et al.*—"Internal Structure of Granitic Pegmatites." *Monograph 2, Economic Geology Publishing Co.*, Urbana, Illinois, 1949. For the person seriously interested in prospecting and mining pegmatites; profusely illustrated.

DANA, E. S., and W. E. FORD—*A Textbook of Mineralogy*. 4th Ed. New York: John Wiley and Sons, 1932. Much information in this splendid work is out of date, but it is still the most complete advanced textbook available to the college student of mineralogy.

DENNEN, W. H.—*Principles of Mineralogy*. Revised printing with determinative tables. New York: The Ronald Press Co., 1960. A fine college-level book providing much interesting reading.

EVANS, R. C.—*An Introduction to Crystal Chemistry*. 2nd Ed. London: Cambridge University Press, 1946. Clearly written treatment of crystals, crystal structures, and the relationships of properties to such structures.

FRONDEL, C.—"Systematic Mineralogy of Uranium and Thorium," *U.S. Geological Survey Bulletin 1064*, 1958. Thorough treatment and a very useful addition to the mineralogical library; extensive bibliography numbering hundreds of items.

——— —*Dana's System of Mineralogy.* 7th Ed. Vol. III—Silica Minerals. New York: John Wiley and Sons, 1962. Deals thoroughly with quartz and other minerals with the formula SiO_2, including opal. Handsomely illustrated and indispensible to the serious amateur.

GLEASON, S.—*Ultraviolet Guide to Minerals.* Princeton, New Jersey: D. Van Nostrand Co., Inc., 1960. The most complete guide on fluorescent minerals available.

GOLDSCHMIDT, V.—*Atlas der Kristallformen.* Heidelberg (Germany): 1913-1923. Nine volumes, atlas and text, of thousands of crystal drawings compiled from many sources and designed to show the habits observed upon crystals of the species treated. Long out-of-print and nearly impossible to obtain. Prized by micromounters.

HEY, M. H.—*Chemical Index of Minerals.* London: British Museum of Natural History, 1962. Name of species, formulas, pronunciations, and many more useful data on all species known at the time of publication. Encyclopedic in scope.

HOLDEN, A., and P. SINGER—*Crystals and Crystal Growing.* Garden City, New York: Doubleday and Co., Inc., 1960. Written for the amateur interested in learning more about crystals through the expedient of growing some at home. Much information on properties of crystals is introduced throughout this interestingly written book.

HONESS, A. P.—*The Nature, Origin and Interpretation of the Etch Figures on Crystals.* New York: John Wiley and Sons, 1927.

HOWELL, J. V.—*Glossary of Geology and Related Sciences.* American Geological Institute, 2101 Constitution Ave., Washington 25, D.C., 1957. Explains many geological terms. *Supplement,* issued in 1960.

HOWELL, J. V., and A. I. Levorsen—*Directory of Geological Materials in North America.* 2nd Ed., revised. American Geological Institute, Washington, D.C., 1957. An invaluable compilation of sources of literature, equipment, maps, photographs, and other materials of interest to geologists and mineralogists.

HURLBUT, C. S.—*Dana's Manual of Mineralogy.* 17th Ed. New York: John Wiley and Sons, Inc., 1959. A standard college-level textbook for many years and now revised to include the latest concepts of mineralogy. Extensive determinative tables.

JAHNS, R. H.—"The Study of Pegmatites," *Economic Geology, 50th Anniversary Volume,* p. 1025-1130; 698 references; 1956.

LINDGREN, W.—*Mineral Deposits.* 4th Ed. New York: McGraw-Hill, 1933. For many years the accepted authority on the formation of mineral deposits.

MASON, B.—*Principles of Geochemistry.* 2nd Ed. New York: John Wiley and Sons, Inc., 1958. How elements combine under various conditions to form minerals and rocks.

NIGGLI, P.—*Rocks and Mineral Deposits.* San Francisco: W. H. Freeman and Co., 1959.

PALACHE, C., et al.—*Dana's System of Mineralogy.* Vol. I, 1944; Vol. II, 1951. New York: John Wiley and Sons, Inc. The authoritative compilation of mineralogical data.

PHILLIPS, F. C.—*An Introduction to Crystallography.* 2nd Ed. London: Longmans, Green and Co., 1956. Numerous diagrams and crystal drawings; one of the best texts available to the advanced amateur.

PIRSSON, L. V.—*Rocks and Rock Minerals.* 3rd Ed., revised by Adolph Knopf. New York: John Wiley and Sons, Inc., 1947.

RICE, C. M.—*Dictionary of Geological Terms.* Princeton, New Jersey: published by the author, 1951.

SINKANKAS, J.—*Gemstones and Minerals—How and Where to Find Them.* Princeton,

New Jersey: D. Van Nostrand Co., Inc., 1961. Explains how minerals form, where they form, how deposits may be recognized, and how to go about extracting specimens.

SMITH, O. C.—*Identification and Qualitative Chemical Analysis of Minerals.* 2nd Ed. Princeton, New Jersey: D. Van Nostrand Co., Inc., 1953. Detailed testing procedures; extensive list of properties.

WOOSTER, W. A.—*A Textbook on Crystal Physics.* London: Cambridge University Press, 1938. An interesting and informative work on the properties of crystals.

PART 2. LOCALITIES AND DEPOSITS

AHLFELD, Fr. and J. MUNOZ REYES—*Mineralogia von Bolivien.* Berlin, 1938. Minerals of Bolivia.

BAIN, H. F.—"The Fluorspar Deposits of Southern Illinois," *U.S. Geological Survey Bulletin 255,* 1905.

BEHIER, J.—"Contribution A La Minéralogie De Madagascar," *Annales Géologiques de Madagascar.* Tananarive: Imprimerie Officielle, République Malgache, 1960. Mineralogy of Madagascar brought up to date.

BØGGILD, O. B.—*The Mineralogy of Greenland.* Copenhagen: C. A. Reitzel, 1953.

BUDDINGTON, A. F., and T. CHAPIN—"Geology and Mineral Deposits of Southeastern Alaska," *U.S. Geological Survey Bulletin 800,* 1929.

BUTLER, B. S., *et al.*—"The Ore Deposits of Utah," *U.S. Geological Survey Professional Paper 111,* 1920.

BUTLER, B. S., and W. S. BURBANK—"The Copper Deposits of Michigan," *U.S. Geological Survey Professional Paper 144,* 1929.

BUTTGENBACH, H.—*Les Minéraux de Belgique et du Congo Belge.* Paris and Liege, 1947. The minerals of Belgium and the Belgian Congo.

CALDERÓN, D. S.—*Los Minerales de España.* Vols. I and II. Madrid, 1910. An excellent, detailed work on the minerals of Spain.

CHHIBBER, H. L.—*The Mineral Resources of Burma.* London: Macmillan and Co., Ltd., 1934.

CAMERON, E. N., *et al.*—"Pegmatite Investigations, 1942-1945, New England," *U.S. Geological Survey Professional Paper 255,* 1954. Describes many mineralogically interesting pegmatites.

CONLEY, J. F.—"Mineral Localities of North Carolina," *North Carolina Division of Mineral Resources Information Circular 16,* Raleigh, 1958.

de SCHMID, H. S.—"Mica," *Canada Department of Mines Report 118,* 2nd. Ed., 1912. Many fine maps of old mines in Ontario and elsewhere in Canada, once productive of excellent specimens.

——— —"Feldspar in Canada," *Canada Department of Mines Report 401,* 1916. Also informative, with fine maps and details on numerous mineralogically interesting pegmatites.

DIETRICH, R. V.—"Virginia Mineral Localities," *Bulletin of the Virginia Polytechnic Institute, Engineering Experiment Station Series 88,* 1953. *Supplement* issued as *Bulletin Series 105,* 1955. An excellent guide.

ECKEL, E. B.—"Minerals of Colorado—A 100 Year Record," *U.S. Geological Survey Bulletin 1114,* 1961. An outstanding compilation of Colorado minerals and their sources.

EMMONS, W. H., and F. F. Grout—"Mineral Resources of Minnesota." *Minnesota Geological Survey Bulletin 30,* Minneapolis, 1943.

ENGEL, A. E. J.—"Quartz Crystal Deposits of Western Arkansas," *U.S. Geological Survey Bulletin 973-E,* 1951. Details on the famous quartz deposits which prove to be very extensive and not likely to be soon exhausted.

FERGUSON, H. G.—"Geology and Ore Deposits of the Manhattan District, Nevada," *U.S. Geological Survey Bulletin 723,* 1924.

FERRAZ, L. C.—*Compendio dos Mineraes do Brazil en Forma de Diccionario.* Rio de Janeiro, 1929. A listing of the minerals of Brazil.

FLAGG, A. L.—*Mineralogical Journeys in Arizona.* Scottsdale, Arizona: A. F. Bitner Co., 1958. Interesting trips for collectors.

GALBRAITH, F. W., and D. J. BRENNAN—*Minerals of Arizona.* 3rd Ed. revised. Tucson: University of Arizona Press, 1959. Brief statements of properties and listings of localities.

GARRELS, R. M., and E. S. LARSEN, 3rd.—"Geochemistry and Mineralogy of the Colorado Plateau Uranium Ores," *U.S. Geological Survey Professional Paper 320,* 1959.

GORDON, S. G.—"Mineralogy of Pennsylvania," *Academy of Natural Sciences of Philadelphia, Special Publication I,* 1922. Very thorough and still very useful.

——— —"The Mineralogy of the Tin Mines of Cerro de Llallagua, Bolivia," *Proceedings of the Academy of Natural Sciences of Philadelphia, Vol. XCVI,* 1944. Detailed treatment of famous deposits; drawings and photos.

GREG, R. P., and W. G. LETTSOM—*Manual of the Mineralogy of Great Britain and Ireland.* London: 1858. Antiquated but contains much locality information which is not available elsewhere.

GRIFFITTS, W. R., and J. C. OLSON—"Mica Deposits of the Southeastern Piedmont, Part 5, Shelby-Hickory District, North Carolina, with Part 6, Outlying Deposits in North Carolina," *U.S. Geological Survey Professional Paper 248-D,* 1953.

HEDDLE, M. F.—*The Mineralogy of Scotland.* 2nd. Ed., Vols. I and II, edited by J. G. Goodchild. Edinburgh: 1901. A fine detailed work on the many prolific localities of Scotland.

JAHNS, R. H., and L. A. WRIGHT—"Gem and Lithium-Bearing Pegmatites of the Pala District, San Diego County, California," *California Division of Mines Special Report 7-A,* 1951. An excellent report on the famous gem mines of this district.

JOHNSTON, R. A. A.—"A List of Canadian Mineral Occurrences," Canada Department of Mines, *Geological Survey Memoir 74,* 1915. An old but still useful compendium of minerals and localities.

KOKSHAROV, N.—*Materialien zur Mineralogie Russlands.* Eleven volumes with atlas, 1853-1891. A monumental compilation of Russian minerals and localities, particularly of the classical occurrences in the Urals.

LACROIX, A.—*Minéralogie de Madagascar.* Vol. I, II, 1922, Vol. III, 1923. Paris. Large and fine work on the mineralogy, geology and petrography of this celebrated island. Excellent drawings and photographs.

LAPHAM, D. M., and A. R. GEYER—"Mineral Collecting in Pennsylvania," *Pennsylvania Geological Survey Bulletin G-33.* Harrisburg, 1959. Prominent localities carefully described.

LEMKE, R. W., *et al.*—"Mica Deposits of the Southeastern Piedmont, Part 2, The Amelia District, Virginia," *U.S. Geological Survey Professional Paper 248-B,* 1952. The classical pegmatite localities near Amelia are mapped and described.

LIEBENER, L., and J. VORHAUSER—*Die Mineralien Tirols.* Innsbruck: 1852. Thorough listing of localities in the famous Austrian Tyrol.

LINDGREN, W., *et al.*—"The Ore Deposits of New Mexico," *U.S. Geological Survey Professional Paper 68,* 1910.

LUEDECKE, O.—*Die Minerale des Harzes.* Berlin: 1896. Minerals of the famous silver mines of the Harz Region, Germany.

LUEDKE, E. M., *et al.*—"Mineral Occurrences in New York State with Selected References to Each Locality," *U.S. Geological Survey Bulletin 1072-F,* 1959. Not very informative as to localities but the references are valuable.

MAINE GEOLOGICAL SURVEY—"Maine Pegmatite Mines and Prospects and Associated Minerals." *Department of Development of Industry and Commerce, Mineral Resources Index I.* Augusta: 1957.

——— —"Maine Mineral Collecting," Augusta: 1960. A small but very fine pamphlet with maps showing noted mineral localities. The previous entry is far more complete and is very helpful to collectors.

MANCHESTER, J. G.—"Minerals of New York and Its Environs," *New York Mineralogical Club Bulletin, Vol. 3, No. I.* New York: 1931. A very fine work with much locality information and fine photographs.

MASON, B.—"Trap Rock Minerals of New Jersey," *New Jersey Geological Survey Bulletin 64,* 1960. Excellent short treatment of zeolites and associated species, with properties, locality data and photographs.

MEEN, V. B., and D. H. GORMAN—"Mineral Occurrences of Wilberforce, Bancroft and Craigmont—Lake Clear Areas, Southeastern Ontario," *Guide Book for Field Trip No. 2,* Geological Society of America and Geological Association of Canada, 1953. How to reach some famous localities and what minerals to find.

MEYERS, T. R., and G. W. STEWART—*The Geology of New Hampshire, Part 3, Minerals and Mines.* New Hampshire State Planning and Development Commission, Concord, 1956.

MORRILL, P.—*New Hampshire Mines and Mineral Localities.* Dartmouth College Museum, Hanover, 1960. Fine for field trips.

MURDOCH, J. and R. W. WEBB—"Minerals of California," *California Division of Mines Bulletin 173,* San Francisco, 1956. *With* supplement, for 1955 through 1957. Complete listings of localities under species, with notes.

NIGGLI, P. *et al.*—*Die Mineralien der Schweizeralpen.* Vols. I, II. Basel: B. Wepf, 1940. A geological and mineralogical literary masterpiece, describing in great detail, the famous Swiss Alpine occurrences.

NORTHRUP, S. A.—*Minerals of New Mexico.* 2nd. Ed. revised. Albuquerque: University of New Mexico Press, 1960. A very thorough work, including much historical data.

OSANN, A.—*Die Mineralien Badens.* Stuttgart, Germany; 1927. Minerals of the Baden Principality, Germany.

OSTRANDER, C. W., and W. E. PRICE—*Minerals of Maryland.* Baltimore: Natural History Society of Maryland, 1940.

PALACHE, C.—"The Minerals of Franklin and Sterling Hill, Sussex County, New Jersey," *U.S. Geological Survey Professional Paper 180,* 1935. An exceptionally thorough treatment of the mineralogy of these famous deposits; numerous illustrations, drawings and maps. Reprinted.

PARKER, R. L.—*Die Mineralfunde der Schweizer Alpen.* Basel: Wepf and Co., 1954. Essentially a condensation of Niggli, Koenigsberger, Parker (see above), but of the same high quality and with many of the crystal drawings and outstanding photographs.

PEMBERTON, E.—*The Minerals of Boron, California.* Montebello, California: The

Mineral Research Society of California, 1960. A fine pamphlet dealing with borates and other minerals found in abundance at this celebrated locality.

PETTERD, W. F.—*Catalogue of the Minerals of Tasmania.* Hobart: 1910.

PITTMAN, E. F.—*The Mineral Resources of New South Wales.* Sydney: Geological Survey of New South Wales, 1901.

POINDEXTER, O. F., *et al.*—"Rocks and Minerals of Michigan," *Michigan Department of Conservation, Geological Survey Division, Publication 42.* Revised Ed., Lansing: 1953. A guide for collectors.

RICHARDSON, C. H.—"The Mineralogy of Kentucky." *Kentucky Geological Survey, Series 6, Geologic Reports, Vol. 33.* Frankfort: 1925.

SCHULZE, E.—*Lithia Hercynica. Verzeichnis der Minerals des Harzes und seines Vorlandes.* Leipzig, Germany: 1896. Mineralogy of the Harz region in Germany.

SHANNON, E. V.—"The Minerals of Idaho," *U.S. National Museum (Smithsonian), Bulletin 131.* Washington: 1926. An exceptionally thorough treatment with numerous crystal drawings and photographs and excellent locality information.

SINKANKAS, J.—*Gemstones of North America.* Princeton, New Jersey: D. Van Nostrand Co., Inc., 1959. Over 2,000 localities, many described in some detail.

SLOAN, E.—"Catalogue of the Mineral Localities of South Carolina," *South Carolina Geological Survey, Series IV, Bulletin 2,* Columbia: 1908.

SOHON, J. A.—"Connecticut Minerals." *Connecticut Geological and Natural History Survey Bulletin 77.* Storrs: 1951.

TRAUBE, H.—*Die Minerale Schlesiens.* Breslau: 1888. Minerals of Silesia.

TRUSHKOVA, N. N., and A. A. KUKHARENKO—*Atlas of Placer Minerals.* Moscow: U.S.S.R. Geological Research Institute, 1961. Numerous photomicrographs of placer minerals from many world localities, including Russian, mostly taken in color. A total of 183 minerals and mineral varieties are treated.

ULRICH, G. H. F.—*Contributions to the Mineralogy of Victoria.* Melbourne: 1870.

UNION OF SOUTH AFRICA—*The Mineral Resources of the Union of South Africa.* Pretoria: Department of Mines, 1940.

Von CALMBACH, W. F.—*Handbuch Brazilianischer Edelsteine und Ihrer Vorkommen.* Rio de Janeiro: N. Medawar, 1938. Brief descriptions of Brazilian gemstones and locality lists.

WEED, W. H.—"Geology and Ore Deposits of the Butte District, Montana," *U.S. Geological Survey Professional Paper 74, 1912.*

WOODFORD, A. O.—"Crestmore Minerals," *California Division of Mines, Report XXXIX of the State Mineralogist, 1943.* Deals in detail with the assemblages of minerals from the famous deposits in Riverside County, California.

ZAMBONINI, F.—*Mineralogia Vesuviana.* 2nd. Ed. Naples: 1935.

ZEPHAROVICH, V. R.—*Mineralogisches Lexicon für das Kaiserthum Oesterreichs.* Vol. I, 1859, Vol. II, 1873. Vienna. Catalog of minerals of Austria.

ZIEGLER, V.—"The Minerals of the Black Hills," *South Dakota School of Mines, Department of Geology and Mineralogy Bulletin 10.* Rapid City: 1914.

PART 3. GEMOLOGY AND LAPIDARY

ANDERSON, B. W.—*Gem Testing.* 6th. Ed. London: Heywood & Co., Ltd., 1958. Thorough treatment designed primarily for use of jewelers and gemologists, but embodying much of value to the amateur mineralogist.

GÜBELIN, E. J.—*Inclusions as a Means of Gemstone Identification.* Los Angeles: Gemological Institute of America, 1953. Characteristic inclusions in certain gem species are explained and identified; numerous photomicrographs.

LIDDICOAT, R. T.—*Handbook of Gem Identification.* 6th. Ed. Los Angeles: Gemological Institute of America, 1962. An excellent book with easily understood text, numerous illustrations and many other features of value to the gemologist and mineralogist.

SINKANKAS, J.—*Gem Cutting—A Lapidary's Manual.* 2nd. Ed., completely revised. Princeton, New Jersey: D. Van Nostrand Co., Inc., 1962. Complete instructions on all phases of lapidary work.

SMITH, G. F. H.—*Gemstones.* 13th. Ed., revised by F. C. Phillips. London: Methuen & Co., Ltd., 1958. The standard authority on the subject in the English language.

WEBSTER, R.—*The Gemmologist's Compendium.* London: N. A. G. Press, Ltd., 1960. A very useful compilation of data on gemstones, including properties.

—— —*Gems, Their Sources, Descriptions and Identification.* Vols. I, II. London: Butterworths, 1962. An extremely thorough treatment of all gemstones, handsomely illustrated.

PART 4. MAGAZINES AND JOURNALS

American Mineralogist—Journal of the Mineralogical Society of America. Editor E. Wm. Heinrich. The journal of professional mineralogists.

Aufschluss—Zeitschrift für die Freunde der Mineralogie und Geologie. Vereinigung der Freunde der Mineralogie und Geologie (VFMG), Heidelberg, Germany. An excellent, non-technical German journal containing numerous articles on mineralogy and geology, usually about localities in Central Europe. Special reports are issued from time to time on famous localities or mineralized districts.

Canadian Mineralogist—Journal of the Mineralogical Association of Canada, 555 Booth Street, Ottawa, Ontario. Primarily for professionals, but often containing articles of interest to amateurs.

Canadian Rockhound—941 Wavertree Rd., North Vancouver, B.C. V7R 1S4. Popular coverage emphasizing Canada.

Earth Science—Box 1815, Colorado Springs, Co. 80901. Popular coverage of all aspects of the earth sciences, lapidary work and gemology.

Gems and Minerals—Box 687, Mentone, Ca. 92359. Popular coverage of all aspects of earth sciences, including lapidary work and gemology. Interesting and informative. Stresses articles on localities.

Lapidary Journal—P.O. Box 80937, San Diego, Ca. 92138. Devoted to gem cutting, lapidary work and associated crafts, although much locality information on gemstones and mineral deposits is frequently included. The specially enlarged April issue, *The Rockhound Buyers Guide,* is particularly valuable because of many locality articles and extensive advertisements.

Mineral Digest—155 East 24th Street, New York, N.Y. 10016. Wide coverage of minerals, gems, and related cultural and craft activities in a large format, beautifully illustrated and designed periodical.

Mineralogical Magazine—The journal of the Mineralogical Society (Great Britain). Editor: M. H. Hey, 41 Queen's Gate, London, SW7, England. The professional mineralogical journal of Great Britain; less formal than its United States counterpart, and of more interest to the amateur.

Mineralogical Record—P.O. Box 783, Bowie, Md. 20715. Devoted to mineralogy, mineral collecting, famous mineral collecting sites, and other aspects of interest to mineral collectors and amateur and professional mineralogists.

Rock & Gem—16001 Ventura Boulevard, Encine, Ca. 91436. General coverage of all aspects of the crafts and earth science hobbies.

Rockhound—320 North Frazier, Conroe, Tx. 77301. Exclusively devoted to field trips.

Rocks and Minerals—4000 Albemarle Street, NW, Washington, D.C. 20016. Popular coverage of all aspects of the earth sciences, including lapidary work. Lays greatest emphasis on amateur mineralogy, and often contains excellent articles on minerals and localities.

Index